SALTERS ADVANCED CHEMISTRY

Chemical Ideas

www.heinemann.co.uk

✓ Free online support
✓ Useful weblinks
✓ 24 hour online ordering

01865 888080

Heinemann is an imprint of Pearson Education Limited, a company incorporated in England and Wales, having its registered office at Edinburgh Gate, Harlow, Essex, CM20 2JE. Registered company number: 872828

www.heinemann.co.uk

Heinemann is a registered trademark of Pearson Education Limited

First published 1994
Second edition published 2000
This edition published 2008

12 11 10 09 08
10 9 8 7 6 5 4

British Library Cataloguing in Publication Data is available from the British Library on request.

ISBN 978 0 435631 49 9

Edited by Tony Clappison
Designed, produced, illustrated and typeset by Wearset Limited, Boldon, Tyne and Wear
Original illustrations © Pearson Education Limited 2008
Cover design by Wearset Limited, Boldon, Tyne and Wear
Picture research by Q2AMedia
Cover photo/illustration © Alfred Pasieka/Science Photo Library
Printed in Malaysia(CTP-VVP)

Acknowledgements
The author and publisher would like to thank the following individuals and organisations for permission to reproduce data.

The infrared, mass and n.m.r. spectra in Chapter 6 are based on data from a variety of sources, including:
Aldrich Chemical Co Ltd (p. 149, Fig. 6)
Schering Agrochemicals (p. 146, Fig. 12)
The NIST Chemistry Webbook (http://webbook.nist.gov/chemistry) (p. 133, Fig. 4; p. 135, Figs. 7, 8; p. 137, Figs. 11, 12; p. 138, Figs. 13, 14, 15; p. 139, Fig. 16; p. 145, Figs. 9, 10; p. 146, Fig. 11)
The Agency of Industrial Science and Technology (AIST), Japan (Research Information Database, http://www.aist.go.jp) (p. 150, Figs. 8, 9; p. 151, Fig. 10; p. 152, Figs. 11, 12, 13; p. 153, Figs. 14, 15, 16)
Department of Chemistry, University of York
Control of Major Accident Hazards (COMAH) Regulations 1999 (http://www.hse.gov.uk/comah) Crown copyright material is reproduced with the permission of the Controller Office of Public Sector Information (OPSI) (p. 355)
Control of Substances Hazardous to Health (COSHH) Regulations 2002 (http://www.hse.gov.uk/coshh) Crown copyright material is reproduced with the permission of the Controller Office of Public Sector Information (OPSI) (p. 355)

Every effort has been made to contact copyright holders of material reproduced in this book. Any omissions will be rectified in subsequent printings if notice is given to the publishers.

CONTRIBUTORS

The following people have contributed to the development of Chemical Ideas (Third Edition) for the Salters Advanced Chemistry Project:

Editors

Chris Otter (Project Director)	University of York Science Education Group (UYSEG)
Kay Stephenson	CLEAPSS

Associate Editors

Adelene Cogill	Idsall School, Shifnal
Frank Harriss	Formerly Malvern College
Gwen Pilling	Formerly University of York Science Education Group (UYSEG)
Gill Saville	Dover Grammar School for Boys
David Waistnidge	King Edward VI College, Totnes
Ashley Wheway	Formerly Oakham School

A list of those who contributed to the first and second editions is given in Chemical Storylines AS.

Assessment adviser

Steven Evans	OCR

Acknowledgement

We would like to thank the following for their advice and contribution to the development of these materials:

Sandra Wilmott (Project Administrator)	University of York Science Education Group (UYSEG)
Richard Lees	University of York

Sponsors

We are grateful for sponsorship from the Salters' Institute, which has continued to support the Salters Advanced Chemistry Project and has enabled the development of these materials.

Dedication

This publication is dedicated to the memory of Don Ainley, a valued contributor to the development of the Salters Advanced Chemistry Project over the years.

CONTENTS

*Sections met for the first time at AS

INTRODUCTION FOR STUDENTS

The Salters Advanced Chemistry course for AS and A2 is made up of 13 teaching modules. *Chemical Storylines AS* and *Chemical Storylines A2* form the backbone of the teaching modules. This *Chemical Ideas* book covers both AS and A2 material and should be used alongside the relevant *Storylines* book. You will also be given sheets to guide you through the **activities**.

Each teaching module is driven by the storyline. You work through each storyline, making 'excursions' to activities and chemical ideas at appropriate points.

The storylines are broken down into numbered sections. You will find that there are **assignments** at intervals. These are designed to help you through each storyline and to check your understanding, and they are best done as you go along.

Excursions to Chemical Ideas

As you work through the storylines, you will also find that there are references to sections in this book. These sections cover the chemical principles that are needed to understand that particular part of the storyline, and you will probably need to study that section of the *Chemical Ideas* book before you can go much further.

As you study *Chemical Ideas* you will find **problems** relating to each section. These are designed to check and consolidate your understanding of the chemical principles involved.

Building up the Chemical Ideas

Salters Advanced Chemistry has been planned so that you build up your understanding of chemical ideas gradually. For example, the idea of chemical equilibrium is introduced in a simple, qualitative way at AS in **The Atmosphere** module. A more detailed, quantitative treatment is given in the A2 teaching modules **Agriculture and Industry** and **The Oceans**. It is important to bear in mind that *Chemical Ideas* is not the only place where chemistry is covered!

Sections in *Chemical Ideas* cover chemical principles that may be needed in more than one module of the course. As *Chemical Ideas* covers both AS and A2 content, those sections met for the first time at AS are clearly marked 'AS' and are also indicated with an asterisk in the contents list. **Note:** Some of these sections may be *revisited* at A2. The context of the chemistry for a particular module is dealt with in the storyline itself and in related activities.

How much do you need to remember?

The specification for OCR GCE Chemistry B (Salters) defines what you have to remember. Each teaching module includes one or more 'Check your knowledge and understanding' activity. These can be used to check that you have mastered all the required knowledge, understanding and skills for the module. Each 'Check your knowledge and understanding' activity lists whether a topic is covered in *Chemical Ideas*, *Chemical Storylines* or in the associated activities.

Salters Advanced Chemistry Project

1 MEASURING AMOUNTS OF SUBSTANCE

1.1 Amount of substance

Relative atomic mass

Imagine you have a bag containing 10 golf balls and 20 table tennis balls. The golf balls would make up most of the mass, but the table tennis balls would make up most of the contents.

Chemists encounter a comparable situation when they try to work out the composition of water, or any other substance. Oxygen makes up nearly 90% of the mass of water, but two-thirds of the atoms are hydrogen atoms. The difference between the two situations is that, whereas golf balls and table tennis balls can be picked up and counted, the chemist is unable to pick up and count the atoms in water molecules – they are far too small.

You know that the formula of water is H_2O, and that this means there are *two* hydrogen atoms combined with *one* oxygen atom in *each* water molecule. The link between the mass of an element and the number of atoms it contains is the **relative atomic mass (A_r)** of the element. It is this link which allows chemists to work out chemical formulae.

The relative atomic mass scale is used to compare the masses of different atoms. The hydrogen atom, the lightest of all, was originally assigned an A_r value of 1, and A_r values of other atoms were compared with this. The reference used now is the carbon-12 isotope (^{12}C) which is assigned a relative atomic mass of exactly 12 (12.000 000 000). (You can learn more about isotopes in **Section 2.1**.)

In chemistry, approximate relative atomic masses are used most of the time, and you can simply think of carbon atoms, set at $A_r = 12$, as being the reference point for the relative atomic mass scale. The approximate relative atomic masses of some elements are listed in Table 1.

Element	Symbol	Approximate relative atomic mass, A_r
hydrogen	H	1
helium	He	4
carbon	**C**	**12**
nitrogen	N	14
oxygen	O	16
magnesium	Mg	24
sulfur	S	32
calcium	Ca	40
iron	Fe	56
copper	Cu	64
iodine	I	127
mercury	Hg	200

◄ **Table 1** Some approximate relative atomic masses.

Note: In this section we are using A_r values rounded to whole numbers. You will usually use A_r values to one decimal place.

Notice that A_r values have no units. Copper atoms *do not* have a mass of 64 g or 64 'anythings'. They are just 64 times heavier than hydrogen atoms, or four times heavier than oxygen atoms, or twice as heavy as sulfur atoms, and so on.

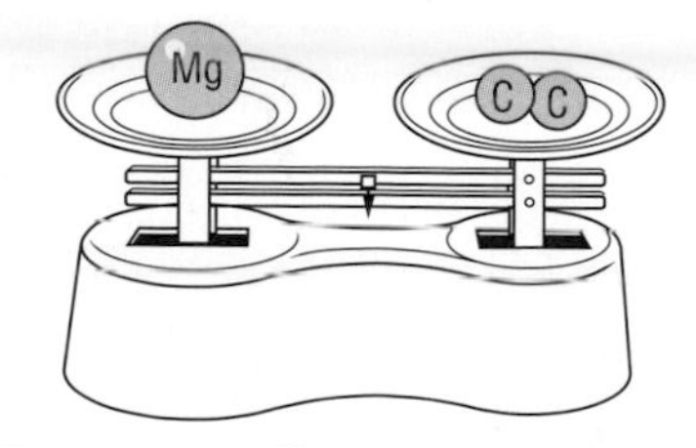

▲ **Figure 1** Two ^{12}C atoms have the same mass as one Mg atom. So the relative atomic mass of Mg is 2 × 12 = 24.

Chemical quantities

Suppose you have two bottles containing equal masses of copper ($A_r = 64$) and sulfur ($A_r = 32$). You would know that you had twice as many sulfur atoms as copper atoms because sulfur atoms have only half the mass of copper atoms. If you had a bottle containing mercury ($A_r = 200$) that was five times heavier than a similar bottle containing calcium ($A_r = 40$), you would know that both bottles contained equal numbers of atoms because each mercury atom has five times the mass of each calcium atom.

12 g of carbon, 1 g of hydrogen and 16 g of oxygen all contain equal numbers of atoms because these masses are in the same ratio as the relative atomic masses.

This amount of each of these elements has a special significance in chemistry. It is called a **mole** (**mol** for short). The mole is the unit that measures **amount of substance** in such a way that equal amounts of elements consist of equal numbers of atoms. The mole is the unit that measures amount in the same way as the kilogram is the unit that measures mass.

Chemical amounts are defined so that the mass of one mole (the **molar mass**) is equal to the relative atomic mass in grams. Thus, the molar mass of carbon is $12\,g\,mol^{-1}$. If you had 6 g of carbon you would have 0.5 mol of carbon atoms; 4 g of oxygen would contain 0.25 mol of oxygen atoms and 3 g of hydrogen would contain 3 mol of hydrogen atoms.

You can work out these amounts like this:

6 g of carbon contains $6\,g \div 12\,g\,mol^{-1} = 0.5\,mol$ of carbon atoms
4 g of oxygen contains $4\,g \div 16\,g\,mol^{-1} = 0.25\,mol$ of oxygen atoms
3 g of hydrogen contains $3\,g \div 1\,g\,mol^{-1} = 3\,mol$ of hydrogen atoms

Each of these calculations uses the relationship:

mass in grams ÷ molar mass = amount in moles of atoms

Relative formula mass

You can use the mole to deal with all substances, compounds as well as elements. A molecule of methane, CH_4, for example, is formed when one carbon atom combines with four hydrogen atoms – this also means that one *mole* of methane is formed when one *mole* of carbon atoms combines with four *moles* of hydrogen atoms.

Just as there are relative atomic masses for atoms, so chemists use **relative formula masses** to compare other substances. The relative formula mass of a substance can be worked out by first writing the formula of the substance and then adding together the relative atomic masses of each of the atoms in the formula. For example:

for methane, CH_4:
relative formula mass $= (1 \times 12) + (4 \times 1)$
$= 16$

for calcium nitrate, $Ca(NO_3)_2$:
relative formula mass $= (1 \times 40) + (2 \times 14) + (6 \times 16)$
$= 164$

Relative formula masses have no units and are on the same scale as relative atomic mass – which has the relative atomic mass of ^{12}C set at 12.000 … .

In substances such as methane, where the formula represents a discrete molecule, the relative formula mass is often called the **relative molecular mass**. Both relative formula mass and relative molecular mass are given the symbol $\boldsymbol{M_r}$.

Formula units

Just as computer memory is made up of individual bits and a box of chalk is made up of individual pieces, the 'bits and pieces' that make up a substance are called **formula units**. They are the basic units, or building blocks, of substances and, not surprisingly, they match the formulae of the substances. Formula units can be single atoms, molecules or groups of ions.

Take copper for example. The formula unit in the metal is simply a Cu atom. Likewise, the formula unit in carbon, a non-metal, is the C atom. In fact, the formula unit in most elements is a single atom and so their relative formula mass is identical to their relative atomic mass. There are some exceptions – in oxygen gas, the formula unit is the O_2 molecule rather than the O atom and so the relative formula mass is *twice* the relative atomic mass of oxygen.

In many covalent compounds, the formula unit is a molecule. In methane, for example, the formula unit is the CH_4 molecule. In ionic compounds, however, the formula unit is a *group of ions*. In calcium nitrate, for example, the formula unit is $Ca^{2+}(NO_3^-)_2$ and contains a group of three ions. These groups of ions are not labelled with any special name and so we just use the general name 'formula unit' when referring to ionic compounds. (You will learn more about ionic and covalent compounds in **Section 3.1**.)

Moles of formula units

The relative formula mass in grams is equal to the molar mass, so the molar mass of methane is 16 g and that of calcium nitrate is 164 g. If you had 8 g of methane you would have 0.5 moles of CH_4 formula units; 41 g of calcium nitrate contains 0.25 moles of $Ca^{2+}(NO_3^-)_2$ formula units. In general:

mass in grams ÷ molar mass = amount in moles of formula units

As well as describing 16 g of methane as containing 1 mole of formula units (or molecules) of CH_4, we can also say that it contains 1 mole of C atoms and $(1 \times 4) = 4$ moles of H atoms. Similarly, 164 g of calcium nitrate contains 1 mole of formula units of $Ca^{2+}(NO_3^-)_2$; it also contains 1 mole of Ca^{2+} ions and 2 moles of NO_3^- ions.

For elements that exist as diatomic gases, such as oxygen, you must be especially careful. For example, 32 g of oxygen gas contains 1 mole of O_2 molecules, but 2 moles of O atoms. To avoid ambiguity when using moles, it is essential to give the formula unit you are referring to (e.g. moles of O atoms or moles of O_2 molecules).

Some definitions

The **relative atomic mass** (A_r) of an element is the mass of its atom relative to $^{12}C = 12$.

The **relative formula mass** (or **relative molecular mass**) (M_r) of a substance is the mass of its formula unit relative to $^{12}C = 12$.

A **mole** (1 **mol**) of a substance is the **amount of substance** which contains as many formula units (atoms, molecules, groups of ions, etc.) as there are atoms in 12 g of ^{12}C.

The **molar mass** of a substance is the mass of substance which contains 1 mol.

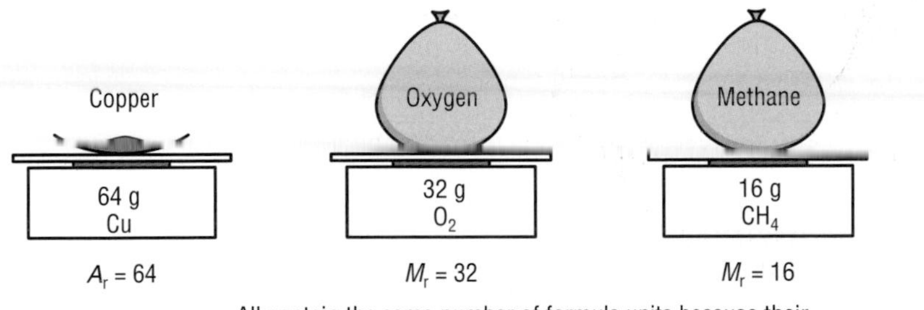

▲ **Figure 2** The important thing to remember is that equimolar amounts of substances contain equal numbers of formula units.

The Avogadro constant

The number of formula units in one mole of a substance is a constant. It is called the **Avogadro constant**, after the Italian scientist Amedeo Avogadro (1776–1856) who did some work which was fundamental to this method of counting formula units. The value of the Avogadro constant (symbol N_A) is 6.02×10^{23} formula units mol^{-1}. If you are counting atoms, it is 6.02×10^{23} atoms mol^{-1}; for molecules it is 6.02×10^{23} molecules mol^{-1}; for electrons it would be 6.02×10^{23} electrons mol^{-1}. The number is so huge that it is difficult to comprehend. Figure 3 shows one way to illustrate how big it is.

The Avogadro constant

One mole of any substance contains 6.02×10^{23} (N_A) formula units.

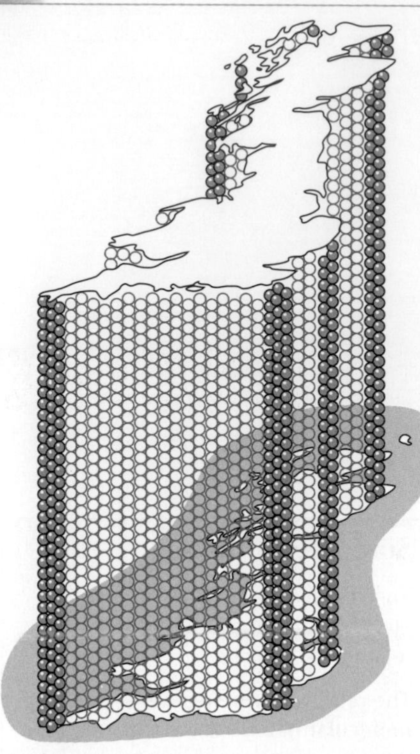

▲ **Figure 3** If the 6.02×10^{23} atoms in 12 g of carbon were turned into marbles, the marbles could cover Great Britain to a depth of 1500 km!

If you want to make sure that you have the same number of formula units of different substances, you don't have to count out incredibly small formula units – neither do you have to work with numbers as big as the Avogadro constant. You just have to measure out a mole – this is easily done by weighing out the molar mass, and is the reason why chemists find the mole so useful.

Chemical formulae

We use moles when we work out chemical formulae. If you did a laboratory experiment to determine the formula of magnesium oxide, you would burn a known mass of magnesium and find out what mass of oxygen combined with it. Here are some specimen results and, from them, a calculation of the formula:

mass of magnesium = 0.84 g	mass of oxygen = 0.56 g
amount of Mg atoms	amount of O atoms
$= 0.84\,g \div 24\,g\,mol^{-1}$	$= 0.56\,g \div 16\,g\,mol^{-1}$
= 0.035 mol	= 0.035 mol

ratio of moles of atoms of Mg : O in magnesium oxide = 1 : 1
formula of magnesium oxide = MgO

Another way of analysing a compound is to find the *percentage mass* of each element it contains. Here are some specimen results for methane:

% mass of carbon = 75%	% mass of hydrogen = 25%
mass of carbon in	mass of hydrogen in
100 g of methane = 75 g	100 g of methane = 25 g
amount of C atoms	amount of H atoms
$= 75\,g \div 12\,g\,mol^{-1}$	$= 25\,g \div 1\,g\,mol^{-1}$
= 6.25 mol	= 25 mol

ratio of moles of atoms of C : H in methane = 1 : 4
formula of methane = CH_4

In a methane molecule there is a central carbon atom surrounded by four hydrogen atoms. The simple ratio of atoms from which it is formed is the same as in the formula of the molecules.

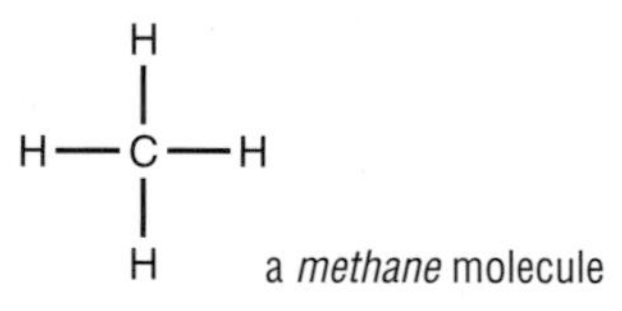

But this isn't the case for all substances. Ethane molecules contain eight atoms – two carbon atoms and six hydrogen atoms.

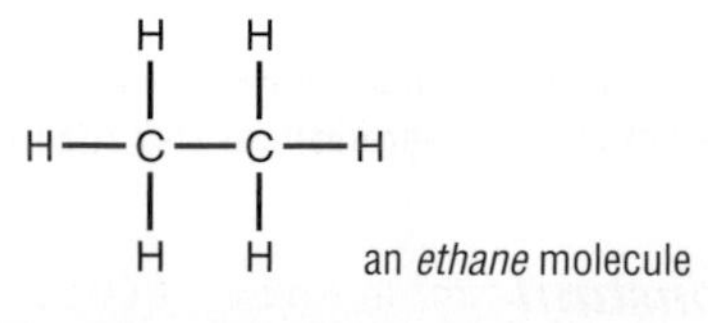

The **empirical formula** of a substance tells you the *simplest ratio* of the numbers of different types of atom in the substance.

The **molecular formula** tells you the *actual numbers* of different types of atom.

The **molecular formula** of ethane is C_2H_6, but the simplest ratio for the moles of atoms of C : H is 1 : 3. So a calculation from percentage masses would lead you to a formula CH_3. Chemists call this type of formula an **empirical formula**. Table 2 shows some more examples of the two types of formulae.

Substance	Molecular formula	Empirical formula
ethene	C_2H_4	CH_2
benzene	C_6H_6	CH
butane	C_4H_{10}	C_2H_5
phosphorus(V) oxide	P_4O_{10}	P_2O_5
oxygen	O_2	O
bromine	Br_2	Br

◀ **Table 2** Some molecular formulae and empirical formulae.

Problems for 1.1

You will need to use a table of relative atomic masses or the Periodic Table in the **Data Sheets** (Table 7) to answer these questions. From now on, you will always see A_r and M_r figures quoted to one decimal place.

1 *Approximately* how many times heavier is
- **a** a magnesium atom than a carbon atom?
- **b** a copper atom than a carbon atom?
- **c** an oxygen atom than a carbon atom?
- **d** a calcium atom than a helium atom?
- **e** a mercury atom than a calcium atom?
- **f** a mercury atom than a helium atom?

2 One atom of element X is approximately 12 times heavier than one carbon atom.
- **a** What is the approximate relative atomic mass of this element?
- **b** Identify X.

3 Copy and complete the table.

Elements	Molar mass ($g\,mol^{-1}$)	Mass of sample (g)	Amount of sample (mol)	Number of atoms
hydrogen	1.0	1.00	1.00	6.02×10^{23}
carbon	12.0	12.00	1.00	?
carbon	12.0	24.00	?	12.04×10^{23}
iron	55.8	?	1.00	?
calcium	40.1	?	2.00	?
iodine	126.9	?	?	3.01×10^{23}

4 The worked example in this problem shows how the empirical formula (simplest formula) of a compound can be calculated when you know the masses of the elements in a sample of it. The calculation is carried out using the general method shown under *Chemical formulae* earlier in this section (page 4). The questions that follow should help you to understand the steps involved in the calculation.

Example
Calculate the empirical formula of a compound if a 16.7 g sample of it contains 12.7 g of iodine and 4.0 g of oxygen.

Steps involved in working out the answer (units have been omitted)

	Iodine	Oxygen
Step 1	12.7	4.0
Step 2	12.7/126.9 = 0.1	4.0/16.0 = 0.25
Step 3	0.1/0.1 = 1	0.25/0.1 = 2.5
Step 4	2	5

The empirical formula of the compound is I_2O_5.

- **a** Why do we need to know the mass of the sample as well as the masses of the elements in it?
- **b** In step 2, what do the numbers 0.1 and 0.25 represent?
- **c** Why, in step 3, do we divide both 0.1 and 0.25 by 0.1?
- **d** Why have we doubled the numbers in moving from step 3 to step 4?
- **e** Write down three possibilities for the molecular formula of this compound based on its empirical formula.
- **f** What additional information do you need to work out the actual molecular formula of the compound?

5 How many moles of atoms are contained in the following masses?
- **a** 32.1 g of sulfur
- **b** 20.0 g of calcium
- **c** 8.0 g of sulfur
- **d** 4.0 g of calcium
- **e** 6.1 g of magnesium
- **f** 31.8 g of copper
- **g** 3.0 g of carbon

h 0.1 g of hydrogen
i 111.6 g of iron
j 1.003 kg of mercury (1 kg = 1000 g)
k Look at your answers to parts **a** and **f**. Explain why they are different, even though each question involves approximately 32 g of an element.

6 There are two different oxides of copper. Black copper oxide has the formula CuO and red copper oxide has the formula Cu_2O. What do these formulae tell us about the relative numbers of particles in each compound?

7 Calculate the empirical formulae of the compounds which contain only the elements shown below.
a 2.0 g of hydrogen and 16.0 g of oxygen
b 12.0 g of carbon and 16.0 g of oxygen
c 12.0 g of carbon and 64.2 g of sulfur
d 12.0 g of carbon and 4.0 g of hydrogen
e 111.6 g of iron and 48.0 g of oxygen
f 31.8 g of copper and 8.0 g of oxygen
g 10.0 g of calcium and 4.0 g of oxygen
h 8.0 g of sulfur and 8.0 g of oxygen
i 1.2 g of magnesium and 0.1 g of hydrogen.

8 A compound contains 92.3% carbon and 7.7% hydrogen by mass.
a In 100 g of the compound, how many grams of carbon are there?
b In 100 g of the compound, how many grams of hydrogen are there?
c Use your answers to parts **a** and **b** to calculate the empirical formula of the compound.

9 Work out the empirical formulae of the compounds with the following percentage compositions by mass.
a 87.5% silicon and 12.5% hydrogen
b 42.8% carbon and 57.2% oxygen
c 27.3% carbon and 72.7% oxygen
d 60.0% magnesium and 40.0% oxygen
e 52.2% carbon, 13.0% hydrogen and 34.8% oxygen
f 40.0% calcium, 12.0% carbon and 48.0% oxygen
g 1.18% hydrogen, 42.00% chlorine and 56.82% oxygen
h 27.4% sodium, 1.2% hydrogen, 14.3% carbon and 57.1% oxygen.

10 What are the empirical formulae of the substances with the following molecular formulae?
a C_3H_6
b P_4O_6
c Al_2Cl_6
d B_2H_6
e C_8H_{10}
f C_3H_4
g $C_6H_{12}O_6$
h $C_{12}H_{22}O_{11}$

11 The table below contains the empirical formulae and relative molecular masses of some compounds. Write down their molecular formulae.

	Empirical formula	Relative molecular mass	Molecular formula
a	HO	34	?
b	CO	28	?
c	CH	26	?
d	CH	78	?
e	CH_2	84	?

12 Approximately how many times heavier is
a an O_2 molecule than an oxygen atom?
b a CO_2 molecule than a helium atom?
c an SO_2 molecule than an O_2 molecule?
d an MgO formula unit than a helium atom?
e a CuO formula unit than an MgO formula unit?

13 Calculate the molar mass, in $g\,mol^{-1}$, of the following compounds.
a ethane, C_2H_6
b benzene, C_6H_6
c cobalt(II) chloride, $CoCl_2$
d calcium carbonate, $CaCO_3$
e potassium manganate(VII), $KMnO_4$
f iron(III) nitrate, $Fe(NO_3)_3$
g ammonium sulfate, $(NH_4)_2SO_4$.

14 Calculate how many moles of each substance are contained in
a 88 g of carbon dioxide
b 234 g of sodium chloride
c 560 g of but-1-ene, C_4H_8
d 2.92 g of sulfur hexafluoride, SF_6
e 0.37 kg of calcium hydroxide, $Ca(OH)_2$
f 18 tonnes of water (1 tonne = 1×10^6 g).

AS

1.2 Balanced equations

A balanced chemical equation tells you the reactants and products in a reaction, and the relative amounts involved. The equation is balanced so that there are equal numbers of each type of atom on both sides. For example, the equation

$$CH_4(g) + 2O_2(g) \rightarrow CO_2(g) + 2H_2O(l)$$

tells you that 1 molecule of methane reacts with 2 molecules of oxygen to form 1 molecule of carbon dioxide and 2 molecules of water. These are also the amounts in moles of the substances involved in the reaction.

The number written in front of each formula in a balanced equation tells you the number of formula units involved in the reaction. Remember that a formula unit may be a molecule or another species such as an atom or an ion. The small subscript numbers are part of the formulae and cannot be changed.

State symbols

State symbols are included in chemical equations to show the physical state of the reactants and products:

(g) gas
(l) liquid
(s) solid
(aq) aqueous solution

Writing balanced equations

The only way to be sure of the balanced equation for a reaction is to do experiments to find out what is formed in the reaction and what quantities are involved. But chemists use equations a lot, and it isn't possible to do experiments every time. Fortunately, if we know the reactants and products we can usually work out their formulae and predict the balanced equation.

The steps for predicting balanced equations

We will illustrate the steps by using the reaction that occurs when calcium reacts with water.

Step 1 Decide what the reactants and products are:
calcium + water → calcium hydroxide + hydrogen

Step 2 Write formulae for the substances involved – state symbols should be included:
$Ca(s) + H_2O(l) \rightarrow Ca(OH)_2(aq) + H_2(g)$
This equation is unbalanced because there are different numbers of each type of atom on each side.

Step 3 Balance the equation so that there are the same numbers of each type of atom on each side:
$Ca(s) + 2H_2O(l) \rightarrow Ca(OH)_2(aq) + H_2(g)$
Equations can only be balanced by putting numbers in front of the formulae. You cannot balance them by altering the formulae because that would create different substances.

The law of conservation of mass

The famous French chemist Antoine Lavoisier noted in 1774 that, if nothing is allowed to enter or leave a reaction vessel, the total mass is the same after a chemical reaction has taken place as it was before the reaction.

Atoms are not created or destroyed in chemical reactions. They are simply rearranged ...

... so equations must balance.

Problems for 1.2

1 Write out the following equations, putting numbers in front of formulae so that the equations are balanced.

a $Mg + O_2 \rightarrow MgO$
b $H_2 + O_2 \rightarrow H_2O$
c $Fe + Cl_2 \rightarrow FeCl_3$
d $CaO + HNO_3 \rightarrow Ca(NO_3)_2 + H_2O$
e $CaCO_3 + HCl \rightarrow CaCl_2 + CO_2 + H_2O$
f $H_2SO_4 + NaOH \rightarrow Na_2SO_4 + H_2O$
g $HCl + Ca(OH)_2 \rightarrow CaCl_2 + H_2O$
h $Na + H_2O \rightarrow NaOH + H_2$
i $CH_4 + O_2 \rightarrow CO_2 + H_2O$
j $CH_3OH + O_2 \rightarrow CO_2 + H_2O$

2 Write out balanced equations for the following reactions.

a calcium + oxygen → calcium oxide (CaO)
b calcium + water → calcium hydroxide ($Ca(OH)_2$) + hydrogen
c carbon + carbon dioxide → carbon monoxide
d nitrogen (N_2) + hydrogen (H_2) → ammonia (NH_3)
e propane (C_3H_8) + oxygen → carbon dioxide + water

3 Write balanced equations, including state symbols, for the following reactions.
  a zinc reacting with sulfuric acid (H_2SO_4) to form zinc sulfate ($ZnSO_4$) and hydrogen
  b magnesium reacting with hydrochloric acid (HCl) to form magnesium chloride ($MgCl_2$) and hydrogen
  c magnesium carbonate ($MgCO_3$) decomposing on heating to form magnesium oxide (MgO) and carbon dioxide
  d ethane (C_2H_6) burning in oxygen to form carbon dioxide and water
  e barium oxide (BaO) reacting with hydrochloric acid (HCl) to form barium chloride ($BaCl_2$) and water.

AS

1.3 Using equations to work out reacting masses

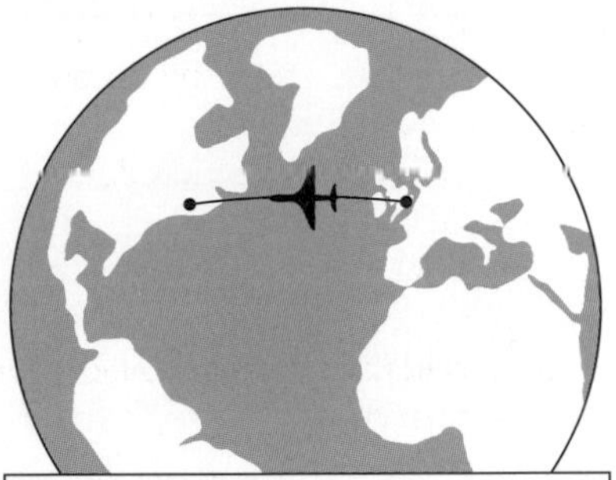

A flight from London to New York takes 7 hours. A jumbo jet carrying 375 people uses 10 tonnes of fuel per hour.

▲ **Figure 1** What mass of CO_2 is produced during one transatlantic flight? (See problem 7 at the end of this section.)

A balanced equation tells you the amount in moles of each substance involved in the reaction. For example, the equation

$$CH_4(g) + 2O_2(g) \rightarrow CO_2(g) + 2H_2O(l)$$

tells you that 1 mole of CH_4 reacts with 2 moles of O_2 to give 1 mole of CO_2 and 2 moles of H_2O.

The mass of 1 mole of CH_4 is $12.0\,g + (4 \times 1.0)\,g = 16.0\,g$
The mass of 1 mole of O_2 is $(2 \times 16.0)\,g = 32.0\,g$
The mass of 1 mole of CO_2 is $12.0\,g + (2 \times 16.0)\,g = 44.0\,g$
The mass of 1 mole of H_2O is $(2 \times 1.0)\,g + 16.0\,g = 18.0\,g$

So, 16.0 g CH_4 reacts with 64.0 g O_2 to give 44.0 g CO_2 and 36.0 g H_2O.
(The total mass on each side of the equation must always be the same.)
This means that chemists can use equations to work out the masses of reactants and products involved in a reaction, without having to do an experiment.

The steps for working out reacting masses

Step 1 Write a balanced equation.
Step 2 In words, state what the equation tells you about the amount in moles of the substances you are interested in.
Step 3 Change amounts in moles to masses in grams.
Step 4 Scale the masses to the ones in the question.

Example

What mass of carbon dioxide is produced when 64 g of methane are burned in a plentiful supply of air?

Step 1 $CH_4(g) + 2O_2(g) \rightarrow CO_2(g) + 2H_2O(l)$
Step 2 1 mole → 1 mole
Step 3 16.0 g → 44.0 g
Step 4 1.0 g → $\frac{44.0}{16.0}\,g$

64.0 g → $\frac{44.0}{16.0} \times 64.0\,g$

$= 176.0\,g$

So, 176 g CO_2 are produced when 64 g CH_4 are burned.
When you are satisfied that you understand these steps, you can try your hand at the problems that follow. They start with another worked example.

Problems for 1.3

You will need to use a table of relative atomic masses or the Periodic Table in the **Data Sheets** (Table 7) to answer these problems.

1 The worked example in this problem shows how a balanced equation can be used to determine the masses of substances involved in a chemical reaction. The questions that follow should help you to understand the steps involved in the calculation.

Example

What mass of magnesium oxide is formed when 6 g of magnesium are burnt in excess oxygen?

Steps involved in working out the answer

Step 1 $2Mg(s) + O_2(g) \rightarrow 2MgO(s)$

Step 2 2 moles 2 moles

Step 3 $2 \times 24.3 = 48.6$ g of magnesium produces $2 \times 40.3 = 80.6$ g of magnesium oxide

Step 4 Burning 1 g of magnesium produces $\frac{80.6}{48.6}$ g of magnesium oxide

Burning 6 g of magnesium produces

$6 \times \frac{80.6}{48.6}$ g = 10 g of magnesium oxide

a Excess oxygen was used. What does this tell you?

b Why do we need the balanced equation for the reaction?

c How do we get the number 40.3 in step 3?

d Why do we multiply 40.3 by 2 in step 3?

e Why do we divide 80.6 by 48.6 in step 4?

f If we had started with 50 g of magnesium, how would step 4 have been different?

2 Copper(II) oxide is reduced by hydrogen to copper and water:

$$CuO(s) + H_2(g) \rightarrow Cu(s) + H_2O(l)$$

What mass of copper(II) oxide will be reduced by excess hydrogen to give 16 g of copper?

3 Calcium carbonate decomposes when heated strongly to give calcium oxide and carbon dioxide:

$$CaCO_3(s) \rightarrow CaO(s) + CO_2(g)$$

a What mass of calcium oxide is produced by heating 5 g of calcium carbonate?

b What mass of carbon dioxide is produced when 7 g of calcium carbonate are heated?

c What mass of calcium carbonate must be decomposed to produce 1.4 g of calcium oxide?

4 1 kg (1000 g) of charcoal is burned in a plentiful supply of air. Assuming charcoal is pure carbon, calculate the mass of carbon dioxide formed.

5 a Write a balanced equation for the complete combustion of the fuel octane (C_8H_{18}) in oxygen.

b What mass of oxygen would be needed to burn 50 kg of octane completely?

c What mass of carbon dioxide would be produced by the complete combustion of 50 kg of octane?

6 Coal used as fuel in a power station contains 1% sulfur by mass. When the coal is burned, the sulfur is also burned and forms sulfur dioxide.

a If the power station burns 5600 tonnes (t) of coal each day, how many tonnes of sulfur will also be burned?

b Write a balanced equation for the burning of sulfur to produce sulfur dioxide.

c How many grams of sulfur dioxide are produced when 32 g of sulfur are burned?

d How many tonnes of sulfur dioxide are produced when 32 tonnes of sulfur are burned?

e How many tonnes of sulfur dioxide are produced when 1 tonne of sulfur is burned?

f How many tonnes of sulfur dioxide are produced by the power station every day?

7 An aircraft flight from London to New York takes 7 hours. A jumbo jet carrying 375 people uses 10 tonnes of fuel per hour. (Jet fuel is a mixture of hydrocarbons, but you can assume it is a hydrocarbon with the formula $C_{12}H_{26}$)

a What is the total mass of carbon dioxide produced by the aircraft during the flight?

b What is the mass of carbon dioxide produced per person during the flight?

8 In the manufacture of iron in the blast furnace, iron(III) oxide in the iron ore is reduced by carbon monoxide to iron. Carbon dioxide is also formed.

a Write a balanced equation for this reaction.

b Calculate the mass of 1 mole of iron(III) oxide.

c Calculate the mass of iron(III) oxide needed to make 1 g of iron.

d How many tonnes of iron(III) oxide are needed to make 1 tonne of iron? (1 tonne (t) = 1000 kg = 1×10^6 g)

e If the iron ore contains 50% iron(III) oxide, how many tonnes of ore are needed to produce 1 tonne of iron?

f If the iron ore contains 12% iron(III) oxide, how much ore is needed to produce 1 tonne of iron?

1.4 Calculations involving gases

Chemists use chemical equations to work out the masses of reactants and products involved in a reaction. If one or more of these is a gas, it is sometimes more useful to know its volume rather than its mass.

Molar volume

> **Avogadro's law**
>
> Equal volumes of all gases at the same temperature and pressure contain an equal number of molecules.

The number of molecules in one mole of any gas is always 6.02×10^{23}. This quantity is known as the **Avogadro constant**, N_A. The molecules in a gas are very far apart so that the actual size of each molecule has a negligible effect on the total volume the gas occupies. So, one mole of any gas always occupies the same volume, no matter which gas it is. Avogadro realised this as long ago as 1811, when he put forward his famous law (sometimes called Avogadro's hypothesis).

The volume occupied by one mole of any gas at a particular temperature and pressure is called the **molar volume**. At standard temperature and pressure (s.t.p.), the molar volume of a gas is 22.4 dm^3. 'Standard temperature and pressure' means a temperature of 0 °C (273 K) and a pressure of 1 atmosphere (101.3 kPa).

At room temperature, around 25 °C (298 K), and 1 atmosphere pressure, the volume of a mole of any gas is about 24 dm^3 (Figure 1).

The idea of molar volume allows you to calculate the amount in moles from the volume of a gas, and vice versa, provided you know the temperature and pressure of the gas. Here are two examples. You will need to assume that the molar volume of a gas at room temperature and pressure (r.t.p.) is 24 dm^3.

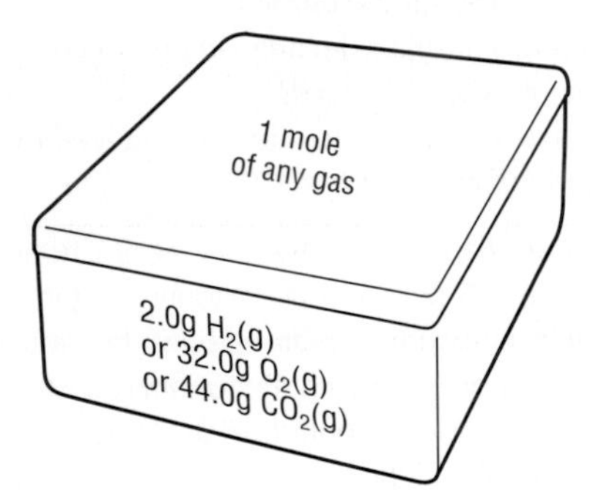

▲ **Figure 1** A mole of any gas at room temperature and pressure occupies 24 dm^3 (about the volume of a large biscuit tin).

Example 1

Calculate the volume occupied by 4.4 g of carbon dioxide at room temperature and pressure.

Step 1 Calculate the amount in moles from the mass.
Relative formula mass of CO_2 = 44.0

$$\text{Amount of } CO_2 = \frac{\text{mass}}{\text{molar mass}}$$

$$= \frac{4.4\,\text{g}}{44.0\,\text{g mol}^{-1}} = 0.1\text{ mol}$$

Step 2 Calculate the volume.
1 mol CO_2 at r.t.p. has a volume of 24 dm^3
0.1 mol CO_2 at r.t.p. has a volume of 2.4 dm^3

Example 2

Calculate the mass of 1.2 dm^3 of methane gas (CH_4) at room temperature and pressure.

Step 1 Calculate the amount in moles.

$$\text{Amount of } CH_4 = \frac{\text{volume}}{\text{molar volume}}$$

$$= \frac{1.2\,\text{dm}^3}{24\,\text{dm}^3\,\text{mol}^{-1}} = 0.05\,\text{mol}$$

Step 2 Calculate the mass.
Relative formula mass of CH_4 = 16.0
Mass of CH_4 = amount in moles × molar mass
= $0.05\,\text{mol} \times 16.0\,\text{g mol}^{-1} = 0.8\,\text{g}$

Reacting volumes of gases

We can use molar volumes to work out the volumes of gases involved in a reaction.

For example, consider the manufacture of ammonia from nitrogen and hydrogen:

$$N_2(g) + 3H_2(g) \rightarrow 2NH_3(g)$$

From the equation:

$$1 \text{ mole } N_2 + 3 \text{ moles } H_2 \rightarrow 2 \text{ moles } NH_3$$

Using the idea that one mole of each gas occupies 24 dm^3 at r.t.p., we can write:

$$24\,dm^3\ N_2 + (3 \times 24)\,dm^3\ H_2 \rightarrow (2 \times 24)\,dm^3\ NH_3$$
$$\text{or } 1\,dm^3\ N_2 + 3\,dm^3\ H_2 \rightarrow 2\,dm^3\ NH_3$$
$$\text{or } 1 \text{ volume } N_2 + 3 \text{ volumes } H_2 \rightarrow 2 \text{ volumes } NH_3$$

So, if we had 10 cm^3 of nitrogen it would react with 30 cm^3 of hydrogen to form 20 cm^3 of ammonia, provided all the measurements were taken at the same temperature and pressure.

For a reaction involving only gases, we can convert a statement about the numbers of moles of each substance involved to the same statement about volumes.

$N_2(g)$ +	$3H_2(g)$ →	$2NH_3(g)$
1 mole	3 moles	2 moles
1 volume	3 volumes	2 volumes

Measuring volumes

The units you will use to measure volume will depend on how big the volume is. In chemistry, large volumes are usually measured in cubic decimetres (dm^3); smaller volumes are measured in cubic centimetres (cm^3). A decimetre is a tenth of a metre, i.e. 10 cm. A cubic decimetre is therefore 10 cm × 10 cm × 10 cm, or 1000 cm^3.

You may also come across other names for these units:

$1\,dm^3 = 1000\,cm^3$

Working with masses and volumes

The methods used in **Section 1.3** can be extended to calculations involving volumes of gases.

Example

What volume of carbon dioxide is produced when 15 g of calcium carbonate completely decompose? Assume that one mole of gas occupies 24 dm^3 at r.t.p.

Step 1	$CaCO_3 \rightarrow CaO + CO_2$	
Step 2	1 mole	1 mole
Step 3	100.1 g	24 dm^3
Step 4	1 g	$\frac{24}{100.1}\,dm^3$
	15 g	$15 \times \frac{24}{100.1}\,dm^3 = 3.6\,dm^3$

Problems for 1.4

Unless stated otherwise, you should assume that the volumes of all gases are measured at the same temperature and pressure.

1 The volumes of one mole of all gases are the same when measured at the same temperature and pressure. The volumes of 1 mole of liquids or solids are almost always different. Why do gases differ from liquids and solids in this way?

2 Methane burns in excess oxygen to form carbon dioxide and water vapour.

a Write a balanced equation, including state symbols, for this reaction.

b How does the volume of oxygen used compare with the volume of methane burned?

c What can you say about the volume of water vapour formed compared with the volume of methane burned?

3 Hydrogen reacts with chlorine to form hydrogen chloride gas.
 a Write a balanced equation, including state symbols, for this reaction.
 b What is the ratio of the volumes of the gases involved?

4 10 cm^3 of hydrogen are burned in oxygen to form water.
 a Write a balanced equation, including state symbols, for this reaction.
 b What volume of oxygen is needed to burn the hydrogen completely?

5 10 cm^3 of a gaseous hydrocarbon reacts completely with 40 cm^3 of oxygen to produce 30 cm^3 of carbon dioxide.
 a How many moles of carbon dioxide must have been formed from one mole of the hydrocarbon?
 b How many carbon atoms must there be in the formula of the hydrocarbon?
 c How many moles of oxygen were used in burning 1 mole of the hydrocarbon?
 d How many moles of water must have been formed in burning 1 mole of the hydrocarbon?
 e What is the formula of the hydrocarbon?

6 10 g of calcium carbonate is reacted with excess hydrochloric acid. Calculate the volume of carbon dioxide formed at room temperature and pressure.

7 1.2 g of magnesium react with excess sulfuric acid. Calculate the volume of hydrogen produced at room temperature and pressure.

8 18 g of pentane (C_5H_{12}) are completely burned in a car engine to form carbon dioxide and water.
 a How many moles of pentane are burned?
 b How many moles of oxygen are needed to burn all the pentane?
 c What volume of oxygen is needed, assuming that 1 mole of gas occupies 24 dm^3?
 d What volume of air is needed (assume that air contains 20% oxygen by volume)?
 e What volume of carbon dioxide is formed?

AS

1.5 *Concentrations of solutions*

Chemists often carry out reactions in solution. When you are using a solution of a substance, it is important to know how much of the substance is dissolved in a particular volume of solution.

Concentrations are sometimes measured in grams per cubic decimetre (see the 'Measuring volumes' box in **Section 1.4**). A solution containing 80 g of sodium hydroxide made up to 1 dm^3 of solution has a concentration of 80 g dm^{-3}.

However, chemists usually prefer to measure out quantities in moles rather than in grams, because working in moles tells you about the number of particles present. So the preferred units for measuring concentrations in chemistry are moles per cubic decimetre, or mol dm^{-3}.

To convert grams per cubic decimetre to moles per cubic decimetre, you need to know the molar mass of the substance involved. For example, the molar mass of sodium hydroxide, NaOH, is 40.0 g mol^{-1}. So a solution containing 80 g dm^{-3} has a concentration of

$$\frac{80\,\text{g}\,\text{dm}^{-3}}{40.0\,\text{g}\,\text{mol}^{-1}} = 2.0\,\text{mol}\,\text{dm}^{-3}$$

In general,

$$\text{concentration (in mol dm}^{-3}\text{)} = \frac{\text{concentration (in g dm}^{-3}\text{)}}{\text{molar mass (in g mol}^{-1}\text{)}}$$

In some books you may see 'mol dm^{-3}' abbreviated to 'M'. This is quick to write, and was once widely used, but nowadays mol dm^{-3} is preferred.

When we make a solution, its concentration will depend on:

- the amount of solute
- the final volume of the solution (Figure 1).

If you know the concentration of a solution, you can work out the amount of solute in a particular volume.

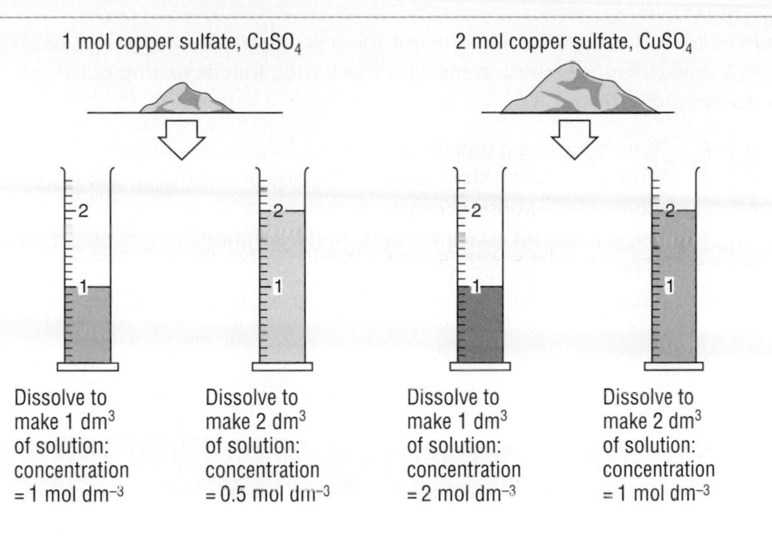

◀ **Figure 1** The concentration of a solution depends on the *amount* of solute and the final *volume* of the solution.

You don't make a 1 mol dm^{-3} solution by adding 1 mol of solute to 1 dm^{-3} of solvent. Instead, you add solvent to the dissolved solute until the volume of the final solution is 1 dm^3.

Example
Suppose we have 250 cm^3 of a solution of sodium hydroxide with a concentration of 2 mol dm^{-3}. How many moles of NaOH do we have?

Answer
250 cm^3 is 0.25 dm^3, and there are 2 mol in each dm^3. So in 0.25 dm^3 we must have (2 mol dm^{-3} × 0.25 dm^3)
= 0.5 mol of NaOH

In general,

amount in mol = (concentration of solution in mol dm^{-3}) × (volume of solution in dm^3)

Using concentrations in calculations

If you are carrying out a chemical reaction in solution, and you know the equation for the reaction, you can use the concentrations of the reacting solutions to predict the volumes you will need.

Example
Consider the reaction of sodium hydroxide with hydrochloric acid. Suppose the concentrations of both solutions are 2 mol dm^{-3}. If you have 0.25 dm^3 of sodium hydroxide solution, what volume of hydrochloric acid would be needed to neutralise it?

Answer
The equation for the reaction is

$$NaOH(aq) + HCl(aq) \rightarrow NaCl(aq) + H_2O(l)$$

The equation shows that 1 mole of NaOH reacts with 1 mole of HCl.

Amount of sodium hydroxide = 2 mol dm^{-3} × 0.25 dm^3 = 0.5 mol

$\Rightarrow$ amount of HCl needed = 0.5 mol

$$\Rightarrow \text{volume of HCl needed} = \frac{0.5\,\text{mol}}{2\,\text{mol dm}^{-3}} = 0.25\,\text{dm}^3$$

You could probably have worked out this simple example without going through these stages, but it isn't so easy when the solutions are of different concentrations, and when they don't react in a 1 : 1 ratio.

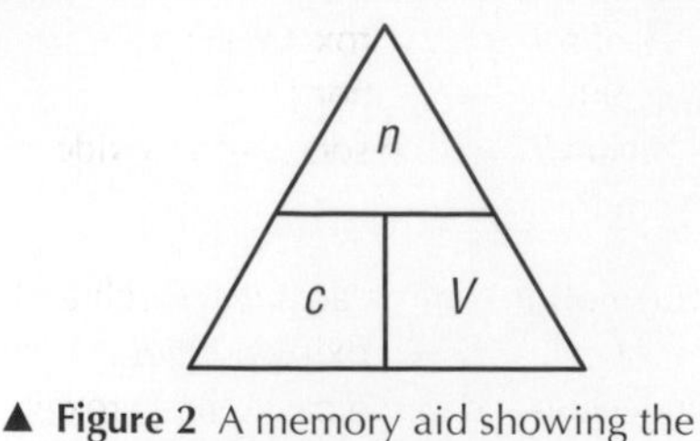

▲ **Figure 2** A memory aid showing the relationship between concentration (*c*) in $mol\,dm^{-3}$, the amount (*n*) in moles and the volume of solution (*V*) in dm^3: $c = \frac{n}{V}$, $n = c \times V$, $V = \frac{n}{c}$

In general, the steps for working out the reacting volumes of solutions are as follows (compare these steps with those used for calculating reacting masses in **Section 1.3**).

Step 1 Write a balanced equation.
Step 2 Write down what the equation tells you about the amounts in moles of the substances you are interested in.
Step 3 Use the known concentrations of the solutions to change amounts in moles to volumes of solutions.
Step 4 Scale the volumes of solutions to the ones in the question.

Problems for 1.5

1 **a** Convert the following volumes into dm^3:
  i $20\,cm^3$ **ii** $1500\,cm^3$
 b Convert the following volumes into cm^3:
  i $220\,dm^3$ **ii** $1.6\,dm^3$

2 A solution of sodium hydroxide is described as having a concentration of $0.4\,mol\,dm^{-3}$. Explain what this means.

3 How many moles of solute are contained in each of the following aqueous solutions?
 a $1\,dm^3$ of $0.5\,mol\,dm^{-3}$ KCl
 b $2\,dm^3$ of $0.2\,mol\,dm^{-3}$ NaOH
 c $250\,cm^3$ of $4\,mol\,dm^{-3}$ HCl
 d $100\,cm^3$ of $2\,mol\,dm^{-3}$ Na_2CO_3
 e $250\,cm^3$ of $0.2\,mol\,dm^{-3}$ H_2SO_4
 f $50\,cm^3$ of $0.04\,mol\,dm^{-3}$ NaI.

4 What is the concentration (in $mol\,dm^{-3}$) of the solutions with the following compositions?
 a 2.0 moles of KOH in $1\,dm^3$ of solution
 b 1.0 mole of NaCl in $500\,cm^3$ solution
 c 0.5 mole HCl in $100\,cm^3$ solution
 d 2.0 mole of HCl in $10\,dm^3$ solution
 e 0.1 mole of HNO_3 in $25\,cm^3$ solution
 f 0.05 mole of $CaCl_2$ in $250\,cm^3$ solution.

5 What is the concentration (in $mol\,dm^{-3}$) of the following solutions?
 a 10 g of NaOH in $1\,dm^3$ of solution
 b 4 g of NaOH in $500\,cm^3$ of solution
 c 100 g of $CaBr_2$ in $100\,cm^3$ of solution
 d 2 g of $CaBr_2$ in $25\,cm^3$ of solution
 e 8 g of $CuSO_4$ in $100\,cm^3$ of solution
 f 40 g of $CuSO_4$ in $2\,dm^3$ of solution.

6 How many grams of sodium hydroxide are needed to make up each of the following solutions?
 a $1\,dm^3$ of a $1\,mol\,dm^{-3}$ solution
 b $100\,cm^3$ of a $1\,mol\,dm^{-3}$ solution
 c $250\,cm^3$ of a $2\,mol\,dm^{-3}$ solution
 d $1\,dm^3$ of a $0.01\,mol\,dm^{-3}$ solution
 e $5\,dm^3$ of a $4\,mol\,dm^{-3}$ solution
 f $50\,cm^3$ of a $0.5\,mol\,dm^{-3}$ solution.

7 Calculate the mass of solute needed to make up the following solutions.
 a $1\,dm^3$ of a $2\,mol\,dm^{-3}$ solution of NaCl
 b $250\,cm^3$ of a $0.1\,mol\,dm^{-3}$ solution of $KMnO_4$
 c $50\,cm^3$ of a $0.5\,mol\,dm^{-3}$ solution of KOH
 d $15\,dm^3$ of a $2\,mol\,dm^{-3}$ solution of $Pb(NO_3)_2$
 e $10\,cm^3$ of a $0.01\,mol\,dm^{-3}$ solution of LiOH
 f $30\,cm^3$ of a $0.75\,mol\,dm^{-3}$ solution of Na_2CO_3
 g $200\,cm^3$ of $0.001\,mol\,dm^{-3}$ HNO_3
 h $2.5\,cm^3$ of $2\,mol\,dm^{-3}$ HCl
 i $1\,dm^3$ of a $0.1\,mol\,dm^{-3}$ solution of copper(II) sulfate, using copper(II) sulfate crystals, $CuSO_4{\cdot}5H_2O$
 j $250\,cm^3$ of a $0.2\,mol\,dm^{-3}$ solution of iron(II) sulfate, using iron(II) sulfate crystals, $FeSO_4{\cdot}7H_2O$.

8 Complete the following table, which shows the concentrations of ions in water from a sample of Dead Sea water.

Ion	Concentration/ $g\,dm^{-3}$	Concentration/ $mol\,dm^{-3}$
Cl^-	183.0	?
Mg^{2+}	36.2	?
Na^+	?	1.37
Ca^{2+}	?	0.335
K^+	6.8	?
Br^-	5.2	?
SO_4^{2-}	?	0.00625

9 Write down the formulae of the ions present in the following solutions.

Solution	Positive ion	Negative ion
copper(II) sulfate	Cu^{2+}	SO_4^{2-}
sodium chloride	?	?
sodium carbonate	?	?
silver nitrate	?	?
magnesium bromide	?	?
sulfuric acid	?	?

10 a What is the concentration of iodide ions in a 1 mol dm^{-3} solution of KI?

b What is the concentration of hydroxide ions in a 0.01 mol dm^{-3} solution of barium hydroxide, $Ba(OH)_2$?

c What is the concentration of hydrogen ions in a 0.1 mol dm^{-3} solution of phosphoric acid, H_3PO_4?

d What is the concentration of iron(III) ions (Fe^{3+}) in a 0.2 mol dm^{-3} solution of iron(III) sulfate, $Fe_2(SO_4)_3$?

11 The concentration of an acid solution can be found by titrating the acid solution with a solution of an alkali of known concentration. In such a titration it was found that 19.00 cm^3 of 0.100 mol dm^{-3} sodium hydroxide were necessary to react with 25.00 cm^3 of a hydrochloric acid solution.

$$NaOH(aq) + HCl(aq) \rightarrow NaCl(aq) + H_2O(l)$$

a How many moles of sodium hydroxide are there in 19.00 cm^3 of 0.100 mol dm^{-3} solution?

b How many moles of hydrochloric acid are, therefore, in 25.00 cm^3 of the acid solution?

c How many moles of hydrochloric acid are, therefore, in 1000 cm^3 of the acid solution?

d What is the concentration of the hydrochloric acid solution?

12 In a titration, 25.00 cm^3 of a sodium hydroxide solution were pipetted into a conical flask. A 0.100 mol dm^{-3} solution of sulfuric acid was run from a burette into the flask. An indicator in the flask changed colour when 22.00 cm^3 of the acid had been added.

$$H_2SO_4(aq) + 2NaOH(aq) \rightarrow Na_2SO_4(aq) + 2H_2O(l)$$

a How many moles of sulfuric acid were added from the burette?

b How many moles of sodium hydroxide must have been in the flask?

c How many moles of sodium hydroxide would, therefore, be in 1000 cm^3 of solution?

d What is the concentration of the sodium hydroxide solution in mol dm^{-3}?

13 15.50 cm^3 of 0.00100 mol dm^{-3} nitric acid react with and neutralise 10.00 cm^3 of calcium hydroxide solution.

a Calculate the concentration of the calcium hydroxide solution in mol dm^{-3}.

b Calculate the concentration in g dm^{-3}.

14 A solution of 0.100 mol dm^{-3} silver nitrate is added to 5.00 cm^3 of 0.500 mol dm^{-3} sodium chloride solution until the reaction is complete.

$$Ag^+(aq) + Cl^-(aq) \rightarrow AgCl(s)$$

a Calculate the number of moles of sodium chloride in 5.00 cm^3 of 0.500 mol dm^{-3} solution.

b How many moles of silver nitrate must have been added to react with the sodium chloride?

c What volume of silver nitrate solution must have been used?

15 A 2.00 mol dm^{-3} solution of sodium hydroxide is used to neutralise 20.00 cm^3 of a 1.00 mol dm^{-3} solution of sulfuric acid.

$$H_2SO_4(aq) + 2NaOH(aq) \rightarrow Na_2SO_4(aq) + 2H_2O(l)$$

a Calculate the number of moles of sulfuric acid that are neutralised.

b What number of moles of sodium hydroxide must have been added to the acid?

c What volume of 2.00 mol dm^{-3} sodium hydroxide solution contains 2 moles of sodium hydroxide?

d What volume of the solution contains 1 mole of sodium hydroxide?

e What volume of the sodium hydroxide solution was needed to neutralise the sulfuric acid?

2 ATOMIC STRUCTURE

2.1 A simple model of the atom

AS

What is inside atoms?

No one has yet been able to look directly inside atoms to see what they are really like, but a considerable amount of experimental evidence has given us a good working model of atomic structure. Scientific models should not be considered as the 'truth' or the 'right answer'. They are useful because they help to explain our observations of nature and they guide our thinking in productive directions.

Sometimes we can explain things by using a simplified version of a model. For example, thinking of atoms as tiny snooker balls is sufficient to explain the states of matter, but is not detailed enough to describe how chemical bonds are formed.

Many chemical and nuclear processes can be explained by a simple model of atomic structure in which atoms are thought to be made from three types of sub-atomic particle: **protons**, **neutrons** and **electrons**. This is not the whole story, but it's enough for present purposes. Protons and neutrons form the dense nucleus (or centre) of atoms. Electrons are much more diffuse and move around the nucleus in a way that is described in **Sections 2.3** and **2.4**. Figure 1 shows a very simple model of the atom. It is not to scale. The nucleus is tiny compared with the volume occupied by the electrons. If you imagined the atom to be the size of Wembley Stadium, the nucleus would be the size of a pea on the centre spot! Since most atoms have a radius of 0.1–0.2 nm (1×10^{-10} to 2×10^{-10} m), the nucleus must be very, very small – about 1×10^{-15} m radius, in fact.

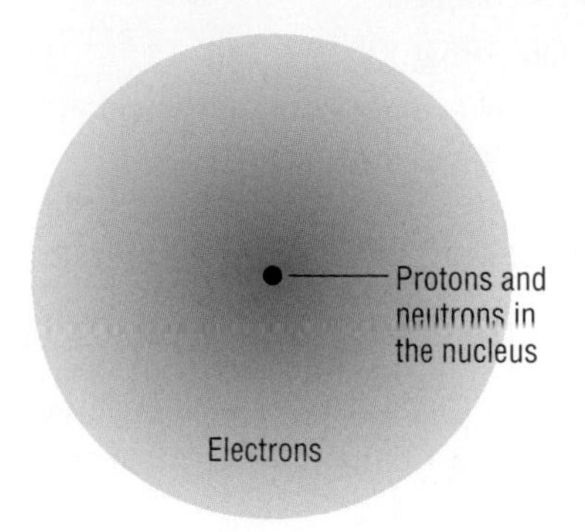

▲ **Figure 1** A simple model of the atom – not to scale.

Sub-atomic particles

Some properties of protons, neutrons and electrons are summarised in Table 1.

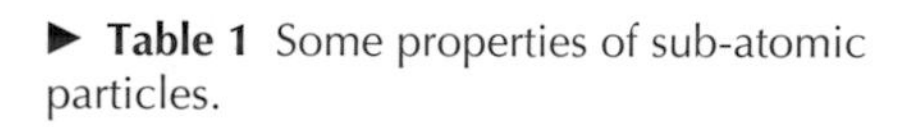

► **Table 1** Some properties of sub-atomic particles.

Particle	Mass on relative atomic mass scale	Charge (relative to proton)	Location in atom
proton	1	+1	in nucleus
neutron	1	0	in nucleus
electron	0.000 55	−1	around nucleus

Protons and electrons have equal but opposite electrical charges. Neutrons have no charge. Protons and neutrons have almost equal masses, and are much more massive than electrons. The nucleus accounts for almost all the mass of the atom and hardly any of its volume. Most of the atom is empty space.

It is the outer parts of atoms which interact together in chemical reactions, so for chemists, electrons are the most important particles.

Nuclear symbols

The nucleus can be described by just two numbers – the **atomic number** (symbol $\boldsymbol{Z}$) and the **mass number** (symbol $\boldsymbol{A}$).

The atomic number is the number of protons in the nucleus. It is also numerically equal to the charge on the nucleus. The atomic number is the same for every atom of an element – for example, $Z = 6$ for all carbon atoms.

The mass number is the number of protons plus neutrons in the nucleus. If the number of neutrons is given the symbol N, then

$$A = Z + N$$

Nuclear symbols identify the mass number and the atomic number as well as the symbol for the element. Figure 2 gives an example.

Sometimes, the atomic number is omitted because the chemical symbol tells us which element it is anyway. So $^{12}_{6}C$ can be simplified to ^{12}C or carbon-12.

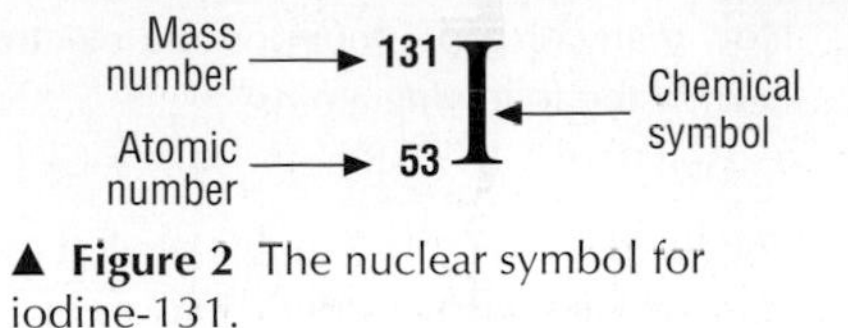

▲ **Figure 2** The nuclear symbol for iodine-131.

What are isotopes?

Atoms of the same element which have different mass numbers are called **isotopes**. Because the number of protons is the same for all atoms of an element, the differences in mass must arise from different numbers of neutrons.

Most elements exist naturally as a mixture of isotopes. The relative atomic mass is an average of the masses of the isotopes (**relative isotopic masses**) taking into account their abundances. Isotopes of some elements are given in Table 2, together with their percentage abundances in a naturally occurring sample.

◀ **Table 2** Isotopes of some elements.

Element	Isotope	Abundance
chlorine	^{35}Cl ^{37}Cl	75% 25%
iron	^{54}Fe ^{56}Fe ^{57}Fe ^{58}Fe	5.8% 91.7% 2.2% 0.3%
bromine	^{79}Br ^{81}Br	50% 50%
calcium	^{40}Ca ^{42}Ca ^{43}Ca ^{44}Ca ^{48}Ca	96.9% 0.7% 0.1% 2.1% 0.2%

The information in Table 2 has been obtained from studies of the elements using mass spectrometry. You can find out more about mass spectrometry in **Section 6.5**.

This type of data can be used to calculate the relative atomic mass of elements. For example, the relative atomic mass of chlorine is

$$\frac{(35.0 \times 75) + (37.0 \times 25)}{100} = \frac{2625 + 925}{100} = \frac{3550}{100} = 35.5$$

The relative atomic mass (A_r) of chlorine = 35.5

Problems for 2.1

1 Copy and complete the table.

Isotope	Symbol	Atomic number	Mass number	Number of neutrons
carbon-12	$^{12}_{6}C$	6	12.0	6
carbon-13	?	?	13.0	7
oxygen-16	$^{16}_{8}O$	8	16.0	?
strontium-90	$^{90}_{38}Sr$	38	?	52
iodine-131	$^{131}_{53}I$	?	?	?
?	$^{123}_{53}I$	?	?	?

2 How many protons, neutrons and electrons are present in each of the following atoms?
  a $^{79}_{35}Br$ b $^{81}_{35}Br$ c $^{35}_{17}Cl$ d $^{37}_{17}Cl$

3 Use the data in Table 2 in this section (page 17) to calculate the relative atomic masses of:
  a bromine b calcium.

4 The relative atomic mass of the element iridium is 192.2. Iridium occurs naturally as a mixture of iridium-191 and iridium-193. Work out the percentage of each isotope in naturally occurring iridium using the following steps.
  a If the percentage of iridium-193 in naturally occurring iridium is x%, there must be x atoms of iridium-193 in every 100 atoms of the element. How many atoms of iridium-191, therefore, must there be in every 100 atoms of iridium?
  b What will be the total relative mass of the atoms of iridium-193 in the 100-atom sample?
  c What will be the total relative mass of the atoms of iridium-191 in the 100-atom sample?
  d Write down the expression for the total relative mass of all 100 iridium atoms.
  e Write down the expression for the relative atomic mass of 1 iridium atom.
  f Use your answer to part **e** and the value of the relative atomic mass of iridium to work out the percentage of each isotope in naturally occurring iridium.

5 The relative atomic mass of antimony is 121.8. Antimony exists as two isotopes, antimony-121 and antimony-123. Calculate the relative abundances of the two isotopes.

6 The relative atomic mass of rubidium is 85.5. Rubidium consists of two isotopes, rubidium-85 and rubidium-87. Calculate the percentage of rubidium-87 in naturally occurring rubidium.

2.2 *Nuclear reactions*

Emissions from radioactive substances

Some isotopes of some elements are unstable. Their nuclei break down *spontaneously*, and they are described as being **radioactive**. These isotopes are called **radioisotopes.** As these nuclei break down, they emit rays and particles called **emissions**. This breakdown (or **radioactive decay**) occurs *of its own accord* – it isn't triggered off by something we do. Some isotopes decay very quickly; for others the process takes thousands of years.

Not all unstable atoms decay in the same way. Three different kinds of emissions have been identified – they are called **alpha (α)**, **beta (β)** and **gamma (γ) emissions.**

All three types of emissions are capable of knocking electrons out of the atoms they collide with – this ionises the atoms. Because of this, these emissions are sometimes referred to as **ionising radiation**. Some of the properties of α, β and γ emissions are summarised in Table 1.

▼ **Table 1** Some properties of α, β and γ emissions.

Property	Type of emission		
	α	β	γ
relative charge	+2	−1	0
relative mass	4	0.000 55	0
nature	2 protons + 2 neutrons (He nucleus)	electron (produced by nuclear changes)	very high frequency electromagnetic radiation
range in air	few centimetres	few metres	very long
stopped by	paper	aluminium foil	lead sheet
deflection by electrical field	low	high	nil

Nuclear equations

Nuclear equations summarise the processes which produce α and β radiation. They include the **mass number** (number of protons plus number of neutrons), **nuclear charge** (usually indicated by the **atomic number**, or number of protons) and **chemical symbol** for each particle involved. Both the mass and the charge must balance in a nuclear equation.

α decay involves the emission of α particles, which are helium nuclei, $^{4}_{2}He$. α decay is common among heavier elements with atomic numbers greater than 83. α decay reduces the mass of these heavy nuclei. The isotope produced by α decay will have a mass number four units lower and a nuclear charge two units lower than the original atom. An example is

$$^{238}_{92}U \rightarrow {}^{234}_{90}Th + {}^{4}_{2}He$$

β decay involves the emission of electrons, written as $^{0}_{-1}e$. β decay is common among lighter elements in which the isotopes contain a relatively large number of neutrons. For example,

$$^{14}_{6}C \rightarrow {}^{14}_{7}N + {}^{0}_{-1}e$$

During β decay, the mass number (protons and neutrons) remains constant but the nuclear charge (number of protons) *increases* by one unit. This means that a neutron, $^{0}_{1}n$, is converted into a proton, $^{1}_{1}p$, plus an electron, which is ejected from the nucleus:

$$^{0}_{1}n \rightarrow {}^{1}_{1}p + {}^{0}_{-1}e$$

Notice that α and β decay result in the production of a different element. For example, when uranium-238 undergoes α decay, it turns into thorium-234; when carbon-14 undergoes β decay, it turns into nitrogen.

γ decay is different – it is the emission of energy from a nucleus which is changing from a high energy level to a lower one. γ radiation often accompanies the emission of α and β particles.

Radiation from radioisotopes can be extremely dangerous, but under carefully controlled conditions some radioisotopes have important uses. For example, you can read about the use of radioactive tracers and radioactive decay in the **Elements of Life** module.

- In both α decay and β decay, new elements are formed.
- In γ decay, no new elements are formed.

Nuclear fusion

In a nuclear fusion reaction, two light atomic nuclei fuse together to form a single heavier nucleus of a new element. The process releases enormous quantities of energy.

Fusion reactions only occur at very high temperatures, such as those found in the Sun and other stars. For two nuclei to fuse, they must come very close together. At the normal temperatures found on the Earth, the positive nuclei repel one another so strongly that this cannot happen. At very high temperatures the nuclei are moving much more quickly and collide with so much energy that this repulsive energy barrier can be overcome. Once the nuclei are close enough, the strong nuclear forces holding protons and neutrons together in the nucleus take over and the nuclei fuse.

Nuclear fusion reactions take place in the gas clouds of stars and result in the formation of new elements. For example, when two hydrogen nuclei join together by nuclear fusion, hydrogen turns into helium. The energy released causes the gas cloud to glow.

Here are two examples of nuclear reactions that take place in the Sun involving the three isotopes of hydrogen ($^{2}_{1}H$ is called deuterium, $^{3}_{1}H$ is called tritium):

$$^{1}_{1}H + ^{2}_{1}H \rightarrow ^{3}_{2}He$$
$$^{2}_{1}H + ^{3}_{1}H \rightarrow ^{4}_{2}He + ^{1}_{0}n$$

All naturally occurring elements, including those found in our bodies, originated in stars as a result of nuclear fusion reactions. You can read about this process in the **Elements of Life** module.

At the high temperatures in the gas clouds of stars, the electrons have enough energy to escape from the nuclei. The gases exist in an ionised form called a **plasma** in which the positive atomic nuclei exist in a 'sea' of delocalised electrons (this means that the electrons are 'spread out' and not attached to any particular nucleus). The gases in the plasma have very different properties from the gases we know at lower temperatures. Although the idea of a plasma may seem strange to us on Earth, the Sun and other stars exist as plasma so it is the most common form of matter in the Universe.

It is hoped that nuclear fusion reactions can be harnessed under carefully controlled conditions to produce electrical energy on Earth. The problem, of course, is to safely generate and contain a plasma at the high temperatures needed for fusion to occur (over 10^{7} °C) and to make the process sustainable and economic. Scientists are still working on this problem. (See the green box 'Fusion on Earth' in the **Elements of Life** module.)

Half-life

Radioactive decay is a random process. Each nucleus in a sample of an isotope decays at random, regardless of what the other nuclei are doing, and regardless of outside conditions such as temperature and pressure. Decay is also independent of the chemical state of the isotope. For example, uranium would decay in the same way whether it was in a compound or in the free state as an element. Similarly, it would make no difference if the radioactive substance was solid, liquid, gas or in solution. Gradually, as more and more of the nuclei in the sample decay, the sample becomes less and less radioactive.

The time for half of the nuclei to decay is called the **half-life**. For any given isotope, the half-life is fixed – it doesn't matter how much of the isotope is present, or the temperature or pressure. In the course of one half-life, half of the radioactivity of the sample always disappears. But the sample never completely disappears – that's why we can still detect radioactive isotopes even though the Earth was formed over four billion years ago.

Table 2 shows some examples of isotopes and their half-lives. For example, strontium-90 has a half-life of 28 years. If you started now with 8 g of strontium-90, in 28 years' time half the nuclei would have decayed, and only 4 g of strontium-90 would be left. In 56 years' time, only 2 g would be left. By that time the isotope would have gone through two half-lives: the original quantity would have been halved and halved again (Figure 1).

► **Table 2** Some isotopes and their half-lives.

Isotope	Half-life
uranium-238	4.5×10^{9} years
carbon-14	5.7×10^{3} years
strontium-90	28 years
iodine-131	8.1 days
bismuth-214	19.7 minutes
polonium-214	1.5×10^{-4} seconds

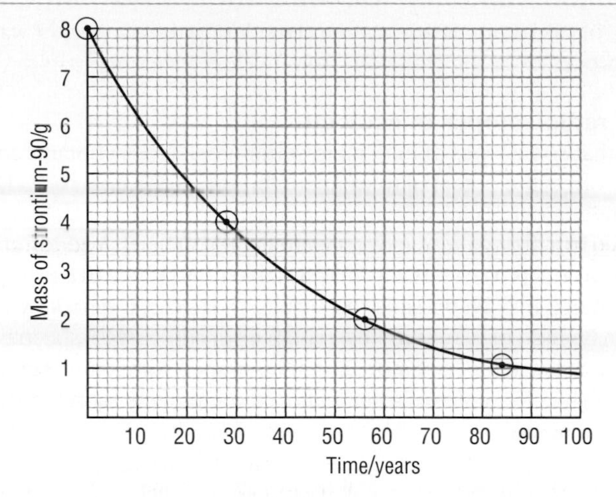

◀ **Figure 1** The half-life of strontium-90 is 28 years. This means that the amount of strontium-90 halves every 28 years.

Calculating and using half-life

Suppose that a sample of a radioisotope has a count rate of 52 counts min^{-1} and 4 hours later the count rate is 13 counts min^{-1}, the half-life of the radioisotope can be calculated as follows.

At the start, the count rate was 52 counts min^{-1}.
So, after 1 half-life the count rate would be 26 counts min^{-1}.
And after 2 half-lives the count rate would be 13 counts min^{-1}.
So the count rate would have fallen to 13 counts min^{-1} after 2 half-lives.
We are told that this actually took 4 hours.
If 2 half-lives is equal to 4 hours then 1 half-life is equal to 2 hours.
The half-life of this radioisotope is 2 hours.

If you know the half-life of a radioactive isotope, you can use it to calculate:

- the time needed for the activity, or the mass of the isotope, to fall to a certain value
- the mass of the isotope, or its activity, after it has decayed for a certain time.

Example 1
The half-life of the radioisotope carbon-14 is 5730 years. How long will it take for the radioactive count rate to drop to 25% of its original value?

At the start it has 100% activity (the original count rate)
After 1 half-life it will be 50% of the original count rate.
After 2 half-lives it will be 25% of the original count rate.

This means that over 2 half-lives the count drops to 25% of the original.

As 1 half-life is 5730 years, the time taken for the isotope to decay to 25% of its original value is 2 × 5730 years = 11460 years.

Calculations similar to this can be used to date objects containing radioisotopes (see below).

Example 2
Another radioisotope, iodine-131, has a half-life of approximately 8 days. If you start with 8 g of the isotope, how much of it will be left after 24 days?

Number of half-lives = $\frac{24}{8} = 3$
At the start we have 8 g of the isotope.
After 1 half-life we have ½ of the original amount, i.e. 8 ÷ 2 = 4 g
After 2 half-lives we have ¼ of the original amount, i.e. 8 ÷ 4 = 2 g
After 3 half-lives we have ⅛ of the original amount, i.e. 8 ÷ 8 = 1 g

This type of calculation is done when considering the dangers of waste from nuclear processes or the emissions from nuclear accidents.

Uses of radioisotopes in geology

The fact that some types of mineral in certain rock types contain radioactive elements offers the potential to estimate the age of the host rock. They can be used as geological clocks.

If the original, or parent, radioisotope decays through a series of daughter products to eventually form a stable non-radioactive isotope of an element then it may be possible to estimate the age of the rock by working out the mass ratio of parent to final daughter product. Mass spectrometry is used here. You can find out more about mass spectrometry in **Section 6.5**.

Table 3 gives examples of some radioactive elements found in naturally occurring minerals.

► **Table 3** Examples of radioactive elements found in natural minerals.

Radioactive parent isotope	Stable daughter isotope	Half-life (years)
potassium-40	argon-40	1.3×10^{9}
rubidium-87	strontium-87	4.5×10^{10}
thorium-232	lead-208	1.4×10^{10}
uranium-235	lead-207	7.0×10^{8}
uranium-238	lead-206	4.5×10^{9}

Most minerals that contain radioactive elements are found in igneous rocks and the dates they give indicate the time when the magma cooled and crystallised. For example, potassium-40 occurs in some feldspar minerals; the ratio of potassium-40 to argon-40 decreases over time and the value of the ratio can be used to date the host rock, providing that the argon gas has been trapped in the rock. This is more likely in igneous rocks.

If $^{40}Ar : ^{40}K = 1:1$ (50% of ^{40}K decayed, 50% left) the rocks are 1.3×10^{9} years old
If $^{40}Ar : ^{40}K = 3:1$ (75% of ^{40}K decayed, 25% left) the rocks are 2.6×10^{9} years old
If $^{40}Ar : ^{40}K = 7:1$ (87.5% of ^{40}K decayed, 12.5% left) the rocks are 3.9×10^{9} years old.

Although the oldest rocks dated on Earth are about 4.0×10^{9} years, rocks from the Moon and meteorites all date at around 4.6×10^{9} years. Meteorites are similar in composition to the Earth and this suggests that they all formed at around the same time as our Solar System, thus suggesting that the Earth formed about 4.6 billion years ago.

A cautionary note

For radiometric systems to work as geological clocks there are some criteria that must be fulfilled:

- The half-life of the radioisotope must be known accurately.
- There must have been no movement of either parent or daughter isotopes out of or into a mineral since the time of crystallisation of the rock. This is a particular problem if a daughter product is gaseous – e.g. argon.
- There must have been no resetting of the radiometric clock – heating and deformation of rocks (often referred to as metamorphism) can cause redeposition or loss of radioisotopes, effectively resetting the 'clock hands'.

Uses of radioisotopes in archaeology – radiocarbon dating

The carbon-14 carbon isotope occurs naturally and is present along with the carbon-12 isotope in all living things. The ratio of carbon-12 to carbon-14 does not change in a living organism, but when an organism dies the carbon-14 is not replenished and the ratio of carbon-12 to carbon-14 gradually increases. If this ratio can be measured it can give an estimate of the age of the material since it died. (See Example 1, page 21.)

The half-life for carbon-14 is 5730 years and this makes it useful for dating archaeological remains up to about 50 000 years old. At ages beyond this, the amount of carbon-14 remaining is so small that measurement is difficult and errors become too large for a sensible age range to be produced.

Radioactive tracers

Radioisotopes can be injected into the body to locate problems including cancerous tumours. Examples include iodine-131 used to detect thyroid problems and technetium-99 used to detect tumours.

Generally, radioisotopes used as tracers should be gamma emitters, enabling detection and minimising the ionisation of cells – however, even alpha emitters have their uses. For example, radium-226 can be used to treat cancer by injecting tiny amounts of radium into a tumorous mass.

Tracer radioisotopes should have a relatively short half-life (typically around a few hours) to allow detection but to minimise harm to the patient.

Problems for 2.2

You will need to refer to the Periodic Table to identify the symbols for the elements produced.

1 Write nuclear equations for the α decay of
a $^{238}_{94}Pu$ **b** $^{221}_{87}Fr$ **c** $^{230}_{90}Th$

2 Write nuclear equations for the β decay of
a $^{90}_{38}Sr$ **b** $^{131}_{53}I$ **c** $^{231}_{90}Th$

3 Write a nuclear equation for each of the following:
a A $^{7}_{3}Li$ nucleus absorbs a colliding proton and then disintegrates into two identical fragments.
b The production of carbon-14 by collision of a neutron with an atom of nitrogen-14.
c The collision of an alpha particle with an atom of nitrogen-14 releasing a proton and forming another element. (Rutherford observed this reaction in 1919. It was the first nuclear reaction ever observed.)
d The collision of an alpha particle with an atom of aluminium-27 to form phosphorus-30.

4 Part of the decay series involving radon-222 is shown below. Copy it out and fill in the missing symbols for atomic number, mass number and type of decay.

$$_{88}Ra \underset{\alpha}{\rightarrow} {}^{222}_{86}Rn \rightarrow Po \underset{\alpha}{\rightarrow} {}^{214}Pb$$

5 $^{232}_{90}Th$ emits a total of six α and four β particles in its natural decay series. What are the atomic number and the mass number of the final product?

6 This problem refers to the isotopes in Table 2 in this section (page 20). Assume that you begin with a 10 g sample of each isotope.
a How much uranium-238 would be left after 4.5×10^9 years?
b How much ^{214}Bi would be left after 78.8 minutes?
c How long would it take for you to be left with 1.25 g of ^{214}Po?
d How much ^{14}C would be left after 4.56×10^4 years?

7 The isotope fermium-257 has a radioactive half-life of 80 days.
a Draw a graph to show how the mass of this isotope changes over the period of one year, starting with 10 g of the element.
b Use your graph to find the mass of fermium-257 remaining after 100 days.
c Use your graph to find the number of days that pass before the mass of the isotope falls to 2 g.

8 Carbon from a piece of wood taken from a beam found in an ancient tomb gave a reading of 9 counts per minute per gram. New wood gives a reading of 36 counts per minute per gram.
a Estimate the year in which the tomb was built, given that the half-life of C-14 is 5730 years.
b Why should you treat your answer with some caution?

2.3 *Electronic structure: shells*

Section 6.1 describes how Neils Bohr's theory of the hydrogen atom explained the appearance of the emission spectrum of hydrogen. However, Bohr's theory only worked for the simple hydrogen atom, with its one electron. The theory needed extending to make sense of the structure of atoms which contain several electrons.

We now know that, for these atoms, energy levels 2, 3, 4 ... have a more complex structure than the single levels which exist in hydrogen. You will find out more about this in **Section 2.4**. It is more appropriate to talk about the first, second, third ... **electron shell** rather than energy level 1, 2, 3 ... The shells are labelled by giving each a **principal quantum number, *n***. For the first shell, $n = 1$; for the second shell $n = 2$, and so on. The higher the value of n, the further the shell is from the nucleus and the higher the energy associated with the shell.

Although each shell can hold more than one electron, there is a limit. The maximum numbers of electrons which can be held in the first four shells are:

first shell	($n = 1$)	2 electrons
second shell	($n = 2$)	8 electrons
third shell	($n = 3$)	18 electrons
fourth shell	($n = 4$)	32 electrons.

A shell which contains its maximum number of electrons is called a **filled shell**. Electrons are arranged so that the lowest energy shells are filled first. Table 1 shows the **electron shell configuration** for elements 1 to 36. The electron shell configuration of sodium, for example, is written as 2.8.1 meaning:

2 electrons in first shell
8 electrons in second shell
1 electron in third shell.

Notice that the filling of shell 3 is not straightforward. This is because of the pattern of energy levels within shells 3 and 4. Electrons 19 and 20 have lower energy if they are placed in shell 4 rather than in shell 3. This is explained in **Section 2.4** and other examples such as copper and chromium are explained in **Section 11.5**.

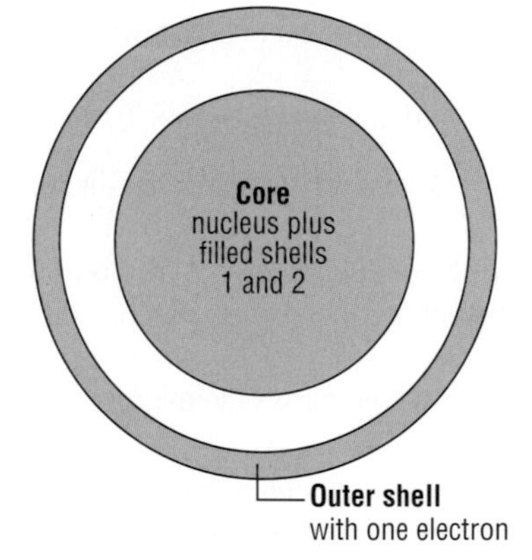

▲ **Figure 1** The core and outer shell model of a sodium atom (2.8.1).

Chemists can explain many of the properties of atoms without needing to use a detailed theory of atomic structure. Much chemistry is decided only by the outer shell electrons, and one very useful model treats the atom as being composed of a **core** of the nucleus plus the inner electrons shells, surrounded by an **outer shell**. Figure 1 shows how a sodium atom would be thought of on this model.

Electronic structure and the Periodic Table

The arrangement of elements by rows and columns in the Periodic Table is a direct result of the electronic structure of atoms.

The *number of outer shell* electrons determines the **group** number. For example Li, Na, K all have *one* outer shell electron and are in Group 1; C and Si have *four* outer shell electrons and are in Group 4. The noble gases all have eight electrons in their outer shells (except helium which has two) and are in Group 0 (the zero refers to the shell *beyond* the filled outer one, which does not have any electrons in it yet).

Element	Atomic number	1st shell	2nd shell	3rd shell	4th shell
hydrogen	1	1			
helium	2	2			
lithium	3	2	1		
beryllium	4	2	2		
boron	5	2	3		
carbon	6	2	4		
nitrogen	7	2	5		
oxygen	8	2	6		
fluorine	9	2	7		
neon	10	2	8		
sodium	11	2	8	1	
magnesium	12	2	8	2	
aluminium	13	2	8	3	
silicon	14	2	8	4	
phosphorus	15	2	8	5	
sulfur	16	2	8	6	
chlorine	17	2	8	7	
argon	18	2	8	8	
potassium	19	2	8	8	1
calcium	20	2	8	8	2
scandium	21	2	8	9	2
titanium	22	2	8	10	2
vanadium	23	2	8	11	2
chromium	24	2	8	13	1
manganese	25	2	8	13	2
iron	26	2	8	14	2
cobalt	27	2	8	15	2
nickel	28	2	8	16	2
copper	29	2	8	18	1
zinc	30	2	8	18	2
gallium	31	2	8	18	3
germanium	32	2	8	18	4
arsenic	33	2	8	18	5
selenium	34	2	8	18	6
bromine	35	2	8	18	7
krypton	36	2	8	18	8

◀ **Table 1** Electron shell configurations for the first 36 elements.

The number of the *shell* which is being filled determines the **period** to which an element belongs. From Li to Ne, the outer electrons are being placed in shell 2, so these elements belong to Period 2. The more detailed description of electronic structure in **Section 2.4** also explains how the elements are assigned to *blocks* in the Periodic Table.

First ionisation enthalpies

Figure 2 shows how **first ionisation enthalpy** varies with atomic number for elements 1–56. (Look at **Section 2.5** for an explanation of ionisation enthalpies.) Notice how the elements at the peaks are all in Group 0 (the noble gases). These elements have high first ionisation enthalpies – they are difficult to ionise and are very unreactive. The elements in the troughs are all in Group 1 (the alkali metals). These elements, with only one outer shell electron, have low ionisation enthalpies – they are easy to ionise, leaving them with a full outer shell. This makes them very reactive.

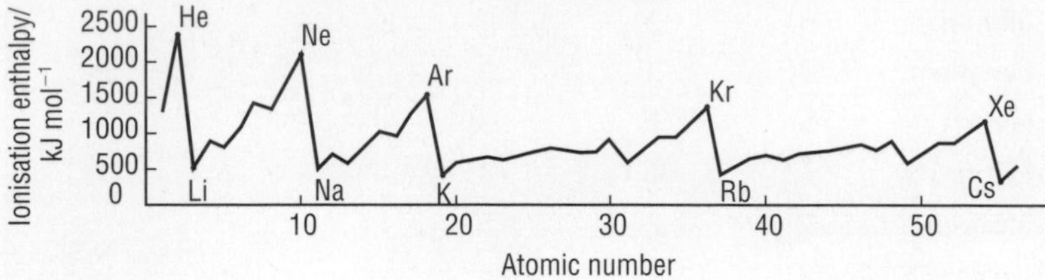

▶ **Figure 2** Variation of first ionisation enthalpy with atomic number for elements with atomic numbers 1 to 56.

These patterns in first ionisation enthalpies provide evidence to support our understanding of the nature of electron shells.

Problems for 2.3

1 Isotopes of elements have different numbers of neutrons. Do they have different electron arrangements? Explain your answer.

2 Look at Figure 1 in this section (page 24) and draw a similar labelled diagram for the core and outer shell model of a fluorine atom.

3 The electron shell configuration for sodium can be written as 2.8.1. Use this notation to write down the electron shell configurations for
  a lithium
  b phosphorus
  c calcium.

4 Copy and complete the following table, which shows the electron shell configurations of some elements and the groups and periods to which they belong.

Electron shell configuration	Group	Period
2.8.7	7	3
2.3	?	?
2.8.6	?	?
?	4	2
?	4	3
2.1	?	?
2.8.1	?	?
2.8.8.1	?	?

5 The electron shell configurations of elements **A** to **E** are given below. Which of these elements are in the same group?

A 2.8.2 **B** 2.6
C 2.8.8.2 **D** 2.7
E 2.2

2.4 *Electronic structure: sub-shells and orbitals*

Sub-shells of electrons

Section 2.3 describes how the electrons in atoms are arranged in shells. Much of our knowledge of electron shells has come from studying the emission spectrum of hydrogen (see **Section 6.1**). The hydrogen atom has only one electron and its spectrum is relatively simple to interpret. When we come to look at elements other than hydrogen, we find their spectra are much more complex – electron shells are not the whole story. The shells are themselves split up into **sub-shells**.

The sub-shells are labelled **s**, **p**, **d** and **f**. The $n = 1$ shell has only an s sub-shell; the $n = 2$ shell has two sub-shells – s and p; the $n = 3$ shell has three sub-shells – s, p and d; and the $n = 4$ shell has four sub-shells – s, p, d and f.

Sub-shell	Maximum number of electrons
s	2
p	6
d	10
f	14

◀ **Table 1** Maximum number of electrons in the s, p, d and f sub-shells

The different types of sub-shells can hold different numbers of electrons. These are given in Table 1. So,

the $n = 1$ shell can hold 2 electrons in the s sub-shell

the $n = 2$ shell can hold 2 electrons in its s sub-shell
6 electrons in its p sub-shell
a total of 8 electrons.

the $n = 3$ shell can hold 2 electrons in its s sub-shell
6 electrons in its p sub-shell
10 electrons in its d sub-shell
a total of 18 electrons.

the $n = 4$ shell can hold 2 electrons in its s sub-shell
6 electrons in its p sub-shell
10 electrons in its d sub-shell
14 electrons in its f sub-shell
a total of 32 electrons.

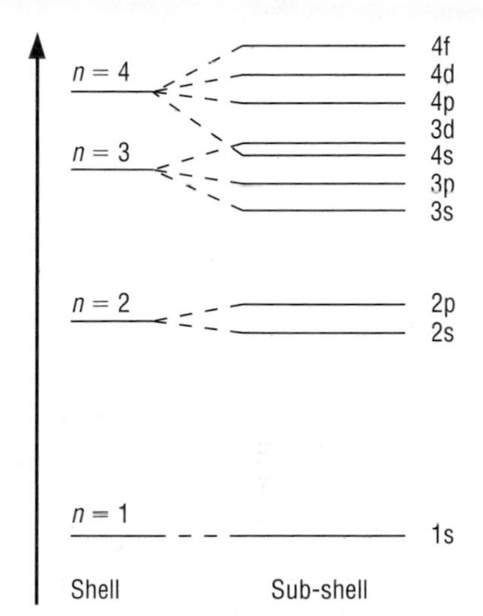

▲ **Figure 1** Energies of electron sub-shells from $n = 1$ to $n = 4$ in a typical many-electron atom. The energy of a sub-shell is not fixed, but falls as the charge on the nucleus increases as you go from one element to the next in the Periodic Table. The order shown in the diagram is correct for the elements in Period 3 and up to nickel in Period 4. After nickel, the 3d sub-shell has lower energy than 4s.

In atoms other than hydrogen, the sub-shells within a shell have different energies. Figure 1 shows the relative energies of the sub-shell for each of the shells $n = 1$ to $n = 4$ in a typical many-electron atom. Note the overlap of energy between the $n = 3$ and $n = 4$ shells in Figure 1. This has important consequences which you will meet later.

Atomic orbitals

The s, p, d and f sub-shells are themselves divided further into **atomic orbitals**. An electron in a given orbital can be found in a particular region of space around the nucleus.

An s sub-shell always contains *one* s atomic orbital
A p sub-shell always contains *three* p atomic orbitals
A d sub-shell always contains *five* d atomic orbitals
An f sub-shell always contains *seven* f atomic orbitals

You can see the energy levels associated with each of these atomic orbitals in Figure 2. In an isolated atom, orbitals in the same sub-shell have the same energy.

Don't think of an atomic orbital as a fixed electron orbit. The position of an electron cannot be mapped exactly. For an electron in a given atomic orbital, we only know the *probability* of finding the electron in any region – Figure 3 shows one way of representing this for a 1s orbital.

Each atomic orbital can hold a maximum of two electrons. Electrons in atoms have a **spin**, which you can picture as a spinning motion in one of two directions. Every electron spins at the same rate in either a clockwise (↑) or anticlockwise (↓) direction. Electrons can only occupy the same orbital if they have opposite or **paired** spins.

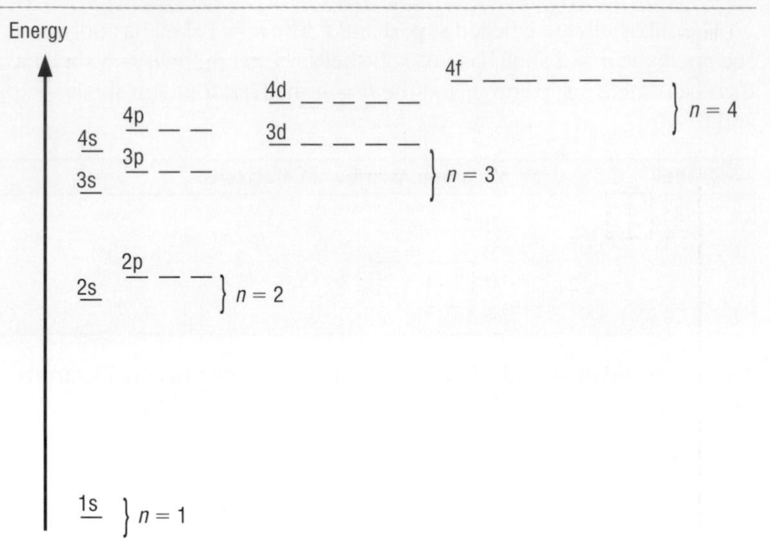

▶ **Figure 2** Energy levels of atomic orbitals in the n = 1 to n = 4 shells.

We can write this as

[↑↓]

where the box represents the atomic orbital and the arrows the electrons. To describe the state of an electron in an atom accurately you need to supply four pieces of information about it – a bit like an address:

- the electron shell it is in
- its sub-shell
- its orbital within the sub-shell
- its spin.

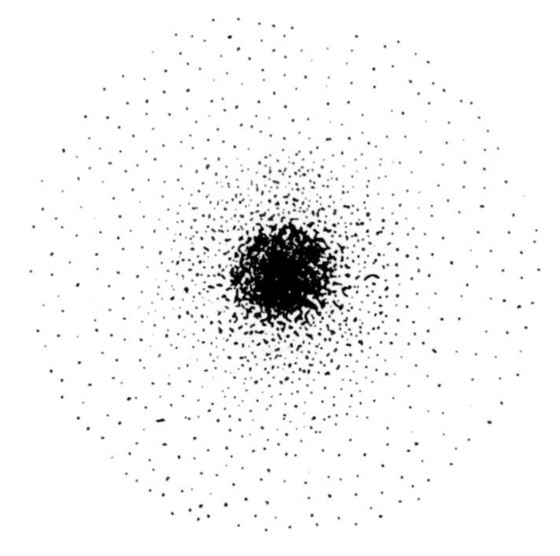

▲ **Figure 3** One way of representing a 1s electronic orbital. The dots represent the probability of finding the electron in that region – the denser the dots, the higher the probability of finding the electron there.

Filling up atomic orbitals

The arrangement of electrons in shells and orbitals is called the **electronic configuration** of an atom. The orbitals are filled in a definite order to produce the lowest energy arrangement possible.

The orbitals are filled in order of increasing energy. Where there is more than one orbital with the same energy, these orbitals are first occupied singly by electrons. This keeps the electrons in an atom as far apart as possible. Only when every orbital is singly occupied do the electrons pair up in orbitals. For the lowest energy arrangement, electrons in singly occupied orbitals have parallel spins.

Using these rules, you can now assign the 11 electrons in a sodium atom to atomic orbitals. This is done in Figure 4.

The electronic configuration of a sodium atom can be represented in shorthand notation as

$1s^2 2s^2 2p^6 3s^1$

The large numbers show the principal quantum number of each shell, the letters show the sub-shells and the small superscripts indicate the numbers of electrons in each sub-shell.

Building up the Periodic Table

Hydrogen is the simplest element, with atomic number Z = 1. It has one electron which will occupy the s orbital of the n = 1 shell.

1s [↑]   H $1s^1$

The next element, helium (Z = 2) has two electrons, which both occupy the 1s orbital with paired spins.

1s [↑↓]   He $1s^2$

◀ **Figure 4** Arrangement of electrons in atomic orbitals in a ground state sodium atom.

Lithium (Z = 3) has three electrons. The third electron cannot fit in the n = 1 shell, and so occupies the next lowest orbital, the 2s orbital,

1s [↑↓] 2s [↑] Li $1s^2 2s^1$

and so on across the first short period.

Nitrogen (Z = 7) has seven electrons

1s [↑↓] 2s [↑↓] 2p [↑ | ↑ | ↑] N $1s^2 2s^2 2p^3$

The three electrons occupying the 2p sub-shell occupy the three separate p orbitals singly and their spins are parallel.

Oxygen (Z = 8) has eight electrons

1s [↑↓] 2s [↑↓] 2p [↑↓ | ↑ | ↑] O $1s^2 2s^2 2p^4$

The electronic configurations of elements with atomic numbers 1 to 36 are shown in Table 2 (page 30).

To write the electronic configuration of the next element, scandium (Z = 21), you need to look back to Figure 1.

The energy level of the 3d sub-shell lies just above that of the 4s sub-shell but just below the 4p sub-shell. This means that, once the 4s level is filled in calcium (Z = 20), the next element, scandium, has the electronic structure $1s^2 2s^2 2p^6 3s^2 3p^6 3d^1 4s^2$. The 3d sub-shell continues to be filled across the period in the elements Sc to Zn. Zinc (Z = 30) has the electronic configuration $1s^2 2s^2 2p^6 3s^2 3p^6 3d^{10} 4s^2$.

Note that the 3d sub-shell is written alongside the other n = 3 sub-shells even though it is filled after the 4s sub-shell. The filling of the 3d sub-shell has an important effect on the chemistry of the elements Sc to Zn.

Once the 3d orbitals are filled, subsequent electrons go into the 4p sub-shell.

In this way we can understand the building up of the Periodic Table (see Figure 5, page 31). Period 2 (Li to Ne) corresponds to the filling of the 2s and 2p orbitals, Period 3 (Na to Ar) to the filling of the 3s and 3p orbitals and so on.

The d-block elements (Sc to Zn, Y to Cd and La to Hg) correspond to the filling of the d orbitals (3d, 4d and 5d, respectively) and the lanthanides and actinides to the filling of the 4f and 5f orbitals, respectively.

▶ **Table 2** Ground state electronic configurations of elements with atomic numbers 1–36.

Atomic number (*Z*)	Element	Electronic configuration
1	hydrogen	$1s^1$
2	helium	$1s^2$
3	lithium	$1s^2 2s^1$
4	beryllium	$1s^2 2s^2$
5	boron	$1s^2 2s^2 2p^1$
6	carbon	$1s^2 2s^2 2p^2$
7	nitrogen	$1s^2 2s^2 2p^3$
8	oxygen	$1s^2 2s^2 2p^4$
9	fluorine	$1s^2 2s^2 2p^5$
10	neon	$1s^2 2s^2 2p^6$
11	sodium	$1s^2 2s^2 2p^6 3s^1$
12	magnesium	$1s^2 2s^2 2p^6 3s^2$
13	aluminium	$1s^2 2s^2 2p^6 3s^2 3p^1$
14	silicon	$1s^2 2s^2 2p^6 3s^2 3p^2$
15	phosphorus	$1s^2 2s^2 2p^6 3s^2 3p^3$
16	sulfur	$1s^2 2s^2 2p^6 3s^2 3p^4$
17	chlorine	$1s^2 2s^2 2p^6 3s^2 3p^5$
18	argon	$1s^2 2s^2 2p^6 3s^2 3p^6$
19	potassium	$1s^2 2s^2 2p^6 3s^2 3p^6 4s^1$
20	calcium	$1s^2 2s^2 2p^6 3s^2 3p^6 4s^2$
21	scandium	$1s^2 2s^2 2p^6 3s^2 3p^6 3d^1 4s^2$
22	titanium	$1s^2 2s^2 2p^6 3s^2 3p^6 3d^2 4s^2$
23	vanadium	$1s^2 2s^2 2p^6 3s^2 3p^6 3d^3 4s^2$
24	chromium	$1s^2 2s^2 2p^6 3s^2 3p^6 3d^5 4s^1$
25	manganese	$1s^2 2s^2 2p^6 3s^2 3p^6 3d^5 4s^2$
26	iron	$1s^2 2s^2 2p^6 3s^2 3p^6 3d^6 4s^2$
27	cobalt	$1s^2 2s^2 2p^6 3s^2 3p^6 3d^7 4s^2$
28	nickel	$1s^2 2s^2 2p^6 3s^2 3p^6 3d^8 4s^2$
29	copper	$1s^2 2s^2 2p^6 3s^2 3p^6 3d^{10} 4s^1$
30	zinc	$1s^2 2s^2 2p^6 3s^2 3p^6 3d^{10} 4s^2$
31	gallium	$1s^2 2s^2 2p^6 3s^2 3p^6 3d^{10} 4s^2 4p^1$
32	germanium	$1s^2 2s^2 2p^6 3s^2 3p^6 3d^{10} 4s^2 4p^2$
33	arsenic	$1s^2 2s^2 2p^6 3s^2 3p^6 3d^{10} 4s^2 4p^3$
34	selenium	$1s^2 2s^2 2p^6 3s^2 3p^6 3d^{10} 4s^2 4p^4$
35	bromine	$1s^2 2s^2 2p^6 3s^2 3p^6 3d^{10} 4s^2 4p^5$
36	krypton	$1s^2 2s^2 2p^6 3s^2 3p^6 3d^{10} 4s^2 4p^6$

Chemical properties and electronic structure

The chemical properties of an element are decided by the electrons in the incomplete outer shells – these are the electrons that are involved in chemical reaction.

Compare the electronic arrangement of the noble gases:

He $1s^2$
Ne $1s^2 2s^2 2p^6$
Ar $1s^2 2s^2 2p^6 3s^2 3p^6$
Kr $1s^2 2s^2 2p^6 3s^2 3p^6 3d^{10} 4s^2 4p^6$

All have sub-shells fully occupied by electrons. Such arrangements are called **closed shell** arrangements. These are particularly stable arrangements, but note that in Ar and Kr the outer shell is only 'temporarily full' and can expand further by filling d and f sub-shells.

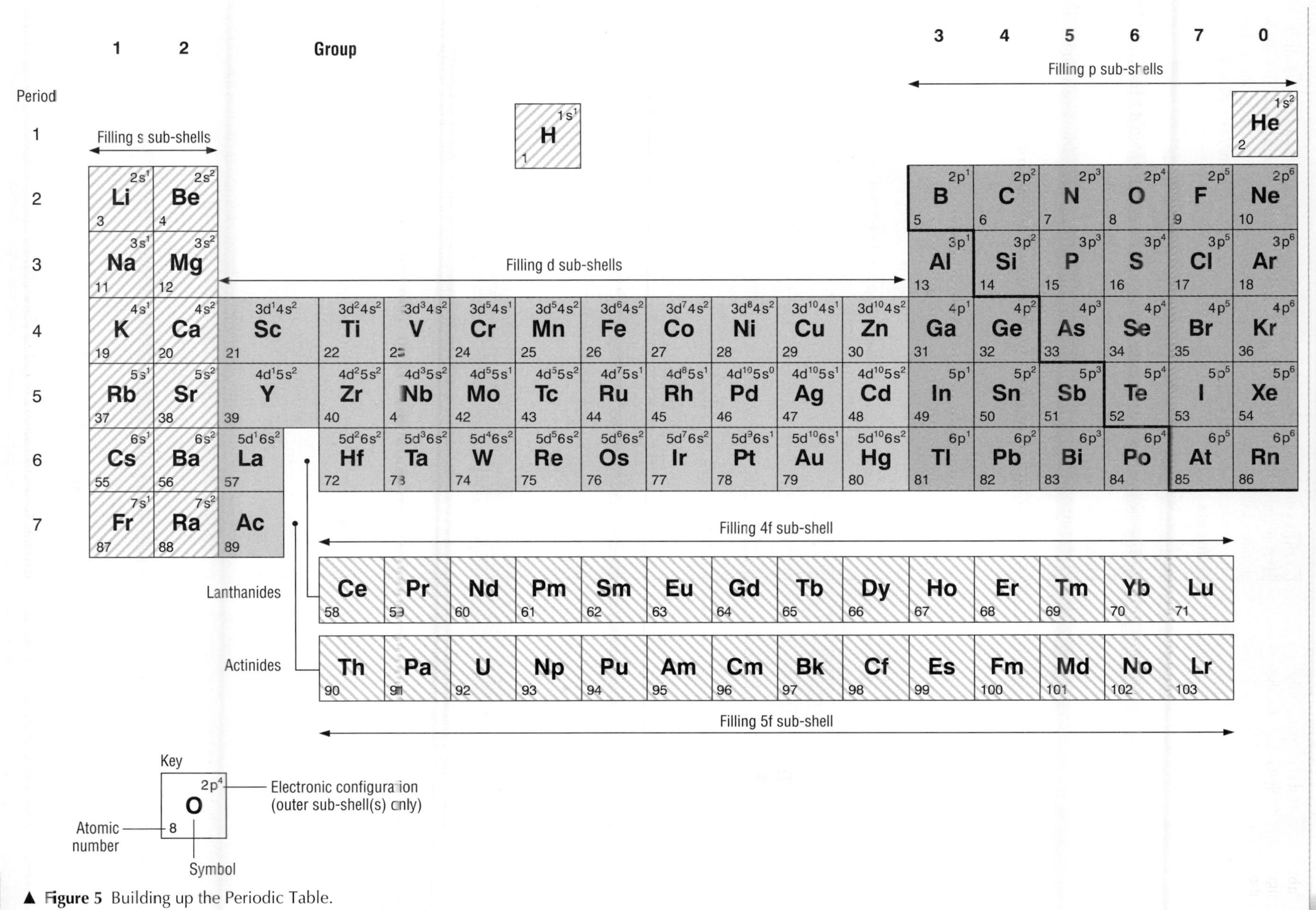

▲ **Figure 5** Building up the Periodic Table.

Now compare the electronic configurations of the elements in Group 1. They all have one electron in the outermost s sub-shell and as a result show similar chemical properties. Group 2 elements all have two electrons in the outermost s sub-shell. Groups 1 and 2 elements are known as **s-block elements**.

In Groups 3, 4, 5, 6, 7 and 0 the outermost p sub-shell is being filled. These elements are known as **p-block elements**.

The elements where a d sub-shell is being filled are called **d-block elements**, and those where an f sub-shell is being filled are called **f-block elements**.

Figure 6 summarises the blocks in the Periodic Table. Dividing it up in this way is very useful to chemists since it groups together elements with similar electronic configurations and similar chemical properties.

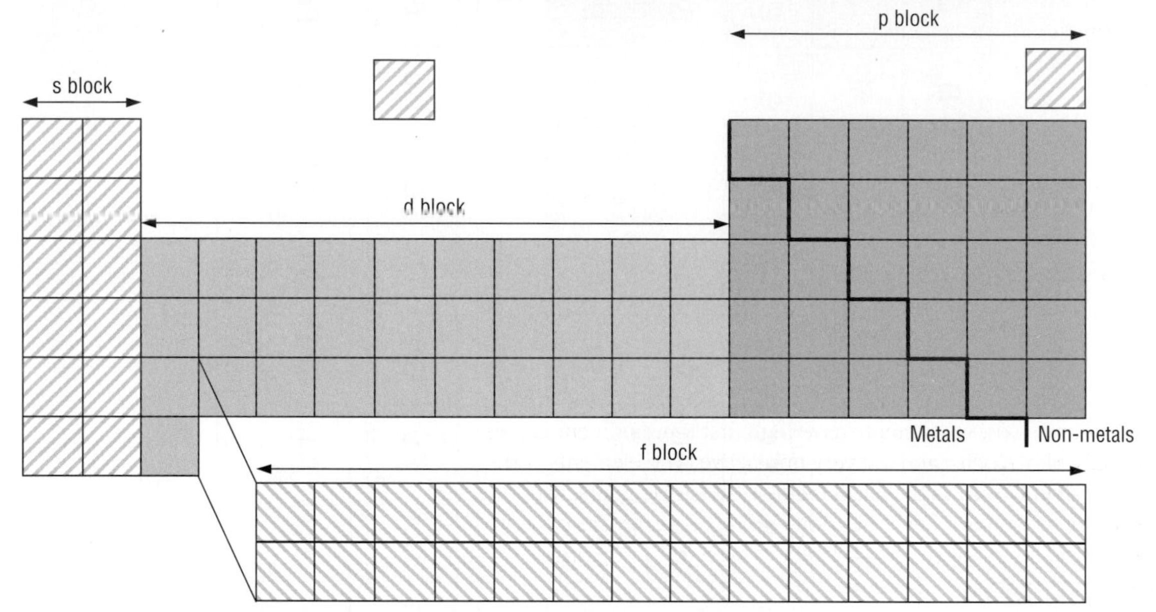

▲ **Figure 6** Dividing up the Periodic Table.

Problems for 2.4

1 Look at Figure 6 and classify the s-, p-, d- and f-blocks as containing metals, non-metals or a mixture of both.

2 When we use the notation $1s^2$ in an electronic configuration:

- **a** what does the '1' in front of the 's' tell us?
- **b** what does the 's' tell us?
- **c** what does the superscript '2' after the 's' tell us?

3 Classify the following elements as s-, p-, d- or f-block elements.

- **a** $1s^2 2s^2 2p^6 3s^2 3p^6 3d^{10} 4s^2 4p^6 5s^1$
- **b** $1s^2 2s^2 2p^6 3s^2 3p^4$
- **c** $1s^2 2s^2 2p^6 3s^2 3p^6 3d^{10} 4s^2 4p^6$
- **d** $1s^2 2s^2 2p^6 3s^2 3p^6 3d^{10} 4s^2 4p^6 4d^{10} 4f^4 5s^2 5p^6 6s^2$
- **e** chromium
- **f** aluminium
- **g** uranium
- **h** strontium

4 The electronic configuration of the outermost shell of an atom of an element X is $3s^2 3p^4$. What is the atomic number and name of the element?

5 Electronic configurations are sometimes abbreviated by labelling the core of filled inner shells as the electronic configuration of the appropriate noble gas. For example, the electronic configuration of neon is $1s^2 2s^2 2p^6$ and that of sodium is $1s^2 2s^2 2p^6 3s^1$ – so we can write the electronic configuration of sodium as $[Ne]3s^1$. Name the elements whose electronic configurations may be written as:

- **a** $[Ne]3s^2 3p^5$
- **b** $[Ar]4s^1$
- **c** $[Ar]3d^2 4s^2$
- **d** $[Kr]4d^{10} 5s^2 5p^2$

6 Write out the electronic configurations of the following atoms.

- **a** boron ($Z = 5$)
- **b** phosphorus ($Z = 15$)
- **c** chlorine ($Z = 17$)
- **d** calcium ($Z = 20$)
- **e** iron ($Z = 26$).

AS

2.5 Atoms and ions

First ionisation enthalpy

If sufficient energy is given to an atom, an electron is lost and the atom becomes a positive ion – **ionisation** has taken place. An input of energy is *always* needed to remove electrons because they are attracted to the nucleus.

When one electron is pulled out of an atom, the energy required is called the **first ionisation enthalpy** (or **first ionisation energy**). We define the first ionisation enthalpy of an element as the energy needed to remove one electron from every atom in one mole of isolated gaseous atoms of the element – a mole of gaseous ions with one positive charge are formed. The general equation for the first ionisation process is

$$X(g) \rightarrow X^+(g) + e^-$$

where X represents the symbol for the element. Let's consider oxygen. Its first ionisation enthalpy is $+1320\,kJ\,mol^{-1}$. This means that an input of 1320 kJ of energy is necessary to bring about the process

$$O(g) \rightarrow O^+(g) + e^-$$

It is the most loosely held electron that is removed on first ionisation. This will be one of the outer shell electrons since they are furthest from the nucleus.

Figure 1 shows how first ionisation enthalpy varies with atomic number for elements 1–56. Notice how the elements at the peaks are all in Group 0 (the noble gases). These elements have high first ionisation enthalpies – they are difficult to ionise and are very unreactive. The elements at the troughs are all in Group 1 (the alkali metals). These elements, with only one outer shell electron, have low ionisation enthalpies – they are easy to ionise and are very reactive. These are examples of *periodicity* – more detail on periodicity can be found in **Section 11.1**.

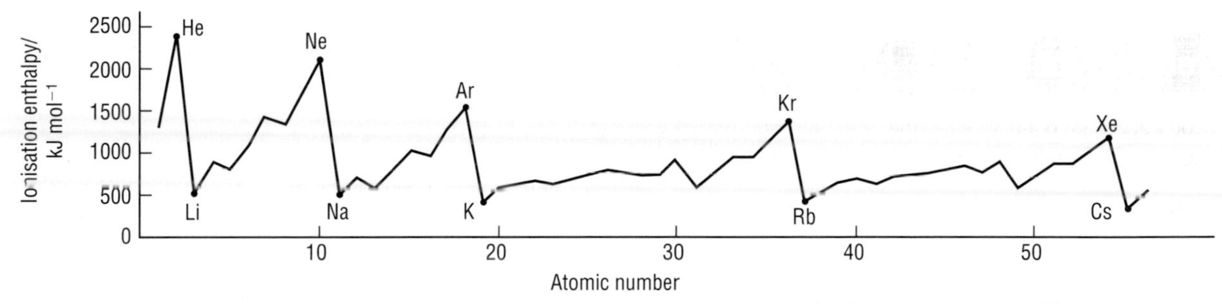

▲ **Figure 1** Variation of first ionisation enthalpy with atomic number for elements with atomic numbers 1 to 56.

Figure 2 shows the first ionisation enthalpies for elements 1–20 in more detail. Why does it become more difficult to remove an electron from an atom as you go across a period in the Periodic Table? Across a period, electrons are being added to the same shell but, at the same time, protons are being added to the nucleus. As the nuclear charge becomes more positive, the electrons will be held more tightly and so it gets harder to pull one from the outer shell. Again, this is an example of periodicity.

If we look at the ionisation enthalpies for the Group 7 elements we can see an example of a *trend* in ionisation enthalpies within the group. Table 1 lists the first ionisation enthalpies in Group 7. (For more information on the properties of elements in this group see **Section 11.4**.)

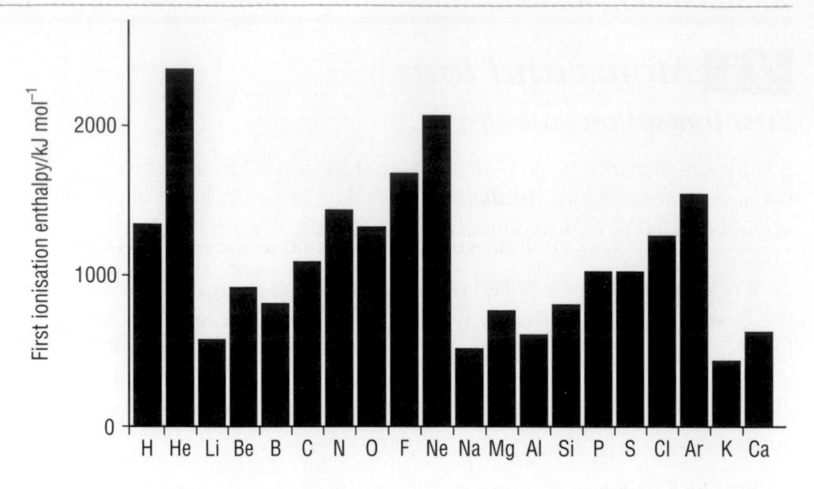

► **Figure 2** First ionisation enthalpies of elements 1–20.

► **Table 1** The first ionisation enthalpies for Group 7 elements.

Group 7 element	1st ionisation enthalpy/kJ mol^{-1}
F	+1687
Cl	+1257
Br	+1146
I	+1010

On going down Group 7, the first ionisation enthalpies *decrease*. This is because the attraction between the nucleus and the outermost electron decreases. This is a result of there being more filled shells of electrons between the nucleus and the outermost electron (see **Section 2.3**). These 'shield' the positively charged nucleus from the outermost electron, thus reducing its attraction to the electron. It is therefore easier for the outermost electron to be removed. A different way of expressing the same thing is to consider the core charge (see **Section 3.1**, page 39). All of the halogens have a core charge of +7. Smaller halogens have this core charge closer to the outermost electron, and therefore are attracted more strongly to the outermost electron. This is an example of a trend in a group property. This is explored further in **Section 11.1**.

Successive ionisation enthalpies

More than one electron can be removed from an atom (except from a hydrogen atom, of course). So, as well as first ionisation enthalpies, there are second, third and fourth (etc.) ionisation enthalpies which refer to the energies required to remove further electrons. The general equations for these ionisation processes are:

first ionisation: $X(g) \rightarrow X^{+}(g) + e^{-}$
second ionisation: $X^{+}(g) \rightarrow X^{2+}(g) + e^{-}$
third ionisation: $X^{2+}(g) \rightarrow X^{3+}(g) + e^{-}$
fourth ionisation: $X^{3+}(g) \rightarrow X^{4+}(g) + e^{-}$

and so on.

Notice that each of the second and subsequent ionisation processes involve the removal of an electron from a *positive ion*. The second ionisation enthalpy, for example, is the energy needed to remove one electron from an $X^{+}(g)$ ion (and *not* the energy required to remove two electrons from an X(g) atom).

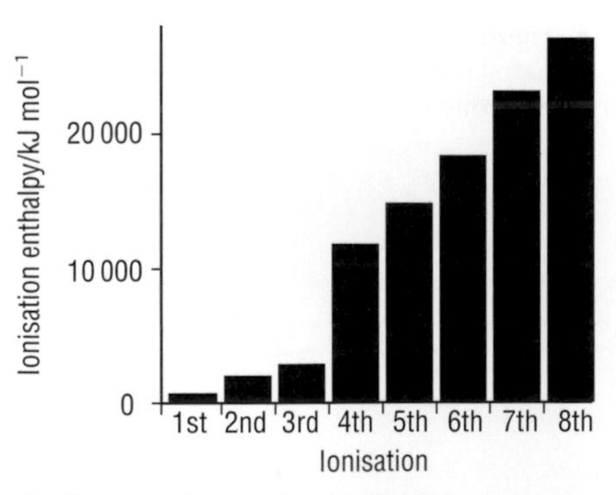

▲ **Figure 3** Successive ionisation enthalpies for aluminium.

Figure 3 shows eight successive ionisation enthalpies for aluminium. The striking feature here is that ionisation enthalpies increase as successive electrons are removed. It is more difficult to remove the second electron than the first; more difficult to remove the third than the second, and so on. This is true for all elements, not just aluminium. It makes sense because once an electron has been removed from an atom or ion, the remaining electrons will be attracted more strongly to the nucleus.

Another important feature of Figure 3 is the sharp increase between the third and fourth ionisation enthalpies of aluminium. For an explanation we need to look at the electron arrangement, 2.8.3, of the aluminium atom. The first, second and third ionisations correspond to the removal of the three electrons from the outer shell. The fourth ionisation, on the other hand, involves pulling an electron from a filled inner shell. Because the second shell is closer to the nucleus, it is much more difficult to remove an electron than from the third shell. Other Group 3 elements also show 'big jumps' between the third and fourth ionisation enthalpies.

If this explanation is correct, then Group 1 elements will have sharp rises between the first and second ionisation enthalpies, Group 2 elements between the second and third ionisation enthalpies and so on. So you would predict that the 'big jump' for phosphorus, a Group 5 element, would come between its fifth and sixth ionisation enthalpies – it is clear from Figure 4 that this indeed is the case.

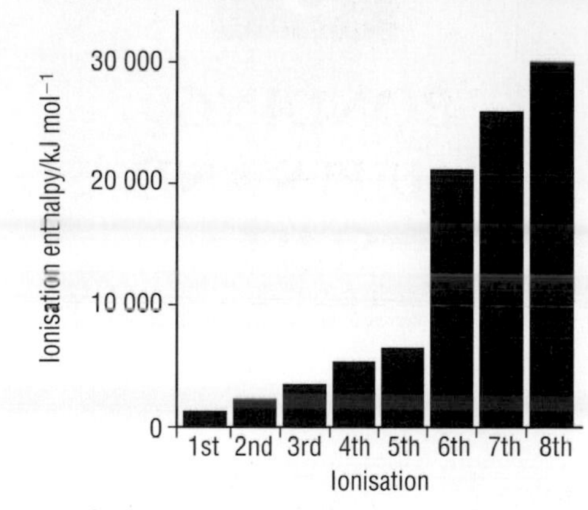

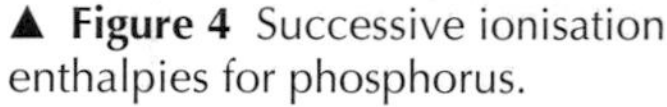
▲ **Figure 4** Successive ionisation enthalpies for phosphorus.

Problems for 2.5

1 The graph below shows the variation in the first ionisation enthalpies with atomic number for 16 elements in the Periodic Table – the element at which the graph starts is not specified.

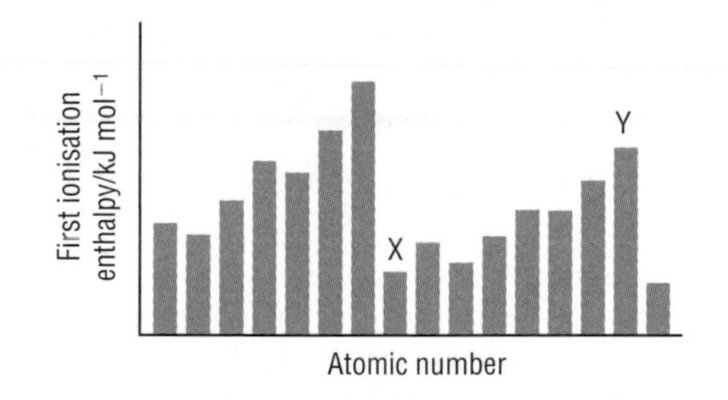

In which group of the Periodic Table is

a element X? **b** element Y?

Explain your answers.

2 The first, second and third ionisation enthalpies of calcium are +596 kJ mol^{-1}, +1160 kJ mol^{-1} and +4930 kJ mol^{-1}, respectively.

a Write equations corresponding to each of these three ionisation enthalpies.

b Explain why the second ionisation enthalpy of calcium is larger than its first ionisation enthalpy.

c Explain why there is a very sharp rise between the second and third ionisation enthalpies of calcium.

3 Explain why the second ionisation enthalpy of sodium (+4560 kJ mol^{-1}) is much larger than the second ionisation enthalpy of magnesium (+1460 kJ mol^{-1}).

4 a Explain why the first ionisation enthalpy values for the Group 1 elements become less positive as you go down the group.

b Why is this called a trend?

3

BONDING, SHAPES AND SIZES

3.1 Chemical bonding

AS

Noble gas electron configurations

In 1916, W. Kossel and G.N. Lewis realised that all the noble gases have eight electrons in their outer shells, with the exception of helium. They linked the chemical stability of the noble gases to this outer shell electron configuration. They suggested that other elements try to achieve eight outer shell electrons by losing or gaining electrons when they react to form compounds. Further, it seemed that some light elements are able to achieve stability by reaching the helium configuration of two outer shell electrons.

This was a very important finding and was the basis for early ideas about why elements combine to form compounds. It is still useful today – though it is not the whole story.

Formation of ions

Atoms of metal elements in Groups 1 and 2 of the Periodic Table have only one or two outer shell electrons. They are able to *lose* these electrons to form positively charged ions, called **cations**.

Most non-metal atoms have more than three outer shell electrons. They are able to gain electrons to form negatively charged ions, called **anions**.

There are limits to how many electrons an atom can pick up. One electron leads to an anion with a single negative charge. A second electron is repelled by this anion because their charges are the same, so making a doubly charged anion is much harder. Getting a third electron to stick, by the process $A^{2-} + e^- \rightarrow A^{3-}$ is very difficult and does not often happen – anions with a charge of 3− are unusual.

It is also hard to remove three or more electrons from atoms – cations with a 4+ charge are almost unknown.

Ionic bonding

When *metals* react with *non-metals* in a chemical reaction, ions are only formed if the *overall* energy change for the reaction is favourable. Electrons are *transferred* from the metal atoms to the non-metal atoms. Often this gives both the metal and the non-metal a stable electronic structure like a noble gas. The cations and anions which are formed are held together by an **electrostatic bond** because of their opposite charges.

We use **electron dot–cross diagrams** to represent the way that atoms bond together. In these, the outer shell electrons of one atom are represented by dots, with crosses for the other. Figure 1 shows dot–cross diagrams for the formation of sodium chloride and magnesium fluoride. The dots and crosses represent only the outer shell electrons. The numbers

▼ **Figure 1** Dot–cross diagrams for two ionic compounds.

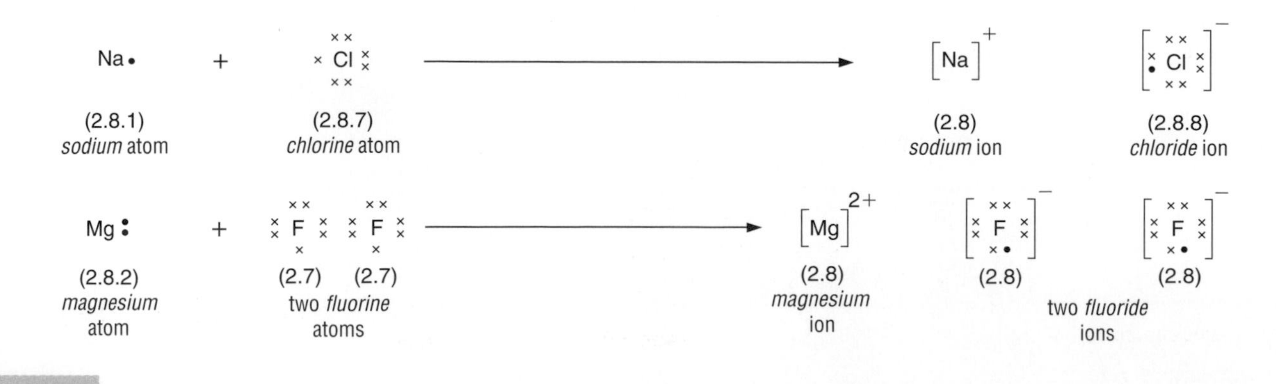

underneath the chemical symbols show the arrangement of electrons in the shells.

Each sodium atom loses one electron and each chlorine atom gains one electron, so the compound formed has the formula NaCl. Each magnesium atom loses two electrons but each fluorine atom gains only one electron, so the formula for magnesium fluoride is MgF_2.

The oppositely charged ions attract each other strongly – this attraction is an **ionic bond**. In the solid compound, each ion attracts many others of opposite charge and the ions build up into a giant lattice like the one shown in Figure 2. There is more about this in **Section 5.1**.

Properties of ionic compounds

Ionic compounds have high melting points, usually dissolve in water (see **Sections 4.5** and **5.1**) and conduct electricity when molten or dissolved in water.

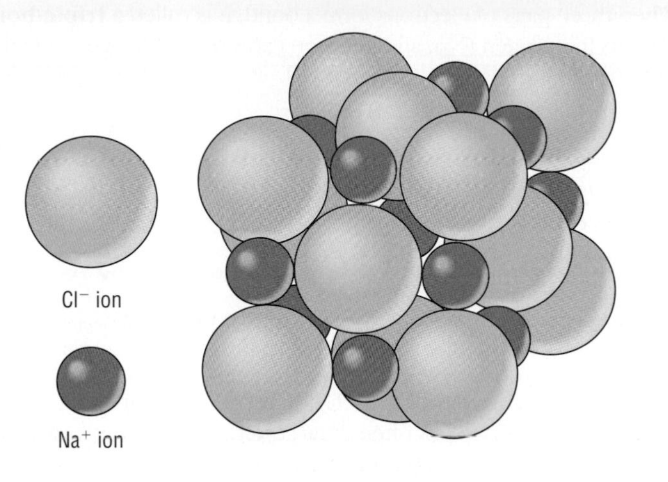

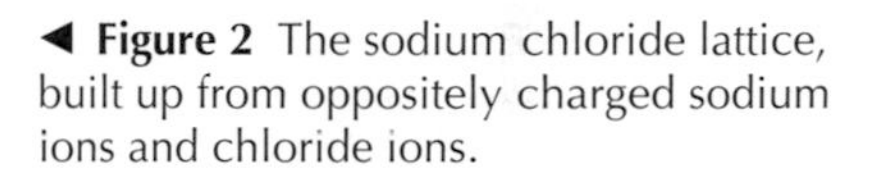

◄ **Figure 2** The sodium chloride lattice, built up from oppositely charged sodium ions and chloride ions.

Covalent bonding

Ionic compounds are usually formed when metals react with non-metals. But there are many compounds containing only non-metallic elements. Here it is not energetically favourable to form ions – electrons are *shared* between the atoms of the elements in these compounds. The resulting compound is more stable than the individual elements. Shared electrons count as part of the outer shell of *both* atoms in the bond. The dot–cross diagram for the H_2 molecule is shown in Figure 3.

Bonds formed by sharing electrons are called **covalent bonds**. The two atoms are held together because their positively charged nuclei are simultaneously attracted to the negatively charged shared electrons. If a pair of electrons is involved, the bond is called a single covalent bond – or more simply a **single bond**.

Examples of dot–cross diagrams for two more simple covalent compounds are shown in Figure 4. Notice that by sharing electrons all the atoms achieve more stable electron structures, like noble gases.

Electron pairs which form bonds are called **bonding pairs**. Pairs of electrons not involved in bonding are called **lone pairs**. Both water (H_2O) and ammonia (NH_3) have lone pair electrons.

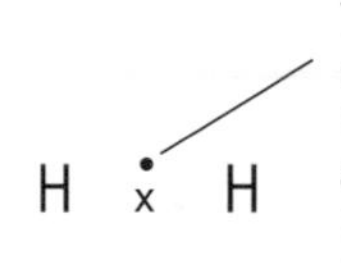

▲ **Figure 3** Electron sharing in the hydrogen molecule.

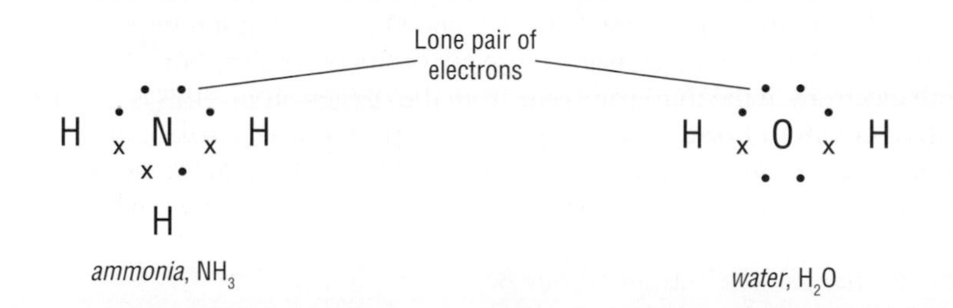

◄ **Figure 4** Dot–cross diagrams for NH_3 and H_2O.

When two pairs of electrons form a covalent bond, it is called a **double bond**. The bonds in molecular oxygen and carbon dioxide are double covalent bonds (Figure 5).

oxygen, O_2

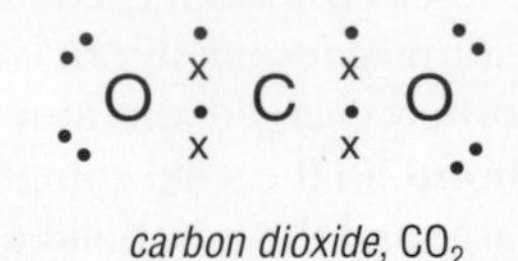

carbon dioxide, CO_2

► **Figure 5** Dot–cross diagrams for O_2 and CO_2.

When three pairs of electrons form a bond, it is called a **triple bond**. The bonds in nitrogen (N_2), and between carbon and nitrogen in hydrogen cyanide (HCN) are examples (Figure 6).

► **Figure 6** Dot–cross diagrams for N_2 and HCN.

Electron dot–cross diagrams are useful for representing individual electrons in chemical bonds. Chemists often draw structures in a simpler way by using *lines* to represent a pair of electrons shared between two atoms. A single line represents a single covalent bond, while double and triple lines represent double and triple covalent bonds, respectively.

Some of the molecules used as examples earlier are shown in this way in Figure 7.

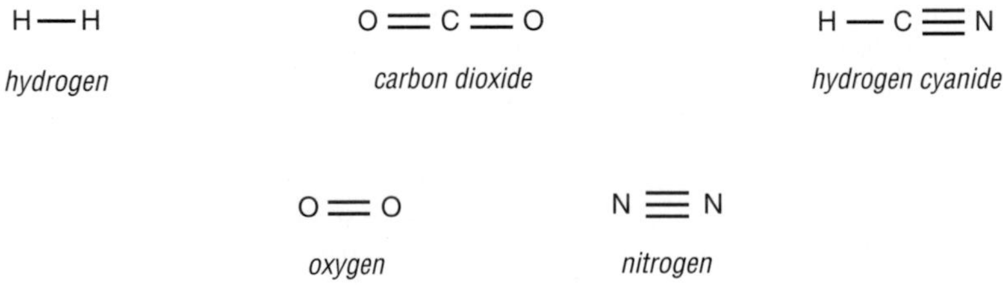

H—H *hydrogen*

O=C=O *carbon dioxide*

H—C≡N *hydrogen cyanide*

O=O *oxygen*

N≡N *nitrogen*

► **Figure 7** Covalently bonded molecules.

Properties of covalently bonded elements and compounds

Covalent network structures (see **Section 5.2**), such as silicon dioxide, have very high melting points, do not normally conduct electricity (one notable exception is graphite) and are insoluble in water.

Elements or compounds with simple molecular structures, such as chlorine and methane, have low melting points, do not conduct electricity and are usually insoluble in water.

You can read more about the links between bonding, structure and properties in **Section 5.8**.

There are three types of covalently bonded structures. These are macromolecular covalent structures (polymers) (**Sections 5.6** and **5.7**), simple molecular structures and covalent network structures. You can find out more about covalent network structures in **Sections 5.2** and **5.8**.

Dative covalent bonds

Figure 8 shows the bonding in carbon monoxide, CO. Look carefully at the three pairs of electrons that make the triple bond between the C and O atoms. Two of these are formed by the C and O atoms contributing an electron each to the shared pair – these are ordinary covalent bonds. But both electrons in the third pair come from the oxygen atom – this is called a **dative covalent bond**. In a dative bond, *both bonding electrons come from the same atom*, unlike an ordinary covalent bond where the atoms each contribute one electron to the pair. A dative covalent bond can be shown by an arrow, with the arrow pointing away from the atom that donates the pair of electrons (Figure 8c).

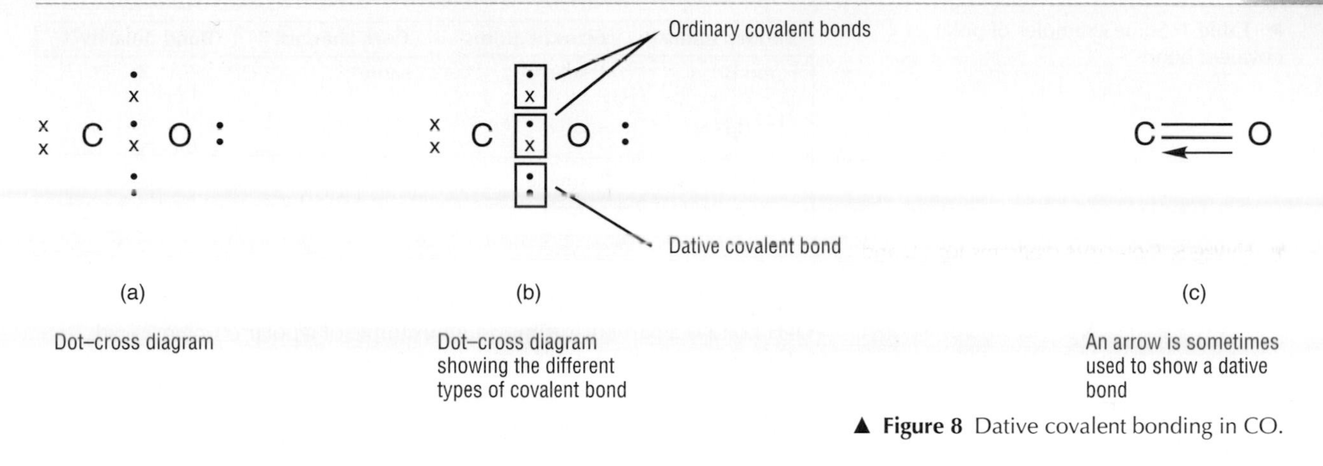

▲ **Figure 8** Dative covalent bonding in CO.

Why are bonds like bears?

Figure 9 shows a more detailed diagram of the way the hydrogen molecule is bonded.

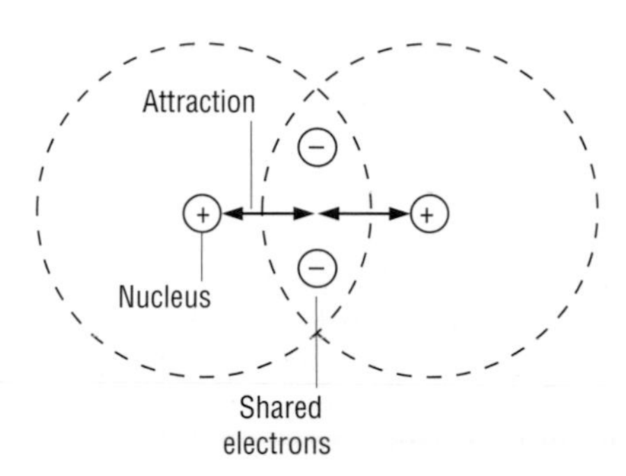

◀ **Figure 9** In a hydrogen molecule, the atoms are held together because their nuclei are both attracted to the shared electrons.

The atoms are held together because their nuclei are both attracted to the electron pair which is shared between them. Both atoms are identical so the electrons are shared equally.

With atoms more complicated than hydrogen, it is the atomic cores which are attracted to the shared electron pairs. The atomic core is made up of everything except the outer shell electrons (see **Section 2.3**). A fluorine atom, for example, has an electron shell configuration of 2.7 and its core is made up from the nucleus with a charge of 9+ and the two electrons in the innermost shell with a charge of 2−. This gives a core with a charge of 7+. The F_2 molecule is held together because the 7+ core charges are attracted to the negative charges on the shared electrons.

Very often the two atoms bonded together have different sizes. In this case, the core of the smaller atom will be closer to the shared electrons and will exert a stronger pull on them (Figure 10a). A similar situation arises when the atoms are from different groups in the Periodic Table and have different core charges (Figure 10b).

In general, different atoms attract bonding electrons unequally. One atom gets a slight negative charge because it has a greater share of the bonding electrons. The other atom becomes slightly positively charged because it has lost some of its share in the bonding electrons. Bonds like this are called **polar bonds**. An example is shown in Figure 11.

The small amounts of electrical charge are shown by δ− and δ+ where 'δ' means 'a small amount of'. Some more examples of **bond polarity** are shown in Table 1.

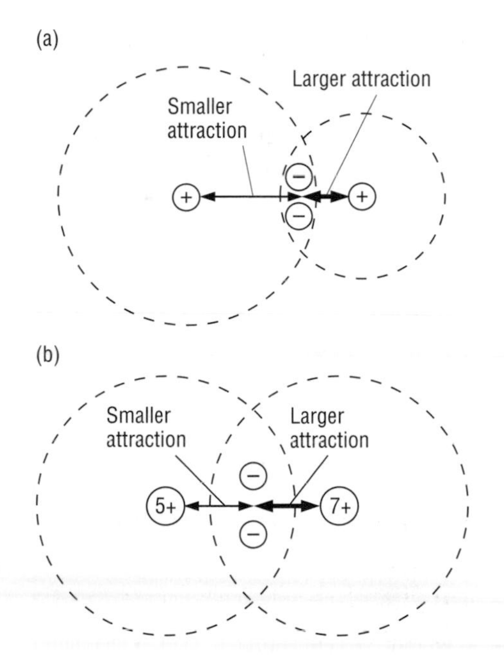

▲ **Figure 10 a** Shared electrons are attracted more strongly by the core of the smaller atom, which is closer.
b Shared electrons are attracted more strongly by the atom with the greater core charge.

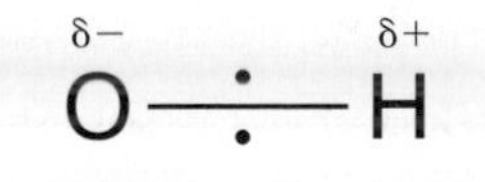

▲ **Figure 11** The O–H covalent bond is polar – the two electrons in the bond are drawn closer to the oxygen atom than the hydrogen atom.

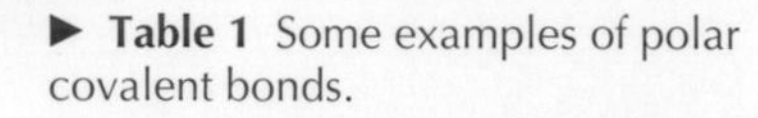

▶ **Table 1** Some examples of polar covalent bonds.

Bonded atoms	Sizes of atoms	Core charges	Bond polarity
F and Br	F smaller	same	$\overset{\delta-}{\text{F}}$—$\overset{\delta+}{\text{Br}}$
O and N	similar	O greater	$\overset{\delta-}{\text{O}}$—$\overset{\delta+}{\text{N}}$
F and P	F smaller	F greater	$\overset{\delta-}{\text{F}}$—$\overset{\delta+}{\text{P}}$

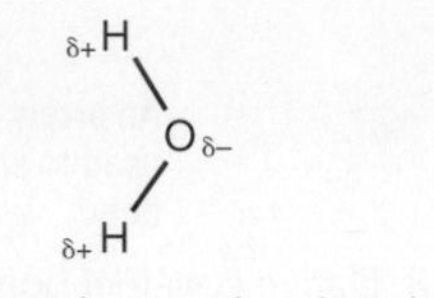

▲ **Figure 12** Polar covalent bonding in the water molecule.

So bonds *are* like bears – some are polar and some are not. The O–H bond is a particularly important example of a polar covalent bond. It has many consequences for the chemistry of water molecules (Figure 12).

The type of bonding in a compound is related to the position of its constituent elements in the Periodic Table:

- when metals on the left-hand side of the Periodic Table combine with non-metals on the right, ionic compounds are formed
- when non-metals on the right-hand side of the Periodic Table combine with each other, covalent compounds are formed.

In non-metallic elements (for example, graphite and chlorine gas) the atoms are bonded together by covalent bonds.

For metallic elements, a different type of bonding, called metallic bonding, gives metals their characteristic properties.

Electronegativity

To decide the polarity of a covalent bond, we need a measure of each atom's attraction for bonding electrons – it's 'electron pulling power'. This is called **electronegativity** – atoms with strong electron pulling power in covalent bonds are said to be highly electronegative.

There are several ways of estimating electronegativity values. Each method gives different numbers but each places the elements in a similar order – they lead to the same relative electronegativity values. Table 2 lists some atoms in order of their electronegativity values.

The electronegativity values in Figure 13 are derived from a method suggested by Linus Pauling, a US chemist and winner of two Nobel Prizes.

If you exclude the noble gases, the elements are more electronegative at the top of a group and at the right-hand side of the Periodic Table. The highest electronegativities correspond to reactive non-metals with small atoms – the lowest electronegativities correspond to reactive metals with large atoms.

▼ **Table 2** Atoms listed in order of electronegativity.

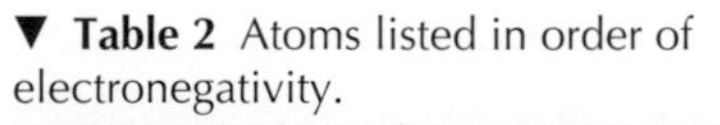

Element	Electronegativity
F	↑
O	
Cl	
N, Br	
I	
S, C	increasing electronegativity
H, P	
Si	
Al	
Mg	
Na	

Electronegativity is a measure of the ability of an atom in a molecule to attract electrons in a chemical bond to itself.

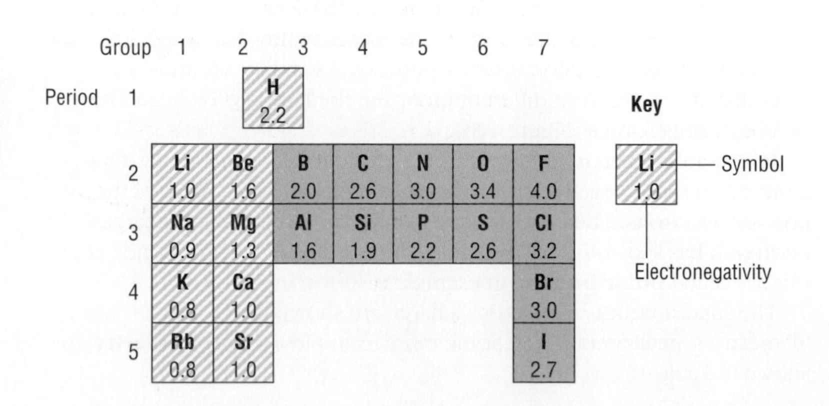

▶ **Figure 13** Pauling electronegativity values for some main group elements in the Periodic Table.

We can use differences in electronegativity values to predict how polar a particular covalent bond will be. For example, in the C–F bond, F has a higher electronegativity value (4.0) than C (2.6) so it attracts the shared electrons more strongly and the polarity of the bond is

$$\overset{\delta+}{C}—\overset{\delta-}{F}$$

Notice that C and H have similar electronegativities, so for practical purposes we can think of the C–H bond as being non-polar. There is more about electronegativity and bond polarity in **Section 5.3**.

Polar bonds are like covalent bonds with a bit of ionic character in them. The ionic and covalent models are extreme forms of bonding – polar bonds are somewhere between the two. The bigger the difference in electronegativity between the atoms, the more polar the bond and the greater the ionic character (Figure 14).

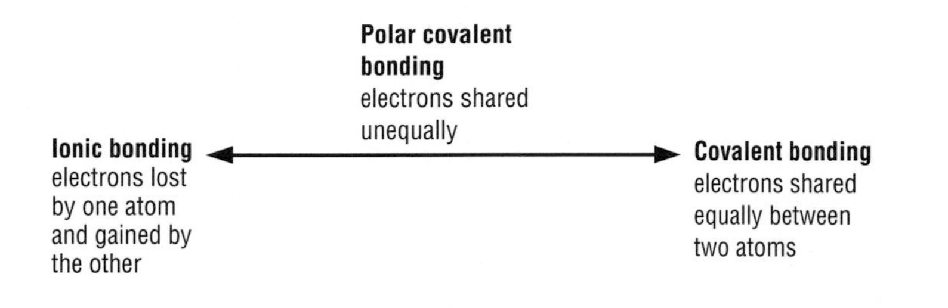

◀ **Figure 14** Bonds can vary from ionic to covalent, with polar covalent bonds in between.

A cautionary note

Theories of bonding are fundamental to an understanding of chemistry. The dot–cross model is a simple approach to bonding. Like the Bohr theory, it is an over-simplified model, but it is nevertheless useful. It needs to be extended before it can do more than explain simple situations. So don't be surprised if dot–cross diagrams sometimes don't work, or if some molecules seem to break the rules. Such an example can be seen in **Section 11.4**, Figure 4 (page 248), where the ClO_3^- ion has 12 electrons in its outer shell.

Metallic bonding

Metal atoms cannot achieve the stable electron arrangement of a noble gas by sharing electrons with each other or by transferring electrons from one to the other – they don't have enough outer shell electrons to allow them to do this. So what holds the metal atoms together in a sample of, say, sodium or iron?

The metal atoms lose their outer electrons to form a **lattice** of regularly spaced positive ions. The outer electrons from each atom contribute to a common 'pool' of electrons which move randomly throughout the lattice of positive ions. Since the electrons in the pool do not belong to any particular metal ion, they are described as being **delocalised** or 'spread out' over the lattice. Each positively charged ion in the metal is attracted to the negatively charged delocalised electrons, and vice versa. It is these electrostatic attractions that are the **metallic bonds** (Figure 15).

The strength of metallic bonding depends on several factors including the number of electrons per atom available for delocalisation in this way. Thus magnesium (two outer shell electrons) has stronger metallic bonding than sodium (one outer shell electron) and this is reflected in the higher melting point and boiling point of magnesium.

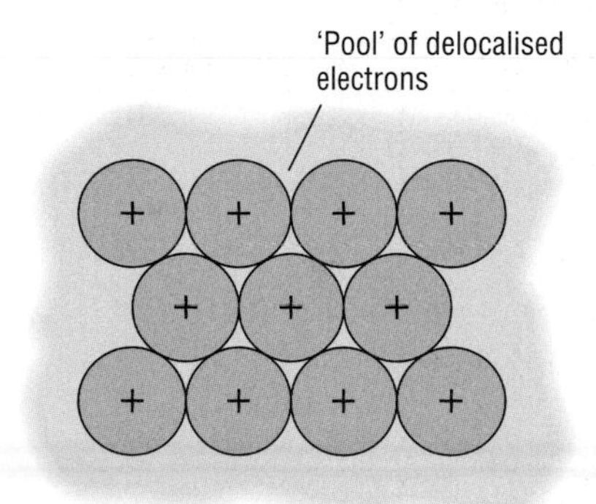

▲ **Figure 15** A model of metallic bonding.

Properties of metals

Metals usually have high melting points. A notable exception is mercury, which is a liquid at room temperature. Metals are insoluble in water and conduct electricity when solid or liquid. You can find out more about metallic properties in **Section 11.5**.

▶ **Table 3** The names and formulae of some common ions. For elements where there is more than one common ion, the oxidation state is given – e.g. iron(II) and iron(III).

Group 1 elements form +1 ions
Group 2 elements form +2 ions
Group 6 elements form –2 ions
Group 7 elements form –1 ions

Positive ions (cations)		Negative ions (anions)	
+1		**–1**	
H^+	hydrogen	F^-	fluoride
Li^+	lithium	Cl^-	chloride
Na^+	sodium	Br^-	bromide
K^+	potassium	I^-	iodide
Cu^+	copper(I)	OH^-	hydroxide
Ag^+	silver	NO_3^-	nitrate(V) or nitrate
NH_4^+	ammonium	NO_2^-	nitrate(III) or nitrite
		HCO_3^-	hydrogencarbonate
		HSO_4^-	hydrogensulfate
+2		**–2**	
Mg^{2+}	magnesium	O^{2-}	oxide
Ca^{2+}	calcium	S^{2-}	sulfide
Sr^{2+}	strontium	CO_3^{2-}	carbonate
Ba^{2+}	barium	SO_4^{2-}	sulfate
Fe^{2+}	iron(II)		
Ni^{2+}	nickel(II)		
Cu^{2+}	copper(II)		
Zn^{2+}	zinc		
Sn^{2+}	tin(II)		
Pb^{2+}	lead(II)		
+3		**–3**	
Al^{3+}	aluminium	N^{3-}	nitride
Fe^{3+}	iron(III)	PO_4^{3-}	phosphate
Cr^{3+}	chromium(III)		

Writing chemical formulae

Table 3 lists some common positive and negative ions.

Notice that it includes some ions that contain more than one type of atom, such as hydroxide (OH^-) and sulfate (SO_4^{2-}). This type of ion is called a complex ion. These ions consist of a group of atoms held together by covalent bonds, but the whole group carries an electric charge. The dot–cross diagram for sodium hydroxide is shown in Figure 16.

$[Na]^+$ $[O H]^-$

• hydrogen electron

× oxygen electron

□ electron transferred from sodium

▶ **Figure 16** Dot–cross diagram for sodium hydroxide.

There is a relationship between simple ions and the position of the element in the Periodic Table. You can see this in Table 4. Transition metals display a range of charges on their ions.

Once you know the formulae of the ions in an ionic compound, it is easy to write the formula of the compound. You need to make sure that the total positive charge is equal to the total negative charge – i.e. when you add all the charges on the ions, they add up to zero.

▼ **Table 4** Table showing the relationship between group in the Periodic Table and the charge on ions. You can find out more about this in **Section 2.5**.

Group	Charge on simple ion	Example
1	+1	Na^+
2	+2	Mg^{2+}
6	–2	S^{2-}
7	–1	Cl^-

Example 1

Sodium chloride is made up of sodium ions (Na^+) and chloride ions (Cl^-). The formula is NaCl because $(+1) + (-1) = 0$

Example 2
Calcium oxide is made up of calcium ions (Ca^{2+}) and oxide ions (O^{2-}). The formula is CaO because $(+2) + (-2) = 0$

Example 3
Magnesium fluoride is made up of magnesium ions (Mg^{2+}) and fluoride ions (F^-). The formula is MgF_2 because $(+2) + (-1) + (-1) = 0$

Example 4
Potassium carbonate is made up of potassium ions (K^+) and carbonate ions (CO_3^{2-}). The formula is K_2CO_3 because $(+1) + (+1) + (-2) = 0$

Example 5
Chromium(III) hydroxide is made up of chromium(III) ions (Cr^{3+}) and hydroxide ions (OH^-). The formula is $Cr(OH)_3$ because $(+3) + (-1) + (-1) + (-1) = 0$. Note that if more than one complex ion is needed in a formula it is always placed in brackets.

For covalent compounds, an indication of the formula is often given in the Greek prefixes used in the names – 'mono' meaning one, 'di' meaning two and 'tri' meaning three. Sulfur dioxide, for example, has the formula SO_2.

Problems for 3.1

1 Draw electron dot–cross diagrams for the following **ionic** compounds.
- **a** lithium hydride, LiH
- **b** potassium fluoride, KF
- **c** magnesium oxide, MgO
- **d** calcium chloride, $CaCl_2$
- **e** calcium sulfide, CaS
- **f** sodium sulfide, Na_2S
- **g** sodium nitride, Na_3N
- **h** aluminium fluoride, AlF_3

2 Draw electron dot–cross diagrams for the following covalent substances.
- **a** chlorine, Cl_2
- **b** hydrogen chloride, HCl
- **c** methane, CH_4
- **d** hydrogen sulfide, H_2S
- **e** aluminium bromide, $AlBr_3$
- **f** silicon chloride, $SiCl_4$
- **g** ethene, $H_2C{=}CH_2$
- **h** ethyne, H—C≡C—H
- **i** methanol, $H_3C{-}O{-}H$

3 An ammonia molecule, NH_3, forms a dative bond with a hydrogen ion, H^+, to produce an ammonium ion, NH_4^+. The other three hydrogen atoms are held to the nitrogen atom by conventional covalent bonds.
- **a** What is the essential difference between a dative bond and a covalent bond?
- **b** Draw dot–cross diagrams for the ammonia molecule and the ammonium ion.

4 Water forms a dative bond with a hydrogen ion to form the oxonium ion, H_3O^+. Draw a dot–cross diagram for the oxonium ion.

5 Ammonia and boron trifluoride, BF_3, combine together to form the molecule NH_3BF_3. This molecule has a dative covalent bond between the nitrogen atom and the boron atom. Draw a dot–cross diagram for this molecule.

6 Explain what is meant by the statement that chlorine is more electronegative than carbon.

7 Predict the polarity of the covalent bonds in the following molecules, using the notation $\delta+$ and $\delta-$ where appropriate.
- **a** HF
- **b** N_2
- **c** HCl
- **d** ClF
- **e** HI
- **f** CS_2

8
- **a** In aluminium, how many electrons will each atom contribute to the 'pool' of delocalised electrons?
- **b** Suggest why metals are good conductors of electricity.

9 Predict the type of bonding present in each of the following substances.
- **a** iron
- **b** chlorine
- **c** silicon tetrafluoride
- **d** sodium bromide
- **e** magnesium oxide
- **f** methane

Which of the above substances would you expect to contain polar covalent bonds?

10 Write down the formulae of the following ionic compounds.
- **a** sodium chloride
- **b** magnesium chloride
- **c** iron(III) chloride
- **d** aluminium oxide
- **e** ammonium chloride
- **f** sodium hydroxide
- **g** potassium carbonate
- **h** magnesium sulfate

11 Write down the formulae of the following covalent compounds.
- **a** ammonia
- **b** hydrogen sulfide
- **c** carbon dioxide
- **d** hydrogen chloride
- **e** carbon monoxide
- **f** sulfur dioxide
- **g** dinitrogen oxide
- **h** sulfur trioxide

12 Copy and complete the table.

Physical appearance at room temperature	Melting point	Electrical conductivity	Solubility in water	Type of bonding
gas	?	negligible	virtually insoluble	?
?	?	does not conduct when solid, but does when molten	?	?
shiny solid	high	?	insoluble	?
solid	?	nil	?	?

AS

3.2 *The shapes of molecules*

Electron pair repulsions

H
x •
H •x C •x H
x •
H

▲ **Figure 1** A dot–cross diagram showing the covalent bonding in methane.

The arrangement of the electrons in methane can be represented by a dot–cross diagram (Figure 1). There are *four groups* of electrons around the central carbon atom – the four covalent bonding pairs. Because similar charges repel, these groups of electrons arrange themselves so that they are as far apart as possible.

The farthest apart they can get is when the H–C–H bond angles are 109°. Clearly, this does not correspond to a flat methane molecule, shaped like a 'plus' sign – in that arrangement the bond angles are only 90°. The 109° angle arises if methane has a **tetrahedral** shape, with the H atoms at the corners and the C atom in the centre (Figure 2).

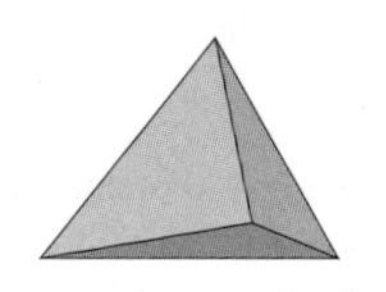

▲ **Figure 2** A regular tetrahedron.

Chemists have developed a system for drawing three-dimensional structures on two-dimensional paper and this has been used in Figure 3. Bonds which lie in the plane of the paper are drawn as solid lines in the normal way. Bonds which come forwards, towards you, are represented as wedge-shaped lines. Bonds which go backwards, away from you, are shown as dashed lines. (Build a model of methane and arrange it so that one C–H bond comes towards you, two go away and the other goes up and down – to correspond with Figure 3.)

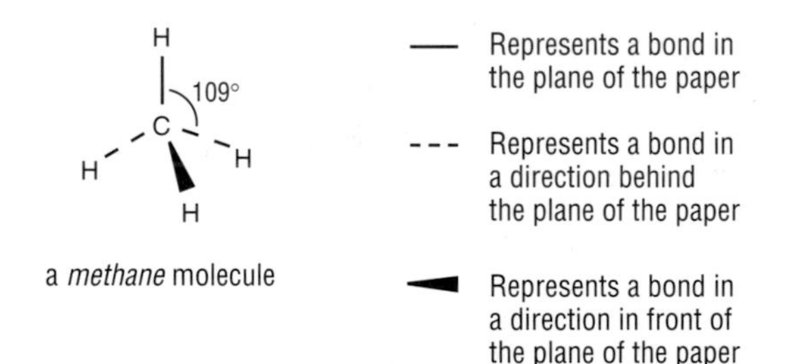

▲ **Figure 3** The methane molecule has a tetrahedral shape.

All carbon atoms which are surrounded by four single bonds have a tetrahedral distribution of bonds. Figure 4 shows ethane, in which both carbon atoms have a tetrahedral arrangement.

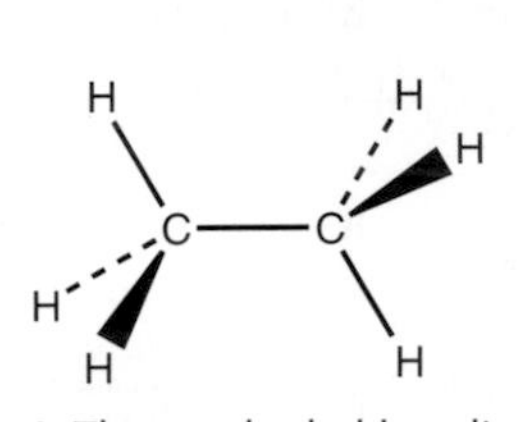

▲ **Figure 4** The tetrahedral bonding around the carbon atoms in ethane.

Lone pairs count too

A molecule of methane has only bonding pairs in the outer shell of carbon – there are no lone pairs. Molecules such as ammonia (NH_3) and water (H_2O) do have lone pairs in the outer shells of nitrogen and oxygen, respectively (Figure 5).

ammonia   water

◄ **Figure 5** Dot–cross diagrams showing the bonding in ammonia and water.

These lone pairs repel the bonding pairs of electrons, as in methane. So, ammonia and water adopt the same tetrahedral shape as methane, but in these molecules one or more of the corners of the tetrahedron are occupied by lone pairs of electrons (Figure 6).

The H–N–H and H–O–H bond angles are both close to 109°. NH_3 is said to have a *pyramidal* shape; H_2O is described as *bent*. (Again, it is best to build models of these molecules to confirm the structures that are drawn.)

The shapes of covalent molecules are decided by a simple rule – *groups of electrons in the outer shell repel one another and move as far apart as possible*. It doesn't matter if they are groups of bonding electrons or lone pairs – they all repel one another.

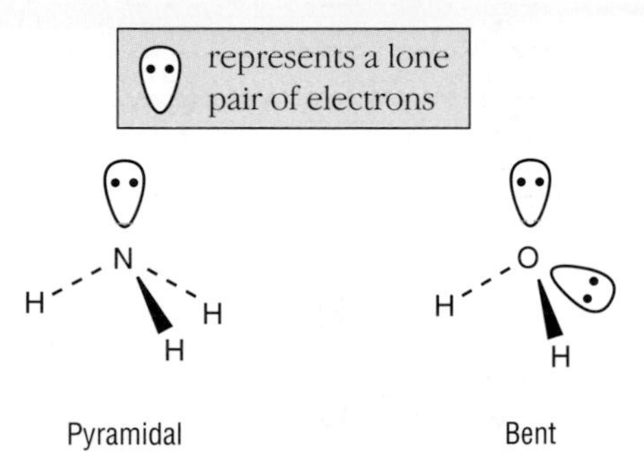

▲ **Figure 6** The shapes of molecules of ammonia and water.

Other shapes

Linear molecules

In $BeCl_2$, there are two groups of electrons around the central atom (Figure 7). Because there are fewer groups of electrons than in methane, they can get further apart. The furthest apart they can get is at an angle of 180°, so $BeCl_2$ is **linear** with a Cl–Be–Cl angle of 180°.

◄ **Figure 7** The shape of the $BeCl_2$ molecule – the lone pairs around the Cl atoms have been omitted for clarity.

Carbon dioxide and ethyne are other examples of linear molecules – there are only two groups of electrons around the central atoms in these molecules (Figure 8).

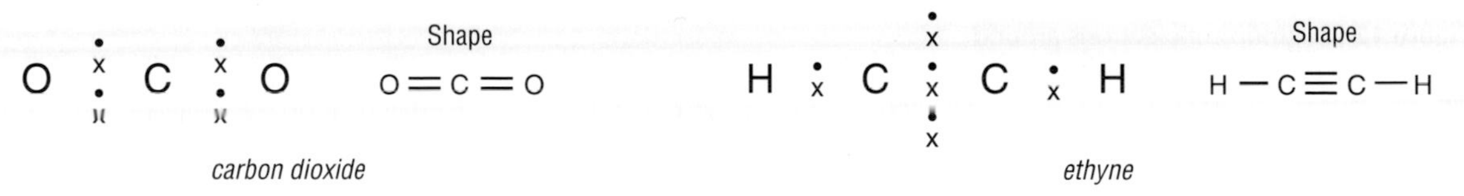

▲ **Figure 8** Carbon dioxide and ethyne both have linear molecules.

Planar molecules

In BF_3 there are *three*, not four, groups of electrons around the central atom. The furthest the three groups of electrons can get apart is at an angle of 120°, so the F–B–F bond angle is 120°. BF_3 is flat and shaped like a triangle with the B atom at the centre and the F atoms at the corners (Figure 9). Its shape is described as **planar triangular**.

There are other molecules with planar structures (Figure 10). In methanal, for example, there are three groups of electrons around the carbon atom, and in ethene *both* carbon atoms have three groups of electrons. Remember it is the number of *groups* of electrons which determine the shape – there are four pairs of electrons around the carbon atoms in these molecules, but the bonds are *not* tetrahedrally directed because there are only three *groups* of electrons.

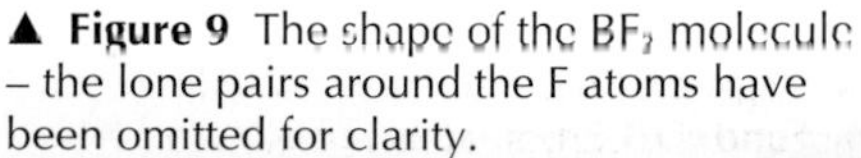

▲ **Figure 9** The shape of the BF_3 molecule – the lone pairs around the F atoms have been omitted for clarity.

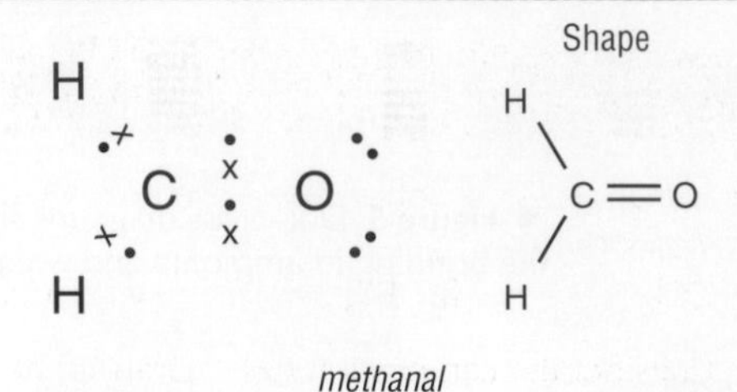

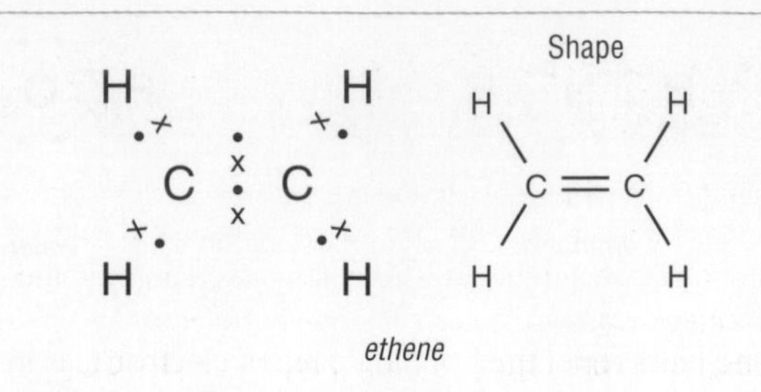

▲ **Figure 10** Methanal and ethene have planar molecules.

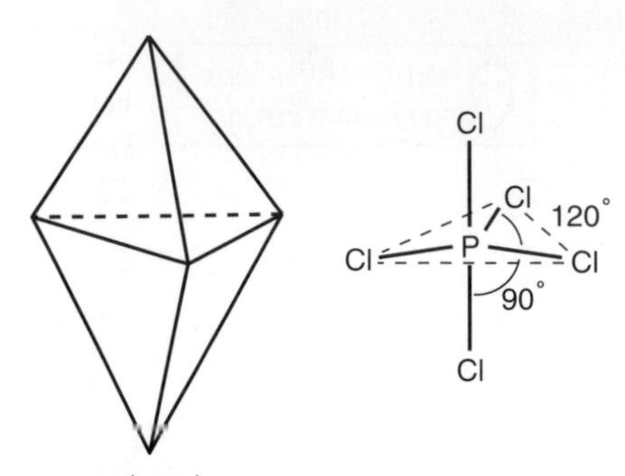

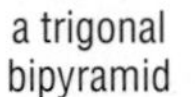

▲ **Figure 11** Five groups of electrons around a central atom or ion gives a trigonal bipyramidal shape.

Bipyramidal molecules

Five groups of electrons around a central atom give rise to a **trigonal bipyramidal** shape, with electrons at the five corners of the shape. An example of such a molecule is phosphorus pentachloride, PCl_5. Bond angles approximate to 120° or 90°, depending on their position within the molecule. Figure 11 helps to explain this.

Octahedral molecules

Six groups of electrons around a central atom gives rise to an **octahedral** shape, where the electrons are directed to the six corners of an octahedron. Some molecules, such as SF_6, adopt this structure but it is more commonly found in the octahedral shapes of complexes of metal ions with six ligands (Figure 12). In the complex ion in Figure 12, the six ligands round the nickel atom are water molecules. (You will find out more about complexes in **Section 11.6**.)

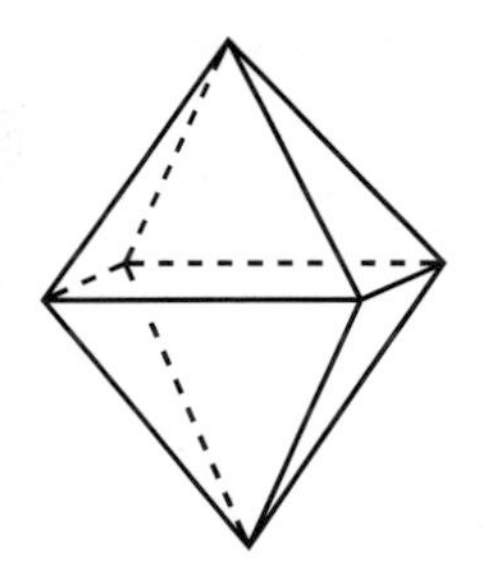

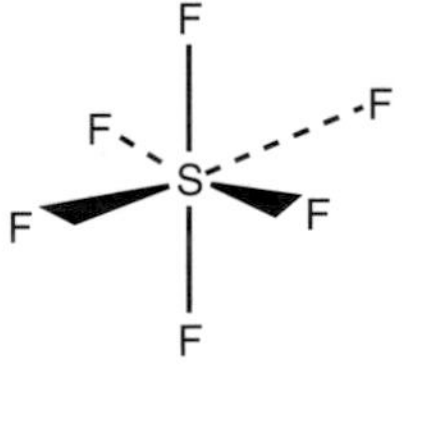

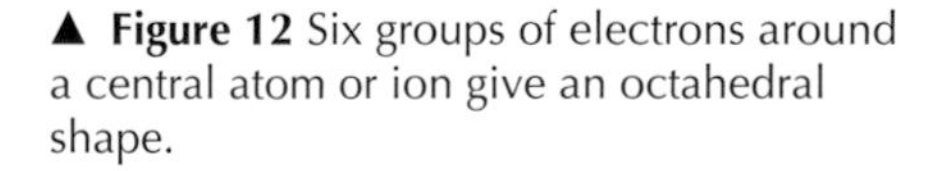
▲ **Figure 12** Six groups of electrons around a central atom or ion give an octahedral shape.

The shapes of ions

The same ideas can be used to work out the shapes of ions. We need to take account of any extra electrons that have been added or any taken away when the ion was formed (see **Section 3.1**). Figure 13 illustrates this idea for two ions based on ammonia – NH_4^+ and NH_2^-.

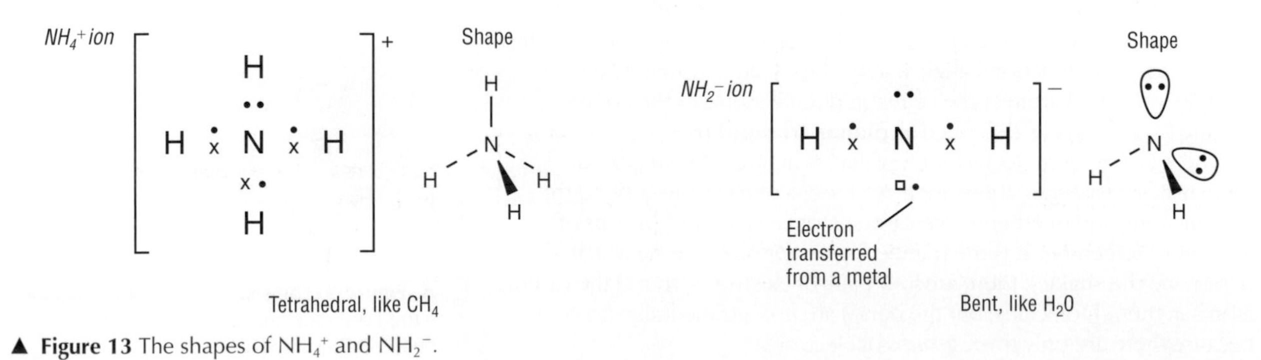

▲ **Figure 13** The shapes of NH_4^+ and NH_2^-.

Problems for 3.2

1 For each of the following molecules, draw an electron dot–cross diagram to show the bonding pairs and lone pairs of electrons in the molecule. Give the approximate bond angles you would expect in each molecule.
  a SiH_4 **b** H_2S **c** PH_3 **d** CO_2
  e SF_2 **f** BCl_3 **g** C_2H_2

2 Draw a diagram to show the bonds and lone pairs of electrons in each of the following molecules. Write in the approximate bond angles you would expect to find in each structure.
  a CH_3CH_3 **b** CH_3OH **c** CH_3NH_2
  d $CH_2{=}CH_2$ **e** $CH_3C{\equiv}N$ **f** NH_2OH
  g $COCl_2$

3 Draw an electron dot–cross diagram for the molecule NF_3.
  a What is the shape of the molecule with respect to the electron pairs?
  b What is the shape of the molecule with respect to the atoms?

4 What shapes are the following molecules (with respect to the atoms)?
  a CCl_4 **b** BF_3 **c** HCN

5 **a** Consider the carbocation CH_3^+. Draw a dot–cross diagram to show the electron structure of this ion (the ion only has six electrons in the outer shell of the C atom) and predict the shape of the ion.
  b Consider the carboanion CH_3^-. Draw a dot–cross diagram to show its electron structure (the ion has eight electrons in the outer shell of the C atom) and predict its shape.

6 Draw electron dot–cross diagrams for the following ions, and predict the bond angles in each of them.
  a NH_2^- **b** NH_2^+

3.3 *Structural isomerism*

AS

Two molecules which have the same molecular formula but differ in the way their atoms are arranged are called **isomers**. Isomers are distinct compounds with different physical properties, and often different chemical properties too.

The occurrence of isomers (**isomerism**) is very common in carbon compounds because of the great variety of ways in which carbon atoms can form chains and rings. However, you will also meet examples of isomerism in inorganic chemistry.

There are two ways in which atoms can be arranged differently in isomers (Figure 1):

- either the atoms are bonded together in a *different order* in each isomer – these are called **structural isomers**;
- or the order of bonding in the isomers is the same, but *the arrangement of the atoms in space is different* in each isomer – these are called **stereoisomers**.

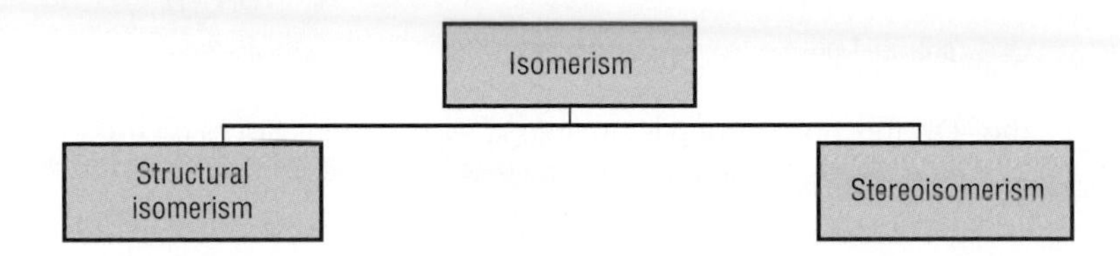

◀ **Figure 1** The two main types of isomerism.

You can find out about stereoisomers in **Sections 3.4** and **3.5**. For the moment, we shall concentrate on structural isomerism.

Structural isomerism

Structural isomers have the same molecular formula but have atoms bonded together in a different order. They have different structural formulae.

There are various ways in which structural isomerism can arise (Figure 2).

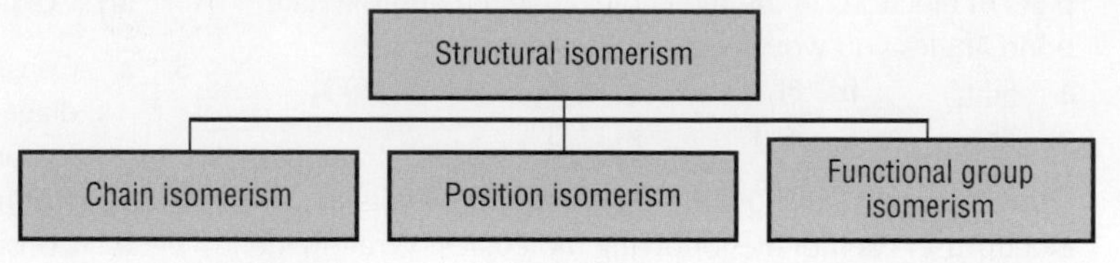

▶ **Figure 2** Types of structural isomerism.

Chain isomerism

There is only one alkane corresponding to each of the molecular formulae CH_4, C_2H_6 and C_3H_8. With 4 or more carbon atoms in a chain, different arrangements are possible. For example,

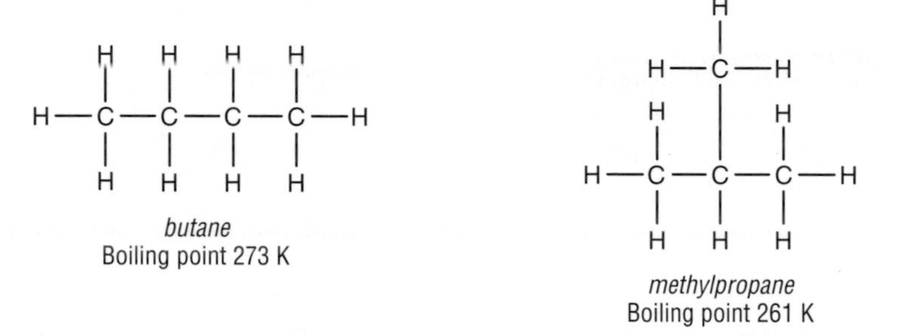

Both these compounds have the same molecular formula, C_4H_{10}. Their different structures lead to different properties. For example, the boiling point of methylpropane is 12 K lower than that of butane.

As the number of carbon atoms in an alkane increases, the number of possible isomers increases. There are over four thousand million isomers with the molecular formula $C_{30}H_{62}$!

Position isomerism

This can occur where there is an atom, or group of atoms, substituted in a carbon chain or ring. These are called **functional groups**.

Isomerism occurs when the functional group is situated in different positions in the molecules.

a Isomers of C_3H_7OH

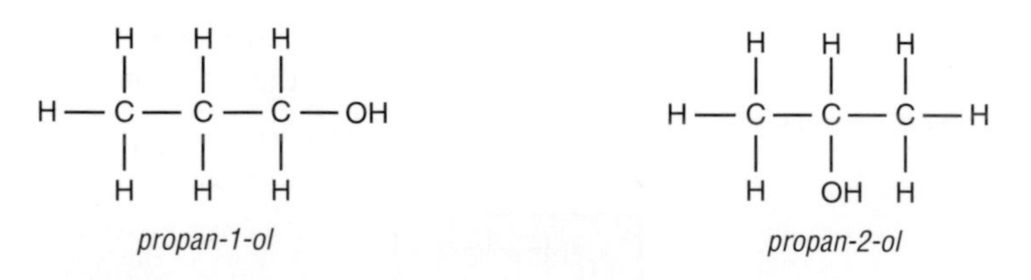

Here the –OH functional group is situated at two different places on the hydrocarbon chains.

b Isomers of $C_6H_4Cl_2$

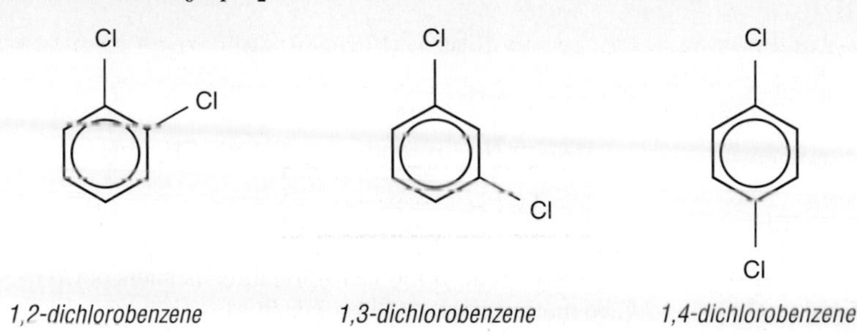

1,2-dichlorobenzene *1,3-dichlorobenzene* *1,4-dichlorobenzene*

In this example, the –Cl functional groups are situated at different positions on the benzene rings.

Functional group isomerism

It is sometimes possible for compounds with the same molecular formula to have different functional groups, and because they have different functional groups they will belong to different homologous series.

a Molecular formula C_3H_8O

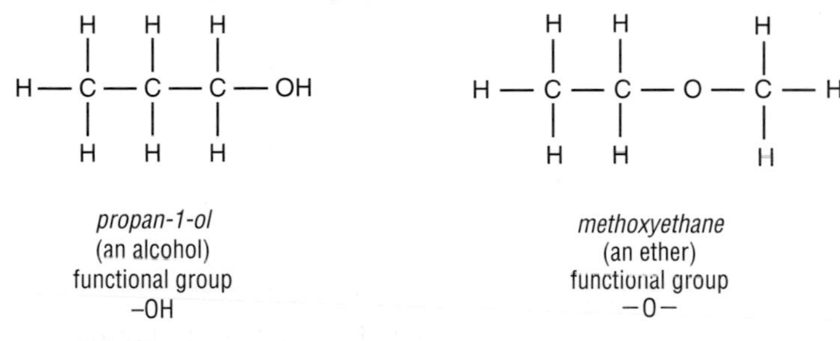

propan-1-ol
(an alcohol)
functional group
–OH

methoxyethane
(an ether)
functional group
–O–

The isomer on the left is called propan-1-ol. Its functional group is –OH and it is a member of the homologous series known as the *alcohols*. The isomer on the right is methoxyethane. Its –O– functional group is quite different from that of propan-1-ol and it belongs to another family of organic compounds, known as the *ethers*. (For more information about alcohols and ethers see **Section 13.2**.)

As well as having different physical properties, propan-1-ol and methoxyethane have very different chemical properties as a result of their differing functional groups.

b Molecular formula C_4H_8

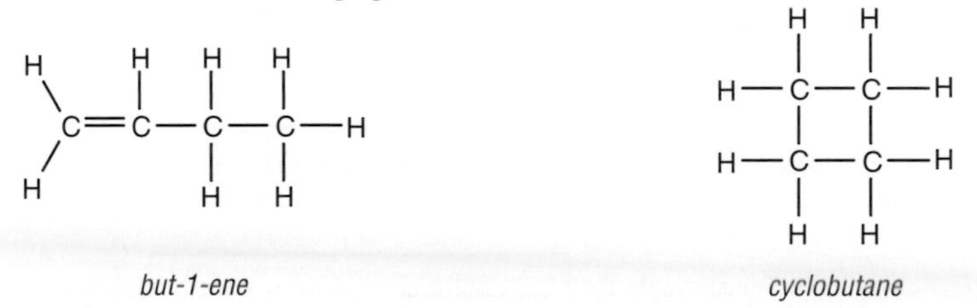

but-1-ene *cyclobutane*

But-1-ene has a C=C functional group and so belongs to the alkene family. Although cyclobutane has no functional group, we can tell from the ring of carbon atoms that it is a member of the cycloalkane series. Since they are not in the same homologous series, but-1-ene and cyclobutane differ significantly in their chemical properties.

Problems for 3.3

1 Write the molecular formula for each of the compounds in the following pairs. Use the molecular formulae to decide whether the compounds in the pair are isomers.

a CH_3–CH_2–CH_2–CH_2–CH_2–CH_3 and (cyclic: CH_2, H_2C, CH_2, H_2C, CH_2, CH_2 ring)

b CH_3–CH_2–CH_2–CH_2–Cl and CH_3–C(–CH_3)(–CH_3)–Cl

c CH_3–CH_2–CH_2–OH and CH_3–C(=O)–CH_3

d (benzene ring with OH and CH_3) and (benzene ring with CH_2–OH)

e

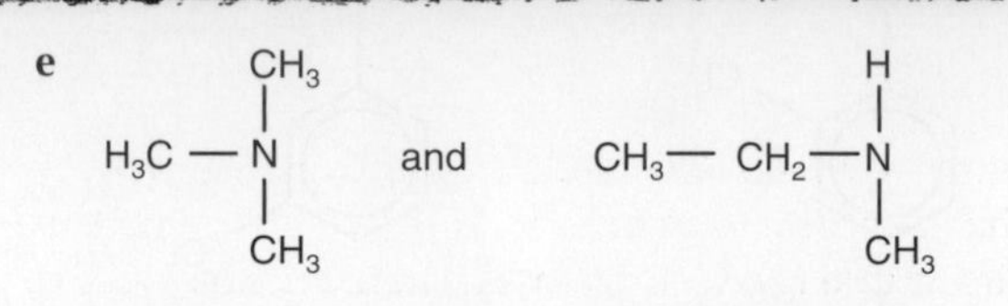

2 Draw structural formulae for the three isomers of C_5H_{12}. Give the systematic name of each isomer.

3 There are four structural isomers with the molecular formula C_4H_9Br. Draw their structures.

4 There are four structural isomers with the molecular formula C_8H_{10} in which each isomer contains a benzene ring and at least one side chain. Draw their structures.

5 Draw the structures of the seven structural isomers which have the molecular formula $C_4H_{10}O$,

3.4 E/Z isomerism

E/Z isomerism, also known as *cis–trans* isomerism, is one type of **stereoisomerism** (Figure 1). **Stereoisomers** have identical molecular formulae and the atoms are bonded together in the same order, but the arrangement of atoms in space is different in each isomer.

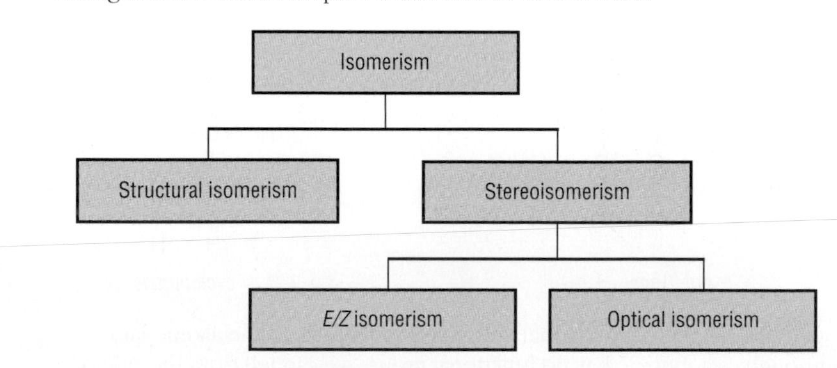

▶ **Figure 1** *E/Z* isomerism is one type of stereoisomerism. (Optical isomerism is covered in **Section 3.5**.)

Modelling E/Z isomers

Build a model of but-2-ene: $CH_3CH{=}CHCH_3$. If a model kit is not available you can draw the structure with correctly represented bond angles.

There is another isomer of but-2-ene with a different shape to the first one you built. Build or draw this other form.

The two forms of but-2-ene look like this:

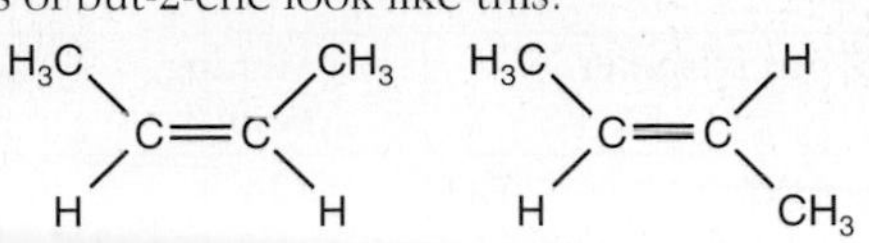

To turn the second form into the first form you would have to spin one end of the molecule round in relation to the other end. This can only be done by first breaking one of the bonds in the double bond. If you have the models available you can easily prove this to yourself. Figure 2 shows what models of the two isomers look like.

The average bond enthalpy for the bond which has to be broken here is about $+270\,\text{kJ mol}^{-1}$ and this much energy is not available at room temperature. A covalent bond has to be broken and another reformed in order to interconvert the forms of but-2-ene. In other words, the process is an example of a chemical reaction and the two forms are different chemicals.

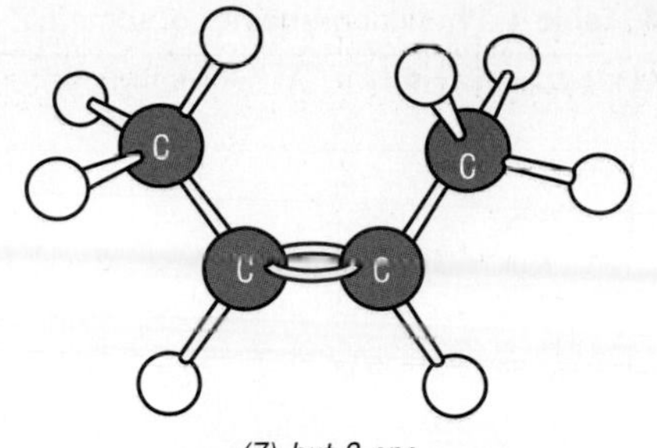

(Z)-but-2-ene

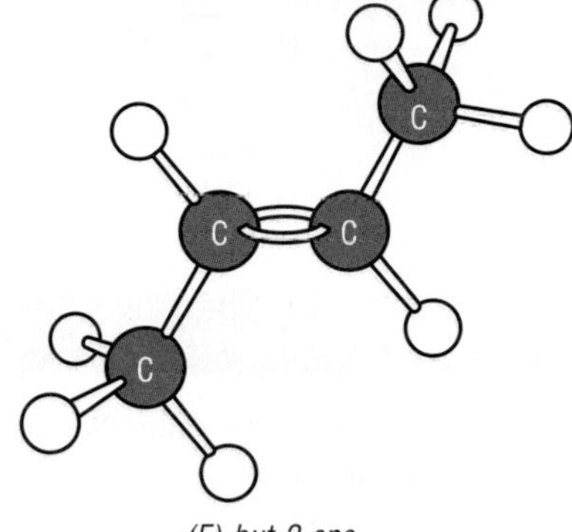

(E)-but-2-ene

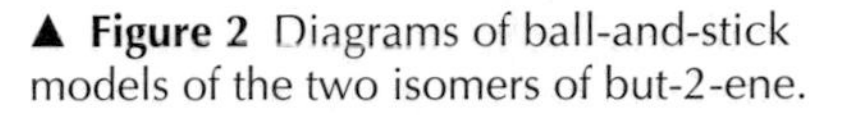

▲ **Figure 2** Diagrams of ball-and-stick models of the two isomers of but-2-ene.

Naming E/Z isomers

The two different but-2-enes need different names. The older way of naming them uses *cis–trans* nomenclature. The form of but-2-ene with the same groups (in this case both are methyl groups) on the same side of the double bond is called *cis*-but-2-ene.

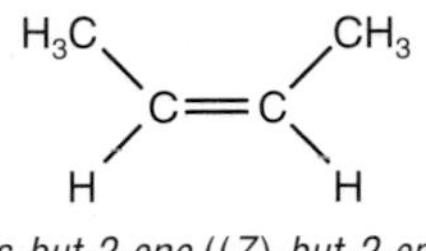

cis-but-2-ene ((Z)-but-2-ene)

Where the substituents are on opposite sides of the double bond the molecule is said to be the *trans* form.

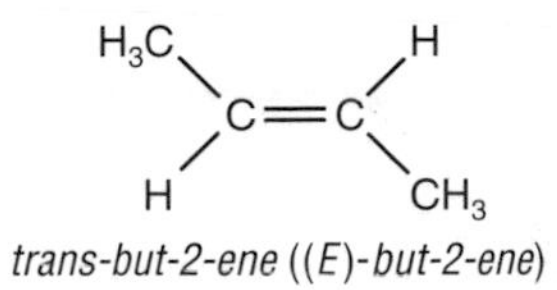

trans-but-2-ene ((E)-but-2-ene)

The more modern *E*/*Z* notation is now more commonly used. '*E*' (from the German *entgegen*) means *opposite* and corresponds to the term '*trans*'; '*Z*' (from the German *zusammen*) means *together* and corresponds to the term '*cis*'.

This means that *cis*-but-2-ene is now called (*Z*)-but-2-ene and *trans*-but-2-ene is now called (*E*)-but-2-ene.

You may be wondering why chemists don't just use the *cis–trans* system. However, the older *cis–trans* system for naming isomers breaks down when there are more than two different substituent groups on a double bond. For example, what about the molecules below?

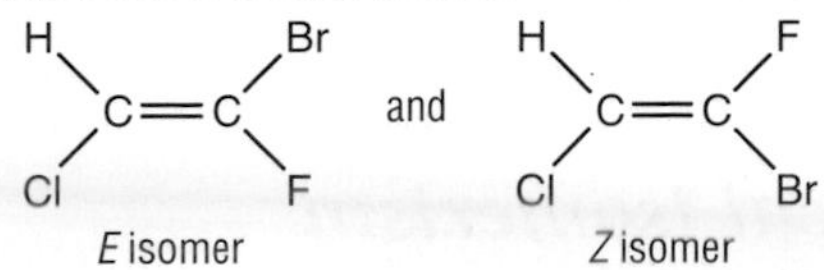

They are clearly isomers of each other, but cannot be named using the *cis–trans* nomenclature. The *E*/*Z* system can cope with this, by having a set of rules called the Cahn–Ingold–Prelog priority rules. You do not need to understand these rules. At this stage, you are only expected to name isomers with the same groups on either end of a double bond and name them as *E* or *Z* isomers.

Z and *E* isomers are different compounds, so they have different properties. Table 1 (page 52) gives some information on some isomers which are clearly different substances.

▼ **Table 1** Physical properties of some *E*/*Z* isomers.

***E* or *trans* isomer**	**Melting point/K**	**Density/g cm^{-3}**	***Z* or *cis* isomer**	**Melting point/K**	**Density/g cm^{-3}**
H_3C H C=C H CH_3	168	0.604	H_3C CH_3 C=C H H	134	0.621
Br H C=C H Br	267	2.23	Br Br C=C H H	220	2.25
HOOC H C=C H COOH	573	1.64	HOOC COOH C=C H H	412	1.59

Problems for 3.4

1 Draw and label the structural formulae of the *E*/*Z* isomers of 1,2-difluoroethene.

2 Draw and label the structural formulae of the *E*/*Z* isomers of pent-2-ene.

3 a Draw and label the *E*/*Z* isomers of 1,2-dichloroethene.
  b Suggest which of the isomers has the higher boiling point.
  c Explain your answer to part **b**.

4 a Draw the structure of 1,1-difluoroethene.
  b Does this compound show *E*/*Z* isomerism?
  c Describe *two* structural features which a molecule must have for it to exist as *E*/*Z* isomers.

5 Nerol, which occurs in bergamot oil, geraniol (in roses), linalool (in lavender) and citronellol (in geraniums) are four compounds from the *terpene* family. Their skeletal structures are shown below.

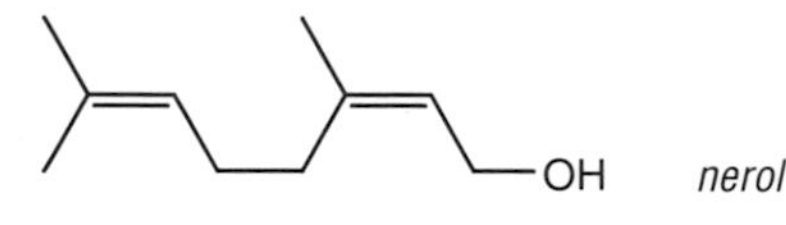

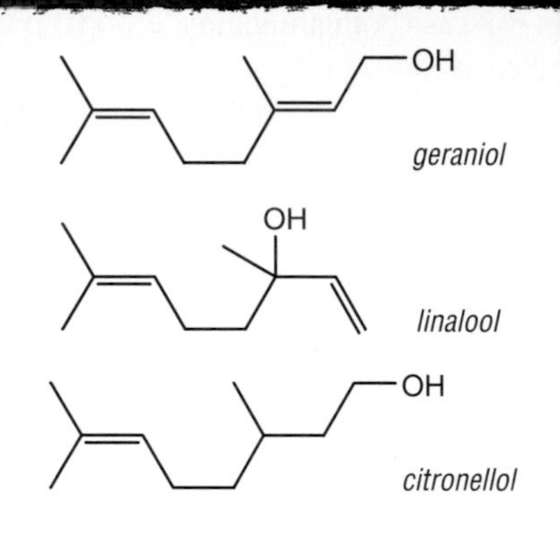

  a How are the structures of nerol and geraniol related?
  b How many moles of hydrogen (H_2) would be required to saturate one mole of geraniol?
  c How are nerol and geraniol related to citronellol?
  d How are the structures of nerol and geraniol related to linalool?

6 Draw and label the structural formulae of the *E*/*Z* isomers of 1-bromo-1,2-dichloroethene.

3.5 *Optical isomerism*

Optical isomerism arises because of the different ways in which you can arrange four groups around a carbon atom. It is a type of stereoisomerism.

It's all done with mirrors

Four single bonds around a carbon atom are arranged tetrahedrally. When four different atoms or groups are attached to these four bonds, the molecule can exist in *two isomeric forms*. Figure 1 illustrates these two forms for the amino acid alanine (2-aminopropanoic acid).

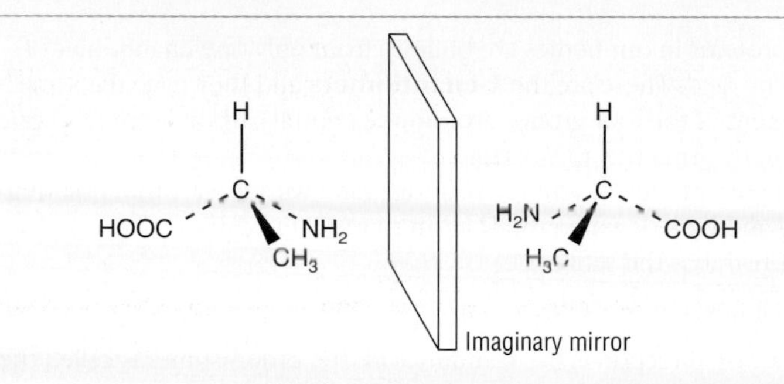

◀ **Figure 1** Two isomers of alanine.

You may need to build models of these two structures to convince yourself that they are different. The way they are different is that the right-hand structure in Figure 1 is the **mirror image** of the left-hand structure. If you have built models then you can use a mirror to prove this to yourself. Otherwise, imagine a mirror placed between the two forms of alanine – the NH_2 group is near to the mirror so it will be at the front of the reflection; the COOH group is furthest from the mirror so it will be at the back of the reflection, and so on.

All molecules have mirror images but they don't all exist as two isomers. For example, glycine (aminoethanoic acid) has only one form (Figure 2).

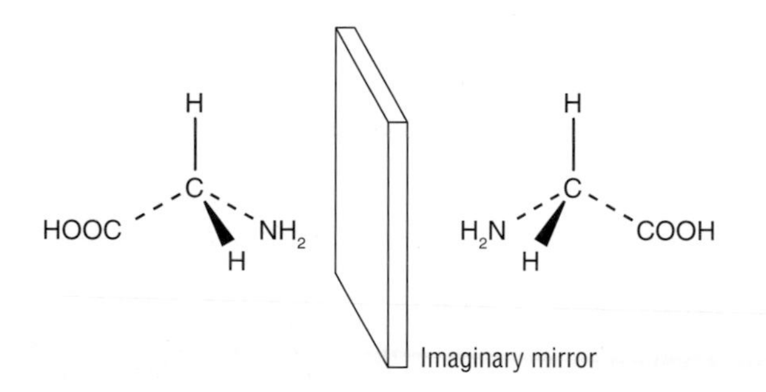

◀ **Figure 2** Glycine drawn to show that the mirror image is identical with the original structure, if you turn it round.

What makes alanine exist in two forms is that the mirror image and the original molecule are **non-superimposable**. If you move the mirror image across to the left then the H atom, the C atom and the CH_3 group will coincide, but NH_2 and COOH will be in the wrong places. No amount of twisting and turning will put things right! The only way you can make them superimpose is to break the C–NH_2 and C–COOH bonds and swap the groups round. Breaking and reforming bonds corresponds to a chemical reaction, and in such a reaction a compound is turned into a new compound. In this example, the new compound is a different isomer of the original one.

On the other hand, the pictures of glycine will coincide if you move them together – they are superimposable, and there is only one form.

Enantiomers

Molecules such as the two forms of alanine are called **optical isomers** or **enantiomers**. You don't always have to build models to find them – whenever a molecule contains a carbon atom that is surrounded by four different atoms or groups there will be optical isomerism.

Molecules that are not superimposable on their mirror images are called **chiral** molecules. A carbon atom that is bonded to four different groups is called a **chiral centre**. There are lots of other chiral things in the world – for example most keys, your shoes and your hands. In fact, chirality comes from a Greek word meaning 'handedness'. If you look at the reflection of your *right* hand in a mirror, the mirror image is superimposable on your *left* hand, but not on your right hand.

The proteins in our bodies are built up from only one enantiomer of each amino acid. These are the **L-enantiomers** and they have the same arrangement of the four groups around the central carbon atom. You can spot them by using the *'CORN rule'*.

Stand your model, or arrange your diagram, with the H atom pointing upwards and look down on the H atom towards the carbon. The optical isomer which has the sequence

COOH, R NH_2 (i.e. 'CORN')

in a *clockwise* direction is the **L**-amino acid; the other isomer is called the **D**-enantiomer (Figure 3).

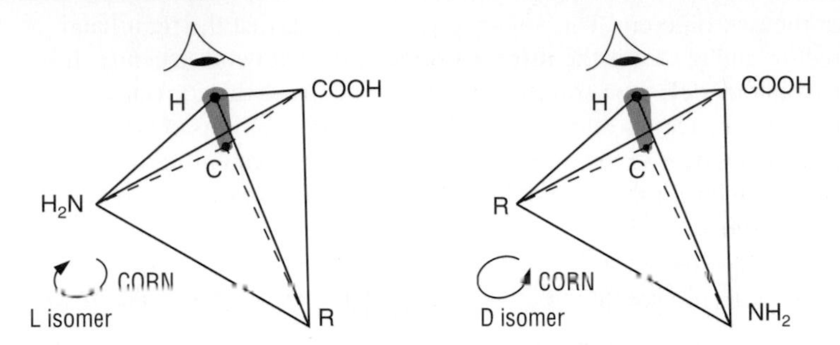

► **Figure 3** The CORN rule – look down the H–C bond from hydrogen towards the central carbon atom.

In Figure 1, the isomer of alanine on the right is the D-isomer; the other form on the left obeys the 'CORN rule' and is the L-isomer.

(You may see an alternative, and more general, naming system for enantiomers using the symbols *R*- and *S*-, in which the groups round the chiral centre are listed in order of a priority based on the relative atomic mass of the attached atom. In this course, we shall use only the *D/L* convention.)

How do enantiomers differ?

Enantiomers behave identically in ordinary test-tube chemical reactions. Most of their physical properties, such as melting point, density and solubility, are also the same. But enantiomers behave differently in the presence of other chiral molecules.

Our proteins are built up from chiral amino acids, and so our enzymes and protein hormones must be chiral too. D- and L-enantiomers will, therefore, react differently with enzymes. It's like your left shoe being made for your left foot – your right foot doesn't really fit into it at all.

Here are some examples of how enantiomers differ:

- Enantiomers can interact differently with the chiral 'taste-buds' on your tongue – D-amino acids all taste sweet; L-amino acids are often tasteless or bitter.
- Enantiomers can smell different. For example, the different smells of oranges and lemons are caused by enantiomers (see problem 6 at the end of this section).
- Many beneficial medicines have enantiomers which have little or no pharmacological effect. In the case of one medicine, thalidomide, the apparently non-active enantiomer was found to damage unborn children when the drug was taken during pregnancy. Medicines have been more thoroughly tested since the thalidomide incident, but it is often expensive to separate enantiomers if no dangers are found, and about 80% of synthetic medicines are marketed as a mixture of isomers.

D-amino acids do exist in nature. For example, penicillin works by breaking peptide links which involve D-alanine. These occur in the cell walls of bacteria but not in humans. When its cell wall is broken, the bacterium is killed. So penicillin is very effective at killing bacteria but cannot have the same effect on us – because we don't use D-amino acids.

Problems for 3.5

1 a What structural feature must a molecule have in order to form optical isomers (enantiomers)?
  b Which of the following compounds can have optical isomers?
   i CH_2Cl_2 ii CH_2ClBr iii $CHClBrI$

2 a Draw the structural formula of 2-bromobutane.
  b Use three-dimensional diagrams to show the structures of the two optical isomers of this compound.

3 a Draw the structural formula of 3-methylheptane.
  b Identify the chiral carbon atom in your structural formula and mark it with an asterisk (*).
  c Use three-dimensional diagrams to show the structures of the two optical isomers of this compound.

4 a Draw the structural formula of the amino acid serine (in which R = $-CH_2OH$).
  b Mark the chiral carbon atom with an asterisk.
  c Use three-dimensional diagrams to show the structures of the two optical isomers of serine.
  d Label your isomers D or L.

5 In the catalogue of a company which makes and sells chemicals, the cost of the amino acid D-cysteine is about 200 times more than the cost of its optical isomer L-cysteine. Suggest why there is such a large difference in price between the isomers.

6 *Limonene* and *carvone* are examples of compounds for which both enantiomers occur naturally. One enantiomer of limonene smells of oranges, the other of lemons; carvone can smell like caraway seeds or like spearmint.

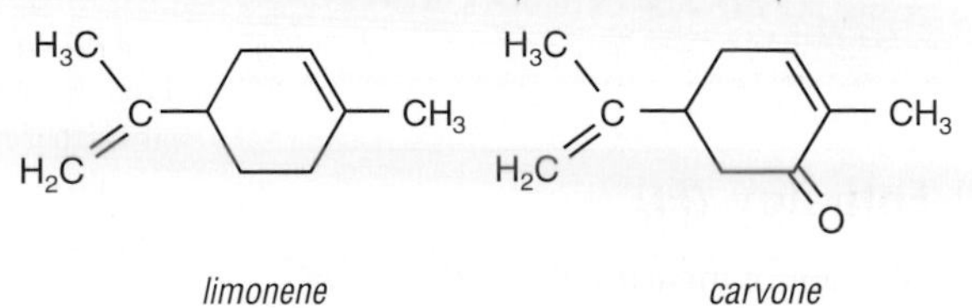

  a Draw out the full structural formulae of limonene and carvone (formulae which show all the bonds and atoms). Use an asterisk to show the chiral carbon atom in each molecule.
  b Draw the structure of the compound which is formed when 1 mole of limonene undergoes addition with 2 moles of hydrogen molecules.
  c Does the product of the reaction in part **b** exhibit optical isomerism? Explain your answer.

ENERGY CHANGES AND CHEMICAL REACTIONS

4.1 *Energy out, energy in*

AS

Energy changes are a characteristic feature of chemical reactions. Many chemical reactions give out energy and a few take energy in. A reaction that gives out energy and heats the surroundings is described as **exothermic**. A reaction that takes in energy and cools the surroundings is **endothermic**.

During an exothermic reaction the chemical reactants are losing energy. This energy is used to heat the surroundings – the air, the test tube, the laboratory, the car engine. The products end up with less energy than the reactants had – but the surroundings end up with more, and get hotter. We measure the energy transferred to and from the surroundings as **enthalpy change**. Enthalpy changes can be shown on an **enthalpy level diagram**, also called an energy level diagram (Figure 1).

Enthalpy *(H)*

We cannot measure the enthalpy, H, of a substance. What we can do is measure the *change* in enthalpy when a reaction occurs:

$$\Delta H = H_{products} - H_{reactants}$$

The enthalpy change in a chemical reaction gives the quantity of energy transferred to or from the surroundings, when the reaction is carried out in an open container.

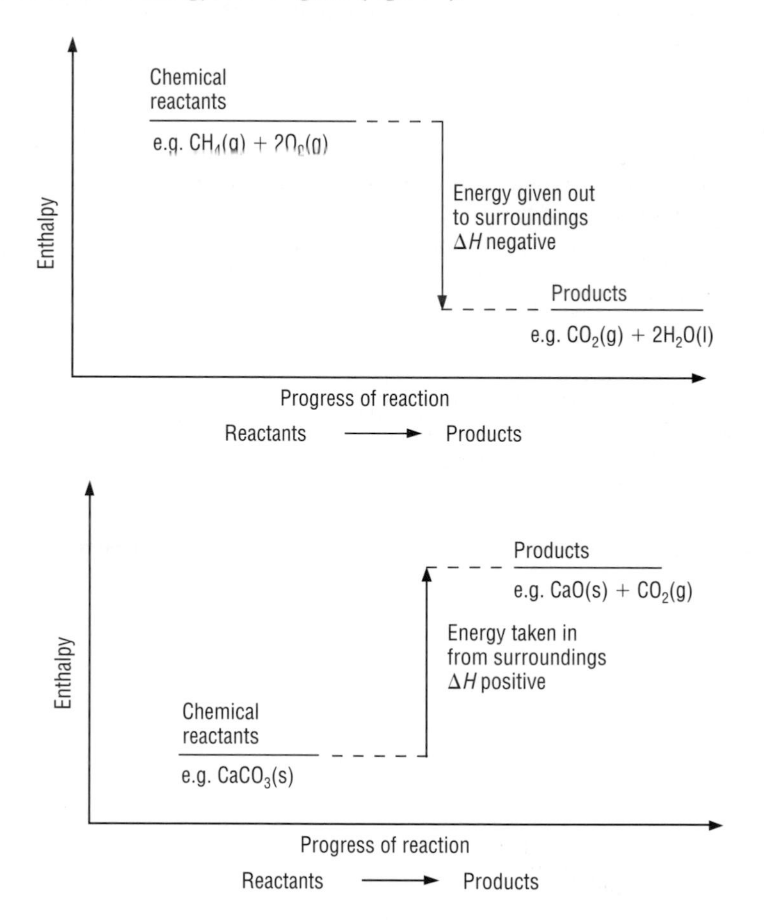

▶ **Figure 1** Enthalpy level diagram for an exothermic reaction – burning methane: $CH_4(g) + 2O_2(g) \rightarrow CO_2(g) + 2H_2O(l)$

▶ **Figure 2** Enthalpy level diagram for an endothermic reaction – decomposing calcium carbonate: $CaCO_3(s) \rightarrow CaO(s) + CO_2(g)$

In an endothermic reaction, the reactants take in energy from the surroundings. This is shown in Figure 2.

In an **exothermic** reaction, the enthalpy of the reacting system decreases.

ΔH is negative

In an **endothermic** reaction, the enthalpy of the reacting system increases.

ΔH is positive

How much?

As your study of chemistry becomes more advanced, you will find you need to put *numbers* to the features of chemical reactions – to make them quantitative. We measure the energy changes associated with chemical reactions by measuring enthalpy changes.

For an exothermic reaction ΔH is *negative*. This is because, from the point of view of the chemical reactants, energy has been *lost* to the surroundings. Conversely, for an endothermic reaction ΔH is *positive* – energy has been *gained* from the surroundings.

The units of ΔH

Enthalpy changes are measured in kilojoules per mole. For example, for the reaction of methane with oxygen, we write

$$CH_4(g) + 2O_2(g) \rightarrow CO_2(g) + 2H_2O(l); \Delta H = -890\,kJ\,mol^{-1}$$

This means that for every mole of methane that reacts in this way, 890 kJ of energy are transferred to heat the surroundings. If we used 2 moles of methane, we would get 2 × 890 = 1780 kJ of energy transferred. This assumes that all the methane is converted into products and that none is left unburned.

When calcium carbonate is heated, it decomposes. Energy is taken in – it is an endothermic reaction. We write

$$CaCO_3(s) \rightarrow CaO(s) + CO_2(g); \Delta H = +179\,kJ\,mol^{-1}$$

For every mole of $CaCO_3$ that is decomposed, 179 kJ of energy is taken in. If we decomposed 0.1 mol of $CaCO_3$, 17.9 kJ would be taken in from the surroundings.

Standard conditions

Like most physical and chemical quantities, ΔH varies according to the conditions. In particular, ΔH is affected by temperature, pressure and the concentration of solutions. So we choose certain *standard* conditions to refer to. We use:

- a *specified* temperature
- a standard pressure of 1 atmosphere ($1.01 \times 10^5\,N\,m^{-2}$)
- a standard concentration of $1\,mol\,dm^{-3}$.

The standard temperature is normally chosen as 298 K (25 °C). If ΔH refers to these standard conditions, it is written as $\Delta H^{\ominus}_{298}$, pronounced 'delta H standard, 298', or just 'delta H standard'. The values of ΔH given in the examples above are all standard values. You can use the **Data Sheets** to look up ΔH values.

Measuring enthalpy changes

Many enthalpy changes can be measured quite simply in the laboratory. We usually do it by arranging for the energy involved in a reaction to be transferred to or from water surrounding the reaction vessel. If it is an exothermic reaction, the water gets hotter; if it is endothermic, the water gets cooler. If we measure the temperature change of the water, and if we know its mass and specific heat capacity, we can work out how much energy was transferred to or from the water during the chemical reaction.

To do this, we need to use the relationship:

$$\text{energy transferred} = cm\Delta T$$

where: c is the specific heat capacity of water ($4.18\,J\,g^{-1}\,K^{-1}$)
m is the mass of water in g
ΔT is the change in temperature of the water.

In enthalpy experiments using simple apparatus, it is difficult to avoid energy being transferred to the surroundings ('heat loss'). Because of this, we need to make a number of simplifying assumptions when calculating energy changes based on results from simple experiments. These assumptions ignore amounts of energy that are small in comparison with that transferred to the water:

- The heat transferred to the calorimeter itself can be ignored. Metals have specific heat capacities of about one-tenth that of water; and plastic cups have a specific heat capacity about half that of water but their masses are very small in comparison to the masses of liquids involved and they are also good insulators.

System or surroundings?

When chemists talk about enthalpy changes they often refer to the **system**. This means the reactants and the products of the reaction they are interested in. The system may lose or gain enthalpy as a result of the reaction.

The **surroundings** means the rest of the world – the test tube, the air ... and so on.

A word about temperature

There are two scales of temperature used in science and you should be familiar with both.

The kelvin (K) is the unit of **absolute temperature** and 0 K is called absolute zero. The kelvin is the SI unit of temperature and should always be used in *calculations* involving temperature.

However, you will usually *measure* temperature using a thermometer marked in degrees Celsius. You can convert temperatures from the Celsius scale to the absolute scale by adding 273. Thus, the boiling point of water = 100 + 273 = 373 K. Note that the kelvin does not have a degree symbol, °.

Note, too, that a *change* in temperature has the same numerical value on both scales.

- The energy transferred to any solids in the calorimeter can be ignored because of the low specific heat capacity the solid will have compared with water.
- When heating or reacting solutions, just the water can be considered – for example, when heating $25\,cm^3$ of $0.2\,mol\,dm^{-3}$ copper sulfate solution, the mass can be taken as 25 g and the specific heat capacity as being the same as that of water.

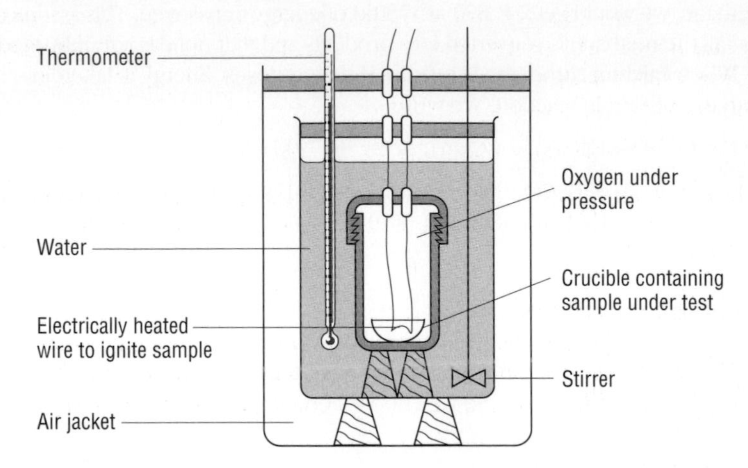

► **Figure 3** A bomb calorimeter for making accurate measurements of energy changes. The fuel is ignited electrically and burns in the oxygen inside the pressurised vessel. Energy is transferred to the surrounding water, whose temperature rise is measured. Note that the experiment is done at constant volume in a closed container. Enthalpy changes are for reactions carried out at constant pressure, so the result needs to be modified accordingly.

For some reactions, measuring ΔH is very straightforward. Take the burning of methane, for example. You could make a rough measurement of ΔH for this reaction using ordinary kitchen equipment and a gas cooker (see problem 6 at the end of this section).

For other reactions it is less simple. For example, decomposing $CaCO_3$ needs a temperature of over 800 °C, which makes it difficult to use water to measure the energy transferred. In cases like this, enthalpy changes can be measured *indirectly*, using enthalpy cycles (see pages 59–61).

Different kinds of enthalpy change

We can define the **standard enthalpy change for a reaction, $\Delta H^{\ominus}_{r,298}$**, as the enthalpy change when molar quantities of reactants *as stated in the equation* react together under standard conditions. This means at 1 atmosphere pressure and 298 K, with all the substances in their standard states.

You must always give an equation when quoting an enthalpy change of reaction – otherwise things can get very muddled. For example,

$$H_2(g) + \tfrac{1}{2}O_2(g) \rightarrow H_2O(l);\ \Delta H^{\ominus}_{r,298} = -286\,kJ\,mol^{-1}$$

but $2H_2(g) + O_2(g) \rightarrow 2H_2O(l);\ \Delta H^{\ominus}_{r,298} = -572\,kJ\,mol^{-1}$

The following kinds of enthalpy change are particularly important and are given special names.

Standard enthalpy change of combustion, $\Delta H^{\ominus}_{c,298}$, is the enthalpy change that occurs when *1 mole* of a substance is burned completely in oxygen. In theory, the substance needs to be burned under standard conditions – that is 1 atmosphere pressure and 298 K. In practice this is impossible, so we burn the substance in the normal way and then make adjustments to allow for non-standard conditions.

For example, the enthalpy change of combustion of octane, one of the alkanes found in petrol, is $-5470\,kJ\,mol^{-1}$. This is much bigger than the $\Delta H^{\ominus}_c$ for methane ($-890\,kJ\,mol^{-1}$) because burning octane involves breaking and making more bonds than burning methane. (Note that if no temperature is given with $\Delta H^{\ominus}_c$ then we assume that the value refers to 298 K.)

Standard state

The standard state of a substance is its most stable state under standard conditions:

- a pressure of 1 atmosphere
- a stated temperature, usually 298 K.

This may be a pure solid, liquid or gas.

All combustion reactions are exothermic

$\therefore \Delta H^{\ominus}_c$ is always negative.

The equations for these combustions are:

$CH_4(g) + 2O_2(g) \rightarrow CO_2(g) + 2H_2O(l)$
$C_8H_{18}(l) + 12\frac{1}{2}O_2(g) \rightarrow 8CO_2(g) + 9H_2O(l)$

When you write an equation to represent an enthalpy change of combustion, the equation must always balance and show *1 mole* of the substance reacting, even if this means having half a mole of oxygen molecules in the equation. You should always include state symbols.

Standard enthalpy change of formation, $\Delta H^\ominus_{f,298}$, is the enthalpy change when *1 mole* of a compound is formed from its elements – again with both the compound and its elements being in their standard states.

For example, the enthalpy change of formation of water, $H_2O(l)$, is $-286\,kJ\,mol^{-1}$. When you make a mole of water from hydrogen and oxygen, 286 kJ are transferred to the surroundings. This is summed up as

$H_2(g) + \frac{1}{2}O_2(g) \rightarrow H_2O(l)$; $\Delta H^\ominus_{f,298} = -286\,kJ\,mol^{-1}$

Notice that the equation refers to 1 mole of H_2O, so only ½ mole of oxygen is needed in the equation.

It is often impossible to measure enthalpy changes of formation directly. For example, the enthalpy change of formation of methane is $-75\,kJ\,mol^{-1}$. This refers to the reaction

$C(s) + 2H_2(g) \rightarrow CH_4(g)$; $\Delta H^\ominus_{f,298} = -75\,kJ\,mol^{-1}$

... but there is a problem. This reaction doesn't actually occur under normal conditions. So how did anyone manage to measure the value of $\Delta H^\ominus_{f,298}$? It has to be done *indirectly*, making use of quantities that *can* be measured and incorporating these into an enthalpy cycle.

$\Delta H^\ominus_f$ may be positive or negative.

Remember too that, by definition, the enthalpy change of formation of a pure element in its standard state is zero.

Enthalpy cycles – enthalpy changes of combustion

Figure 4 shows an **enthalpy cycle**, also known as an energy cycle. There is both a direct and an indirect way to turn graphite (C) and hydrogen (H_2) into methane (CH_4). We can't measure the enthalpy change for the *direct* route. The *indirect* route goes via CO_2 and H_2O and involves two enthalpy changes both of which we *can* measure. Since most organic compounds burn easily, we can often use cycles such as this, based on enthalpy changes of combustion, to work out indirectly the enthalpy change of an organic reaction (see problem 11 at the end of this section).

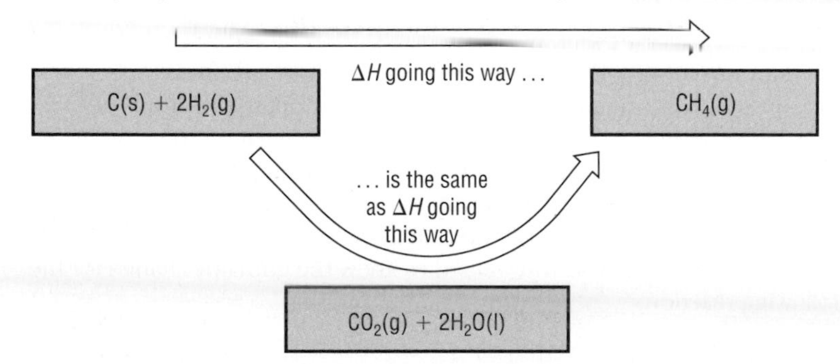

◀ **Figure 4** An enthalpy cycle for finding the enthalpy change of formation of methane, CH_4.

The key idea is that the total enthalpy change for the indirect route is the same as the enthalpy change via the direct route. This makes sense if you think about it; energy cannot be created or destroyed – this is the law of conservation of energy. So, *as long as your starting and finishing points are the same, the enthalpy change will always be the same no matter how you get from start to finish*. This is one way of stating **Hess's law**, and an enthalpy cycle like the one in Figure 4 is called a **Hess cycle** or a **thermochemical cycle**.

Hess's law

The enthalpy change for any chemical reaction is independent of the intermediate stages, so long as the initial and final conditions are the same for each route.

If you know the enthalpy changes involved in two parts of the cycle, you can work out the enthalpy change in the third. So, referring to Figure 5, if we can measure ΔH_2 and ΔH_3, we can find ΔH_1, which is the enthalpy change that cannot be measured directly. Quite simply,

$$\Delta H_1 = \Delta H_2 - \Delta H_3$$

(It has to be minus ΔH_3 because the reaction to which ΔH_3 applies actually goes in the opposite direction to the way we want it to go in order to produce CH_4. To find the enthalpy change for the reverse reaction, we must reverse the sign.)

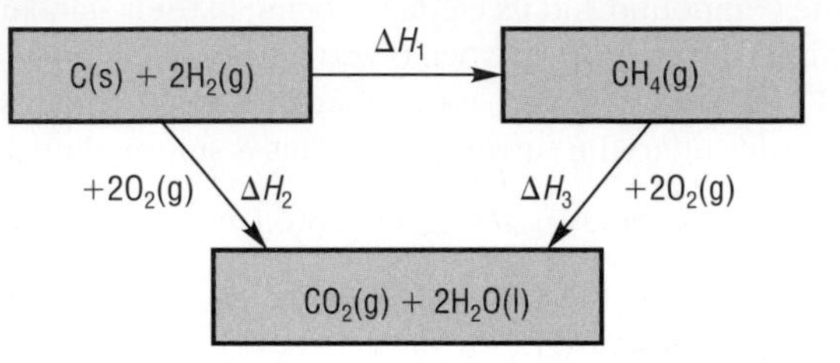

▶ **Figure 5** Using an enthalpy cycle to find ΔH_1.
$\Delta H_1 = \Delta H_2 - \Delta H_3$

Now ΔH_2 is the sum of the enthalpy changes of combustion of 1 mole of carbon and 2 moles of hydrogen, and ΔH_3 is the enthalpy change of combustion of methane. ΔH_1 is the enthalpy change of formation of methane, $\Delta H_f^{\ominus}(CH_4)$, which is the quantity we are trying to find.

$$\Delta H_1 = \Delta H_f^{\ominus}(CH_4)$$
$$\Delta H_2 = \Delta H_c^{\ominus}(C) + 2\Delta H_c^{\ominus}(H_2)$$
$$\Delta H_3 = \Delta H_c^{\ominus}(CH_4)$$

$\Delta H_c^{\ominus}(C) = -393\,kJ\,mol^{-1}$
$\Delta H_c^{\ominus}(H_2) = -286\,kJ\,mol^{-1}$
$\Delta H_c^{\ominus}(CH_4) = -890\,kJ\,mol^{-1}$

Putting in the values of the enthalpies of combustion, which we can measure, we get

$$\begin{aligned}\Delta H_f^{\ominus}(CH_4) &= \Delta H_c^{\ominus}(C) + 2\Delta H_c^{\ominus}(H_2) - \Delta H_c^{\ominus}(CH_4) \\ &= -393\,kJ\,mol^{-1} + 2(-286)\,kJ\,mol^{-1} - (-890)\,kJ\,mol^{-1} \\ &= -75\,kJ\,mol^{-1}\end{aligned}$$

Calculations look cluttered if $\Delta H^{\ominus}_{c,298}$ is written in full each time it occurs. So, we often use just $\Delta H_c^{\ominus}$ for the standard enthalpy change at 298 K.

Using the enthalpy cycle has enabled us to find a value for an enthalpy change which we could not find directly. You will find these cycles very useful in other parts of your chemistry course. Another example is shown below.

Using enthalpy changes of formation in Hess cycles to measure enthalpy changes of reaction

Enthalpy changes for the formation of many compounds from their elements in their standard states are available in data books. Some of these have been measured directly; others have been calculated from enthalpy changes for other reactions, as shown above for methane. These data are very useful as they can be used to calculate the enthalpy change for a reaction, rather than attempt to do an experiment to measure it.

For example, suppose you wanted to know the enthalpy change for the following reaction:

$$NH_3(g) + HCl(g) \rightarrow NH_4Cl(s)$$

The reaction can be included in an enthalpy cycle, as shown in Figure 6.

In this case, ΔH_2 is the sum of the enthalpy changes of formation of ammonia and hydrogen chloride. ΔH_3 is the enthalpy change of formation of ammonium chloride.

$$\Delta H_1 = -\Delta H_2 + \Delta H_3$$
$$\Delta H_2 = \Delta H_f^{\ominus}(NH_3) + \Delta H_f^{\ominus}(HCl)$$
$$\Delta H_3 = \Delta H_f^{\ominus}(NH_4Cl)$$

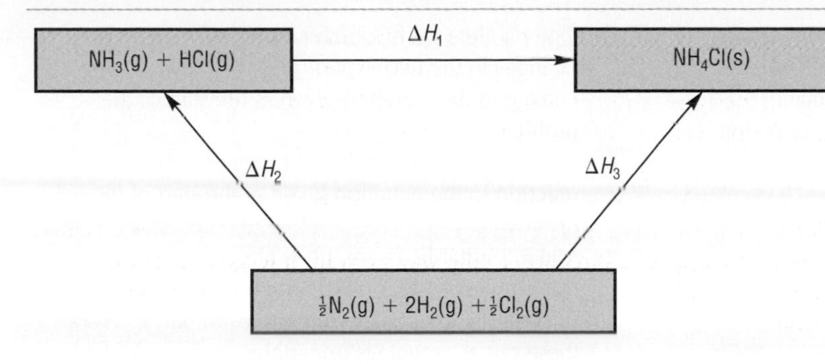

◀ **Figure 6** Using enthalpy changes of formation to find the enthalpy change for a reaction.

Putting values for the enthalpy changes of formation into the equation, we get

$$\begin{aligned}\Delta H_1 &= -\Delta H_f^\ominus(NH_3) - \Delta H_f^\ominus(HCl) + \Delta H_f^\ominus(NH_4Cl) \\ &= -(-46.1)\,kJ\,mol^{-1} - (-92.3)\,kJ\,mol^{-1} + (-315)\,kJ\,mol^{-1} \\ &= -176.6\,kJ\,mol^{-1}\end{aligned}$$

$\Delta H_f^\ominus(NH_3) = -46.1\,kJ\,mol^{-1}$
$\Delta H_f^\ominus(HCl) = -92.3\,kJ\,mol^{-1}$
$\Delta H_f^\ominus(NH_4Cl) = -315\,kJ\,mol^{-1}$

In **Section 4.2** you will discover that 'bond enthalpies' can also be used in Hess cycles to determine the enthalpy changes of reactions.

Problems for 4.1

You will need to consult the **Data Sheets** (Table 18) to find values for standard enthalpy changes when doing these problems.

1 Define the following enthalpy changes:
  a standard enthalpy change of combustion
  b standard enthalpy change of formation.

2 Explain why standard enthalpy changes of formation may have a positive sign, but standard enthalpy changes of combustion are always negative.

3 The standard enthalpy change of formation of hydrogen chloride is $-92.3\,kJ\,mol^{-1}$ and that of hydrogen iodide is $+26.5\,kJ\,mol^{-1}$. Draw labelled enthalpy level diagrams to represent the reactions which occur when each of these compounds is formed from its elements.

4 Write the equations, including state symbols, to which the following enthalpy changes apply.
  a standard enthalpy change of formation of ethanol, C_2H_5OH
  b standard enthalpy change of combustion of ethanol
  c standard enthalpy change of formation of butane, C_4H_{10}
  d standard enthalpy change of combustion of butane
  e standard enthalpy change of formation of glucose, $C_6H_{12}O_6$
  f standard enthalpy change of combustion of glucose.

5 The standard enthalpy change of combustion of carbon is equal to the standard enthalpy change of formation of carbon dioxide. Explain why this is so by referring to the equations for the two reactions.

6 Suppose you were asked to measure the enthalpy change of combustion of methane and you were given a gas cooker and a saucepan.
  a What other equipment would you need?
  b What measurements would you make?
  c What would be the main sources of error?

7 **a** In a student's experiment, 0.16 g of methanol (CH_3OH) is used up when a spirit burner heats 100 g of water from 17 °C to 24 °C. Calculate the enthalpy change of combustion of methanol.
  b Calculate the percentage uncertainties involved in making the measurements in the experiment. (Make appropriate assumptions about the nature of the apparatus by looking at the number of significant figures to which the data is quoted.) Add these uncertainties to give an estimate of the overall uncertainty. Apply this percentage uncertainty to the value you calculated in part **a** to give an overall '±' on your calculated result.
  c Which piece of apparatus should the student change to reduce the percentage uncertainty by the greatest value?
  d Look up a value for the standard enthalpy change of combustion of methane in the **Data Sheets** (Table 18) and record it.
  e Does the Data Sheet value lie within or outside the uncertainty limits of your calculated value? Comment on the significance of this.

8 The standard enthalpy change of combustion of heptane, C_7H_{16}, is $-4817\,kJ\,mol^{-1}$.
  a What is the relative molecular mass of heptane?
  b Calculate the energy transferred when the following quantities of heptane are burnt.
    i 10 g   ii 10 kg
    What assumptions have you made in these calculations?
  c What further information would you need in order to calculate the energy transferred when $1\,dm^3$ of heptane is burned?

9 a Write the equation to represent the formation of 1 mole of water from its elements in their standard states.
  b Look up and write down the standard enthalpy change of formation of water.
  c Calculate the enthalpy change when 1.0 g of hydrogen burns in oxygen. What assumptions have you made?
  d What is the standard enthalpy change for the following reaction?

$$H_2O(l) \rightarrow H_2(g) + \frac{1}{2}O_2(g)$$

10 A student wishes to measure the enthalpy change of the following reaction:

$$CuSO_4(s) + 5H_2O \rightarrow CuSO_4{\cdot}5H_2O$$

anhydrous copper sulfate → *hydrated copper sulfate*

She decides that it will not be possible to do the experiment in one step but must use two steps and apply Hess's law to work out her answer.

Experiment 1
She added 3.99 g (0.0250 mol) of anhydrous copper sulfate to $50\,cm^3$ water in a plastic cup, and stirred the contents until all the solid had dissolved. The temperature rose from 19.52 °C to 27.40 °C.

Experiment 2
She then added 6.24 g (0.0250 mol) of hydrated copper sulfate to $48\,cm^3$ water ($48\,cm^3$ to allow for the '$5H_2O$' making up the volume to $50\,cm^3$). This was again done by stirring the water and copper sulfate in a plastic cup. The temperature changed from 19.56 °C to 18.28 °C.
  a Suggest why the enthalpy change of the reaction given at the start of the problem cannot be measured directly.
  b Show, by calculation, that $48\,cm^3$ was the correct volume to use for the second experiment.
  c Calculate the energy change in each experiment in $kJ\,mol^{-1}$. (The specific heat capacity of water is $4.18\,J\,g^{-1}\,K^{-1}$)
  d Explain the assumptions you made in doing these calculations.
  e Draw a Hess cycle which connects the enthalpy changes in the two experiments with the enthalpy change in the equation given at the start of the problem.
  f Calculate a value for the enthalpy change of the reaction in the equation given at the start of the problem.
  g Making suitable assumptions about the apparatus used by the student, calculate the percentage uncertainty in each of your values in part **c**. Work these out as actual (±) uncertainty values. Add these uncertainties to give the overall uncertainty in your answer to part **f**.
  h Which piece of apparatus generates the greatest percentage uncertainty?
  i The 'accepted' value for the enthalpy change is $-77.4\,kJ\,mol^{-1}$. Comment on this compared with your value and uncertainty limits.

11 In this problem, you can calculate a value for the enthalpy change of formation of butane using a Hess cycle and the enthalpy changes of combustion of butane and its elements.
  a Write the equation representing the formation of 1 mole of butane, $C_4H_{10}(g)$, from its elements in their standard states.
  b Draw an enthalpy cycle to show the relationship between the formation of butane from carbon and hydrogen and the combustion of these elements to give carbon dioxide and water (see Figure 5 if you need help).
  c Use your enthalpy cycle to calculate a value for the standard enthalpy change of formation of butane. You will also need the following data:

$$\Delta H_c^{\ominus}(C) = -393\,kJ\,mol^{-1}$$
$$\Delta H_c^{\ominus}(H_2) = -286\,kJ\,mol^{-1}$$
$$\Delta H_c^{\ominus}(C_4H_{10}) = -2877\,kJ\,mol^{-1}$$

12 In this problem you use enthalpy changes of formation in a Hess cycle to measure the enthalpy change of a reaction.
Silicon chloride reacts with water as shown in the equation:

$$SiCl_4(l) + 2H_2O(l) \rightarrow SiO_2(s) + 4HCl(g)$$

Use the data below to draw a Hess cycle, and use it to work out a value for the enthalpy change of the reaction above (see Figure 6 if you need help).

Substance	Enthalpy change of formation, $\Delta H_f^{\ominus}/kJ\,mol^{-1}$
$SiCl_4(l)$	−640
$H_2O(l)$	−286
$SiO_2(s)$	−910
$HCl(g)$	−92

4.2 Where does the energy come from?

AS

All chemical reactions involve breaking and making chemical bonds. Bonds break in the reactants and new bonds form in the products. The energy changes in chemical reactions come from the energy changes that happen when bonds are broken and made.

You can remind yourself of the basic ideas of chemical bonding by reading **Section 3.1**.

Bond enthalpies

A chemical bond is basically an electrical attraction between atoms or ions. When you break a bond, you have to do work in order to overcome these attractive forces. To break the bond completely, you need (theoretically) to separate the atoms or ions so they are an infinite distance apart. Figure 1 illustrates this for the H–H bond in a molecule of hydrogen, H_2.

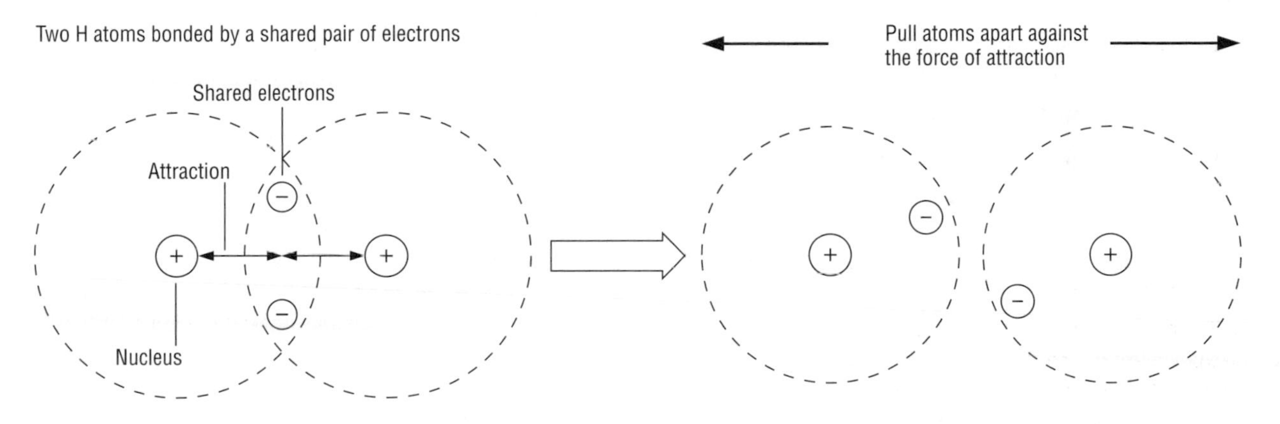

▲ **Figure 1** Breaking a bond involves using energy to overcome the force of attraction.

The quantity of energy needed to break a particular bond in a molecule is called the **bond dissociation enthalpy**, or **bond enthalpy** for short. For the bond shown in Figure 1, the process involved is

$$H_2(g) \rightarrow 2H(g);\ \Delta H = +436\,kJ\,mol^{-1}$$

So the bond enthalpy of the H–H bond is $+436\,kJ\,mol^{-1}$. Notice that this is a *positive* ΔH value because *breaking* a bond is an endothermic process – it needs energy. When you *make* a new bond, you get energy given out.

Bond enthalpies are very useful because they tell you how strong bonds are. The stronger a bond, the more difficult it is to break and the higher its bond enthalpy.

Bond enthalpy and bond length

When a bond like the one in Figure 1 forms, the atoms move together because of the attractive forces between nuclei and electrons. But there are also **repulsive** forces, between the nuclei of the two atoms, and these get bigger as the atoms approach until the atoms stop moving together. The distance between them is now the equilibrium bond length (Figure 2). The shorter the bond length, the stronger the attraction between the atoms.

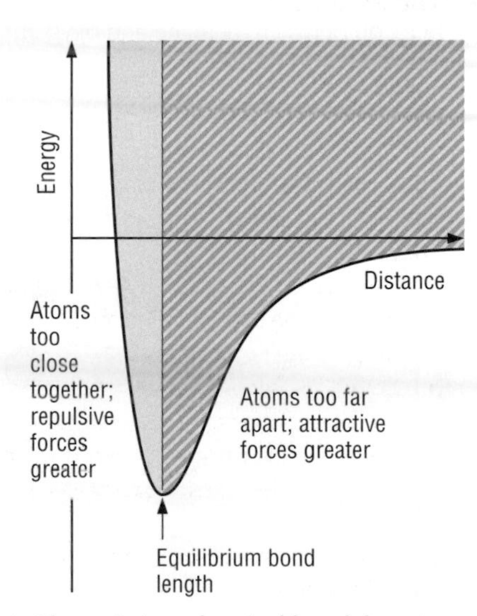

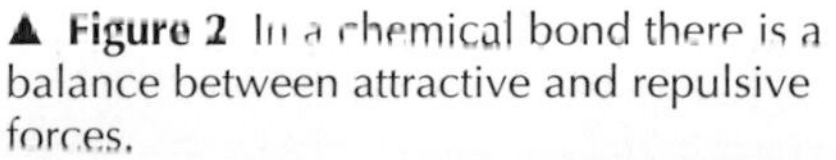
▲ **Figure 2** In a chemical bond there is a balance between attractive and repulsive forces.

▼ Table 1 Average bond enthalpies and bond lengths (*except for C=O where the data are for CO_2).

Bond	Average bond enthalpy/kJ mol^{-1}	Bond length/nm
C–C	+347	0.154
C=C	+612	0.134
C≡C	+838	0.120
C–H	+413	0.108
O–H	+464	0.096
C–O	+358	0.143
C=O*	+805	0.116
O=O	+498	0.121
N≡N	+945	0.110

Table 1 gives some values for bond enthalpies and bond lengths. These are all *average* bond enthalpies because the exact value of a bond enthalpy actually depends on the particular compound in which the bond is found. Notice these points:

- Double bonds have much higher bond enthalpies than single bonds; triple bond enthalpies are even higher.
- In general, the higher the bond enthalpy, the shorter the bond – you can see this if you compare the lengths of the single, double and triple bonds between carbon atoms.

Bond enthalpies always refer to breaking a bond in the *gaseous* compound. This means we can make fair comparisons between different bonds.

Measuring bond enthalpies

It isn't easy to measure bond enthalpies because there is usually more than one bond in a compound. Also, it is very difficult to make measurements when everything is in the gaseous state. For this reason, bond enthalpies are measured indirectly using enthalpy cycles.

Breaking and making bonds in a chemical reaction

Let's look again at the reaction that occurs when methane burns:

$$CH_4(g) + 2O_2(g) \rightarrow CO_2(g) + 2H_2O(l); \Delta H = -890\,kJ\,mol^{-1}$$

The reaction involves both breaking bonds and making new bonds. First you have to break four bonds between C and H in a methane molecule, and bonds between O and O in two O_2 molecules. This bond breaking requires energy. But once the bonds have been broken, the atoms can join together again to form new bonds: two C=O bonds in a CO_2 molecule and four O–H bonds in two H_2O molecules.

Figure 3 illustrates this – but note that you do not have to break *all* the old bonds before you can make new ones. New bonds start forming as soon as the first of the old bonds have broken.

In this reaction, the energy taken in during the bond-breaking steps is less than the energy given out during the bond-making steps, so the overall reaction is exothermic. (If the reverse is true, the reaction is endothermic.)

Bond breaking is an *endothermic* process, so bond enthalpies are always *positive*.

▼ Figure 3 Breaking and making bonds in the reaction between methane and oxygen.

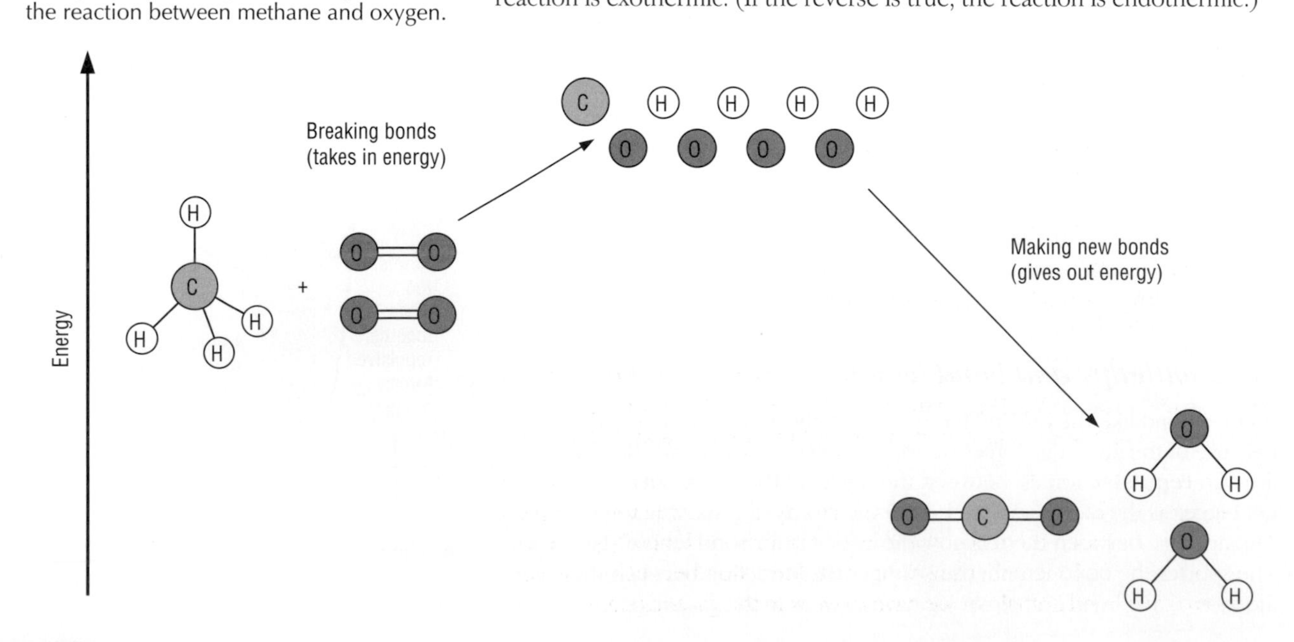

Because you need to break bonds before product molecules can begin to form, many reactions need heating to get them started. All reactions need energy to stretch and break bonds to start them off. Some reactions need only a little energy and there is enough energy available in the surroundings at room temperature to get them started. The reaction of acids with alkalis is one example.

Other reactions need heating to get them started – the burning of fuels is an example. When you use a match to ignite methane, or a spark to ignite petrol vapour in a car engine, you are supplying the energy that is needed to stretch and break bonds so that new bonds can begin to form.

It isn't necessary for *all* the bonds to break before a reaction gets going. If it was, you would have to heat things to very high temperatures to make them react. Once one or two bonds have broken, new bonds can start to form and this usually gives out enough energy to keep the reaction going, as in burning fuels. Many other reactions need continuous heating – for example, reactions that are only slightly exothermic.

Bonds and enthalpy cycles

We can represent bond breaking and bond making in an enthalpy cycle, such as the one given in Figure 4.

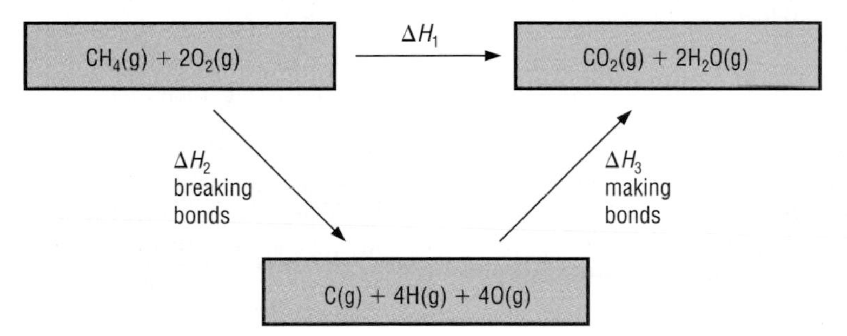

◀ **Figure 4** An enthalpy cycle to show bond breaking and bond making in the combustion of methane.

We can use the enthalpy cycle in Figure 4 to work out a value for the enthalpy change of combustion of methane, represented by ΔH_1. The calculation looks like this:

ΔH_2 = enthalpy change when bonds are broken
$= 4 \times E(\text{C–H}) + 2 \times E(\text{O=O})$
$= +2648\,\text{kJ mol}^{-1}$

ΔH_3 = enthalpy change when bonds are made
$= -[2 \times E(\text{C=O}) + 4 \times E(\text{O–H})]$ (the minus sign is there because energy is *released* when bonds are made)
$= -3466\,\text{kJ mol}^{-1}$

So the enthalpy change of combustion, ΔH_1, is given by:

$\Delta H_1 = \Delta H_2 + \Delta H_3$
$= +2648\,\text{kJ mol}^{-1} + (-3466\,\text{kJ mol}^{-1})$
$= -818\,\text{kJ mol}^{-1}$

Are these exact values? Notice that this value is a little different from the $-890\,\text{kJ mol}^{-1}$ given in the **Data Sheets** (Table 18) for the *standard* enthalpy change of combustion of methane. There are two main reasons for this:

- The value of ΔH_1 calculated here is not actually the standard value. In the equation we have been using, the water in the products is $H_2O(g)$, not $H_2O(l)$ as it would be under standard conditions. We used $H_2O(g)$ because when using bond enthalpies we have to work in the gaseous state.

- The bond enthalpies given are often averages from several compounds. This makes them very useful as a 'toolkit' of values but it does mean that the results of such calculations are not always precise. However, bond enthalpies are useful because they enable enthalpy changes to be measured when there is little specific data for a compound (see problems 5 and 6 in this section).

Problems for 4.2

You will need to look up values of bond enthalpies for some of these problems. You can use Table 1 in this section (page 64) or the **Data Sheets** (Table 19).

1 **a** Write an equation (with state symbols) to represent the combustion of 1 mole of propane, C_3H_8, to give carbon dioxide and water vapour.

b Write this equation again using full structural formulae, showing all the bonds within each molecule.

c Make a list of the types and numbers of bonds that are broken in the reaction.

d Make a list of the types and numbers of bonds that are made in the reaction.

e Use a table of bond enthalpies to calculate a value for the total enthalpy change involved in *breaking* all the bonds in the reaction of propane and oxygen. (You will need to decide whether bond breaking is an exothermic or endothermic process so that you can put a negative or positive sign in front of the number you have calculated.)

f Use a table of bond enthalpies to calculate a value for the total enthalpy change involved in *making* all the new bonds in the reaction of propane and oxygen. (You will need to decide whether bond making is an exothermic or endothermic process so that you can put a negative or positive sign in front of the number you have calculated.)

g Add together the enthalpy changes involved in breaking and making these bonds to calculate the overall enthalpy change of the reaction.

2 Repeat steps **a** to **g** in problem 1 for the combustion of 1 mole of methanol, $CH_3OH(g)$, to give carbon dioxide and water vapour. Hence find the overall enthalpy change for the reaction. Suggest two reasons why this value is different from the standard enthalpy change of combustion.

3 Hydrazine, $H_2N\text{–}NH_2(g)$, has been used as a rocket fuel because it reacts highly exothermically with oxygen to form gaseous nitrogen and water vapour. Use bond enthalpies to calculate a value for the enthalpy change of the reaction

$$H_2N\text{–}NH_2(g) + O_2(g) \rightarrow N_2(g) + 2H_2O(g)$$

The bond enthalpy of N–N is + 158 kJ mol^{-1} and that of N–H is +391 kJ mol^{-1}.

4 Ethene reacts with bromine to form 1,2-dibromoethane according to the equation

$$H_2C\text{=}CH_2(g) + Br_2(g) \rightarrow BrH_2C\text{–}CH_2Br(g)$$

Use bond enthalpies to calculate a value for the enthalpy change of the reaction. The bond enthalpy of Br–Br is +193 kJ mol^{-1} and that of C–Br is +290 kJ mol^{-1}.

The problems above could have been solved by using enthalpy changes of formation as in the problems for **Section 4.1**. The two that follow can only be solved using bond enthalpies!

5 Some apparently simple organic molecules do not exist! This is because they are unstable and form another compound or compounds very easily. One such compound is ethenol, CH_2CHOH, which is unstable with respect to ethanal, CH_3CHO.
Use bond enthalpies to calculate the enthalpy change of the reaction

$$CH_2{=}CHOH \rightarrow CH_3\text{—}C(=O)H$$

6 Inspection of the **Data Sheets** (Table 18) will reveal that the difference between the enthalpy change of combustion of successive straight-chain alkanes is around −650 kJ mol^{-1}. This is because each alkane differs from the previous one by the fragment

```
H
|
C—
|
H
```

(the 'spare' bond is to another carbon atom)

a Write an equation for the combustion of the one carbon and two hydrogen atoms in this fragment.

b Use bond enthalpies to calculate a value for the enthalpy change of the reaction in part **a** (include the breaking of the C–C bond).

c Suggest why this value is not −650 kJ mol^{-1}.

AS

4.3 *Entropy and the direction of change*

Things happen by chance

If you spill some petrol in an enclosed space, such as a garage, you can soon smell it all over the place. The petrol vaporises, and the vapour **diffuses** (spreads out) to occupy all the available space. This is why petrol is such a serious fire risk; as the vapour spreads out it mixes with the air to make a highly flammable mixture.

But *why* does the vapour diffuse? Why doesn't it all stay in one part of the room? It is the laws of chance and probability that say it must diffuse.

Look at Figure 1. We have simplified the situation so that the petrol vapour starts off in one container and can diffuse into the other when the partition is removed. We have ignored the presence of air molecules and have shown just five 'petrol' molecules. (Of course, petrol is really a mixture of different hydrocarbons with many different kinds of molecules.)

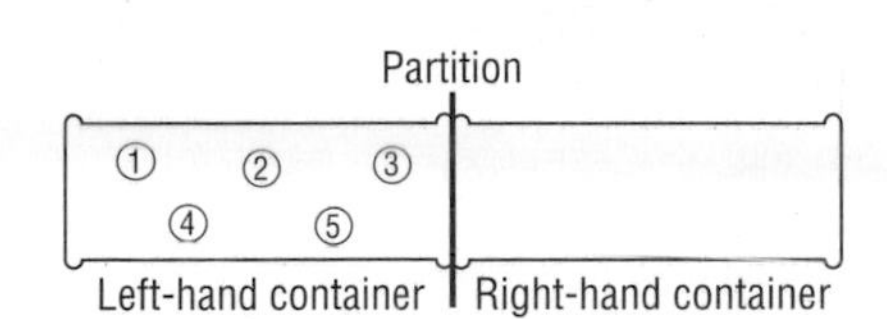

▲ **Figure 1** A simplified situation. Imagine five molecules in a container. What will happen when the partition is removed?

Each molecule moves in a straight line, until it collides with another gas molecule or the wall of the container, when it changes direction.

Figure 2 shows some of the things that can happen once the partition is removed. The molecules move around at random, and it is pure chance which container they end up in after a given length of time. Each of the molecules could end up in one of two places: the left-hand or the right-hand container. There are five molecules altogether, each with two places to be, so the total number of ways the molecules could arrange themselves once the partition is removed is $2 \times 2 \times 2 \times 2 \times 2 = 2^5 = 32$. Each of these ways is equally likely.

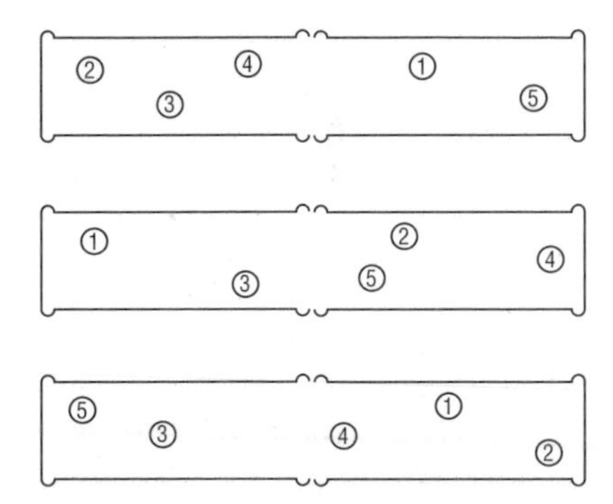

▲ **Figure 2** Possible arrangements of molecules after the partition has been removed.

Only one of these 32 arrangements has all the molecules where they started, in the left-hand container. So the chance that they will all stay all in one container, instead of spreading out between the two, is 1 in 32. The molecules diffuse because there are more ways of being spread out than ways of being all in one place.

Now consider the real-life situation, when there are billions of billions of molecules instead of just five. The number of ways all these molecules can spread out to fill the two containers is unimaginably large, so the chance that they will all remain in one container is unimaginably tiny.

The idea of 'number of ways' is very important in chemistry (and in physics and biology for that matter) because it decides whether changes are likely to take place or not. The basic idea is that *the events that happen are the ones that are most likely to*. The more ways an event can occur, the more likely it is to happen.

Why do liquids mix?

The mixing of liquids is another example of this rule. Figure 3 illustrates the point. If you have a jar half-full of roasted peanuts and you carefully pour onto them half a jar of roasted cashew nuts, you get two layers of different nuts. If you shake the jar, they will mix and you will have a jar of mixed nuts.

However much you shake them, the nuts will never unmix and give you two layers again. There are far more ways that the nuts can be mixed than unmixed. Each time you shake the jar, you produce another way of mixing them.

Unmixing could happen in theory – the different nuts could by chance get shaken to the top and bottom. But this is so unlikely that in practice it never happens. The nuts stay mixed.

If the different nuts represent molecules of different liquids then you

▲ **Figure 3** Peanuts at the bottom, cashews on the top.

have the situation when two liquids are mixed – for example, when two different hydrocarbons are blended in petrol. The liquids mix because there are more ways of being mixed than unmixed.

It doesn't always happen like this. Some pairs of liquids don't mix – for example, petrol and water. This is because there is something to prevent the natural mixing process from happening. There are weak bonds between all molecules, but if the bonds between the molecules of one liquid are stronger than those between the molecules of the other liquid (and stronger than the bonds between the two different types of molecules) then mixing is unlikely. It is as if the peanuts in Figure 3 had a sticky coating that makes them stay attracting one another rather than getting mixed up with the cashew nuts.

The general rule about mixing is that *substances always tend to mix unless there is something stopping them*, such as strong attractive forces holding one set of molecules together so they cannot easily break away from each other and mix.

Entropy: measuring the number of ways

Clearly the 'number of ways' is very important in deciding how things change. It is useful to have a measure of the 'number of ways' a chemical system can be arranged, then we can tell how the system is likely to change. We use a quantity called **entropy** to measure 'number of ways' – the higher the number of ways, the higher the entropy. In general, the more spread-out, mixed-up or disordered a system is then the higher its entropy will be.

We will see later that entropy is a precise physical quantity that can be measured and tabulated. The symbol for entropy is *S*. Table 1 gives some standard molar entropies for different substances. For the time being, you need not worry about how these numbers were obtained, but it is interesting to compare them with one another.

▼ **Table 1** Some values of standard molar entropy.

Substance	Standard molar entropy, $S^{\ominus}$/J K^{-1} mol^{-1}
diamond, C(s)	2.4
hydrogen, H_2(g)	130.7
iron, Fe(s)	27.3
sodium chloride, NaCl(s)	72.4
water (solid), H_2O(s)	48.0
water (liquid), H_2O(l)	70.0
water (gas), H_2O(g)	188.7
methane, CH_4(g)	186.2
ethane, C_2H_6(g)	229.5
propane, C_3H_8(g)	269.9

In general, gases have higher entropies than liquids, which have higher entropies than solids. You can see this by looking at the standard molar entropies of the three states of water. In gases the molecules are arranged completely at random, so there are more ways they can be arranged than in liquids or solids. Liquids have a more random arrangement than solids, and so more ways of arranging the molecules. The more regular and crystalline a solid is then the lower its entropy. Notice the very low entropy of diamond, which has one of the most perfectly regular structures of all.

Notice too that substances with larger molecules, such as ethane, have higher entropies than substances with smaller molecules, such as methane.

How do we predict the entropy change in a reaction?

The entropy of the products will be greater than the entropy of the reactants if:

- solids turn to liquids, or liquids turn to gases, in the reaction;
- there are more moles of gaseous products than gaseous reactants.

But it's not as simple as that... In this treatment of entropy we have talked about the number of ways of arranging the molecules of substances. But entropy does not simply measure arrangements of molecules – it also measures the number of ways that *energy* can be distributed among these molecules. This is a very important contribution to the entropy of a substance and you will meet it later in **Section 4.4**.

Problems for 4.3

1 For each of the following changes, state whether the entropy of the system described would increase, decrease or stay the same.
 a Petrol vaporises
 b Petrol vapour condenses
 c Sugar dissolves in water
 d Oil mixes with petrol
 e A suspension of oil in water separates into two layers
 f Car exhaust gases are adsorbed onto the surface of a catalyst in a catalytic converter.

2 For each of the following pairs of substances, which substance do you believe would have the higher standard molar entropy? Give your reasons.
 a Solid wax or molten wax
 b $Br_2(l)$ or $Br_2(g)$
 c Separate samples of copper and zinc, or a sample of brass (an alloy of copper and zinc)
 d Pentane $C_5H_{12}(l)$ or octane $C_8H_{18}(l)$.

3 Look at Figure 1 in this section (page 67). Suppose there were eight molecules in the left-hand container instead of five. What would be the chance that they would all end up in the left-hand container after the partition had been removed?

4 Explain the differences between the entropies of the following pairs of substances.

	Substance	Standard molar entropy, $S^\ominus$/J K^{-1} mol^{-1}
a	He(g)	126
	Ar(g)	155
b	Hg(l)	76
	Hg(g)	175
c	$F_2(g)$	203
	$Cl_2(g)$	223

5 For each of the following reactions, state whether you would expect the entropy of the products to be greater or less than the entropy of the reactants. In each case explain your answer.
 a $CaCO_3(s) \rightarrow CaO(s) + CO_2(g)$
 b $2SO_2(g) + O_2(g) \rightarrow 2SO_3(g)$
 c $2Mg(s) + O_2(g) \rightarrow 2MgO(s)$
 d $N_2(g) + 3H_2(g) \rightarrow 2NH_3(g)$

4.4 *Energy, entropy and equilibrium*

Inside solids, liquids and gases

You should be familiar with the simple molecular kinetic model for the structures of solids, liquids and gases. This is summarised in Figure 1.

Solids are *rigid* because the molecules are held together in a *lattice*. (We will use the word 'molecule' for the particles involved, though they could also be atoms or ions.) Because the molecules in liquids and gases are moving around, we say these substances are *fluids*. They can take up any shape – usually the shape of the container they are in.

▼ **Figure 1** A comparison of solids, liquids and gases.

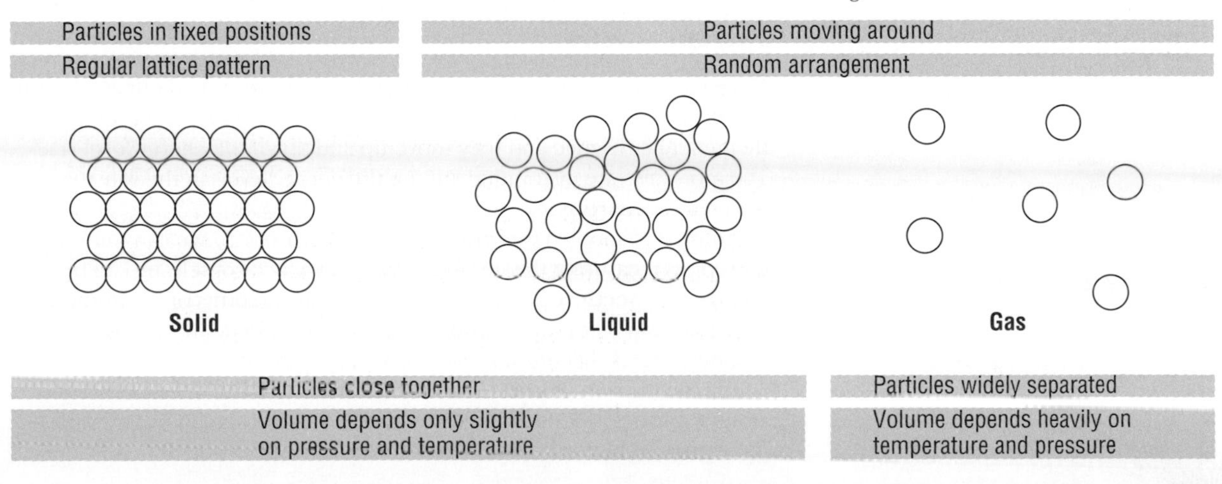

Solids and liquids do not expand much when they are heated but, at a given pressure, gases expand a lot on heating. What is more, gases can be easily compressed by increasing the pressure. Solids and liquids, in contrast, have low *compressibilities*.

If a solid, liquid or gas is heated than its temperature increases. There are only *two* exceptions to this rule. If the solid happens to be at its melting point, or the liquid at its boiling point, the temperature does not increase – it stays constant while the solid melts or the liquid boils.

Stacks of energy

We use **specific heat capacity** (symbol c_p when the measurement is made at constant pressure) as a measure of how much energy is required to warm something up. It tells us how many joules are needed to raise the temperature of 1 g of a substance by 1 K. So c_p has units of $J\,g^{-1}K^{-1}$.

The temperature of a substance is related to the *kinetic energy* of the molecules. A substance feels hotter when its molecules are moving more energetically. That is because movement energy can easily be passed on to the molecules in you, or the thermometer or whatever.

There are three forms of kinetic energy which molecules can possess:

- **translation** – movement of the whole molecule from one place to another
- **rotation** – spinning around
- **vibration** – stretching and compressing bonds.

You can read more about these forms of energy in **Section 6.2**.

If we continue putting energy into a substance, we may increase the **electronic energy** of the molecules, or break bonds between them or even within them. But none of these processes alter the temperature because changes in electronic energy and **bonding energy** do not affect the *motion* of the molecules.

Energy is quantised and molecules are restricted to particular levels of electronic, vibrational, rotational and translational energy. The size of the quanta (i.e. the gaps between the energy levels) increase in the order

translation < rotation < vibration < electronic.

If we think of just one electronic level of a simple molecule, such as H_2, the energy levels can be illustrated as shown in Figure 2.

Overall, there are a large number of closely spaced energy levels for the H_2 molecules to be in. For larger and more complex molecules, there are even more energy levels available.

Entropy matters

Molecules don't all have the same energy – they are spread out among the energy levels. As molecules collide and exchange energy with each other, their energies change and they move up and down the energy level stack. The molecules are distributed among the energy levels in the way that gives the greatest **entropy**.

Entropy (symbol *S*) was introduced in **Section 4.3**. We can think of entropy as measuring the number of ways in which something can be arranged. In **Section 4.3** we discussed entropy in terms of the number of ways that *molecules* can be arranged. Entropy is also a measure of the number of ways that *quanta of energy* can be arranged.

We can explain diffusion and mixing by saying there are more ways of arranging the positions of the molecules if they are spread out in space – the entropy is higher when the molecules are spread out. Similarly, there

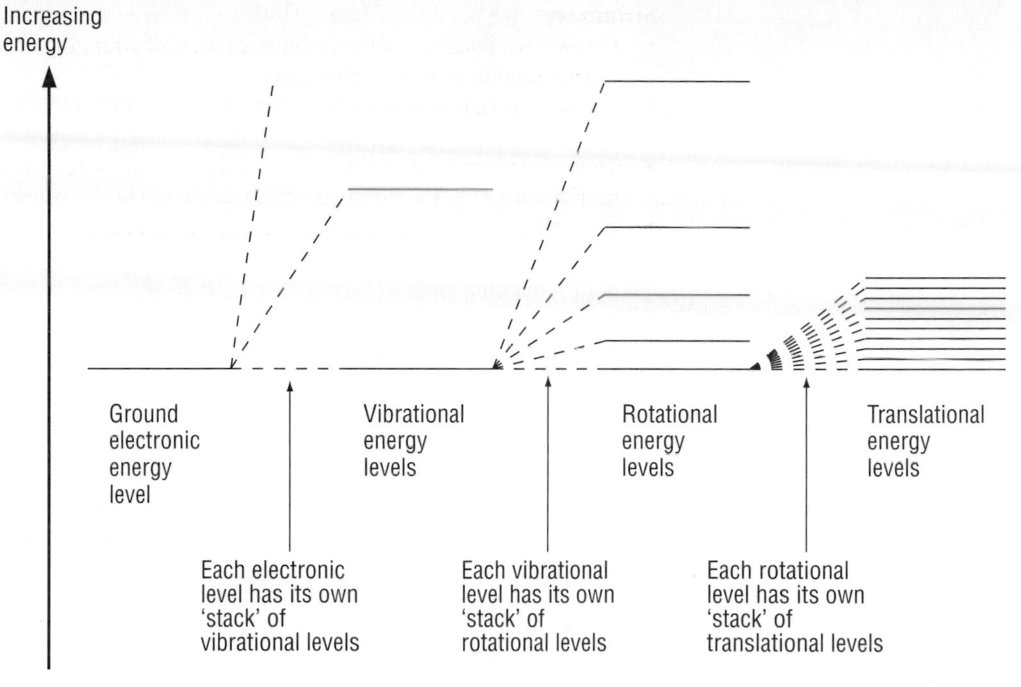

▲ **Figure 2** Each electronic energy level has within it several vibrational, rotational and translational energy levels – note that the levels are not to scale.

are more ways of arranging the quanta of energy in molecules if they are spread out in an energy stack.

Imagine two molecules with four quanta of energy between them. They could each have two quanta, but there are several other possibilities too. In fact, there are five ways in which the energy can be divided between the molecules, as shown in Table 1.

Notice that, even though the energy is spread out and the molecules can change from one level to another when they collide, the *total energy* must always equal four quanta – energy cannot be created or destroyed. That's one way of expressing the **First Law of Thermodynamics** – the law you make use of whenever you do Hess's law calculations.

What happens if we give the molecules six quanta instead of four? There are now *seven* ways of distributing the energy, as shown in Table 2.

So, when we heat something we increase the total number of quanta, and we increase the entropy. This idea will be very useful – we shall come back to it later.

The entropy also increases if we increase the number of molecules sharing the energy. For example, there are 15 ways in which three particles can share four quanta. See if you can work out why.

If the energy levels are closely spaced then the quanta are smaller, but there are more of them at a given temperature. The more closely spaced the energy levels are, the greater the number of quanta and the higher the entropy.

Generally, solids have lower entropies than liquids and gases. This is because solids have a regular lattice structure, whereas the molecules in liquids and gases are arranged randomly. Gases tend to have greater entropies than liquids because gas molecules occupy a greater volume and there is more randomness in their distribution.

Energy levels are more closely spaced for molecules which contain heavier atoms. The number of energy levels tends to increase with the number of atoms in a molecule, thus making adjacent levels closer together in larger molecules. So molecules with heavier atoms, and molecules with larger numbers of atoms, have higher entropies.

▼ **Table 1** Ways of dividing four quanta of energy between two molecules.

Quanta possessed by molecule 1	Quanta possessed by molecule 2
0	4
1	3
2	2
3	1
4	0

▼ **Table 2** Ways of dividing six quanta of energy between two molecules.

Quanta possessed by molecule 1	Quanta possessed by molecule 2
0	6
1	5
2	4
3	3
4	2
5	1
6	0

Summary

- Entropy is a measure of the number of ways of arranging molecules and distributing their quanta of energy.
- A collection of molecules has higher entropy if the molecules are spread out more.
- There are more ways of arranging the energy of a collection of molecules if they spread out among the energy levels available to them.
- The entropy is increased if the energy is shared among more molecules.
- The entropy depends on the number of quanta of energy available. This in turn depends on:
  - the temperature
  - the spacing of the energy levels.
- Substances have higher entropies if their molecules contain:
  - heavier atoms
  - larger numbers of atoms.
- In general, gases have higher entropies than liquids, and liquids have higher entropies than solids.

The whole Universe if necessary

Let's look at melting and freezing in more detail. Why should 0 °C mark the change from water to ice?

$$H_2O(l) \rightarrow H_2O(s) \quad \Delta S = -22.0\,J\,K^{-1}\,mol^{-1}$$
$$\Delta H = -6.01\,kJ\,mol^{-1}$$

ΔS is the **entropy change** for this process. It is negative because the entropy *decreases* when liquid water becomes solid. ΔH is also negative – the change is exothermic, and energy is transferred from the water to the surroundings by heating.

So far we have only considered the molecules in the *system* we are studying – the water and the ice – but changes to the system also affect the *surroundings*. These changes in the surroundings are very important in determining what happens.

Freezing is an exothermic process. Energy is transferred from the chemical system to the surroundings by heating. When ice freezes on a window pane, energy is transferred to the glass but it soon spreads out – by conduction, convection and radiation – into things around the glass. Lots of things – such as the frame, the wall, the air and the plant beside the window – get hotter, and therefore increase in *entropy*.

We can't work out how much each individual substance in the surroundings has increased in entropy because it is impossible to say exactly how the energy has been shared out. Fortunately, we don't need to know this; there is a very simple relationship which allows us to think just in terms of *surroundings* – whatever they are.

The relationship tells us that the entropy change in the surroundings is equal to the energy transferred (the enthalpy change in the surroundings) divided by the temperature. Since the energy *gained* by the *surroundings* is the same as the energy *lost* by the *chemical system*, and vice versa, the enthalpy change in the surroundings is equal to $-\Delta H$.

So the entropy change in the surroundings, ΔS_{surr}, is given by

$$\Delta S_{surr} = \frac{-\Delta H}{T}$$

The powerful thing about this relationship is that it allows us to make a prediction about the surroundings of a chemical process – whatever they are (the rest of the Universe if necessary) – using measurements that we can make on the chemical system in the laboratory.

Will it or won't it?

To find the **total entropy change**, ΔS_{total}, for a process, we need to combine the entropy change for the chemical system, ΔS_{sys}, with the entropy change in the surroundings, ΔS_{surr}.

$$\Delta S_{total} = \Delta S_{sys} + \Delta S_{surr}$$

We can use this to find the total entropy change when water freezes at, say, –10 °C (263 K).

$$\Delta S_{sys} = -22.0\,J\,K^{-1}\,mol^{-1}$$

$$\Delta S_{surr} = \frac{-\Delta H}{T} = \frac{-(-6010\,J\,mol^{-1})}{263\,K} = +22.9\,J\,K^{-1}mol^{-1}$$

$$\Delta S_{total} = \Delta S_{sys} + \Delta S_{surr} = (-22.0\,J\,K^{-1}\,mol^{-1}) + (+22.9\,J\,K^{-1}\,mol^{-1}) = +0.9\,J\,K^{-1}\,mol^{-1}$$

Overall, the process leads to an increase in entropy.

Now let's work at a higher temperature and see what happens if ice *melts* at 10 °C (283 K). We will assume that ΔS and ΔH do not change with temperature.

$$\Delta S_{sys} = +22.0\,J\,K^{-1}\,mol^{-1}$$

Notice that the sign of ΔS_{sys} has changed because we are now producing a liquid from a solid.

$$\Delta S_{surr} = \frac{-\Delta H}{T} = \frac{-(+6010\,J\,mol^{-1})}{283\,K} = -21.2\,J\,K^{-1}\,mol^{-1}$$

We are melting ice this time, so notice another sign change: ΔH is *endothermic*.

$$\Delta S_{total} = \Delta S_{sys} + \Delta S_{surr} = (+22.0\,J\,K^{-1}\,mol^{-1}) + (-21.2\,J\,K^{-1}\,mol^{-1}) = +0.8\,J\,K^{-1}\,mol^{-1}$$

Again, the process leads to an increase in entropy.

These entropy data are summarised in Figure 3.

Both these changes are **spontaneous** – ice melts at 10 °C and water freezes at –10 °C of their own accord and without any help from us. Both processes result in an *increase* in ΔS_{total}. This is part of a general rule – we can tell if a process will be spontaneous or not by finding the value of ΔS_{total}.

ΔS_{total} is positive for a spontaneous process

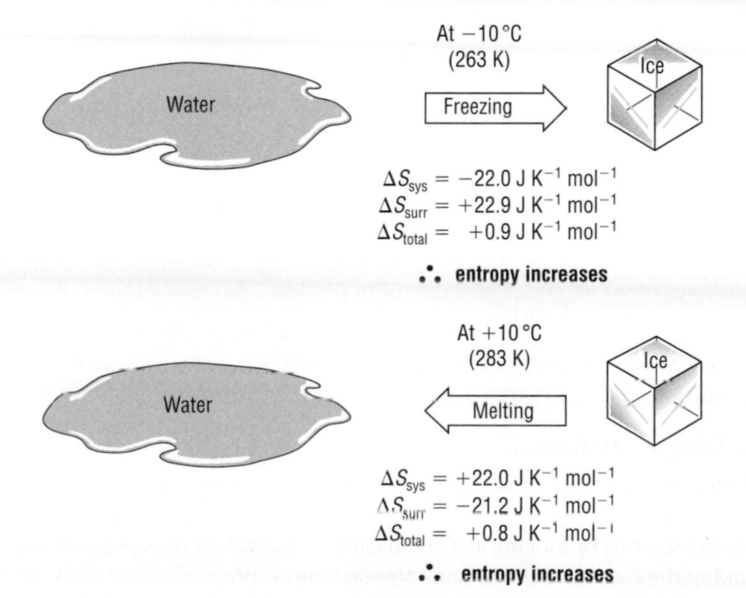

◄ **Figure 3** Entropy changes for water freezing at −10 °C and ice melting at 10 °C.

More about entropy

You have learned a lot more about entropy in the last few pages – here is a summary of the key points.

- When changes take place in a chemical *system* there are nearly always accompanying changes in the *surroundings*.
- To predict whether or not a change will take place spontaneously, we need to take account of the *entropy changes* in the system *and* its surroundings.
- The total entropy change for a process is given by

$$\Delta S_{total} = \Delta S_{sys} + \Delta S_{surr}$$

- ΔS_{surr} can be found using the relationship

$$\Delta S_{surr} = \frac{-\Delta H}{T}$$

- The Second Law of Thermodynamics tells us that for a spontaneous chemical change, the total entropy must increase

$$\Delta S_{total} > 0$$

This is one way of expressing the **Second Law of Thermodynamics**. It is a simple but powerful law which allows us to predict whether or not something *should* happen. Chemistry is all about making things happen and explaining why they happen, so the 'Second Law' is one of the foundations of chemistry. Indeed, the Second Law of Thermodynamics is fundamental to the whole of science. It has been said that ignorance of the Second Law is equivalent to never having read a word of Shakespeare.

Here are some other ways in which famous scientists have expressed the Second Law:

> 'You can never restore everything to its original condition after a change has occurred' (based on *G.N. Lewis*)
> 'Entropy is time's arrow' (*A. Eddington*)
> 'Gain in information is loss in entropy' (*G.N. Lewis*)

Freezing sea water

Let's now put the Second Law of Thermodynamics to work to explain why sea water freezes at a lower temperature than pure water, and why putting salt on the roads helps to keep them clear of ice in winter.

The freezing process is

$$H_2O(l) \rightarrow H_2O(s) \quad \Delta S = -22.0\,J\,K^{-1}\,mol^{-1}$$
$$\Delta H = -6.01\,kJ\,mol^{-1}$$

We will once again assume that ΔH and ΔS do not vary with temperature, though in fact they do vary a bit.

When sea water freezes, pure ice is produced. The energy released will be the same as when molecules from pure water fit together to form a lattice in an ice crystal; ΔH will still be $-6.01\,kJ\,mol^{-1}$.

However, ΔS is different for the system. The entropy of salt solution is greater than the entropy of pure water because the ions are spread out in the solution. There are more ways of arranging water molecules *and* the ions from the salt than water molecules alone. Therefore ΔS is *more negative* when salt solution freezes (Figure 4).

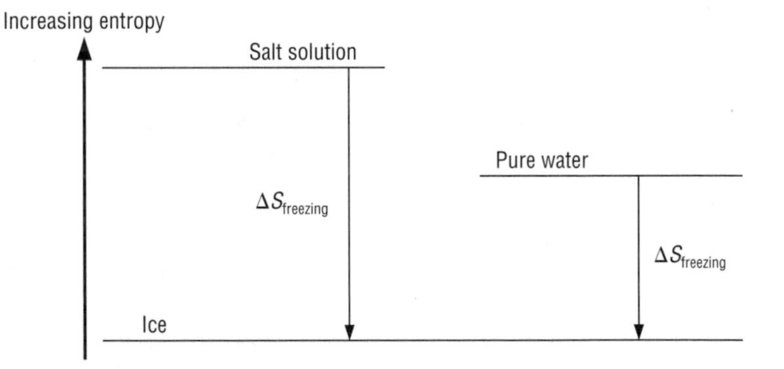

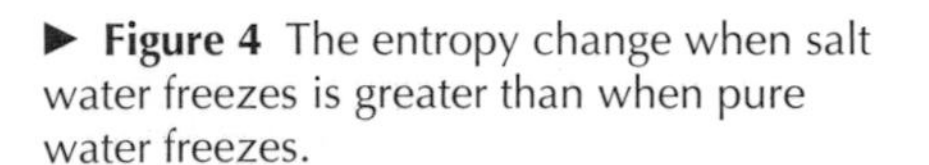

► **Figure 4** The entropy change when salt water freezes is greater than when pure water freezes.

When sea water freezes, ΔS_{sys} is more negative than $-22.0\,J\,K^{-1}\,mol^{-1}$. Before freezing can occur, ΔS_{surr} has to be more positive than $+22.0\,J\,K^{-1}\,mol^{-1}$.

Because $\Delta S_{surr} = \frac{-\Delta H}{T}$, this can only happen if we divide ΔH by a smaller value of T. So salt solution will *not* freeze at 0 °C, so a *lower* temperature is required.

Taking it further ...

Melting and freezing do not involve changing one compound into another – but the ΔS_{total} rule applies to the production of new chemicals just as much. We can see this by looking at a chemical reaction which is responsible for putting the fizz into Perrier water, the decomposition of calcium carbonate.

Limestone rocks have existed for a long time. At the temperatures on the Earth's surface, the calcium carbonate has not broken down into calcium oxide and carbon dioxide – but this is exactly what happens when limestone gets very hot, for example when it is next to lava which has forced its way into the Earth's crust. The process we need to consider is

$$CaCO_3(s) \rightarrow CaO(s) + CO_2(g) \quad \Delta S = +159\,J\,K^{-1}\,mol^{-1}$$
$$\Delta H = +179\,kJ\,mol^{-1}$$

What are the total entropy changes at, say, 298 K (25 °C) and 1273 K (1000 °C)?

At 298 K

$$\Delta S_{sys} = +159\,J\,K^{-1}\,mol^{-1}$$
$$\Delta S_{surr} = \frac{-(+179\,000\,J\,mol^{-1})}{298\,K} = -601\,J\,K^{-1}\,mol^{-1}$$
$$\Delta S_{total} = (+159\,J\,K^{-1}\,mol^{-1}) + (-601\,J\,K^{-1}\,mol^{-1}) = -442\,J\,K^{-1}\,mol^{-1}$$

At 1273 K

$$\Delta S_{sys} = +159\,J\,K^{-1}\,mol^{-1}$$
$$\Delta S_{surr} = \frac{-(+179\,000\,J\,mol^{-1})}{1273\,K} = -141\,J\,K^{-1}\,mol^{-1}$$
$$\Delta S_{total} = (+159\,J\,K^{-1}\,mol^{-1}) + (-141\,J\,K^{-1}\,mol^{-1}) = +18\,J\,K^{-1}\,mol^{-1}$$

At the Earth's surface at 298 K, the decomposition of limestone would result in a large entropy *decrease*. Therefore it does not take place. At higher temperatures, near the hot lava, the total entropy change is positive and the reaction is favourable.

It is the temperature which has made the difference. When we divide ΔH by a small temperature, we get a large negative value for ΔS_{surr}, which dominates the situation and makes it unfavourable. Dividing ΔH by a much bigger value of T doesn't change the sign of ΔS_{surr}, but it does reduce the significance of ΔS_{surr} by making it smaller.

The same relationship between ΔS_{sys} and ΔS_{surr} allows us to explain dissolving and crystallisation. The arguments are slightly different because the temperature remains constant. Why should salt dissolve to produce a solution with a concentration like that of sea water, but crystallise from solution in the salt works when most of the water has been evaporated?

The process involved is

$$NaCl(s) + aq \rightarrow Na^+(aq) + Cl^-(aq)$$

When a 1 mol dm^{-3} salt solution is produced, $\Delta S = +39\,J\,K^{-1}\,mol^{-1}$ and $\Delta H = +3.9\,kJ\,mol^{-1}$.

At 298 K

$$\Delta S_{sys} = +39\,J\,K^{-1}\,mol^{-1}$$
$$\Delta S_{surr} = \frac{-\Delta H}{T} = \frac{-(+3900\,J\,mol^{-1})}{298\,K} = -13\,J\,K^{-1}\,mol^{-1}$$
$$\Delta S_{total} = (+39\,J\,K^{-1}\,mol^{-1}) + (-13\,J\,K^{-1}\,mol^{-1}) = +26\,J\,K^{-1}\,mol^{-1}$$

ΔS_{total} is positive and so salt dissolves to produce a 1 mol dm^{-3} solution at 298 K.

When the solution is more concentrated, ΔH may be very slightly different. However, we can neglect this and assume that ΔH and, therefore, ΔS_{surr} remain the same. We cannot, however, neglect the change to ΔS_{sys}. This changes because, among other things, there is a smaller volume of water for the ions to be dispersed into, so there are fewer ways of arranging their positions.

This causes ΔS_{sys} to be less positive for the formation of a more concentrated solution. Eventually ΔS_{sys} becomes less than $+13\,J\,K^{-1}\,mol^{-1}$, which leads to a *negative* value of ΔS_{total} for the dissolving process. Dissolving becomes unfavourable, so salt crystallises from the solution, rather than dissolving.

You may need to calculate the entropy change for a chemical reaction given the entropies of reactants and products.

You can do this easily using the following equation:

$$\Delta S_{reaction} = \Sigma\Delta S_{products} - \Sigma\Delta S_{reactants}$$

What about equilibrium?

Let us return to the melting of ice. A spontaneous change always takes place in the direction which corresponds to an increase in ΔS_{total}. This is the case for $H_2O(s) \rightarrow H_2O(l)$ at $+10\,°C$.

At $-10\,°C$, however, ΔS_{total} for the melting of ice is negative. Therefore the reverse process occurs, and water freezes.

What happens when ΔS_{total} is equal to zero? There is no net change in either direction – liquid and solid are in **equilibrium**. We can check this by looking at the figures again – this time at 0 °C (273 K).

At 273 K

$H_2O(s) \rightleftharpoons H_2O(l) \quad \Delta S_{sys} = +22.0\,J\,K^{-1}\,mol^{-1}$

$\Delta H = +6.01\,kJ\,mol^{-1}$

$$\Delta S_{sys} = +22.0\,J\,K^{-1}\,mol^{-1}$$

$$\Delta S_{surr} = \frac{-\Delta H}{T} = \frac{-(+6010\,J\,mol^{-1})}{273\,K} = -22.0\,J\,K^{-1}\,mol^{-1}$$

$$\Delta S_{total} = (+22.0\,J\,K^{-1}\,mol^{-1}) + (-22.0\,J\,K^{-1}\,mol^{-1}) = 0\,J\,K^{-1}\,mol^{-1}$$

Liquid water and ice exist together at 0 °C – they are in equilibrium – and at this temperature $\Delta S_{total} = 0$.

The requirement for equilibrium is that ΔS_{total} must be zero.

The very important point is that it is the total entropy change *under the actual experimental conditions* that must be zero at equilibrium. If conditions happen to be standard, the standard entropy change will be zero, but the standard entropy change need not be zero at equilibrium – in fact it rarely is.

Problems for 4.4

1 Explain the pattern in the entropies of the first five alkanes at 298 K, given in the following table.

Alkane	$CH_4(g)$	$C_2H_6(g)$	$C_3H_8(g)$	$C_4H_{10}(g)$	$C_5H_{12}(l)$
S/J K^{-1} mol^{-1}	186	230	270	310	261

2 Calculate the entropy change for each of the following reactions using the standard molar entropies provided. In each case, comment on the values you have obtained.

a $Hg(l) \rightarrow Hg(g)$

b $C(s) + H_2O(g) \rightarrow CO(g) + H_2(g)$

c $2NO_2(g) \rightarrow N_2O_4(g)$

d $N_2(g) + 3H_2(g) \rightarrow 2NH_3(g)$

Substance	S/J K^{-1} mol^{-1}
Hg(l)	76.0
Hg(g)	174.8
C(s) (graphite)	5.7
$H_2O(g)$	188.7
CO(g)	197.6
$H_2(g)$	130.6
$NO_2(g)$	240.0
$N_2O_4(g)$	304.2
$N_2(g)$	191.6
$NH_3(g)$	192.3

3 The values of ΔH and ΔS_{sys} that follow refer to changes at 298 K under standard conditions. Predict whether or not the following changes will be spontaneous at 298 K. Explain your answer in each case.

a $Ca^{2+}(aq) + CO_3^{2-}(aq) \rightarrow CaCO_3(s)$
$\Delta S_{sys} = +203\,J\,K^{-1}\,mol^{-1}$
$\Delta H = +13\,kJ\,mol^{-1}$

b $H_2O_2(l) \rightarrow H_2O(l) + \frac{1}{2}O_2(g)$
$\Delta S_{sys} = +63\,J\,K^{-1}\,mol^{-1}$
$\Delta H = -98\,kJ\,mol^{-1}$

c $N_2(g) + O_2(g) \rightarrow 2NO(g)$
$\Delta S_{sys} = +25\,J\,K^{-1}\,mol^{-1}$
$\Delta H = +180\,kJ\,mol^{-1}$

d $NH_4NO_3(s) \rightarrow N_2O(g) + 2H_2O(l)$
$\Delta S_{sys} = +209\,J\,K^{-1}\,mol^{-1}$
$\Delta H = -124\,kJ\,mol^{-1}$

e C(graphite) $\rightarrow$ C(diamond)
$\Delta S_{sys} = -4\,J\,K^{-1}\,mol^{-1}$
$\Delta H = +2\,kJ\,mol^{-1}$

4 Use the data given for the process in problem 3 **e** to explain why you would not expect to be able to make diamonds from graphite at any temperature at atmospheric pressure.

4.5 *Energy changes in solutions*

Dissolving ionic solids

Many ionic substances dissolve in water, but others do not. What decides whether an ionic substance will dissolve? One important factor is the energy changes that are involved.

Before an ionic solid such as sodium chloride can dissolve, the ions must be separated from the lattice so they can spread out in the solution – an endothermic process. The ions in solution are hydrated – an exothermic process (see **Section 5.1**). This section looks at each of these processes in detail, and then uses them to explain why some substances dissolve and others do not.

Lattice enthalpy

Before an ionic solid can dissolve, the ions must be separated from their lattice so they can spread out in the solution (Figure 1). This means supplying energy to overcome the electrical attraction between the oppositely charged ions.

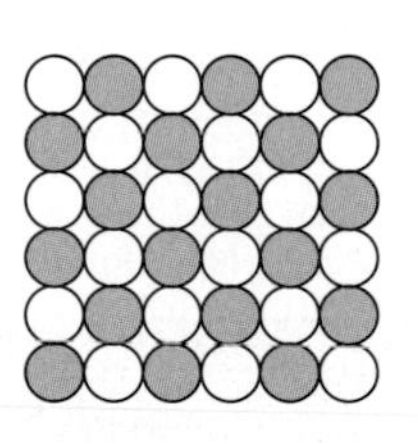

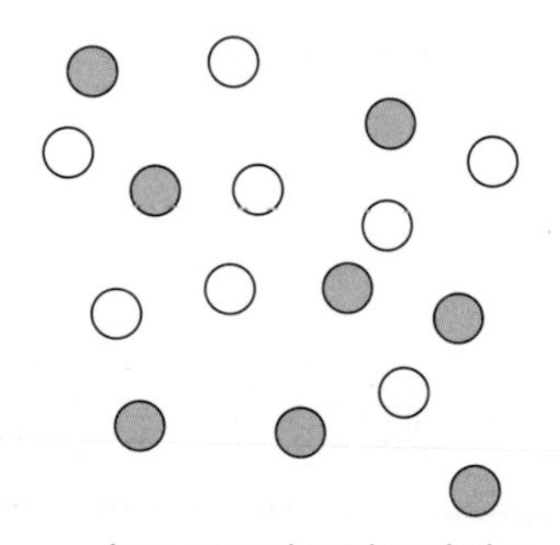

Ions in solid lattice

Ions spread out in solution

◄ **Figure 1** When an ionic solid dissolves, the ions in the lattice get spread out through the solution.

We measure the strength of the ionic attractions in a lattice by the **lattice enthalpy** of the solid. The lattice enthalpy, ΔH_{LE}, is the enthalpy change when 1 mole of solid is *formed* by the coming together of the separate ions. When ions are separated from one another, we can think of them as being in the gaseous state – when they are together in the lattice they are in the solid state. So we define lattice enthalpy as the enthalpy change involved in processes such as

$Na^+(g) + Cl^-(g) \rightarrow NaCl(s)$ for NaCl, $\Delta H_{LE} = -788\,kJ\,mol^{-1}$
$Mg^{2+}(g) + 2Br^-(g) \rightarrow MgBr_2(s)$ for $MgBr_2$, $\Delta H_{LE} = -2434\,kJ\,mol^{-1}$

All lattice enthalpies are large *negative* quantities. If we want to break down a lattice, we have to put in energy equal to $-\Delta H_{LE}$ (which now becomes a *positive* quantity because we are putting energy *in*). This tends to stop substances dissolving – unless the energy is 'paid back' later. Table 1 sets out some lattice enthalpy values for some simple ionic compounds.

Forming an ionic lattice from gaseous ions is always an exothermic process.

∴ ΔH_{LE} is always negative.

The *stronger* the ionic bonding in the lattice, the *more negative* is ΔH_{LE}.

Compound	ΔH_{LE}/ kJ mol^{-1}	Compound	ΔH_{LE}/ kJ mol^{-1}	Compound	ΔH_{LE}/ kJ mol^{-1}
Li_2O	−2806	MgO	−3800	Al_2O_3	−15916
Na_2O	−2488	CaO	−3419		
K_2O	−2245	SrO	−3222		
LiF	−1047	MgF_2	−2961		
NaF	−928	CaF_2	−2634		
KF	−826				

◄ **Table 1** Lattice enthalpies for some ionic compounds.

▶ **Table 2** Ionic radii for some cations.

Ion	Radius/nm	Ion	Radius/nm	Ion	Radius/nm
Li^+	0.078	Mg^{2+}	0.078	Al^{3+}	0.057
Na^+	0.098	Ca^{2+}	0.106		
K^+	0.133	Sr^{2+}	0.127		
Rb^+	0.149				

Table 2 lists values for the radii of the positive ions present in the compounds in Table 1. Notice how lattice enthalpy depends on the *size* and *charge* of the ions. Lattice enthalpies become more negative (i.e. more energy is given out) when:

- the ionic charges increase
- the ionic radii decrease.

This is because ions with a higher charge attract one another more strongly. They also attract more strongly if they are closer together. Stronger attractions mean more negative lattice enthalpies (Figure 2). Substances with large negative lattice enthalpies, such as Al_2O_3, are usually insoluble.

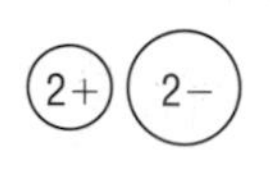

Small, highly charged ions get close together and attract strongly: ΔH_{LE} more negative, more energy given out

+ –

Large, singly charged ions are further apart and attract less strongly: ΔH_{LE} less negative, less energy given out

▶ **Figure 2** The factors deciding the size of lattice enthalpies.

When an ionic substance breaks up, the amount of energy needed corresponds to the reverse of the lattice energy ($-\Delta H_{LE}$).

Hydration and solvation

Despite the need to supply energy to break up the lattice, many ionic substances *do* dissolve. Something else must happen to supply the energy needed.

For the moment, let us restrict our thinking to aqueous solutions where the solvent is water. The covalent bonds in water molecules are polar because of the difference in electronegativity between oxygen and hydrogen (see **Section 3.1**). Because the water molecule has a bent shape (see **Section 3.2**), the whole molecule is polar and behaves as a tiny **dipole**.

The tiny charges on the water molecules are attracted to the charges on the ions. This happens on the surface of an ionic solid which is placed in water, so the ions are pulled into solution. In solution, the positive ions are surrounded by water molecules with the negative end of the dipole facing towards them. The negative ions are surrounded by water molecules with the positive end of the dipole facing towards them (Figure 3). The ions in solution are **hydrated** – they have water molecules bound to them.

Water molecules bind weakly to some ions, with the result that they are not extensively hydrated. Other ions are extensively hydrated with water molecules bound very strongly to them. When these ions crystallise out of solution, the strongly bound water molecules crystallise with them to give hydrated crystals, such as blue hydrated copper(II) sulfate, $CuSO_4{\cdot}5H_2O$.

Table 3 shows approximately how many water molecules are likely to be attached to particular positive ions in solution (though the ion will have an effect on all the water molecules surrounding it – not just the closest ones). The higher the charge density of the ion, the more water

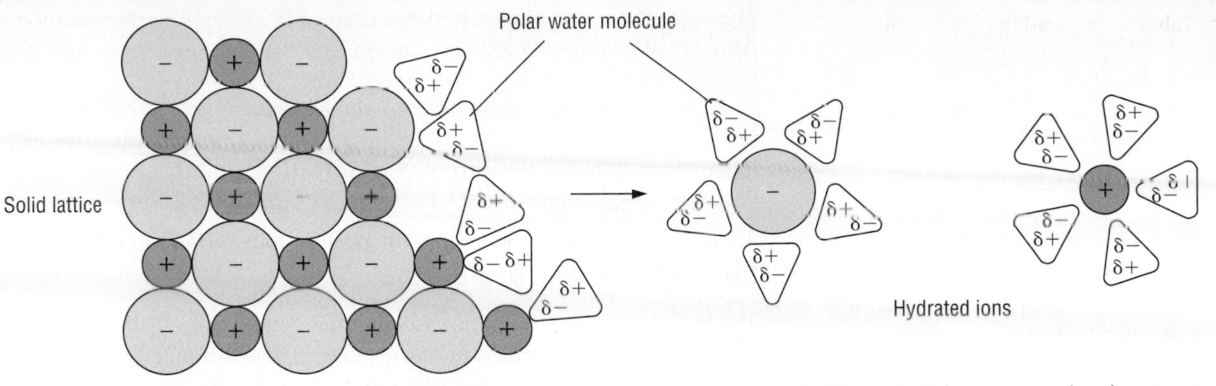

▲ **Figure 3** Polar water molecules attract the ions in a solid lattice to give hydrated ions.

molecules it attracts, and the bigger it becomes. So you have the situation in which an ion that is small in the absence of water becomes large when it is in the aqueous phase.

◀ **Table 3** Approximate extent of hydration for some positive ions.

Ion	Average number of attached water molecules	Ion	Average number of attached water molecules
Li^+	5	Mg^{2+}	15
Na^+	5	Ca^{2+}	13
K^+	4	Al^{3+}	26

When bonds form between ions and water molecules, energy is released and this may supply enough energy to pull ions out of the lattice. It is not quite as simple as this, though. Water molecules are strongly attracted to each other by intermolecular bonds, called *hydrogen bonds*. (You can find out more about hydrogen bonds in **Sections 5.4** and **5.5**). When ions dissolve in water, some of the water molecules must be pulled apart so that they can regroup around the ions. This process requires energy too.

The strength of the attractions between ions and water molecules is measured by the **enthalpy of hydration**, or ΔH_{hyd}. This is the enthalpy change for the formation of a solution of ions from 1 mole of gaseous ions. For example,

$$Na^+(g) + aq \rightarrow Na^+(aq) \quad \Delta H_{hyd} = -406\,kJ\,mol^{-1}$$
$$Br^-(g) + aq \rightarrow Br^-(aq) \quad \Delta H_{hyd} = -337\,kJ\,mol^{-1}$$

(The symbol 'aq' is used in an equation to represent water when it is acting as a solvent.)

Enthalpies of hydration depend on the concentration of the solution produced. Values quoted refer to an 'infinitely' dilute solution, where we can assume interactions between the ions are negligible.

Enthalpies of hydration are always negative – i.e. hydration is *exothermic* and energy is given out. Some values are listed in Table 4. If you compare Table 4 with Table 2 you will see that the most exothermic values occur for the ions with:

- the greatest charge
- the smallest radii.

◀ **Table 4** Enthalpies of hydration of some ions.

Ion	ΔH_{hyd}/kJ mol^{-1}	Ion	ΔH_{hyd}/kJ mol^{-1}	Ion	ΔH_{hyd}/kJ mol^{-1}
Li^+	−520	Mg^{2+}	−1926	Al^{3+}	−4680
Na^+	−406	Ca^{2+}	−1579		
K^+	−320	Sr^{2+}	−1446		
Rb^+	−296				

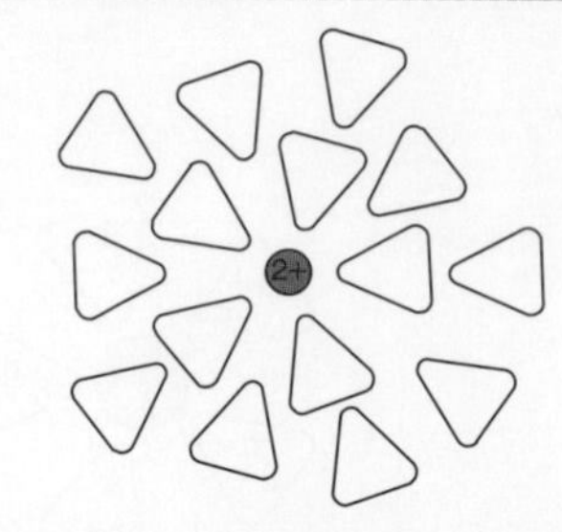

Mg^{2+}(aq)
Small, doubly charged Mg^{2+} ion attracts H_2O molecules strongly. $\Delta H_{hyd} = -1926$ kJ mol^{-1}.

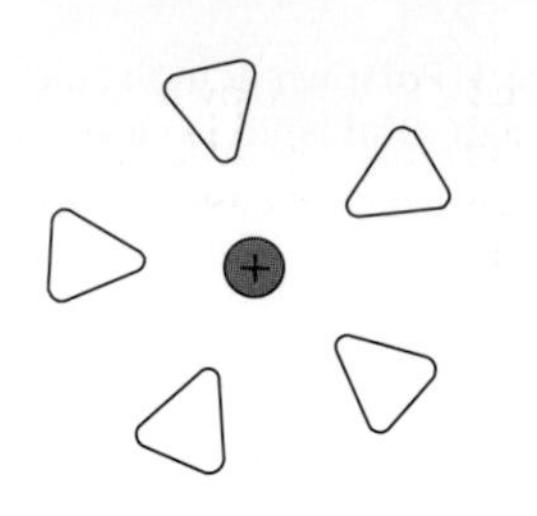

Na^{+}(aq)
Larger Na^{+} ion with smaller charge attracts H_2O molecules less strongly. $\Delta H_{hyd} = -406$ kJ mol^{-1}.

▲ **Figure 4** The magnitude of enthalpy of hydration depends on the size and charge of the ion.

> Hydration of ions is an exothermic process.
>
> ∴ ΔH_{hyd} is always negative.

The reasons are very similar to those used earlier to explain the variation in ΔH_{LE}. Small, highly charged ions can get close to water molecules and attract them strongly (Figure 4).

Molecules of some other solvents, such as ethanol, are also polar and can bind to ions. When we are dealing with solvents other than water, we talk more generally about the **enthalpy of solvation**, or ΔH_{solv}, rather than enthalpy of hydration.

Enthalpy change of solution

The hydration of ions favours dissolving and helps to supply the energy needed to separate the ions from the lattice. The difference between the enthalpies of hydration of the ions and the lattice enthalpy gives the **enthalpy change of solution**, or $\Delta H_{solution}$. $\Delta H_{solution}$ can be measured experimentally – it is the enthalpy change when 1 mole of a solute dissolves to form an 'infinitely' dilute solution:

$$\Delta H_{solution} = \Delta H_{hyd}(\text{cation}) + \Delta H_{hyd}(\text{anion}) - \Delta H_{LE}$$

We can represent the enthalpy changes involved using an enthalpy cycle like the ones you met in **Section 4.1**.

This enthalpy cycle is shown in Figure 5. Notice that the enthalpy change for breaking up the ionic lattice is *minus* ΔH_{LE}, because ΔH_{LE} is defined as the enthalpy change when the lattice is *created* – the opposite of breaking it up.

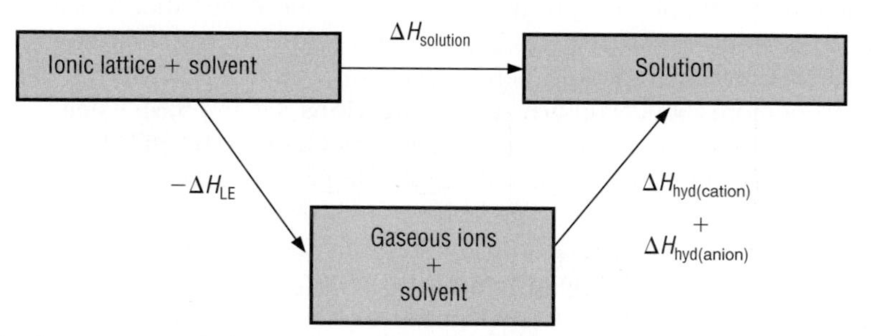

▲ **Figure 5** An enthalpy cycle to show the dissolving of an ionic solid.

Figures 6, 7 and 8 show three different examples for this enthalpy cycle, but this time in the form of an enthalpy level diagram. This makes it easier to compare the sizes of the different enthalpy changes involved.

Figure 6 represents a solute for which $\Delta H_{solution}$ is *negative*. The hydration of the ions provides slightly more energy than is needed to break up the lattice. This type of solute normally dissolves, giving out a little energy in the process.

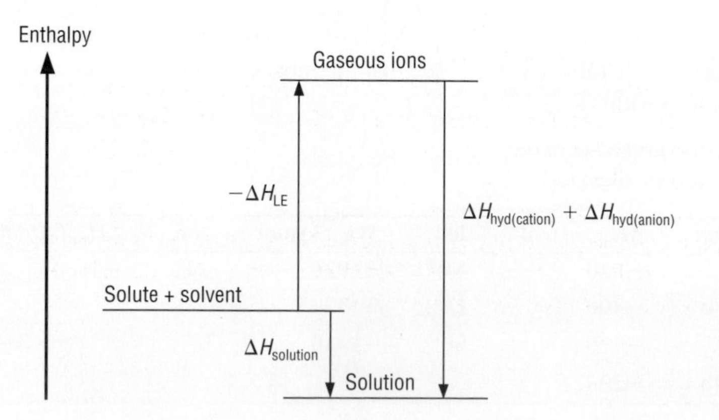

► **Figure 6** An enthalpy level diagram for a solute with a negative $\Delta H_{solution}$ – this type of solute will normally dissolve.

Figure 7 represents a solute for which $\Delta H_{solution}$ has a large *positive* value. The hydration of the ions does not provide as much energy as is needed to break up the lattice. This solute does not dissolve.

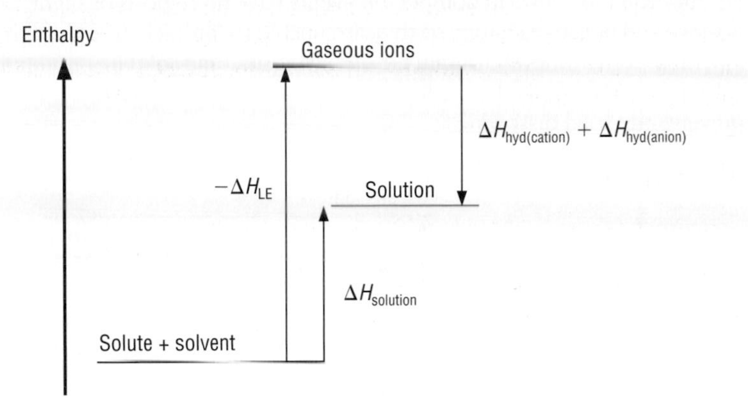

◀ **Figure 7** An enthalpy level diagram for a solute with a positive $\Delta H_{solution}$ – the solute will not dissolve because too much energy is needed.

Figure 8 represents a solute for which $\Delta H_{solution}$ is slightly positive, *but which dissolves nevertheless*. Many ionic solutes are like this – they dissolve even though they need a little energy from the surroundings to do so. Just why this can happen concerns *entropy*.

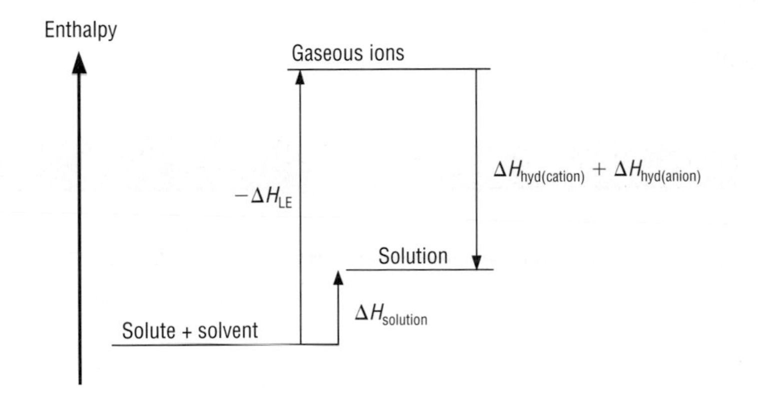

◀ **Figure 8** An enthalpy level diagram for a solute with a *slightly* positive $\Delta H_{solution}$ – if the entropy increase is favourable, this solute may dissolve despite needing energy from the surroundings.

Entropy and dissolving

When a solute dissolves there is often an entropy increase because the solute becomes more disordered as it spreads out through the solvent. (You can read more about entropy in **Section 4.3**.)

An increase in entropy favours dissolving – even if a little energy is needed. So substances with a small positive $\Delta H_{solution}$ can still dissolve, provided there is a favourable increase in entropy. But if $\Delta H_{solution}$ is *very* large and positive then the substance will not dissolve, even if the entropy change is favourable. We can use these ideas to explain why salt dissolves in the sea but the White Cliffs of Dover do not.

NaCl has a small positive $\Delta H_{solution}$ but there is an entropy increase when it dissolves – NaCl is soluble in water. $CaCO_3$ has a small negative $\Delta H_{solution}$ but there is a large entropy *decrease* when it dissolves – $CaCO_3$ is insoluble in water.

But why is there a decrease in entropy when calcium carbonate dissolves? The answer is that, even though the solute becomes more 'disordered', the solvent is becoming more 'ordered' as water molecules cluster round the doubly charged ions.

Non-polar solvents

Ionic solids, such as NaCl, are insoluble in **non-polar solvents**, such as hexane. The molecules in non-polar solvents have no regions of slight positive and negative charge, so they are unable to interact strongly with ions. The enthalpy level diagram for dissolving an ionic solid in a non-polar solvent would look like Figure 9 – the large positive value of $\Delta H_{solution}$ prevents the solid dissolving.

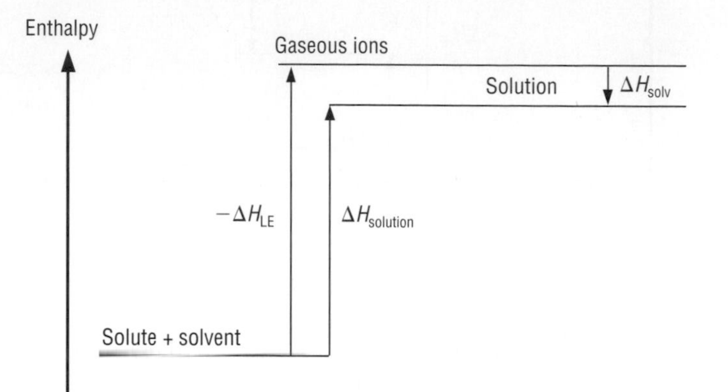

► **Figure 9** The situation when you try to dissolve an ionic solid in a non-polar solvent such as hexane.

The enthalpy change of solvation is so small that $\Delta H_{solution}$ is large and positive, and dissolving is unlikely to occur.

Problems for 4.5

1 a What is meant by the statement that *the lattice enthalpy of sodium fluoride is –915 kJ mol⁻¹*?
  b How do the sizes of the ions in an ionic compound affect its lattice enthalpy?

2 The lattice enthalpy of an ionic lattice is influenced by the sizes and charges of the ions. Use these ideas to decide which compound in each of the following pairs has the more negative (i.e. the more exothermic) lattice enthalpy.
  a LiF and NaF
  b Rb_2O and Na_2O
  c MgO and Na_2O
  d KF and KCl

3 Which compound in each of the following pairs has the more negative lattice enthalpy? Explain your answer in each case.
  a RbF and SrF_2
  b Cs_2O and BaO
  c CuO and Cu_2O (consider the charge on the copper ion in each compound)

4 The enthalpy of hydration of an ion is influenced by its size (small ions can attract water molecules more effectively than big ions) and the charge on the ion. Use these ideas to decide which ion in each of the following pairs has the most negative (i.e. most exothermic) enthalpy of hydration.
  a Li^+(g) and Na^+(g)
  b Mg^{2+}(g) and Ca^{2+}(g)
  c Na^+(g) and Ca^{2+}(g)

5 The lattice enthalpy of AgF is $-958\,kJ\,mol^{-1}$ and that of AgCl is $-905\,kJ\,mol^{-1}$. The enthalpy of hydration of Ag^+(g) is $-446\,kJ\,mol^{-1}$, that of F^-(g) is $-506\,kJ\,mol^{-1}$, and that of Cl^-(g) is $-364\,kJ\,mol^{-1}$.
  a Explain the difference in the lattice enthalpies of the two silver halides.
  b Explain the difference in the enthalpies of hydration of the two halide ions.
  c Draw separate enthalpy level diagrams for dissolving silver fluoride and silver chloride in water, linking together the lattice enthalpy, the sum of the enthalpies of hydration of the silver and the halide ions, and the endothermic enthalpy changes of solution of the silver halides.
  d Use your enthalpy level diagrams to help you to calculate values for the enthalpy changes of solution of silver fluoride and silver chloride.
  e What do the values of the enthalpy changes of solution of the silver halides suggest about the solubilities of the solids in water?

6 a Use the following table of lattice enthalpies and enthalpies of hydration to draw enthalpy level diagrams to determine the enthalpy change of solution of
 i magnesium hydroxide
 ii calcium hydroxide.

Compound	Lattice enthalpy ΔH_{LE}/kJ mol^{-1}	Ion	Enthalpy of hydration ΔH_{hyd}/kJ mol^{-1}
$Mg(OH)_2$	−2998	Mg^{2+}	−1926
$Ca(OH)_2$	−2506	Ca^{2+} OH^-	−1579 −460

b Use the enthalpies of solution you have calculated in part **a** to suggest which of the two hydroxides is more soluble in water.

c What other factor must be taken into account when explaining the relative solubilities of the two compounds?

STRUCTURE AND PROPERTIES

5.1 *Ions in solids and solutions*

Ionic solids

In ionic solids, ions are held together by their opposite electrical charges. Each positive ion (called a **cation**) attracts several negative ions (called **anions**), and vice versa. The ions build up into a giant **ionic lattice**, in which very large numbers of ions are arranged in fixed positions.

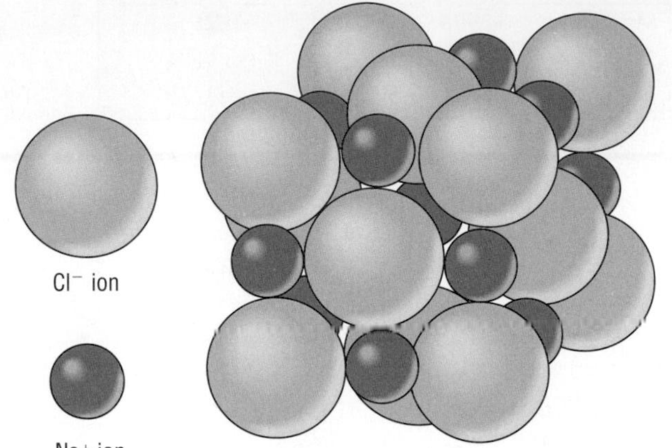

► **Figure 1** The sodium chloride lattice.

One of the simplest examples is sodium chloride, Na^+Cl^-(s) (Figure 1). In the sodium chloride lattice, each Na^+ ion is surrounded by six Cl^- ions, and each Cl^- ion is surrounded by six Na^+ ions. Each Na^+ is attracted to the six Cl^- round it, but repelled by other Na^+ ions which are a bit further away (there are 12 of these), and attracted to the next lot of Cl^- ions (eight of them) which are further away still … and so on. It adds up to an infinite series of attractions and repulsions but overall the attractions are stronger than the repulsions, which is why the lattice holds together. Indeed, it holds together very strongly, which is why ionic solids are hard and have high melting and boiling points.

It is important to realise that an ionic solid like Na^+Cl^- isn't just pairs of Na^+ and Cl^- ions, but a huge lattice. You can see from Figure 1 that the lattice has a cubic shape, which accounts for the fact that sodium chloride crystals are cubic.

The simple cubic lattice structure of sodium chloride is also found in some other Group 1 halides, such as potassium fluoride and lithium chloride. However, other ionic solids may have more complicated lattices, depending on the number and size of the different ions present.

Hydrated crystals

The crystals of some ionic solids include molecules of water. For example, the formula of magnesium chloride crystals is $MgCl_2{\cdot}6H_2O$. The water is not just mixed with the magnesium chloride crystals – that would just make them damp. Instead, the H_2O molecules are *fitted in the lattice* in the same regular way as the ions.

The lattice is, therefore, an interweaving regular pattern of three different particles: Mg^{2+} ions, Cl^- ions and H_2O molecules. The water in compounds like this is called **water of crystallisation** and the crystals are called **hydrated crystals**.

When hydrated crystals are heated, the water is driven off as steam leaving behind an **anhydrous** solid. You are probably familiar with blue, hydrated copper(II) sulfate crystals, $CuSO_4{\cdot}5H_2O$, which form white, anhydrous copper(II) sulfate, $CuSO_4$, on heating.

Ionic substances in solution

Many ionic substances dissolve readily in water. When they do, the ions become surrounded by water molecules and spread out through the water. Figure 2 illustrates this for sodium chloride. The dissolved ions, $Na^+(aq)$ and $Cl^-(aq)$, are no longer regularly arranged – they are scattered through the water at random. What's more, now the Na^+ and Cl^- ions are separated they behave independently of each other. It is as if each has forgotten the other exists.

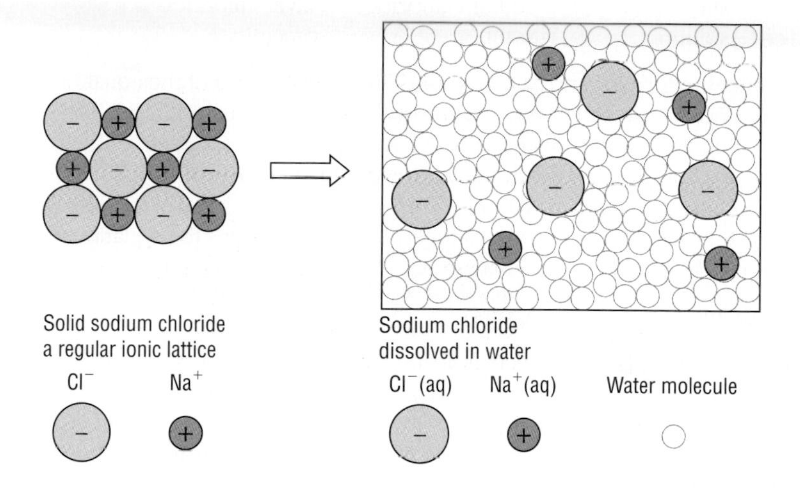

◄ **Figure 2** What happens when an ionic substance such as sodium chloride dissolves in water.

Each ion in solution is surrounded by water molecules. Water is a polar molecule. You have seen in **Section 3.2** that water has a bent shape. You have also seen in **Section 3.1** that oxygen is more electronegative than hydrogen. The polarity of the water molecule is shown in Figure 3.

The positive ends of water molecules are attracted to negative ions, and the negative ends of water molecules are attracted to positive ions. Each ion in solution is surrounded by its own sphere of water molecules – this is known as hydration. It takes a lot of energy to break the bonds within an ionic lattice but hydration releases energy to compensate for this. You can find out more about this in **Section 4.5**.

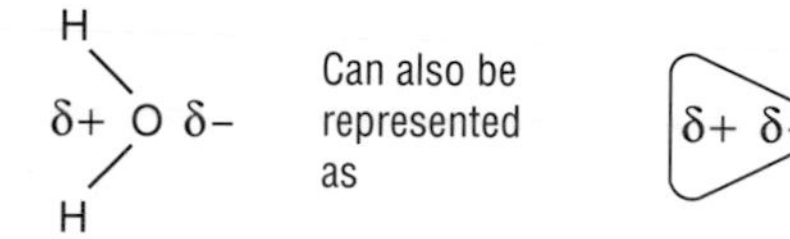

▲ **Figure 3** Water molecules are polar.

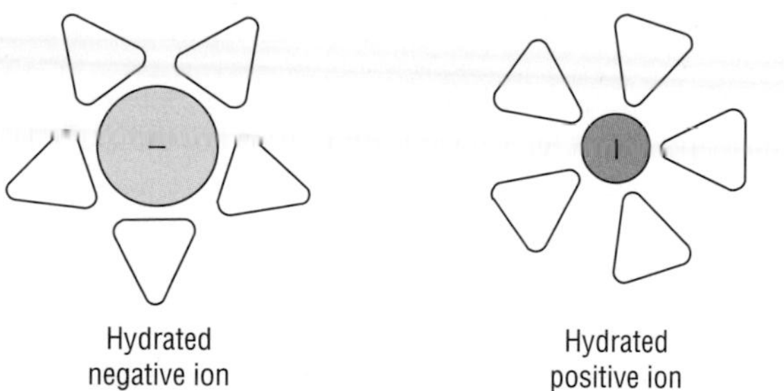

◄ **Figure 4** Hydrated ions – each ion in solution is surrounded by water molecules.

This applies to all ionic substances. As soon as they are dissolved, the positive and negative ions separate and behave independently. So it is best to regard sea water, for example, as a mixture of positive and negative ions dissolved in water, rather than as a solution of ionic compounds.

Ionic equations

You now know that ions in solution behave independently – and this includes their chemical reactions. The reactions of an ionic substance, such as sodium chloride, quite often involve only one of the two types of ion – the other ion does not get involved in the reaction.

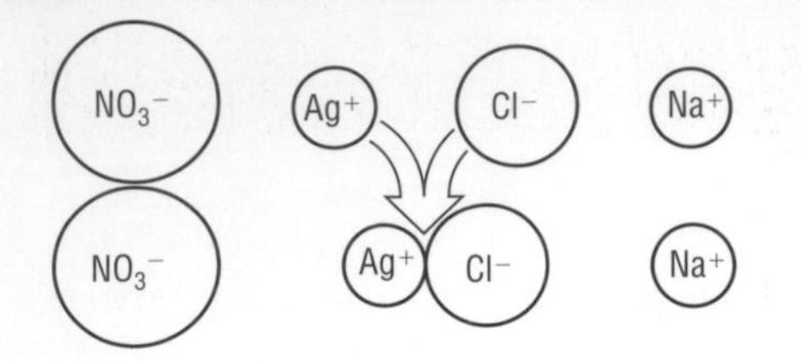

▲ **Figure 5** Making a precipitate of silver chloride by mixing silver nitrate and sodium chloride solutions – only the $Ag^+(aq)$ and $Cl^-(aq)$ ions react.

For example, if you add a solution of sodium chloride to silver nitrate solution, you get a white **precipitate** of silver chloride. Silver ions, Ag^+, and chloride ions, Cl^-, have come together to form insoluble silver chloride, which precipitates out (Figure 5).

We can write an equation for this reaction. Sodium chloride solution contains $Na^+(aq)$ and $Cl^-(aq)$, and silver nitrate solution contains $Ag^+(aq)$ and $NO_3^-(aq)$, so we write

$$\underbrace{Ag^+(aq) + NO_3^-(aq)}_{\text{silver nitrate solution}} + \underbrace{Na^+(aq) + Cl^-(aq)}_{\text{sodium chloride solution}} \rightarrow \underbrace{AgCl(s)}_{\text{silver chloride precipitate}} + Na^+(aq) + NO_3^-(aq)$$

Notice that $Na^+(aq)$ and $NO_3^-(aq)$ ions are on both sides of the equation. They do not take part in the reaction – it is as if they are on the sidelines, watching the action going on between $Ag^+(aq)$ and $Cl^-(aq)$ ions without getting involved. They are described as **spectator ions**.

Because $Na^+(aq)$ and $NO_3^-(aq)$ are not involved in the reaction, we can leave them out of the equation altogether. This simplifies the equation so that it only shows the ions that are actually taking part in the reaction:

$$Ag^+(aq) + Cl^-(aq) \rightarrow AgCl(s)$$

This type of equation is called an **ionic equation** – it shows only the ions that take part in the reaction and excludes the spectator ions.

State symbols are very important in ionic equations. They will help you, for example, to identify that the above reaction involves **ionic precipitation**.

In order to help predict whether an ionic precipitation reaction will occur when two solutions are mixed together it is useful to have some rules. Table 1 gives the solubilities of some common ionic compounds. **Section 4.5** can help you understand why some ionic compounds are soluble in water and others are not.

► **Table 1** Solubilities of some common ionic compounds.

All nitrates are soluble
All chlorides are soluble except for $AgCl$ and $PbCl_2$
Most sulfates are soluble except for $BaSO_4$, $PbSO_4$ and $SrSO_4$
All carbonates are insoluble except NH_4^+ and those of the Group 1 elements
All sodium, potassium and ammonium salts are soluble

Sometimes, when solutions of ions are mixed, a covalent compound is formed. This happens when an acid is neutralised by an alkali. For example, hydrochloric acid contains $H^+(aq)$ and $Cl^-(aq)$, so we write

$$\underbrace{H^+(aq) + Cl^-(aq)}_{\text{hydrochloric acid}} + \underbrace{Na^+(aq) + OH^-(aq)}_{\text{sodium hydroxide solution}} \rightarrow Na^+(aq) + Cl^-(aq) + \underbrace{H_2O(l)}_{\text{water}}$$

Excluding the spectator ions, $Na^+(aq)$ and $Cl^-(aq)$, the ionic equation becomes

$$H^+(aq) + OH^-(aq) \rightarrow H_2O(l)$$

This is the same for all **neutralisation** reactions, whatever acid and alkali are involved.

Problems for 5.1

1 Write down the separate ions present in solutions of the following compounds. For example, in a solution of $Mg(NO_3)_2$ the ions present are $Mg^{2+}(aq) + 2NO_3^-(aq)$.

a $Ca(OH)_2$
b $MgSO_4$
c Na_2O
d KOH
e $AgNO_3$
f $Al_2(SO_4)_3$

2 Write formulae for the following ionic compounds. (You may find it helpful to look at the table of common ions in **Section 3.1**.)

a sodium bromide
b magnesium hydroxide
c sodium sulfide
d barium oxide
e calcium carbonate
f calcium nitrate
g potassium carbonate

3 Many of the compounds of Group 2 elements are insoluble in water. Write ionic equations (with state symbols) for the precipitates formed when the following solutions are mixed.

	Aqueous solutions mixed	Precipitate (colour)
a	barium chloride, $BaCl_2$, and sodium sulfate, Na_2SO_4	barium sulfate (white)
b	magnesium sulfate, $MgSO_4$, and sodium hydroxide, NaOH	magnesium hydroxide (white)
c	calcium chloride, $CaCl_2$, and sodium carbonate, Na_2CO_3	calcium carbonate (white)
d	barium nitrate, $Ba(NO_3)_2$, and potassium chromate, K_2CrO_4	barium chromate (yellow)

4 When a saturated solution of sodium sulfate is cooled crystals with the formula $Na_2SO_4{\cdot}10H_2O$ are formed. Write an ionic equation, with state symbols, for the formation of the crystals from the solution.

5 Write ionic equations (with state symbols) for the following reactions:

a nitric acid with potassium hydroxide solution to form potassium nitrate solution and water
b zinc metal with sulfuric acid to form zinc sulfate solution and hydrogen
c copper(II) oxide powder with hydrochloric acid to form copper(II) chloride solution and water
d marble chips (calcium carbonate) with hydrochloric acid to form calcium chloride solution, carbon dioxide and water.

Refer back to page 84 where you learned about ionic solids before you answer problem 6.

6 The lattice structure of caesium chloride is shown below. A caesium ion lies at the centre of a cube of chloride ions, and a chloride ion lies at the centre of a cube of caesium ions.

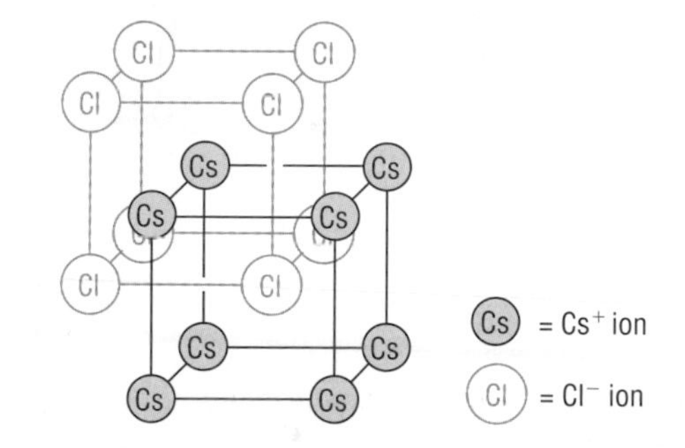

a i Predict the shape of caesium chloride crystals.
ii How many chloride ions closely surround one caesium ion?
iii How many caesium ions closely surround one chloride ion?

b Suggest why caesium chloride
i is hard
ii has a high melting point (918 K).

5.2 *Molecules and networks*

AS

Carbon and silicon oxides

We breathe out CO_2 and walk on SiO_2. Both are essential for life on Earth – one is an atmospheric gas used in photosynthesis, and the other is found in rocks in the Earth's crust and produces the soil which supports growing plants.

Carbon and silicon are both in Group 4 of the Periodic Table and you might expect their compounds to show similar properties. Yet you could hardly imagine two more physically different substances. CO_2 is a gas at

Why the difference?

When carbon dioxide turns from a solid to a gas, the gas, like the solid, contains CO_2 molecules. Energy is needed only to pull the CO_2 molecules apart.

When silica turns from a solid to liquid and then to a gas, the network of Si and O atoms breaks up. This involves breaking strong Si–O covalent bonds and requires a large energy input.

room temperature; it turns directly from a solid to a gas (sublimes) at −78 °C (195 K). SiO_2 is a hard solid with a high melting point – sand, for example, is impure SiO_2.

The reason for this dramatic difference in physical properties is the difference in bonding between carbon and oxygen on one hand, and silicon and oxygen on the other. Both are covalent compounds, but the small size of the carbon atom makes it possible for carbon to form double bonds with oxygen, so that carbon dioxide is composed of individual molecules:

O=C=O

Bonds *between* the carbon dioxide molecules are weak. Little energy is needed to separate individual molecules in the solid and liquid phases to form a gas, so CO_2 is a gas at room temperature. It freezes at 195 K to a white solid (called dry ice). It is soluble in water, giving an acidic solution.

Silicon atoms are larger than carbon atoms and they normally bond to four oxygen atoms. Quartz, SiO_2, is an extended network of SiO_4 units in which the central silicon is covalently bonded to each of four oxygen atoms. Every Si atom has a half-share in four oxygen atoms (Figure 1).

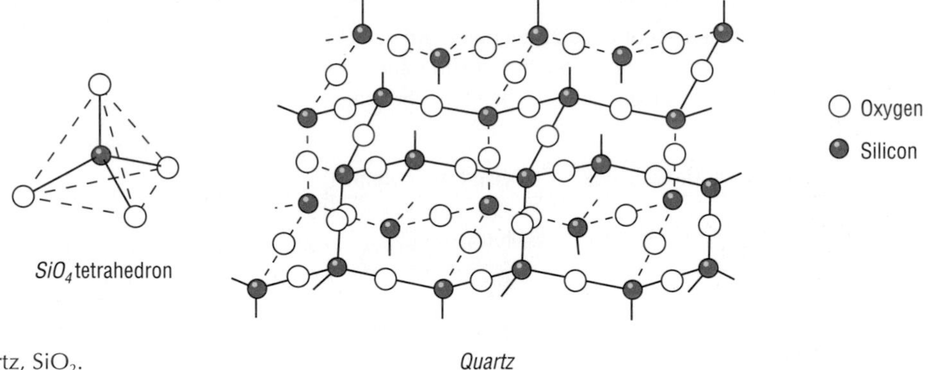

► **Figure 1** The structure of quartz, SiO_2.

Because of its extended **network structure** (sometimes called a **giant structure**), SiO_2 is insoluble in water and has high melting and boiling points (1883 K and 2503 K, respectively). The reason for these properties is the strong covalent bonding which exists throughout the SiO_2 structure. Considerable energy is needed to break bonds within the structure so that a very high temperature is needed to melt quartz.

Note that CO_2 and SiO_2 are named differently. CO_2 is called carbon dioxide because this describes its simple molecules, made up of one carbon and two oxygen atoms. SiO_2 is sometimes called silicon dioxide, but this can be misleading because it implies that the substance contains simple molecules. It is better to use the systematic name, silicon(IV) oxide.

Two types of covalent structures

So covalent substances can exist as two types of structures, depending on how the atoms are bonded. The two types of structures show very different properties.

- **Covalent molecular structures** (for example, CO_2) consist of discrete molecules. The covalent bonds within the molecules are strong. There are also weak bonds *between* the molecules. The strength of these intermolecular bonds determines whether the substance is a gas, solid or liquid at normal temperatures. (You can learn more about this in **Sections 5.3**, **5.4** and **5.5**.) Covalent molecular substances often dissolve in organic solvents. Some, like carbon dioxide, dissolve in water.

- **Covalent network structures** (for example, SiO_2) consist of giant repeating lattices of covalently bonded atoms. Substances with network structures are insoluble solids with high melting and boiling points.

The examples above are covalent *compounds* containing two different elements. But the same types of structures are shown by non-metal *elements* – some exist as covalent molecules, whereas others exist as covalent networks.

Elements with molecular structures

Some non-metal elements that exist as molecules are shown in Figure 2. Hydrogen, oxygen, nitrogen, fluorine and chlorine all exist as **diatomic molecules**, and so their chemical formulae are written as X_2. They are all gases at room temperature.

Phosphorus and sulfur are both soft, low-melting solids. Phosphorus molecules, P_4, are made up of four phosphorus atoms, whereas sulfur molecules, S_8, contain eight sulfur atoms in a ring. The structures of these molecules are shown in Figure 3. For the sake of simplicity, when writing chemical equations, phosphorus and sulfur are often simply written as P and S, respectively.

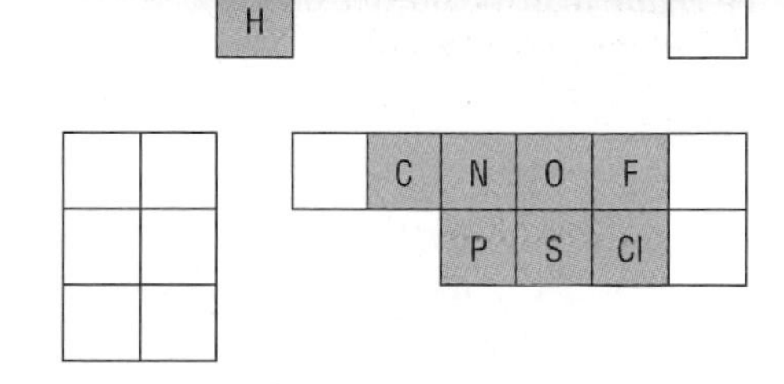

▲ **Figure 2** The position in the Periodic Table of some covalent elements with molecular structures.

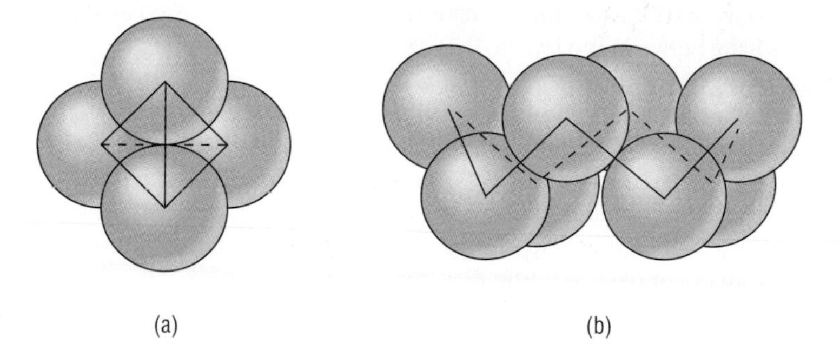

◄ **Figure 3 a** Structure of a phosphorus molecule, P_4. **b** Structure of a sulfur molecule, S_8.

Elements with network structures

Figure 4 shows some common elements that exist as covalent network structures. These elements are solids with high melting points and boiling points. The bonds between the individual atoms are strong covalent bonds and a lot of energy is needed to pull the atoms apart to form a liquid or a gas.

Figure 5 shows the structure of silicon. Every silicon atom is bonded to four other silicon atoms in a tetrahedral arrangement. The giant lattice repeats the structure indefinitely in all directions – even a small lump of silicon would contain billions of atoms joined up in this way.

You may have noticed that carbon appears in both Figure 2, among the molecular elements, and in Figure 4, among the network elements. This is not a misprint. In fact, carbon can exist in several different forms. Two of these have network structures – the other forms are molecular.

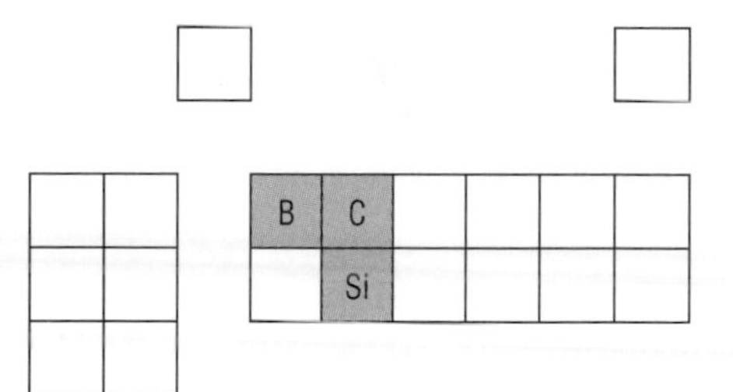

▲ **Figure 4** The position in the Periodic Table of covalent elements with network structures.

Diamond and graphite

Diamond and graphite are both giant lattices made up of carbon atoms, but the atoms are arranged very differently in the two forms. Look at the structures of diamond and graphite in Figure 6 (page 90).

Note the similarity between the structure of diamond in Figure 6a and the structure of silicon in Figure 5. Carbon, like silicon, is in Group 4 of the Periodic Table and can form four covalent bonds.

In **diamond**, every carbon atom is joined tetrahedrally to four other carbon atoms by strong covalent bonds forming a giant lattice. This highly

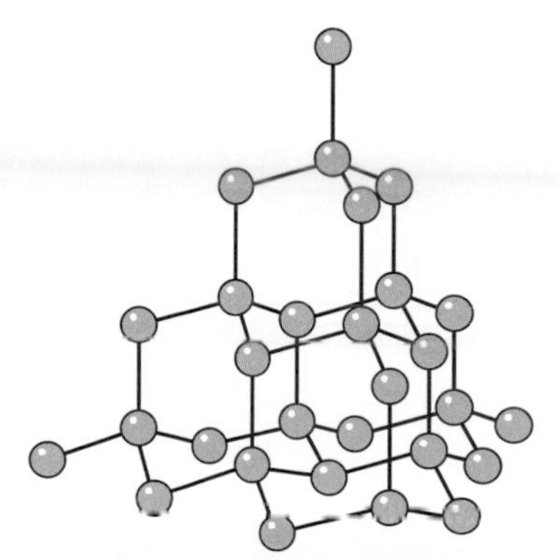

▲ **Figure 5** The network structure of silicon.

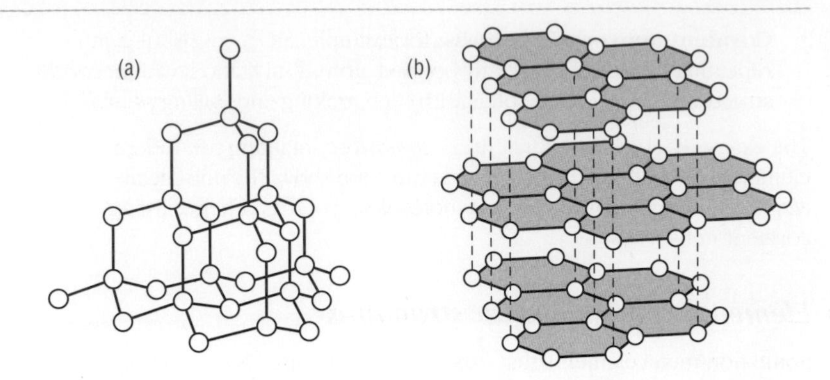

► **Figure 6 a** The structure of diamond. **b** The structure of graphite.

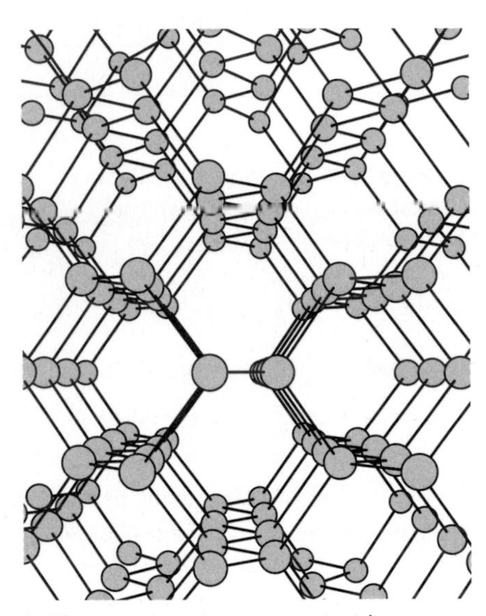

▲ **Figure 7** Imagine you are inside a diamond. The regular network structure would repeat in all directions – as far as the edge of the diamond.

symmetrical structure and the strong C–C bonds make diamond the hardest naturally occurring substance. Industrial grade diamonds are used on the tips of drills, and crushed diamonds are used on high quality grinding and cutting wheels.

The structure of **graphite** is quite different (Figure 6b). It is made up of giant flat layers of carbon atoms. Each layer is a two-dimensional network. In the layers, each carbon atom is joined by strong covalent bonds to three others forming rings containing six carbon atoms. This uses up only three of the four outer shell electrons in each carbon atom. The fourth electron of each atom contributes to the formation of extended delocalised electron clouds lying between the layers. This causes the layers to be held together relatively weakly.

The delocalised electrons are not attached to particular carbon atoms and are free to move over the layer. This allows graphite to be a good conductor of electricity, a property not normally associated with covalent substances. Most covalent substances do not conduct electricity because all the electrons are tightly held in covalent bonds so that they are not free to move from one molecule to the next through the structure.

Graphite is a soft, brittle solid. Because the layers are bonded weakly to each other, they can slide over one another. This is why a lump of graphite has a slippery feel to it, almost like wet soap, and also why graphite is used as a lubricant. Pencil 'leads' are made by mixing powdered graphite with clay.

Fullerenes

The existence of a new form of carbon was announced in 1985. Trace amounts of the substance were discovered by accident in the soot produced when a high-power laser was used to evaporate graphite in an inert atmosphere. Five years later, larger quantities were made by passing an electric arc between graphite rods.

The new form of carbon dissolved in benzene to give a red solution, which suggested that it has a molecular rather than a network structure. The molecules were shown to have the formula C_{60}.

The new substance was named **buckminsterfullerene** because the structure of the molecule was similar to a geodesic dome – a type of building designed by the architect Buckminster Fuller. Figures 8a and 8b show the structure of buckminsterfullerene, which looks like a football.

Other carbon molecules with similar structures have since been discovered and they are known collectively as **fullerenes**. Figure 8c shows a C_{70} molecule, which is shaped like a rugby ball. The fullerenes are sometimes referred to as 'buckyballs'. Sir Harry Kroto at the University of Sussex was awarded the Nobel Prize in Chemistry in 1996 (along with Robert Curl and Richard Smalley at the University of Texas) for the discovery of the fullerenes.

Recently, elongated fullerenes looking like tiny molecular tubes have been discovered. These have been called 'buckytubes' or 'nanotubes'.

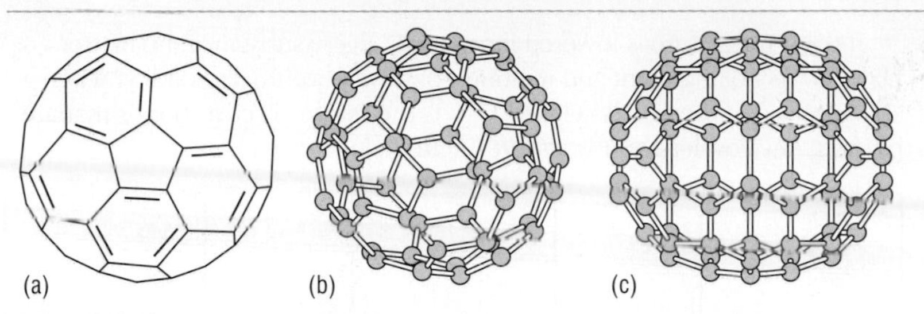

◀ **Figure 8** The fullerenes are a molecular form of carbon.
a C_{60}, named buckminsterfullerene – it is made up of a mixture of 5-membered and 6-membered rings and looks like a football.
b Another way of presenting C_{60}, showing the positions of the carbon atoms.
c C_{70}, which is shaped like a rugby ball.

If long versions of these could be manufactured, the possibility exists of creating incredibly strong, lightweight fibres.

Fullerenes were discovered so recently that their properties have not been fully established. Research work is taking place to investigate possible large-scale uses and there are already hundreds of fullerene patents. It has been discovered, for example, that C_{60} is an *optical limiter*. Shining light on a solution of C_{60} causes it to turn darker instantly, and the more light there is the darker it gets, so that the intensity of the transmitted light is limited to a maximum value. This property could be put to practical use in the manufacture of protective goggles for people working with lasers.

Problems for 5.2

1. In terms of bonds between particles, explain why CO_2 is a gas at room temperature and pressure, while SiO_2 is solid.

2. Explain the following observations in terms of particles, structure and bonding:
   - **a** substances with covalent network structures are insoluble in all solvents
   - **b** substances that are gases at room temperature always contain simple molecules or isolated atoms.

3. Silicon carbide, SiC, is a covalent compound that is almost as hard as diamond.
   - **a** Draw part of the structure of silicon carbide, showing a total of about 10–15 atoms.
   - **b** Suggest why silicon carbide is such a hard substance.

4. Boron nitride, BN, has a similar structure to graphite but it does not conduct electricity.
   - **a** Draw part of the structure of boron nitride.
   - **b** Suggest why boron nitride does not conduct electricity.

5. Buckminsterfullerene was obtained by adding benzene to the soot obtained when an electric discharge is passed between graphite electrodes. Buckminsterfullerene dissolves in benzene to give a red solution. Explain why this was a startling observation at the time, and why it was taken as evidence that buckminsterfullerene has a molecular structure.

5.3 *Bonds between molecules: temporary and permanent dipoles*

Solid, liquid or gas?

In **Sections 3.1** and **5.2** you learned how covalent bonds hold the atoms together within a molecule. At room temperature, some substances made up of covalent molecules are gases, others are liquids and the remainder are solids. What is it that determines whether a substance is a gas, a liquid or a solid?

Most of the gases in the Earth's atmosphere exist as *small* covalent molecules (e.g. N_2, O_2 and CO_2). The noble gases (e.g. Ar, Ne and He) exist as isolated atoms.

In a liquid or a solid there must be forces *between the molecules* causing them to be attracted to one another, otherwise they would move apart and become a gas. These forces are called **intermolecular bonds**.

If the temperature is lowered far enough, every substance, no matter how low its boiling point and melting point, will eventually solidify. When a solid melts and then boils (Figure 1), it is the intermolecular bonds that are broken. Any covalent bonds *within* the molecules remain intact.

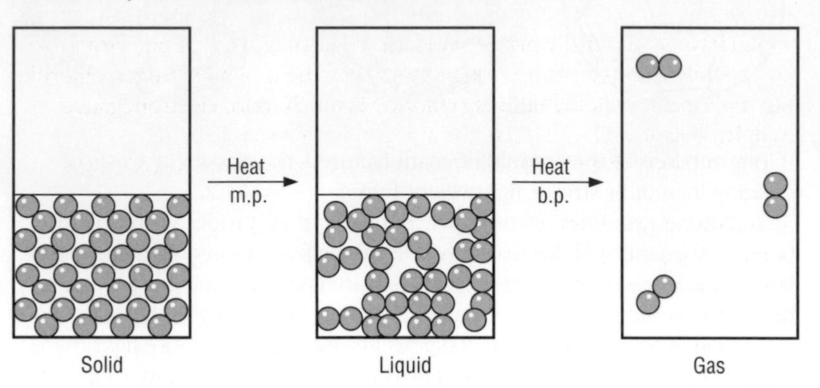

► **Figure 1** On heating, the molecular substance (represented by ⚭) changes from a solid to a liquid and then to a gas. Energy must be supplied to overcome the intermolecular bonds. Note that the covalent bonds within the molecules remain intact.

So, when a substance has a structure consisting of covalent molecules, the boiling point gives a good indication of how strong the bonds are *between* the molecules.

The boiling points of noble gases and alkanes

Look at the boiling points given in Table 1 for the noble gases.

When the noble gases change from liquid to gas, energy is only required to separate the atoms. There are no covalent bonds in the noble gases, but there must be some bonds holding the atoms together.

The low boiling points of the noble gases show that the bonds between the isolated atoms in the liquid state are very weak. However, the boiling points get higher as the atoms get bigger. This suggests that the larger the atoms, the stronger the bonds between them.

Now look at the boiling points of the alkanes in Table 2.

▼ **Table 1** Boiling points for the noble gases.

Element	Boiling point/K
helium	4
neon	27
argon	87
krypton	121
xenon	161

► **Table 2** Boiling points of some alkanes.

Alkane	Structural formula	Boiling point/K
methane	CH_4	111
ethane	CH_3CH_3	184
propane	$CH_3CH_2CH_3$	231
butane	$CH_3CH_2CH_2CH_3$	273
pentane	$CH_3CH_2CH_2CH_2CH_3$	309

There is a pattern to these boiling points also – the longer the alkane chain, the higher the boiling point. There must be stronger bonds between the long molecules than between the short molecules. More energy and higher temperatures are needed to overcome these stronger bonds and set the alkane molecules free.

To understand how the intermolecular bonds between molecules such as those of alkanes arise, and what affects their relative sizes, you need to know about the charges on molecules, called dipoles.

Dipoles

Polar bonds are described in **Section 3.1**. A **dipole** is simply a molecule (or part of a molecule) with a positive end and a negative end. Hydrogen chloride has a dipole – we show the molecule as

$$\overset{\delta+}{H}—\overset{\delta-}{Cl}$$

When a molecule has a dipole we say it is **polarised**. There are several ways a molecule can become polarised.

Permanent dipoles

Permanent dipoles occur when a molecule has two atoms bonded together which have substantially different electronegativities, so that one atom attracts the shared electrons much more than the other. Hydrogen chloride has a permanent dipole because chlorine is much more electronegative than hydrogen, and so attracts the shared electrons more.

Molecules with a permanent dipole are said to be **polar molecules**.

Instantaneous dipoles

Some molecules do not possess a permanent dipole because the atoms that are bonded together have the same, or very similar, electronegativities, so that the electrons are evenly shared. The chlorine molecule, Cl_2, is an example.

However, even though a molecule does not have a permanent dipole, a **temporary**, or **instantaneous**, **dipole** can arise. We can think of the electrons in a chlorine molecule as forming a negatively charged cloud. The electrons in this cloud are in constant motion, and at a particular instant they may not be evenly distributed over the two atoms. This means that one end of the chlorine molecule has a greater negative charge than the other end – instantaneously, the molecule has developed a dipole (Figure 2).

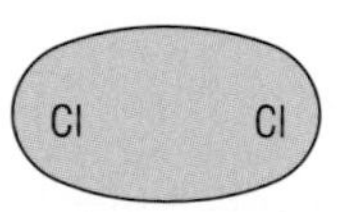

Electron cloud evenly distributed; no dipole

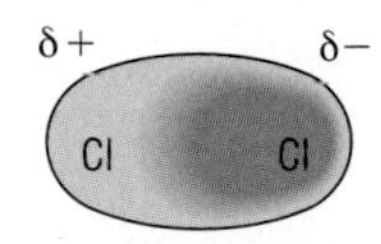

At some instant, more of the electron cloud happens to be at one end of the molecule than the other; molecule has an instantaneous dipole

◀ **Figure 2** How a dipole forms in a chlorine molecule.

Left on its own, this dipole only lasts for an instant before the swirling electron cloud changes its position, cancelling out or even reversing the dipole. However, if there are other molecules nearby then the instantaneous dipole may affect them and produce **induced dipoles**.

Induced dipoles

If an unpolarised molecule finds itself next to a dipole, the unpolarised molecule may get a dipole induced in it. The dipole attracts or repels electrons in the charge cloud of the unpolarised molecule, **inducing** a dipole in it.

Figure 3 illustrates this by showing what happens if an unpolarised chlorine molecule finds itself next to an HCl dipole.

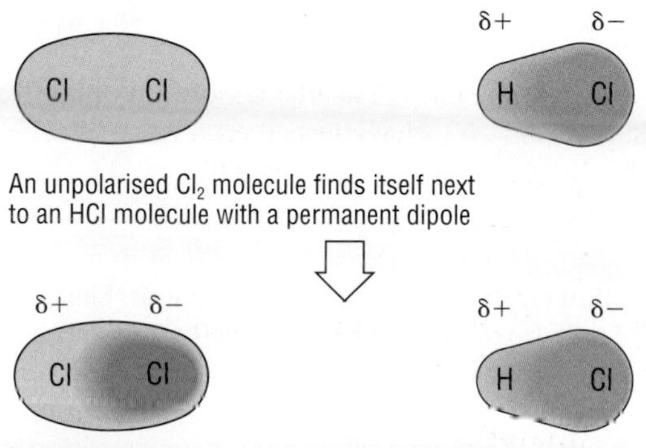

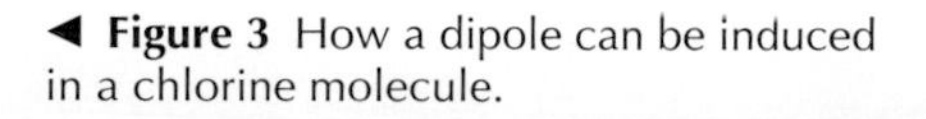

◀ **Figure 3** How a dipole can be induced in a chlorine molecule.

In Figure 3, the dipole has been induced by the effect of a *permanent* dipole. A dipole can also be induced by the effect of an *instantaneous* dipole. This makes it possible for a whole series of dipoles to be set up in a substance that contains no permanent dipoles. How this occurs in a noble gas like xenon is explained later in this section.

Dipoles and intermolecular bonds

If a molecular substance contains dipoles, they can attract each other (Figure 4).

$$\begin{array}{c} \delta+ \ \ \delta- \quad \delta+ \ \ \delta- \quad \delta+ \ \ \delta- \quad \delta+ \ \ \delta- \\ \text{-----H—Cl-----H—Cl-----H—Cl-----H—Cl-----} \end{array}$$

► **Figure 4** Intermolecular bonds arise from the attractions between dipoles. The dotted lines represent the bonds *between* HCl molecules; the solid lines represent the stronger covalent bonds *inside* each molecule.

All intermolecular bonds arise from the attractive forces between dipoles. There are three kinds of bond:

- **permanent dipole–permanent dipole**, where two or more permanent dipoles attract one another. This kind of bonding occurs between the permanent dipoles in HCl (as shown in Figure 4).
- **permanent dipole–induced dipole**, in which a permanent dipole induces a dipole in another molecule, then the two attract one another. This kind of bonding occurs between HCl and Cl_2 molecules (as shown in Figure 3).
- **instantaneous dipole–induced dipole**, in which an instantaneous dipole induces a dipole in another molecule, then attracts it. This kind of bonding occurs between Cl_2 molecules and between the atoms in a noble gas.

In the next sections we will look in more detail at two of these types of bonding: instantaneous dipole–induced dipole bonds and the bonds between permanent dipoles.

Instantaneous dipole–induced dipole bonds

These bonds are the weakest type of intermolecular bonding. They act between *all* molecules because instantaneous dipoles can arise in molecules that already have a permanent dipole. However, you notice them most clearly in substances such as the noble gases and the alkanes, because there are no other intermolecular bonds present.

Figure 5 shows how an instantaneous dipole in a xenon atom induces polarisation in a neighbouring atom.

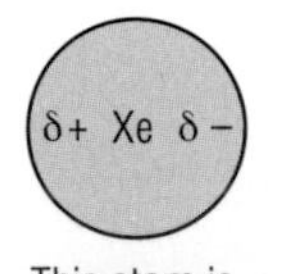

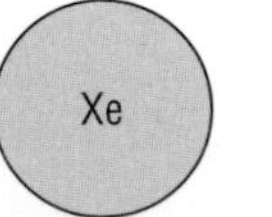

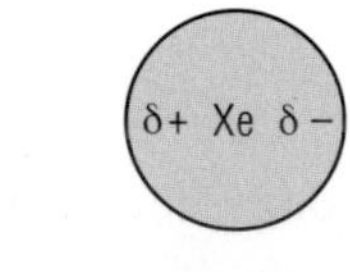

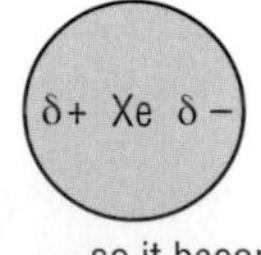

▲ **Figure 5** How an induced dipole is formed in a xenon atom.

The polarisation illustrated here is not permanent. Electrons are always moving around and the original polarity will be lost and replaced by another, which will be different – possibly the other way round. In the gaseous and liquid states, the atoms are always moving around. As they move past each other the dipoles constantly change position, although they are always lined up so that they attract rather than repel one another.

The bonds between atoms in xenon are stronger than those between helium atoms. The xenon atom is bigger, and has more electrons. The outer electrons are less strongly held by the positive nucleus and more easily pushed around from outside. We say that xenon atoms are more **polarisable** than helium atoms. This leads to bigger induced dipoles, which in turn lead to bigger bonds, which is why xenon has a higher boiling point than helium.

The importance of molecular shape

In the straight-chain and branched alkanes in Table 3, the atoms in the different isomers are identical, but in straight-chain alkanes there are more contacts between atoms of different molecules. Therefore, there are more opportunities for induced dipole bonds to form. That is why straight-chain alkanes have higher boiling points than their branched isomers.

Alkane	Skeletal formula	Boiling point/K
hexane		342
3-methylpentane	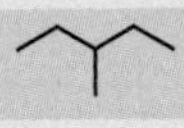	336
2-methylpentane	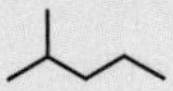	333
2,3-dimethylbutane	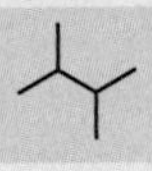	331
2,2-dimethylbutane	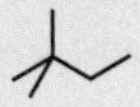	323

◀ **Table 3** Boiling points of the isomers of C_6H_{14}.

Polymers are molecules with very long chains. If the chains can line up neatly then the intermolecular bonds are quite strong, making the polymer material strong too.

If you are not sure about the amount of contact which is possible between molecules of different shape, building models will help. Try building two models each of hexane and 2,2-dimethylbutane and compare how closely they fit together.

To sum up so far …

Instantaneous dipole–induced dipole bonds arise because electron movements in molecules cause instantaneous dipoles, which then induce polarity in neighbouring molecules. The bonds between molecules are the result of electrostatic interactions between the dipoles. These interactions are normally strongest for large atoms, which have more electrons and are more polarisable than small atoms. For molecules, the bonds become stronger as the number of atoms in the molecule increases. This is because larger molecules have more electrons and this increases the possibility of greater numbers of instantaneous dipoles being in existence at any given time.

Although instantaneous dipole–induced dipole bonds are relatively weak, the *overall* bonds between large molecules (for example, between the chains of a polymer) can be quite strong. This is why poly(ethene) is a solid at room temperature, even though instantaneous dipole–induced dipole bonds are the only type of intermolecular bonds present.

Remember that these bonds are present in *all* substances.

Permanent dipole–permanent dipole bonds

In this section, we will look in more detail at the additional bonds which can arise when molecules have **permanent dipoles**.

Bond polarity and polar molecules

Section 3.1 covers bond polarity and describes how you can use electronegativity values to decide how polar a particular bond will be.

Let's look at an important example of a polar molecule, H_2O. Oxygen is much more electronegative than hydrogen, so the O–H bond is polar:

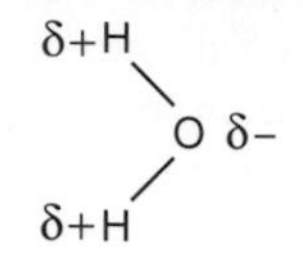

One side of the molecule is positive, the other is negative. The water molecule has a dipole with the positive charge centred between the two hydrogen atoms. We could also draw it like this:

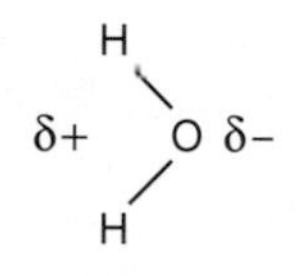

An example of a more complicated polar molecule is 1,1,1-trichloroethane. The three chlorine atoms make one end of the molecule negatively charged – the three hydrogen atoms at the other end are positively charged. The dipole can be seen more clearly if we draw the structure as

$$\overset{\delta+}{H_3C} - \overset{\delta-}{CCl_3}$$

When the atoms which form a molecule have no electronegativity difference, or when the electronegativity differences are small e.g. CH_4 (in other words, when the bonds are non-polar), any dipole will be very small. It is also possible for a molecule to have no overall dipole even though the bonds are polar. CCl_4 is an example of such a molecule. Each Cl atom in CCl_4 carries a small negative charge and the central carbon is positive:

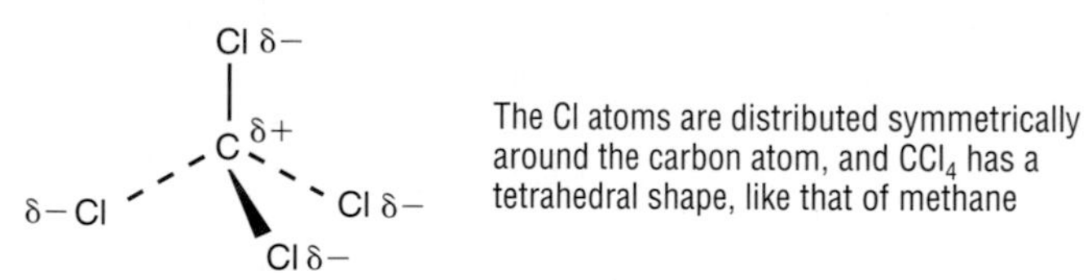

The Cl atoms are distributed symmetrically around the carbon atom, and CCl_4 has a tetrahedral shape, like that of methane

Because the chlorine atoms are distributed tetrahedrally around the carbon, the centre of negative charge is midway between all the chlorines. It is at the centre of the molecule and is superimposed on the positive charge on the carbon. The molecule has no overall dipole – we say it is **non-polar**.

So, *bond* polarity depends on electronegativity differences. A *molecular* dipole depends on electronegativity differences *and* the shape of the molecule. You may need to build a model to help you to decide if a molecule has a dipole.

You need a table of electronegativity values before you can do problems and make decisions about whether or not molecules contain dipoles. Table 4 gives the Pauling electronegativity values for some common elements. You can read more about electronegativity in **Section 3.1**.

Atom	Electronegativity
F	4.0
O	3.4
Cl	3.2
N	3.0
Br	3.0
I	2.7
S	2.6
C	2.6
H	2.2
Si	1.9

◀ **Table 4** Pauling electronegativity values for some common elements.

Dipole–dipole bonds

Let's look at the bonds between molecular dipoles, like those in water and 1,1,1-trichloroethane. These substances are liquid at room temperature, and in the liquid phase the molecules are constantly moving and tumbling around. Sometimes the negative end of one molecule will be lined up with the positive end of another, making them attract, but often there will be two negative ends or two positive ends together, in which case they will repel (Figure 6). Overall, there will be more attraction than repulsion between the permanent dipoles, which is one reason why the molecules stay together as a liquid.

In the liquid state, the molecular dipoles are constantly tumbling around.

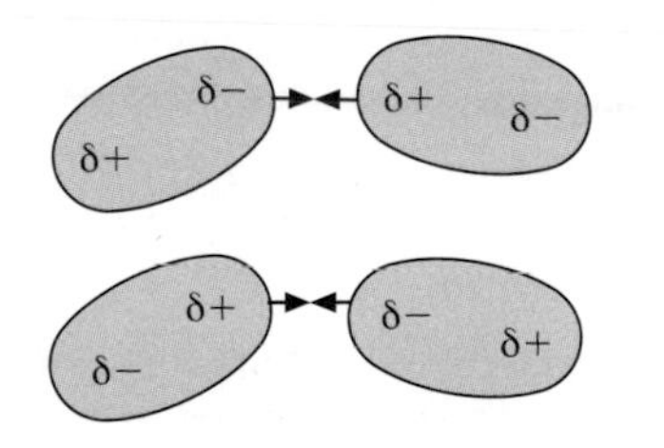

Sometimes opposite charges are next to one another, causing attraction . . .

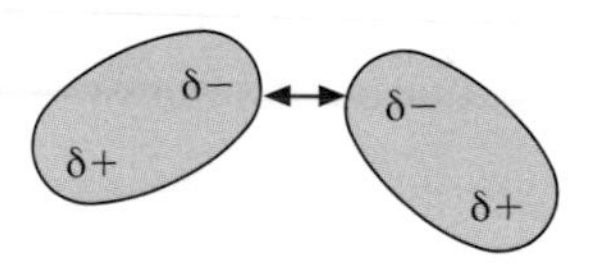

. . . sometimes like charges are next to one another, causing repulsion

◀ **Figure 6** Molecular dipoles in the liquid state.

In addition to the bonds between the permanent dipoles, there will also be some bonding between the molecules due to permanent dipole–induced dipole bonds and instantaneous dipole–induced dipole bonds. For a molecule of any given size, these will be weaker than the permanent dipole–permanent dipole bonds. However, these bonds are always attractive and they play an important part in holding the molecules in the liquid together.

In the solid state things are rather different. In a crystalline solid the molecules are in fixed positions rather than tumbling around, and it is possible for the permanent dipoles to line up with opposite charges next to one another. The bonds between these aligned dipoles help to hold the molecules together in the solid state.

Permanent dipole bonds are present in many molecular substances – for example, chloromethane (CH_3Cl), propanone (CH_3COCH_3) and poly(chloroethene) (PVC). The bonds are stronger if the size of the dipoles increase, or if the interacting dipoles are able to approach each other closely.

Summary

Instantaneous dipole–induced dipole bonds:

- are bonds which operate between all molecules and between the isolated atoms in noble gases
- are much weaker than other types of intermolecular bonds
- are the result of electrostatic attraction between temporary dipoles and induced dipoles caused by movement of electrons.

The strength of the bonds is related to the number of electrons present and hence to the size of the molecule (or isolated atom).

If we have two substances, one polar and one non-polar, whose molecules have the same relative molecular mass (and so approximately the same number of electrons), the polar substance will have a higher boiling point than the non-polar substance.

Permanent dipole–permanent dipole bonds:

- are additional electrostatic bonds between polar molecules
- are stronger than instantaneous dipole–induced dipole bonds for molecules of equivalent size.

A polar molecule is one that has a permanent dipole. The spatial arrangement of polar covalent bonds is important in determining whether or not a molecule as a whole is polar.

Problems for 5.3

1 For each pair of chemicals given below, arrange the formulae in the order in which the strength of the instantaneous dipole–induced dipole bonds increases. In each case, indicate which has the higher boiling point.

a Xe and Kr

b C_8H_{18} and C_6H_{14}

c CCl_4 and CH_4

d $CH_3—CH_2—CH_2—CH_3$ and $CH_3—CH(CH_3)—CH_3$

e $CH_3—CH_2—CH_2—CH_2—CH_2—CH_3$ and $CH_3—CH(CH_3)—CH(CH_3)—CH_3$

2 Explain why noble gases have very low boiling points.

3 Draw skeletal formulae showing how two molecules of pentane can approach close to one another. Now do the same for both of its structural isomers, 2-methylbutane and 2,2-dimethylpropane. The boiling points of the three isomers are given in the table below.

Isomer	Boiling point/K
pentane	309
2-methylbutane	301
2,2-dimethylpropane	283

a Explain the variation in boiling points of the three isomers in terms of the strength of the intermolecular bonds present.

b Account for the differences in strengths of the intermolecular bonds.

4 For a molecular substance, measuring the boiling point can give an indication of the strength of the bonds between molecules. When comparing the strength of intermolecular bonds in polar and non-polar substances, explain why it is important to compare substances of similar relative molecular mass.

5 The boiling points of some molecular substances are shown in the table below.

Substance	Relative molecular mass	Boiling point/K
A	74	390
B	98	371
C	30	252
D	78	353
E	110	513
F	56	281
G	32	338

a Which pairs of molecules could be used to compare the strength of their intermolecular bonds?

b For each pair, state the substance that has the stronger bonds between its molecules.

6 Which of the covalent bonds in the following list will be significantly polar? In cases where there is a polar bond, show which atom is positive and which is negative, using the $\delta+$ and $\delta-$ convention.

a C–F

b C–H

c C–S

d H–Cl

e H–N

f S–Br

g C–O

7 Which of the molecules listed below possesses a permanent dipole? You will need to think about the shape of each molecule, and look at the electronegativity values in Table 4 (page 97).

a CO_2
b $CHCl_3$
c C_6H_{12} (cyclohexane)
d CH_3OH
e $(CH_3)_2CO$
f benzene
g (*Z*)-1,2-difluoroethene
h (*E*)-1,2-difluoroethene
i 1,2-dichlorobenzene
j 1,4-dichlorobenzene

8 Two compounds of silicon and sulfur have the following formulae and boiling points:

SiH_4, 161 K; H_2S, 213 K

a i How many electrons are present in each molecule?
ii How will the strengths of instantaneous dipole–induced dipole bonds compare in the two compounds?
iii Does either molecule possess a permanent dipole? Explain your answer.
b Use your answers in part **a** to explain the relative sizes of the boiling points of these two compounds.

9 Of the types of intermolecular bonds listed below, identify those present in the listed substances.
- instantaneous dipole–induced dipole
- permanent dipole–permanent dipole.

Remember that more than one type of intermolecular bond may operate for each substance.

a Ra
b Cl_2
c HCl
d CH_4
e CH_3Cl
f poly(ethene)
g poly(propene)
h poly(chloroethene) (PVC)

AS

5.4 *Bonds between molecules: hydrogen bonding*

What is special about hydrogen bonding?

Hydrogen bonding is another type of intermolecular bond, similar to those that you studied in **Section 5.3**. However, it is also the strongest type of intermolecular bond and can be thought of as a special case of permanent dipole–permanent dipole bonding.

For hydrogen bonding to occur, the molecules involved must have the following three features:

- a *large dipole* between an H atom and a highly electronegative atom such as O, N or F
- the *small H atom* which can get very close to O, N or F atoms in nearby molecules
- a *lone pair of electrons* on the O, N or F atom, with which the positively charged H atom can line up.

Figure 1 illustrates hydrogen bonding in liquid hydrogen fluoride, HF.

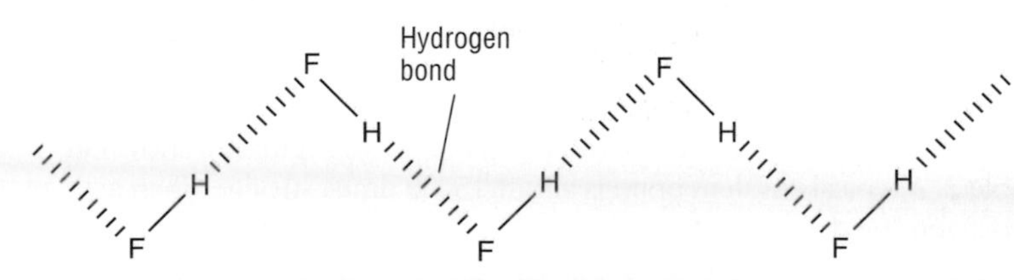

▲ **Figure 1** Hydrogen bonding in liquid hydrogen fluoride.

The H atoms have a strong positive charge because of the highly electronegative F atom to which they are bonded. This positive charge lines up with a lone pair on another F atom because it provides a region of concentrated negative charge (Figure 2). The H and F atoms can get very close, and therefore attract very strongly, because the H atom is so small.

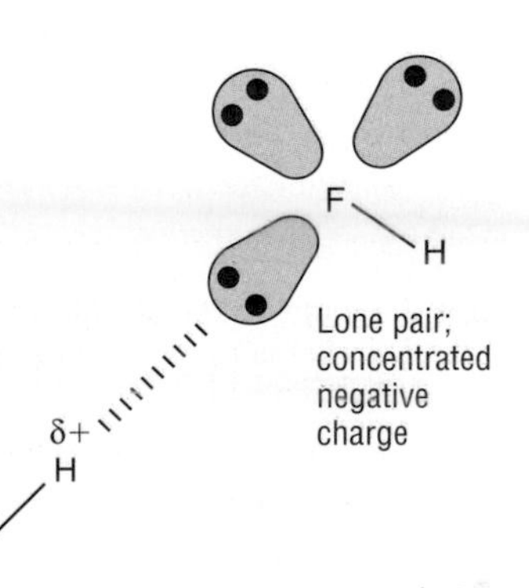

▲ **Figure 2** The positively charged H atom lines up with a lone pair on an F atom.

▼ **Table 1** Boiling points of the hydrogen halides.

Compound	Boiling point/K
HF	292
HCl	188
HBr	206
HI	238

The lining up of HF molecules is important because it means that positive and negative charges are always lined up with one another, so the forces are always attractive. As a result, the HF molecules are held together in the molecular chains shown in Figure 1.

Hydrogen bonds are much stronger than the other kinds of intermolecular bonds that we have met (instantaneous dipole–induced dipole and permanent dipole–permanent dipole bonds).

Dipole–dipole bonds are present in all the hydrogen halides (HF, HCl, HBr and HI) but the greater strength of the hydrogen bonds in HF can be seen from a comparison of the boiling points. Look at Table 1. HF has the highest boiling point despite having the lowest relative molecular mass.

Intermolecular bonds affect the melting point and boiling point of a covalent substance because energy is needed to overcome these bonds and to separate molecules when the substance melts and then boils. If the bonds *between molecules* are strong, then the melting point and boiling point of the substance tend to be high – whereas if the bonds are weak, the melting point and boiling point tend to be low. In the case of hydrogen fluoride, the strong hydrogen bonds give it an exceptionally high boiling point compared with the other hydrogen halides.

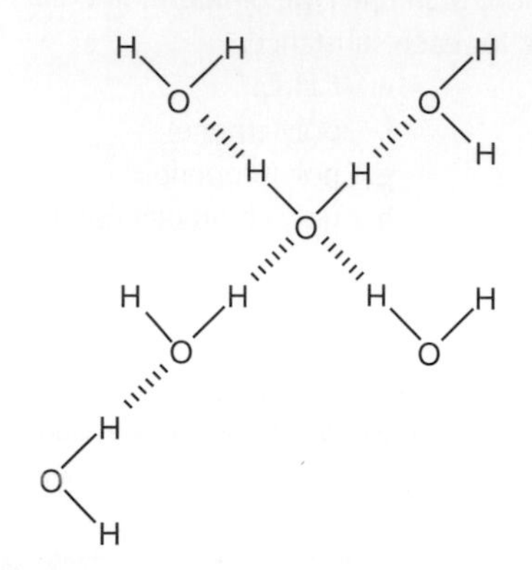

▲ **Figure 3** Hydrogen bonding between water molecules.

The hydrogen bonding in water is shown in Figure 3. Notice that water molecules can form twice as many hydrogen bonds as hydrogen fluoride. The oxygen atom possesses two lone pairs of electrons for positive hydrogen atoms to interact with. There are twice as many hydrogens as oxygens – just the right number to maximise the bonding to the lone pairs (Figure 4).

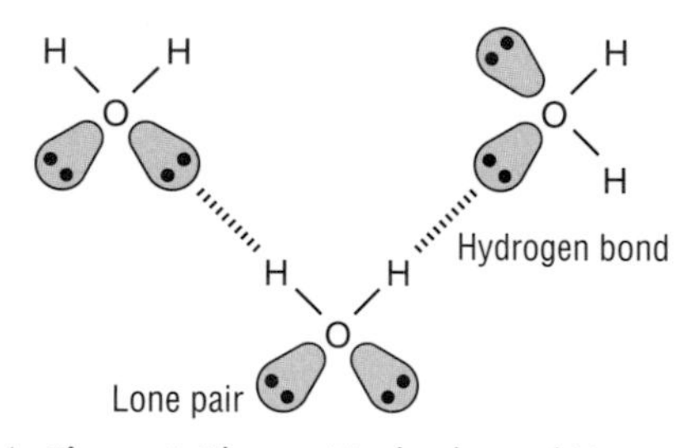

▲ **Figure 4** The positively charged H atoms line up with the lone pairs on the O atoms.

Water is unique in this respect. In HF the fluorine has three lone pairs but there are only as many H atoms as F atoms – so only one-third of the available lone pairs can be used. In NH_3, another hydrogen-bonded substance, there is only one lone pair on the N so, on average, only one of the three hydrogen atoms can form hydrogen bonds.

Summary

For hydrogen bonding to occur there has to be:

- a hydrogen atom made partially positively charged by being attached to a highly electronegative atom or group of atoms (e.g. N, F, O)
- a small, highly electronegative atom with at least one lone pair of electrons for the H atom to interact with.

N, O and F atoms are electronegative enough to satisfy these conditions (Figure 5).

Y— H ||||||||| X—

Hydrogen bond

X = F, O or N
Y = F, O or N

► **Figure 5** Hydrogen bonds only form between particular kinds of atoms.

Hydrogen bonds are stronger than other intermolecular bonds, but *much weaker* than covalent bonds. Some typical bond strengths are shown in Table 2. A typical covalent bond is about 15–20 times stronger than a hydrogen bond.

► **Table 2** Relative strengths of covalent bonds and some intermolecular forces.

Type of attraction	Bond enthalpy/kJ mol^{-1}
O–H covalent bond	+464
hydrogen bond	+10 to +40
instantaneous dipole–induced dipole bonds	$< +10$

The effects of hydrogen bonding

Many covalent substances have N–H and O–H bonds in their molecules. For example, alcohols and organic acids have O–H bonds, while amines and amides have N–H bonds. Amino acids and proteins have both types of bonds. All these substances form hydrogen bonds and hydrogen bonding plays a major role in determining their properties.

Look at Table 3. The two compounds with hydrogen bonding have higher boiling points than compounds with similar relative molecular masses that do not. Propane ($CH_3CH_2CH_3$) is a non-polar molecule and there are only instantaneous dipole–induced dipole bonds between its molecules. Ethanol (CH_3CH_2OH) is a polar molecule but it has hydrogen bonding between its molecules, in addition to instantaneous dipole–induced dipole bonds and permanent dipole–permanent dipole bonds. More energy is needed to overcome the intermolecular bonds between ethanol molecules than between propane molecules and so the boiling point of ethanol is much higher.

Compound	Formula	Relative molecular mass	Hydrogen bonding?	Boiling point/K
propane	$CH_3CH_2CH_3$	44	no	231
ethanol	CH_3CH_2OH	46	yes	351
heptane	$CH_3(CH_2)_5CH_3$	100	no	371
glycerol	CH_2OH | $CHOH$ | CH_2OH	92	yes	563

◀ **Table 3** The effect of hydrogen bonding on boiling point.

Liquids which have hydrogen bonding have a higher than expected *viscosity* – this is a measure of how easily a liquid flows. For a liquid to flow, the molecules must be able to move past each other and this requires the constant breaking and forming of intermolecular bonds. The stronger the bonds between the molecules, the more difficult this becomes. Glycerol is an example of a viscous liquid containing a lot of hydrogen bonding – each glycerol molecule contains three –OH groups.

Substances with hydrogen bonding, such as glycerol and ethanol, are often soluble in water. Hydrogen bonds can form between water molecules and molecules of the substance and this helps the dissolving process.

Hydrogen bonds are responsible for the strong intermolecular bonds between the polymer chains of many fibres. In nylon, for example, there are bonds between the H atoms of the N–H groups and the O atoms of the C=O groups.

Hydrogen bonding also helps fibres to absorb water. The protein chains in wool contain lots of O and N atoms that can hydrogen bond to water, which means that wool can absorb sweat and doesn't feel sticky to wear. Polythene, on the other hand, cannot form hydrogen bonds at all, so polythene clothes would feel very damp and sticky.

For example, in poly(ethenol)

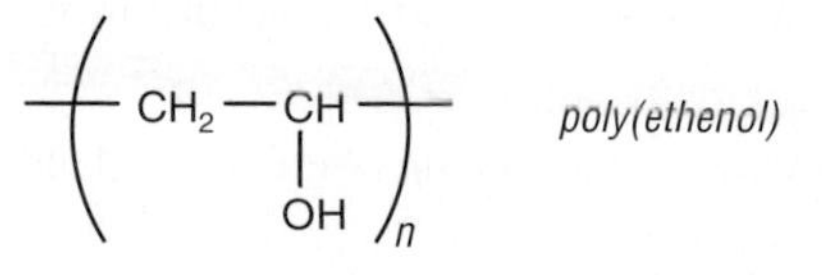

the –OH groups on the chain can hydrogen bond with water molecules and the polymer is water soluble. You can read more about how hydrogen bonding affects the solubility of polymers in the **Polymer Revolution** module.

Problems for 5.4

1 Look carefully at the structural formulae of the substances **a** to **g** given below – some of them will be unfamiliar to you. Suggest which of these substances will be able to form hydrogen bonds.

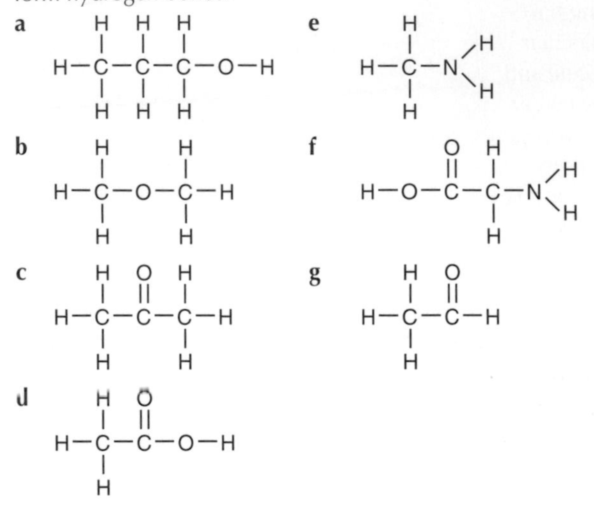

2 **a** For each of the compounds listed below state whether hydrogen bonding will be present or not.

i CH_4
ii NH_3
iii HI
iv H_2S
v CH_3OH
vi C_6H_6
vii HF
viii CH_3Cl

b Wherever you decide that hydrogen bonding is present, draw a diagram to show a hydrogen bond between two molecules of the compound.

3 Hydrogen bonding can occur between *different* molecules *in mixtures.* Draw diagrams to show where the hydrogen bonds form in the following mixtures.

a NH_3 and H_2O
b CH_3CH_2OH and H_2O
c poly(ethenol) and H_2O

4 In this problem you will need to use your knowledge of all the types of intermolecular bonds you have met in **Sections 5.3** and **5.4**. List the types of intermolecular bonds that exist in each of the following compounds – remember that more than one type of intermolecular bond may operate for each molecule. The types of bonds to choose from are:

- instantaneous dipole–induced dipole bonds
- permanent dipole–permanent dipole bonds
- hydrogen bonding.

a NH_3
b Br_2

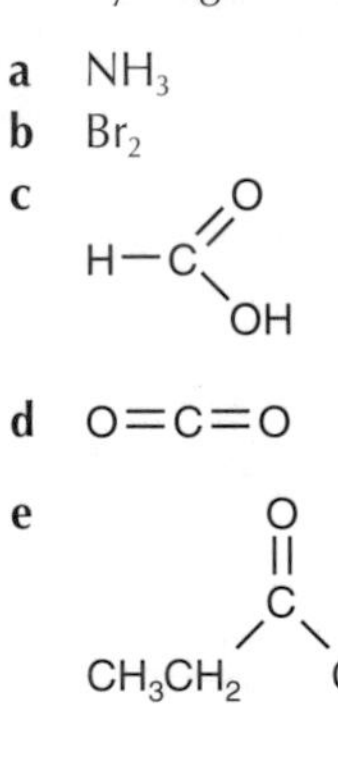

5.5 *Hydrogen bonding and water*

Unusual properties of water

You were introduced to the idea of hydrogen bonding in **Section 5.4** – we now concentrate specifically on hydrogen bonding in water. Hydrogen bonding in water is particularly strong because there are two lone pairs of electrons *and* two positively charged hydrogen atoms per oxygen atom, so intermolecular bonds are maximised (see Figures 3 and 4 in **Section 5.4**). Many of the properties of water are unusual when we compare them with those of other similar compounds of low molecular mass. In particular, water shows:

- unusually high values for its boiling point and enthalpy change of vaporisation
- a greater specific heat capacity than almost any other liquid
- a *decrease* in density when it freezes.

The graph in Figure 1 shows how the boiling points of hydrides change as you go down a group in the Periodic Table.

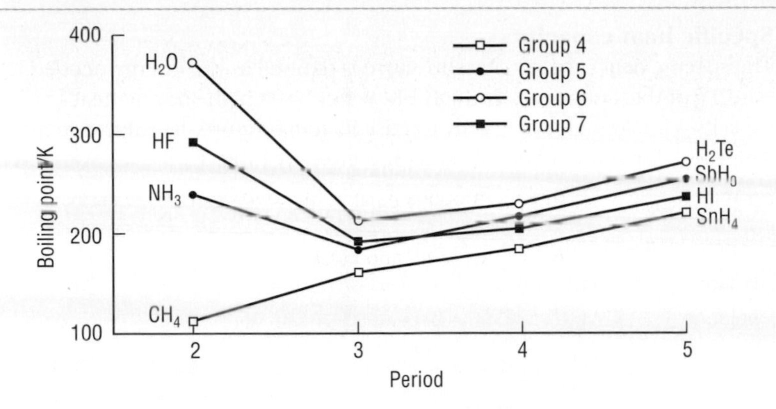

◀ **Figure 1** Variation in the boiling points of the hydrides of some Group 4, 5, 6 and 7 elements.

The data for Group 4 (CH_4 to SnH_4) show how we might *expect* the boiling points to behave. The pattern corresponds to what we know about instantaneous dipole–induced dipole intermolecular bonds and the polarisability of atoms. (These were discussed in **Section 5.3**.) The points for H_2O, NH_3 and HF are clearly out of line.

There is a similar situation for the **enthalpy changes of vaporisation**, as shown in Figure 2. The enthalpy change of vaporisation is a measure of the energy we have to put into a liquid to overcome the intermolecular bonds and turn one mole of molecules from liquid to vapour.

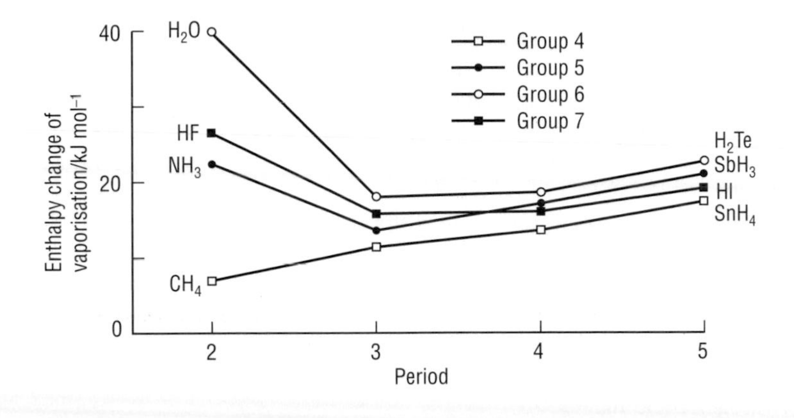

◀ **Figure 2** Variation in the enthalpy changes of vaporisation of some Group 4, 5, 6 and 7 elements.

The unusual behaviour of water, ammonia and hydrogen fluoride can be explained in terms of their ability to form hydrogen bonds between molecules. Hydrogen bonding holds molecules together more strongly than other types of intermolecular bonds, and so more energy is needed to overcome them before boiling or vaporisation can occur. This means that for compounds whose molecules are involved in hydrogen bonding, the boiling point and enthalpy change of vaporisation values are higher than might otherwise be expected. As we saw in **Section 5.4**, each water molecule can form twice as many hydrogen bonds as ammonia or hydrogen fluoride molecules and so the elevation of the boiling point above what we might expect is particularly pronounced.

Fluorine is highly electronegative, so the H–F bond is highly polarised and the molecules can form strong hydrogen bonds. Nitrogen is highly electronegative, but not as much as fluorine or oxygen so the hydrogen bonds that form (only one per molecule of ammonia) are not as effective as in water or hydrogen fluoride. Carbon and hydrogen have similar electronegativity values so the bond between them is barely polarised. No hydrogen bonding occurs in methane, so the boiling point and enthalpy of vaporisation values are low.

Specific heat capacity

The specific heat capacity of a substance is defined as the energy needed to raise 1 g of that substance through 1 K. Water has a high specific heat capacity – it takes a lot of energy to raise its temperature. Its value, together with those for some other substances, is given in Table 1.

▶ **Table 1** Some specific heat capacities.

Compound		Specific heat capacity/$J\,g^{-1}\,K^{-1}$
water	$H_2O(l)$	4.18
ethanol	$C_2H_5OH(l)$	2.45
heptane	$C_7H_{16}(l)$	2.44
copper	$Cu(s)$	0.38
mercury	$Hg(l)$	0.14

To raise the temperature of 1 g of copper by 1 K requires 0.38 J. Doing the same for water requires over 10 times this quantity of energy. Water is an excellent substance for absorbing and storing energy, which is just as well for life in and around the oceans – and for the manufacturers of hot water bottles!

Water's unusually high specific heat capacity is another consequence of its hydrogen bonding. A lot of energy is used in overcoming hydrogen bonding within clusters of water molecules. So this energy is not available for increasing the kinetic energy of the molecules – it is not available for raising the temperature. In liquids with no hydrogen bonding between molecules, a greater proportion of the energy can be used for increasing the kinetic energy of the molecules, so raising the temperature – such liquids have lower specific heat capacities.

The density of water and ice

If we take some water at room temperature and cool it to 4 °C (277 K), the water contracts and its volume decreases. Now, most substances contract on cooling so this is not unexpected. What is unusual is what happens next – the water *expands* as its temperature approaches 0 °C (273 K) and there is a further expansion as water freezes.

Figure 3a shows how the *density* of water changes with temperature around the freezing point. A similar graph for heptane is shown in Figure 3b. The changes in the density of heptane are typical of most solids and liquids. The changes for water are very unusual and are due to the effects of hydrogen bonding.

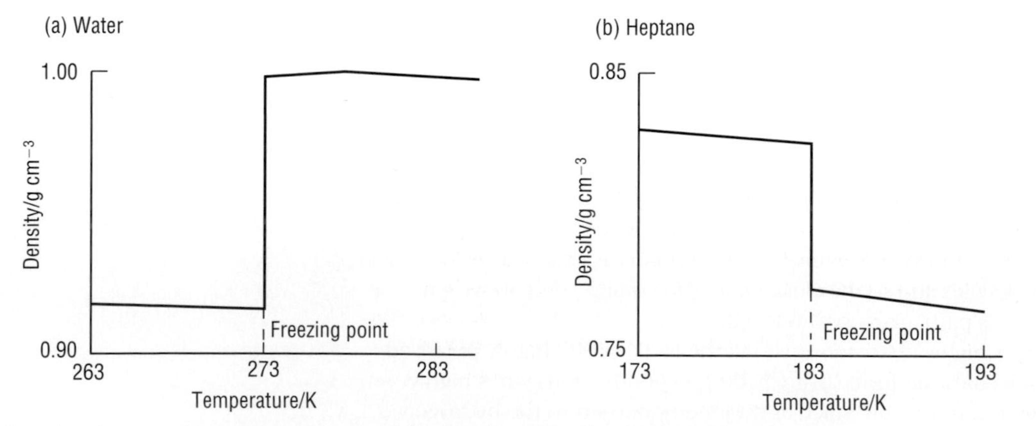

▲ **Figure 3** Density changes at temperatures around the freezing point for water and heptane.

The structure of ice is shown in Figure 4. Note that the four groups around each oxygen atom are arranged tetrahedrally in three dimensions. This arrangement of water molecules maximises the hydrogen bonding between them, but leads to a very 'open' structure with large spaces in it. Therefore, the density of ice at 273 K is less than the density of water at the same temperature. When ice melts, the open structure collapses and water molecules fall into some of the open spaces, thus giving water a greater density than the ice it came from.

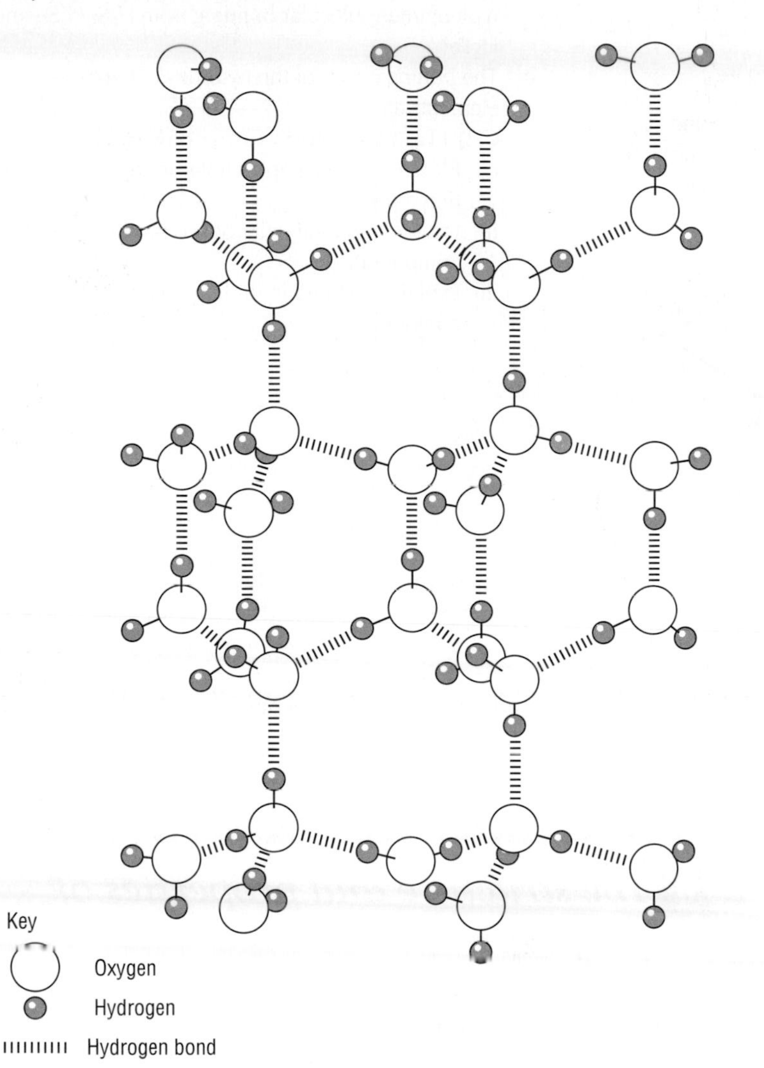

◀ **Figure 4** The arrangement of water molecules in ice.

When water is cooled, it seems that regions in the liquid begin to adopt the open structure of ice before the freezing point at 273 K is reached. This explains why the fall in density begins as the liquid is cooled below 277 K.

The lower density of ice is bad news for people who fail to insulate exposed water pipes in cold weather, or who forget to add antifreeze to the coolant system of their car – but it is good news for fish! Icebergs float and ponds freeze from the top down. The layer of surface ice insulates the water underneath. If ice sank then there would be no insulation – ponds and lakes would be more likely to freeze solid, killing fish and other creatures. An occasional burst pipe is a small price to pay for the environmental advantages of the expansion of freezing water.

Problems for 5.5

1 a Why does the density of most solids and liquids decrease with increasing temperature?

b Explain the following properties of water in terms of hydrogen bonding:

i the greater density of liquid water compared with ice

ii its relatively high boiling point

iii its relatively high specific heat capacity.

2 a The graph below shows the boiling points for the hydrides of Group 6 elements – H_2O, H_2S, H_2Se, H_2Te.

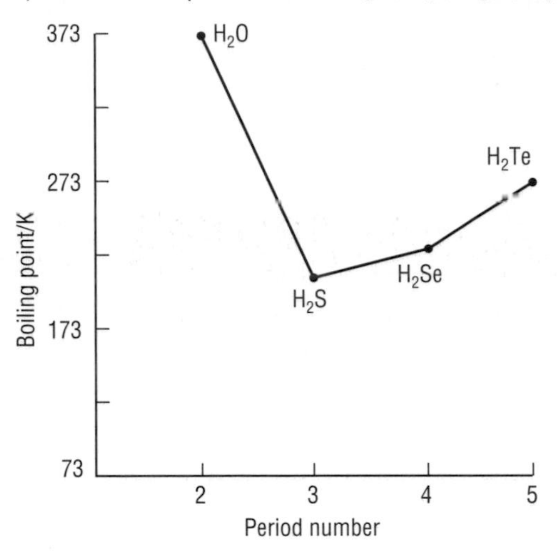

i List the intermolecular bonds present in each of the compounds.

ii Explain why the boiling point of water is so high.

b Why does the shape of the graph for the hydrides of Group 6 elements suggest that H_2O has a different type of intermolecular bonding from H_2S, H_2Se and H_2Te?

c The boiling points of the hydrides of Group 4 elements are

CH_4 112 K; SiH_4 161 K; GeH_4 185 K; SnH_4 221 K

i Plot these on a graph similar to that for the Group 6 hydrides.

ii List the intermolecular bonds present in each compound.

iii Explain the trend in boiling points for these compounds.

5.6 *The structure and properties of polymers*

To support your work in the **Polymer Revolution** module, you will need to read this section about addition polymers.

What is a polymer?

A **polymer** molecule is a long molecule made up from lots of small molecules called **monomers**. If all the monomer molecules are the same, and they are represented by the letter A, an A–A polymer forms:

$$-----A + A + A + A----- \rightarrow -----A\text{–}A\text{–}A\text{–}A-----$$

Poly(ethene) and poly(chloroethene), PVC, are examples of A–A polymers. If two different monomers are used, an A–B polymer may be formed, in which A and B monomers alternate along the chain:

$$-----A + B + A + B----- \rightarrow -----A\text{–}B\text{–}A\text{–}B-----$$

Polyamides (nylons) and polyesters are examples of this type of A–B polymer.

Addition polymerisation

Many polymers are formed in a reaction known as **addition polymerisation**. The monomers usually contain C=C double bonds, for example in alkenes. The addition polymerisation of propene is typical:

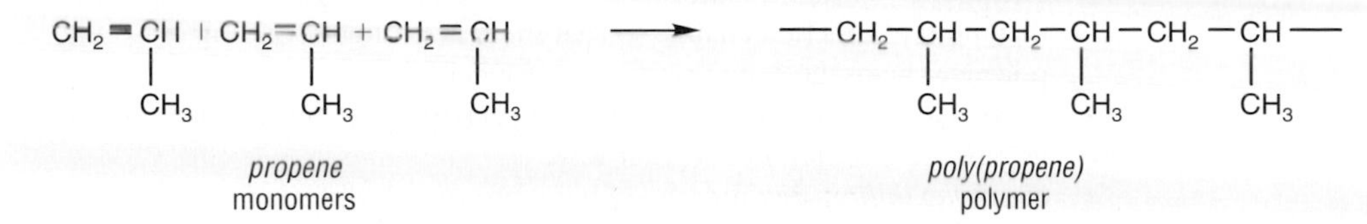

In the chain, the same basic unit is repeated over and over again, so the polymer structure can be shown simply as:

$$\left(CH_2 - \underset{\displaystyle CH_3}{\underset{|}{CH}} \right)_n$$

n = a very large number

The repeating unit of *poly(propene)*

Sometimes more than one monomer is used in addition polymerisation. For example, if some ethene is added to the propene during the polymerisation process, both monomers become incorporated into the final polymer. This is called **copolymerisation**. A section of the copolymer chain could look like this:

$$-CH_2-\underset{\displaystyle CH_3}{\underset{|}{CH}}-CH_2-\underset{\displaystyle CH_3}{\underset{|}{CH}}-CH_2-CH_2-CH_2-\underset{\displaystyle CH_3}{\underset{|}{CH}}-CH_2-\underset{\displaystyle CH_3}{\underset{|}{CH}}-$$

Elastomers, plastics and fibres

Polymer properties vary widely. Polymers that are soft and springy, which can be deformed and then go back to their original shape, are called **elastomers**. Rubber is an elastomer.

Poly(ethene) is not so springy and when it is deformed it tends to stay out of shape, undergoing permanent or plastic deformation. Polymers like this can be incorporated in materials called **plastics**.

Stronger polymers, which do not deform easily, are just what is needed for making clothing materials. Some can be made into strong, thin threads which can then be woven together. These polymers, such as nylon, can be used as **fibres**.

Poly(propene) is on the edge of the plastic/fibre boundary. It can be used as a plastic, like poly(ethene), but it can also be made into fibres for use in carpets.

What decides the properties of a polymer?

The physical properties of a polymer, such as its strength and flexibility, are decided by the characteristics of its molecules. In general, the stronger the attractive bonds are between polymer chains then the stronger, and less flexible, the polymer will be. For a polymer to be flexible the polymer chains must be able to slide past each other easily.

In poly(ethene) the attractive bonds between the polymer chains are weak instantaneous dipole–induced dipole bonds. In poly(chloroethene) there are additional permanent dipole–permanent dipole bonds due to the polar side groups. In nylons hydrogen bonding can also occur between polymer chains.

The particularly important characteristics of polymers are:

- *chain length* – in general, the longer the chains the stronger the polymer
- *side groups* – polar side groups (such as –Cl) give stronger bonding between polymer chains making the polymer stronger
- *branching* – straight, unbranched chains can pack together more closely than highly branched chains, so that there is stronger bonding between polymer chains, and the polymer is stronger
- *stereoregularity* – polymer chains can pack together more closely if side groups are oriented in a regular way
- *chain flexibility* – hydrocarbon chains are very flexible and can take up many different orientations; changes that make the chain more rigid lead to stronger polymers. (You will meet examples of this at A2 – where benzene rings are part of the polymer chain, for example.)
- *cross-linking* – if polymer chains are linked together extensively by covalent bonds then the polymer is harder and more difficult to melt; thermosetting polymers have extensive cross-linking.

Thermoplastics and thermosets

We can put polymers into two groups, depending on how they behave when they are heated.

Thermoplastics

These are materials which contain polymers with no cross-links between the chains. The intermolecular bonds between the chains are much weaker than the covalent cross-links in a **thermoset** and so the bonds can be broken by warming. The chains can then move relative to one another (Figure 1a).

When this happens the polymer changes its shape – it is deformed – and thermoplastics can be moulded. When the polymer is cool the intermolecular bonds reform, but between different atoms.

Over 80% of all the plastics produced are thermoplastics. This includes all poly(alkenes) such as poly(ethene), polyesters such as Terylene, and polyamides such as nylon.

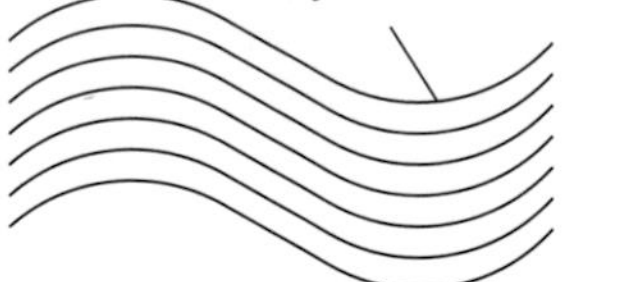

HEAT

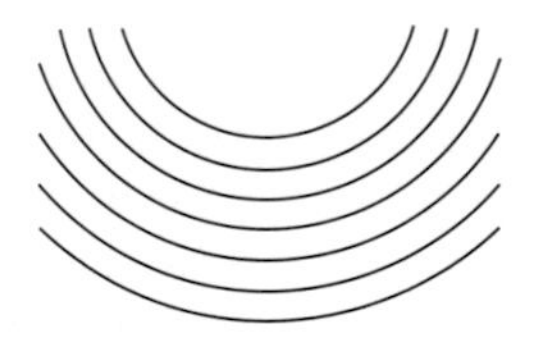

(a) Thermoplastic: no cross-linking

Weak forces between polymer chains easily broken by heating; polymer can be moulded into new shape

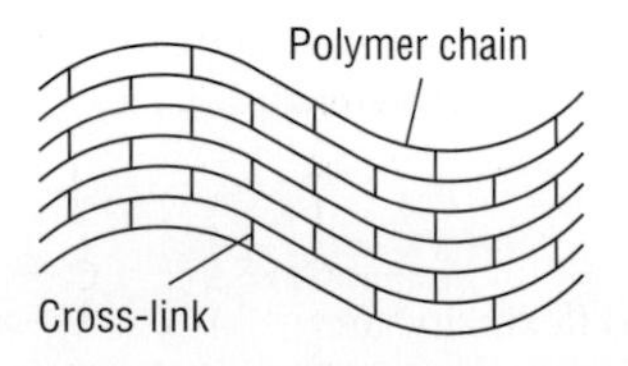

HEAT

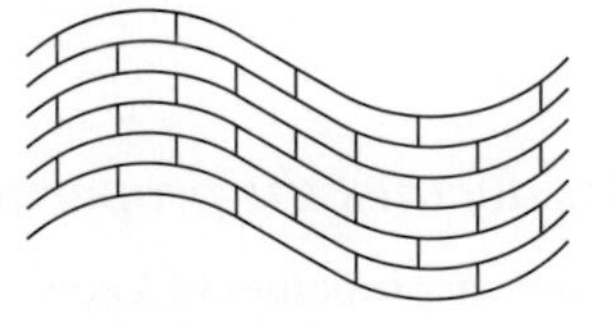

(b) Thermoset: extensive cross-linking

Strong covalent bonds between polymer chains cannot be easily broken; polymer keeps shape on heating

► **Figure 1** Thermoplastics and thermosets.

Thermosetting polymers

Also called **thermosets**, these materials have extensive cross-linking. The cross-links prevent the chains moving very much relative to one another, so the polymer stays set in the same shape when it is heated (Figure 1b).

Bakelite is an example of a thermosetting polymer. It is made from phenol and methanal. When heated together, a three-dimensional structure is formed. This is a network with strong covalent bonds (Figure 2).

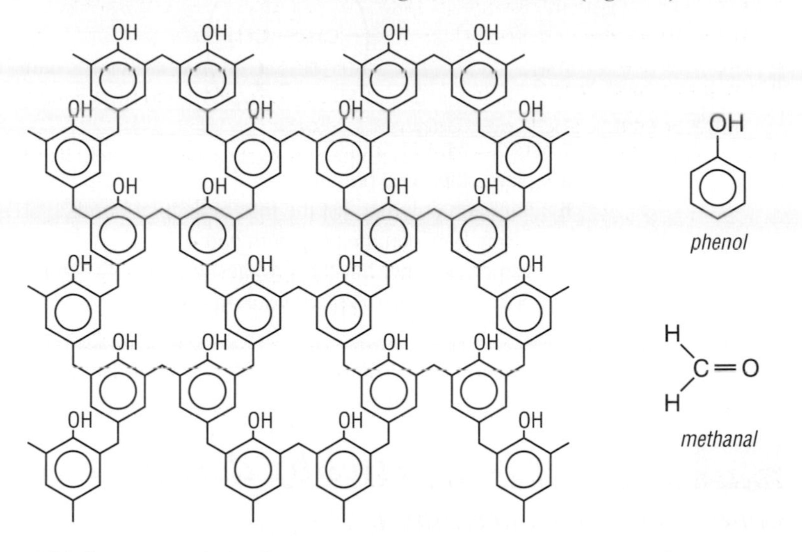

◄ **Figure 2** The structure of Bakelite, a thermosetting polymer. This is only one possible arrangement. Remember that ∧ is a shorthand way of writing $-CH_2-$.

This huge cross-linked structure requires a massive amount of energy to break it down because covalent bonds would have to be broken. Instead of melting when it is heated, Bakelite chars and decomposes.

The structure would also have to be broken down into much smaller pieces before it could dissolve. Bakelite is a very insoluble substance because the energy needed to do this is not normally available.

Bakelite was the first synthetic polymer to be made. It is still used occasionally – for example in the brown plastic used in some electrical insulators.

Problems for 5.6

1 Write down the structure of a length of polymer formed from 6 monomer units for:

a poly(ethene) **b** poly(propene)

2 The polymer PVC is made from a monomer with the structure $H_2C{=}CHCl$. Write down the structure of a length of polymer formed from 6 monomer units.

3 In the section of the copolymer chain shown on page 107 how many monomer units are there of:

a ethene? **b** propene?

4 From the structures of the polymers shown below, draw the structural formula of the monomer used in each case.

a

```
     H   H   H   H
     |   |   |   |
  —C—C—C—C—
     |   |   |   |
     H  CN   H  CN
```

b

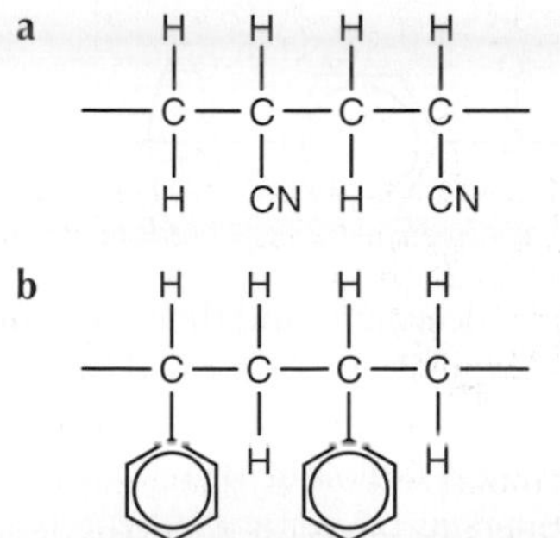

c

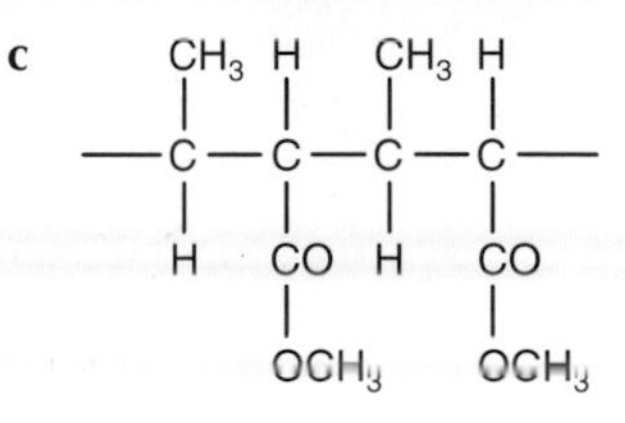

5 For each of the polymers in problem 4, draw the structure of the repeating unit.

6 Name the two alkenes used to make the following copolymer:

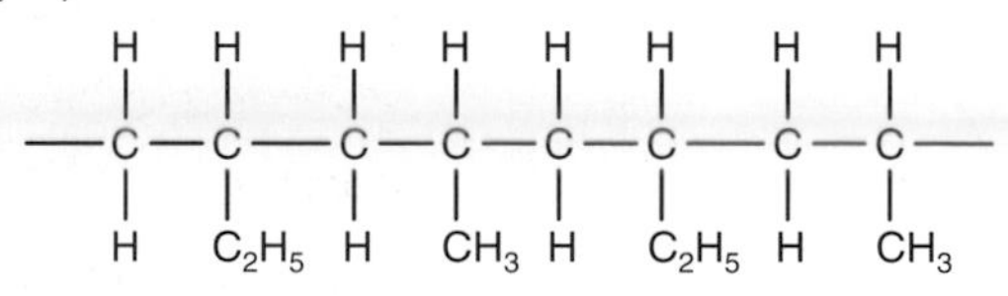

7 Draw a section of the polymer chain formed when chloroethene and ethenyl ethanoate copolymerise in an addition polymerisation reaction to form an A–B polymer.

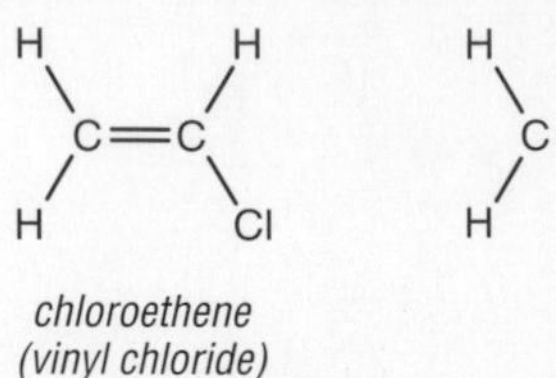

chloroethene (vinyl chloride)

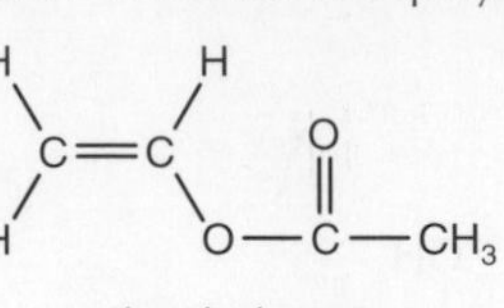

ethenyl ethanoate (vinyl acetate)

8 Three addition polymers have the general structure shown below.

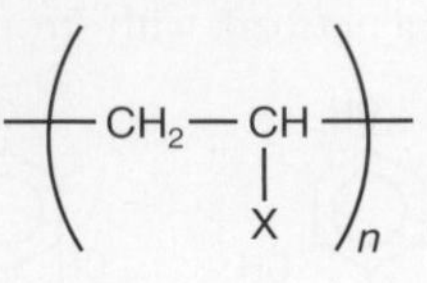

where X = H, CH_3 and Cl.

a Name the three polymers.

b For each polymer, list the intermolecular bonds acting between the polymer chains.

c Which polymer has the strongest bonds between the polymer chains? Explain your answer.

5.7 Polymer properties by design

Condensation polymerisation

In addition polymerisation, the monomers join together to give a polymer *only*. In a condensation reaction the monomers join together to give a polymer *and* another small molecule.

Polymers can be formed by addition reactions (**Section 5.6**) and also by condensation reactions – this is the same type of reaction that is used to make esters. Polymers formed from condensation reactions are known as *condensation polymers*. The monomers must have at least two suitable functional groups per molecule for a condensation polymer to be produced. During condensation polymerisation, the small monomers react together to give a longer chain polymer *and* a small stable molecule – usually water or hydrogen chloride.

Terylene, a **polyester**, is a typical condensation polymer. It is formed from two monomers, as shown below:

n HO — CH_2CH_2 — OH + n

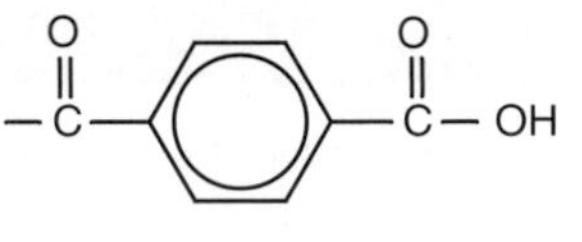

ethane-1,2-diol *benzene-1,4-dicarboxylic acid (terephthalic acid)*

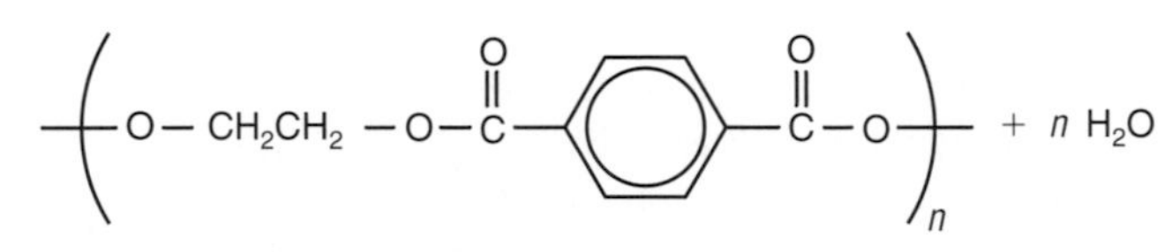

Terylene

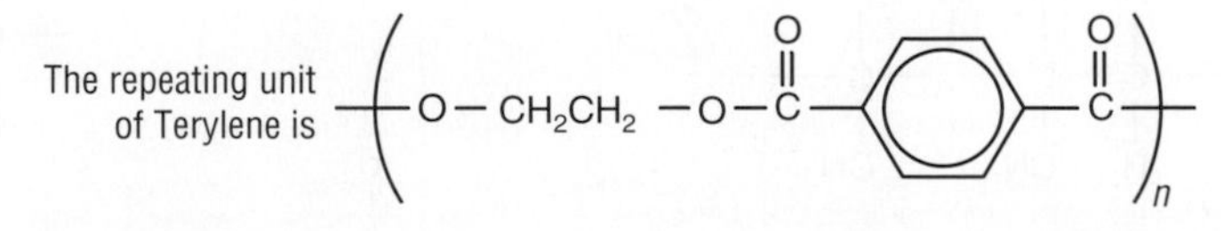

Nylon and Terylene are examples of linear polymers, and their structure makes them ideal for use as **fibres**. Nylons are polyamides formed by the condensation polymerisation of a diamine with a dicarboxylic acid. For example, when 1,6-diaminohexane and hexane-1,6-dioic acid are polymerised, nylon-6,6, a polyamide, is formed and water is eliminated.

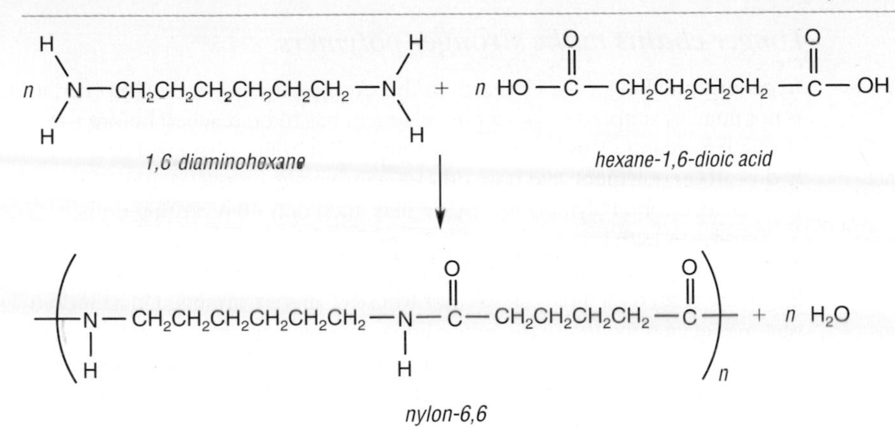

How are polymers affected by temperature changes?

When you heat solids made of small molecules, they simply melt to form free flowing liquids. Carry on heating and eventually the liquid boils.

With polymers it is not so simple. You may have seen a piece of rubber tubing cooled in liquid nitrogen and then hit with a hammer. The rubber shatters – it has become brittle or **glassy**. Or, you may have used a margarine tub for storing food in the freezer. Some of these tubs are made of poly(propene), which becomes brittle at around −10 °C. When you try to get the frozen food out, the tub splits apart – quite unlike a similar tub taken from the fridge.

The structure of many polymers is a mixture of ordered (crystalline) regions and random (amorphous) regions. In the glassy state, the tangled polymer chains in the amorphous regions are 'frozen' so that easy movement of the chains is not possible. If the polymer has to change shape it does so by breaking. At lower temperatures, the intermolecular bonds have a greater effect on properties than at higher temperatures.

If you heat up the glassy material, the polymer chains will reach a temperature at which they can move relative to each other. This temperature is called the **glass transition temperature** (T_g). When the polymer is warmer than this, it becomes flexible and shows the typical plastic properties we expect of polymers.

On further heating, the **melting temperature** (T_m) is reached, the crystalline areas break down and the polymer becomes a viscous fluid. These processes are reversible for thermoplastics. A graph of strength against temperature is shown in Figure 1.

Today's polymers are designed so that they have T_g and T_m values which are suitable for the particular properties needed by a manufacturer.

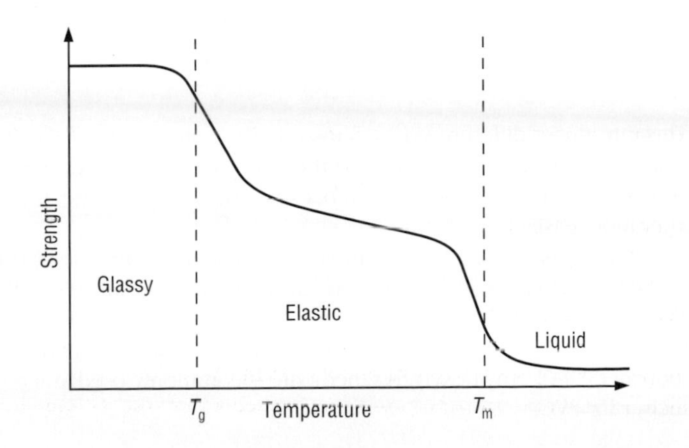

Figure 1 A graph of strength against temperature for a typical part-crystalline polymer.

Longer chains make stronger polymers

In general, the longer the chains then the stronger the polymer. However, it is not quite as simple as this – a critical length has to be reached before the strength increases. This length is different for different polymers. For hydrocarbon polymers, like poly(ethene), an average of at least 100 repeating units is necessary. Polymers like nylon may need only 40 repeating units.

Tensile strength is a measure of how much you have to pull on a sample of a polymer before it snaps. Figure 2 shows how tensile strength and chain length are related for a typical polymer whose chains are arranged in a random way. You get a reminder of the rather limited tensile strength of poly(ethene) whenever you overload your polythene carrier bag with heavy shopping.

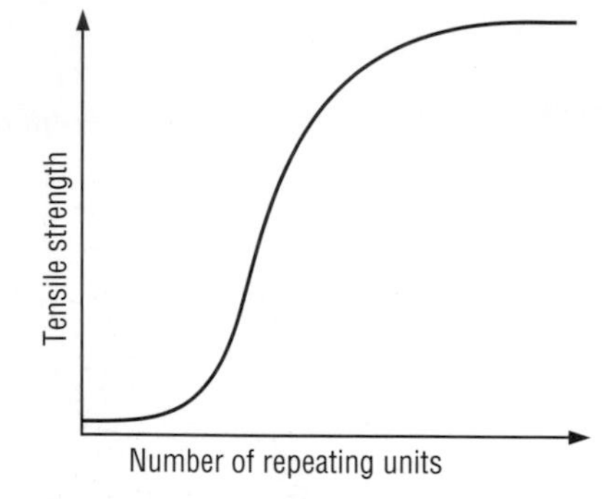

▲ **Figure 2** The relationship between tensile strength and chain length for a polymer.

There are two factors that lead to this increase in tensile strength of a polymer with increasing chain length:

- longer chains are more tangled together – this makes it difficult for the chains to slide over each other, so reducing flexibility.
- when chains are longer they have more points of contact with chains of neighbouring polymer molecules. There are, therefore, more intermolecular bonds to hold the chains to one another. These bonds *between molecules* are quite weak for poly(ethene) but are much stronger for some other polymers, such as nylon. An increase in the number of intermolecular bonds means that the chains are attracted to each other more strongly. This means it is more difficult for the chains to slide over each other – this means that the polymer is less flexible.

You can remind yourself about the relationship between structure and strength of polymers in **Section 5.6**.

Matching polymer properties to needs

The wide range of uses of polymers means that polymers with different glass transition temperatures are needed. Two important ways of changing T_g involve *copolymerisation* and *plasticisers*.

Pure poly(chloroethene) – commonly called PVC – has a T_g of about 80 °C. It is rigid and quite brittle at room temperature. This is the form used to make items such as drainpipes. It is sometimes called *unplasticised* PVC, or uPVC.

If it is to be made more flexible, its T_g value must be lowered. One way of doing this is to **copolymerise** the chloroethene with a small amount of ethenyl ethanoate.

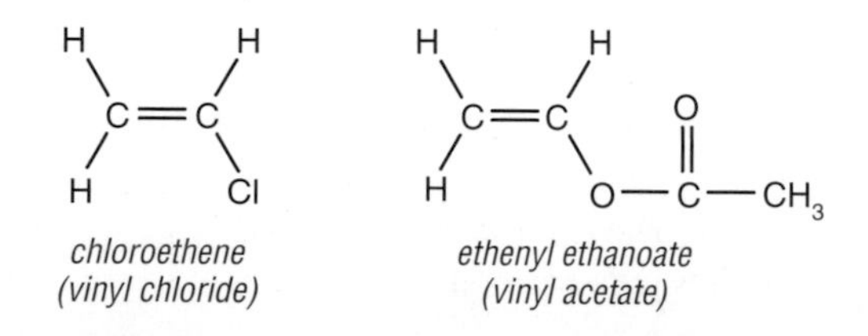

This introduces different side groups into the polymer chain. As a result, the chains pack together less well and the bonds between the chains are weaker. The polymer is more flexible because the chains can move over one another more easily.

Another way is to use a kind of 'molecular lubricant'. This allows the PVC chains to slide around each other more easily. Such a substance is called a **plasticiser**. Figure 3 shows a plasticiser in place between polymer chains.

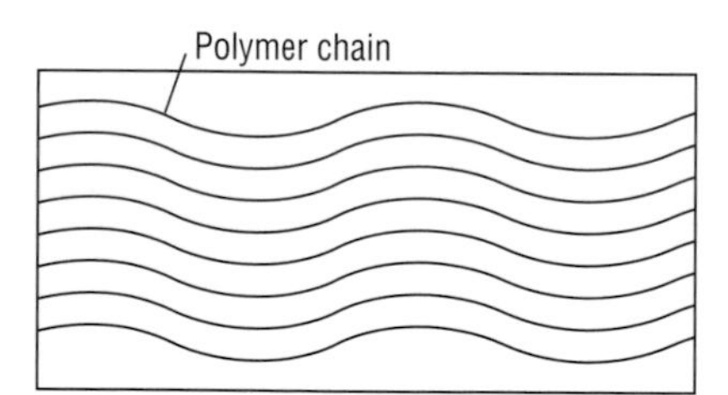

(a) Unplasticised
PVC chains attract strongly and cannot slide past each other. The polymer is rigid

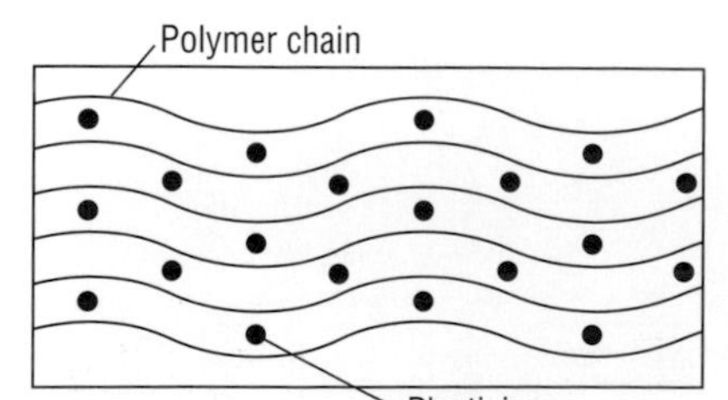

(b) Plasticised
Plasticiser molecules push PVC chains apart and help them slide. The polymer is flexible

▲ **Figure 3** Using a plasticiser to increase flexibility.

Plasticisers have to be chosen carefully so that they are compatible with the polymer. Di-(2-ethylhexyl) hexanedioate is commonly used as a plasticiser for PVC.

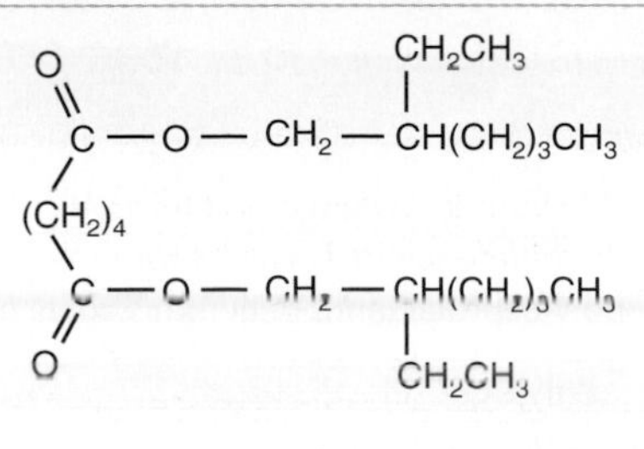

di-(2-ethylhexyl) hexanedioate

It is important that the plasticisers added to PVC film used for wrapping food do not dissolve into fatty foods as they may be harmful to health.

Crystalline polymers

You probably think of *crystals* as regular, flat sided solids, such as copper(II) sulfate and sodium chloride crystals. These crystals have a regular shape externally because the ions are packed inside in a very ordered way. Polymer chemists use the term **crystalline** to describe *the areas in a polymer* where the chains are closely packed in a regular way.

Many polymers contain mixtures of crystalline (ordered) regions and amorphous (random) regions, where the chains are further apart and have more freedom of movement. Any one polymer chain may be involved in both crystalline and amorphous regions along its length. Figure 4 shows the crystalline and amorphous regions in a polymer.

The polymers most likely to contain extensive crystalline regions are those with regular chain structures, such as isotactic poly(propene), and without bulky side groups or extensive chain branching, such as high density poly(ethene).

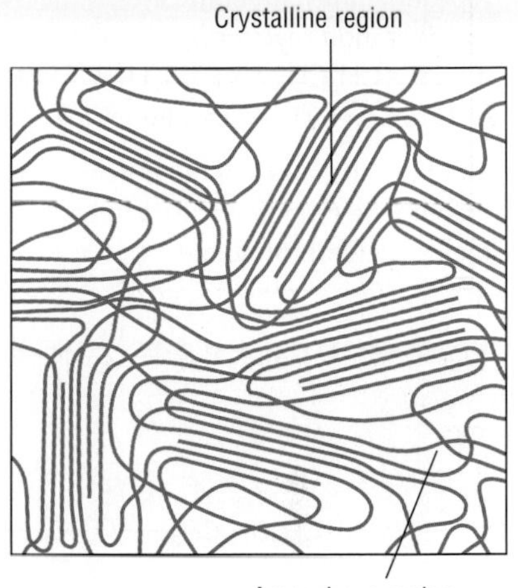

▲ **Figure 4** Crystalline and amorphous regions in a polymer.

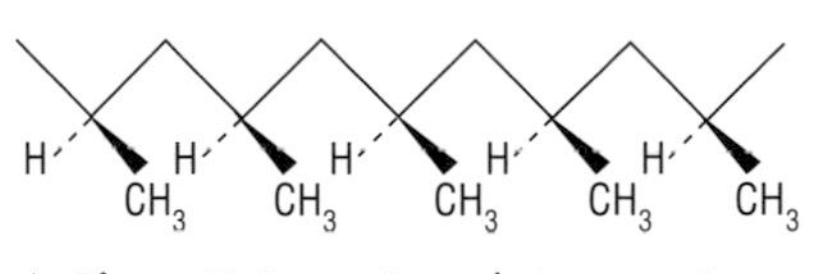

▲ **Figure 5** Isotactic *poly(propene).*

▲ **Figure 6** High density *poly(ethene).*

The proportion of crystallinity in a polymer is very important in determining its properties. The more crystalline the polymer, the stronger and less flexible it becomes. The reason for this is that in the crystalline form the polymer chains pack closely together, maximising intermolecular bonds. This means that the chains will slide over each other less easily and the polymer will be more rigid than polymers with large amorphous regions. In these amorphous regions the polymer chains do not pack closely together and the intermolecular bonds are very weak. This means the polymer chains will slide over each other easily and this type of polymer is flexible.

Cold-drawing

All polymers contain some crystalline regions. When the polymer is stretched (cold-drawn) a *neck* forms (see Figure 7). In the neck the polymer chains are lined up to form a more crystalline region. The process carries on until all the polymer, except the ends which are being pulled, has aligned chains.

Since the proportion of crystalline regions is increased, cold-drawing leads to a significant increase in the polymer's strength. It is used to produce tough fibres.

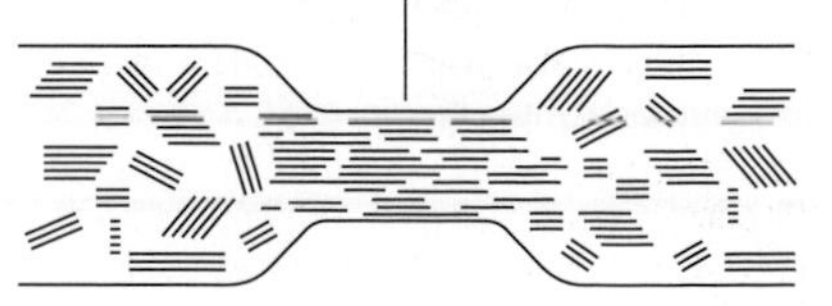

▲ **Figure 7** When a polymer is cold-drawn, a neck forms in which the polymer chains become aligned.

Problems for 5.7

To answer the problems in this section, you will need to know about both addition polymerisation (**Section 5.6**) and condensation polymerisation. You may also need to look back to **Sections 5.3** and **5.4** on intermolecular bonds.

1 Draw the structures of the condensation polymers formed from the following pairs of monomers, showing *two* repeating units. (In both cases, water is eliminated in the condensation reaction.)

a $HO–CH_2CH_2–OH$ and $HOOC–CH_2–COOH$

b $H_2N–CH_2CH_2–NH_2$ and $HOOC–CH_2CH_2–COOH$

2 From the structures of the polymers shown below, identify the monomers used in each case (by drawing their structural formulae).

a

b

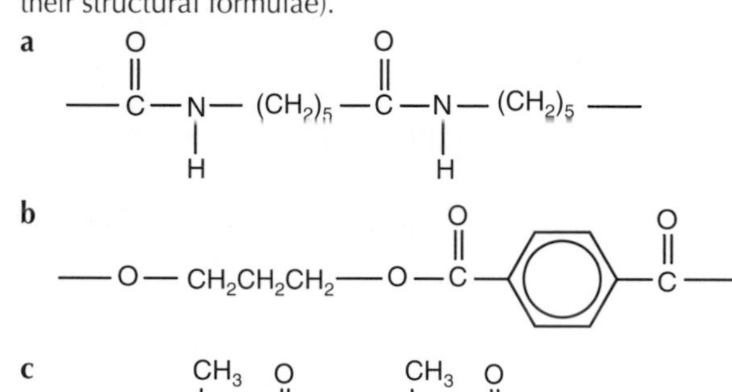

c

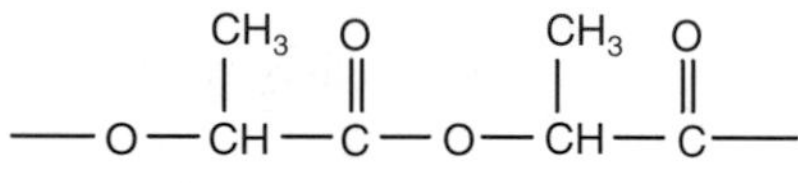

3 Poly(propene), poly(chloroethene) and poly(propenenitrile) all have side chains of similar size:

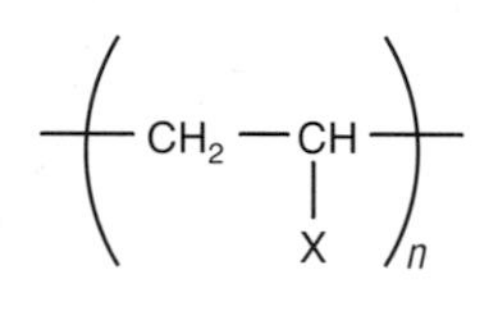

where X is CH_3, Cl or C≡N

Explain the differences in their T_g values in terms of the intermolecular bonds in the polymers.

Polymer	T_g/K
poly(propene) (X is CH_3)	258
poly(chloroethene) (X is Cl)	353
poly(propenenitrile) (X is C≡N)	378

4 a List the intermolecular bonds present in the two polymers below.

i $-[CH_2-CH_2]_n-$

ii $-[CH_2O]_n-$

b Use your answer to part **a** to decide which polymer has the higher T_g value.

5 Use your knowledge of intermolecular bonds to explain the difference in T_g values for poly(caprolactone) and poly(caprolactam), better known as nylon-6.

Polymer	T_g/K
poly(caprolactone) $-[(CH_2)_5COO]_n-$	213
Poly(caprolactam) $-[(CH_2)_5CONH]_n-$	333

6 You may find it helpful to make models to assist you in answering this problem.

a Explain the difference in T_m values for the following two nylons by referring to their structures and the bonds between the polymer chains.

	Polymer	T_m/K
i	nylon-6	486
ii	nylon-11	457

b Explain why nylon-6,10 is more flexible than nylon-6,6.

5.8 *Bonding, structure and properties: a summary*

The properties of substances are decided by their **bonding** and their **structure**.

- Bonding means the way the atoms are held together – you met ionic, covalent and metallic bonding in **Section 3.1**, and learned more about intermolecular bonding in **Sections 5.3** and **5.4**.
- Structure means the way the atoms are arranged relative to one another – you have already met the major types of structure in different parts of this course (see Table 1); they are summarised in Figure 1.

Type of structure	Where to look
ionic lattice	Section 5.1
metallic lattice	Sections 3.1 and 11.5
covalent molecules and networks	Section 5.2
macromolecules (e.g. polymers)	Sections 5.6 and 5.7

◀ **Table 1** Where to find out about types of structures.

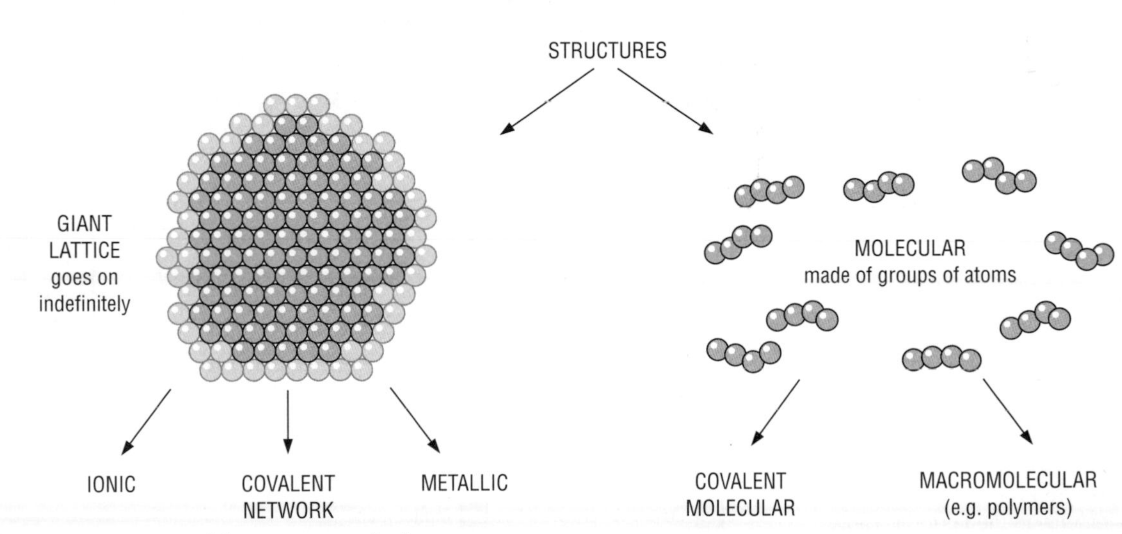

▲ **Figure 1** A summary of the structures of substances.

How do bonding and structure decide properties?

The properties of a solid substance are decided by three main factors.

- **The type of particles it contains** – these may be atoms, ions or molecules. For example, if the substance contains ions, as in an ionic substance such as sodium chloride, it will conduct electricity when molten or dissolved. If a substance contains ions or polar molecules, it may dissolve in water.
- **The way the particles are bonded together** – this may be ionic, covalent, metallic bonding or involve weak intermolecular bonds. The stronger the bonds then the higher the melting and boiling points of the substance and the greater its hardness.

For example, silica, SiO_2, has strong covalent bonds linking every atom to several others in a network structure that goes on indefinitely. This makes the atoms difficult to separate – silica is hard and very difficult to melt.

On the other hand, although carbon dioxide, CO_2, has strong covalent bonds between the C and the O atoms – so it doesn't decompose into carbon and oxygen – it has only weak intermolecular bonds between the individual molecules. This makes the *molecules* easy to separate, so CO_2 has low melting and boiling points and is a gas at room temperature.

- **The way the particles are arranged relative to one another** (the structure) – the particles may be arranged in one-dimensional chains (like poly(ethene)), two-dimensional sheets (like clays) and many different kinds of three-dimensional arrangements. For example, polymers (with their one-dimensional molecules) are often flexible, while silica (with its three-dimensional network structure) is very hard and gritty.

Table 2 summarises the different types of structures and their main properties. It also indicates where you can find out more details about each type.

Structure and the Periodic Table

As you go across a period of the Periodic Table, there is a trend evident in the structures of the elements.

Figure 2 shows the first 20 elements as they are arranged in the Periodic Table, and the type of structure shown by each element.

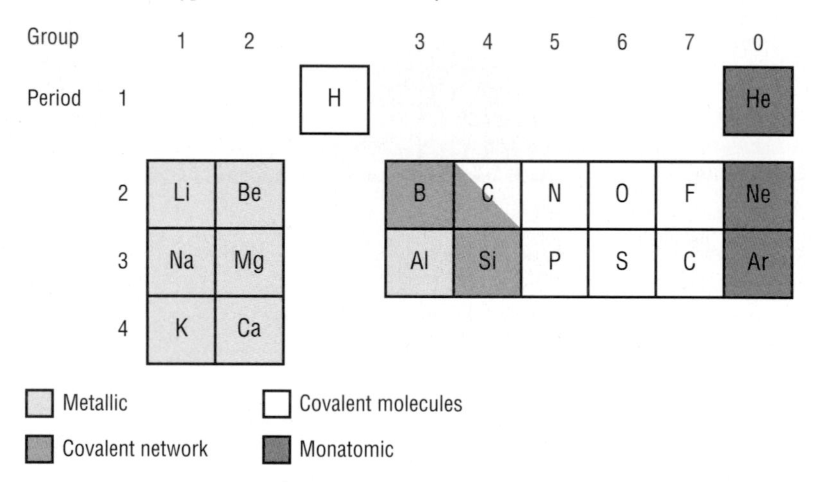

▶ **Figure 2** Trends in the structures of the first 20 elements in the Periodic Table.

Remember that carbon exists in different forms. Two of these (diamond and graphite) are covalent network structures, whereas the fullerenes are molecular. Note the trend across each period from metallic structures on the left, to covalent network structures in the centre, to covalent molecular structures, and finally to the monatomic noble gases.

▼ **Table 2** A summary of the various types of structures and bonding.

	Giant lattice			Covalent molecular	
	Ionic	**Covalent network**	**Metallic**	**Simple molecular**	**Macromolecular**
What substances have this type of structure?	compounds of metals with non-metals	some elements in Group 4 and some of their compounds	metals	some non-metal elements and some non-metal/non-metal compounds	polymers
Examples	sodium chloride, NaCl; calcium oxide, CaO	diamond C; graphite, C; silica, SiO_2	sodium, Na; copper, Cu; iron, Fe	carbon dioxide, CO_2; chlorine, Cl_2; water, H_2O	poly(ethene), nylon, proteins, DNA
What type of particles does it contain?	ions	atoms	positive ions surrounded by delocalised electrons	small molecules	long-chain molecules
How are the particles bonded together?	strong ionic bonds; attraction between oppositely charged ions	strong covalent bonds; attraction of atoms' nuclei for shared electrons	strong metallic bonds; attraction of atoms' nuclei for delocalised electrons	weak intermolecular bonds between molecules; strong covalent bonds between the atoms within molecules	weak intermolecular bonds between molecules; strong covalent bonds between the atoms within molecules
What are the typical properties?					
Melting point and boiling point	high	very high	generally high (except mercury)	low	moderate – often decompose on heating
Hardness	hard but brittle	very hard (if three-dimensional)	hard but malleable (except mercury)	soft	variable; many are soft but often flexible
Electrical conductivity	electrolytes – conduct when molten or dissolved in water	do not normally conduct (except graphite)	conduct when solid or liquid	do not conduct	do not normally conduct
Solubility in water	often soluble	insoluble	insoluble (but some react)	usually insoluble, unless molecules contain groups which can hydrogen bond with water	usually insoluble
Solubility in non-polar solvents (e.g. hexane)	insoluble	insoluble	insoluble	usually soluble	sometimes soluble
Where can you find out more?	Sections 3.1, 5.1 and 4.5	Sections 3.1 and 5.2	Sections 3.1 and 11.5	Sections 3.1 and 5.2	Sections 3.1, 5.6 and 5.7

Problems for 5.8

1 a Magnesium oxide has a similar structure to sodium chloride. Refer back to the sodium chloride structure in Figure 1 at the start of **Section 5.1** (page 84) and draw a section of the magnesium oxide structure. What words are used to describe this type of structure?

b Draw diagrams of the structures of the following substances.

i Ne iii CH_4 ii H_2S iv CO_2

What words are used to describe structures like these?

c The diagram shows a section of the structure of diamond, in which carbon is linked covalently to four other carbon atoms at the corners of a tetrahedron.

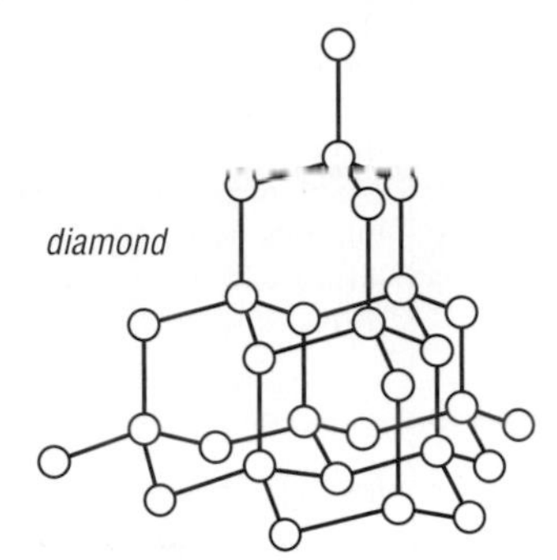

Silica, SiO_2, has the same kind of structure with each silicon bonded to four oxygens, and each oxygen to two silicons.

i Draw a section of this structure.

ii What name is used to describe this type of structure?

d Draw a diagram to show the metallic bonding in magnesium and explain why the structure is called a 'giant lattice'.

2 Use the information given in Table 2 (page 117) in this section to predict the properties of the substances **a–h**. In each case, state whether you expect the substance to be a solid, a liquid or a gas at room temperature, whether you expect the substance to dissolve in water or not, and what you predict about its electrical conductivity.

a sodium iodide
b carbon monoxide
c diamond
d tetrachloromethane
e ethanol
f copper(II) chloride
g vanadium
h poly(propene)

3 Which type of structure (from Table 2 in this section, page 117) would you expect each of the following substances to have in the solid state?

a argon
b an alloy of nickel and cobalt
c silicon carbide
d silk
e rubidium bromide
f xenon tetrafluoride

4 Which type of structure (from Table 2 in this section, page 117) would you expect each of substances **a–e** to have?

a A white solid which starts to soften at 200 °C and can be drawn into fibres.
b A liquid which conducts electricity and solidifies at 39 °C.
c A white solid which melts at 770 °C and conducts electricity when molten, but not in the solid state.
d A hard grey solid which conducts electricity and melts at 3410 °C.
e A white solid which melts at −190 °C.

5 Explain the following in terms of particles, structure and bonding.

a Ionic substances do not conduct electricity when they are solid, but metallic substances do.
b Molten ionic substances are decomposed when they conduct electricity.
c Substances with covalent network structures are insoluble in all solvents.
d Substances that are gases at room temperature always contain simple covalent molecules.

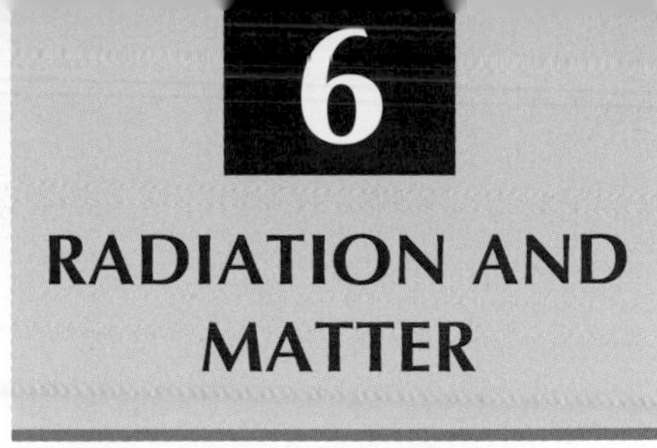

6.1 *Light and electrons*

AS

Most of our understanding of the electronic structure of atoms has come from the area of science known as **spectroscopy** – the study of how light and matter interact. Before you go on to learn about this, you need to know more about the nature of light.

Chemists use two models to describe the behaviour of light – the **wave model** and the **particle model**. Neither fully explains all the properties of light. Some are best described by the wave model – the particle model is better for others. We choose the theory which is most appropriate to the situation.

The wave theory of light

Light is one form of electromagnetic radiation. Like all electromagnetic radiation, it behaves like a wave with a characteristic wavelength and frequency.

A wave of light will travel the distance between two points in a certain time – it doesn't matter what kind of light it is, the time is always the same. The speed at which the wave moves, the *speed of light* (symbolised by c), is the same for all kinds of light, and indeed for all kinds of electromagnetic radiation. It has a value of $3.00 \times 10^8\,\mathrm{m\,s^{-1}}$ when the light is travelling in a vacuum.

Like all waves, the light wave has a **wavelength** (symbol λ) and a **frequency** (symbol ν). Different colours of light have different wavelengths. Figure 1 illustrates this for two different waves.

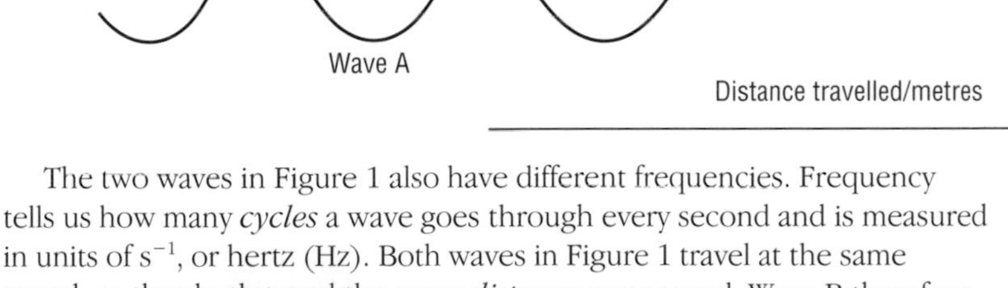

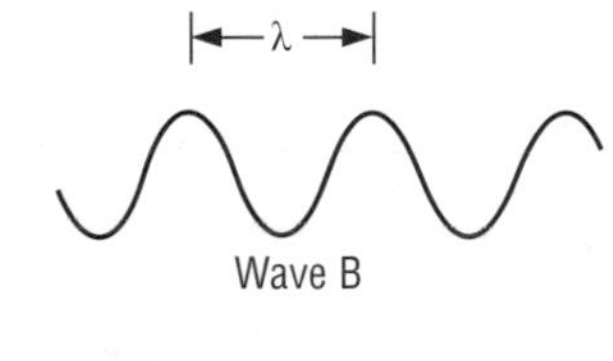

▲ **Figure 1** Wavelength measures the distance (in metres) travelled by the wave during one cycle. Wave A has twice the wavelength of wave B.

The two waves in Figure 1 also have different frequencies. Frequency tells us how many *cycles* a wave goes through every second and is measured in units of $\mathrm{s^{-1}}$, or hertz (Hz). Both waves in Figure 1 travel at the same speed, so they both travel the same *distance* per second. Wave B therefore has twice the frequency of wave A.

Frequency and wavelength are very simply related. In Figure 1, wave B has twice the frequency but half the wavelength of wave A. If you multiply wavelength and frequency together you get a constant – this constant is the speed of light, c. So c, λ and ν are related by the very simple equation

$$c = \lambda\nu$$

When we use the term 'light' we normally mean the visible light to which our eyes respond. But visible light is only a small part of the **electromagnetic spectrum**, which includes all the different forms of electromagnetic radiation. There are other regions – for example radio waves, ultraviolet, infrared and gamma rays. A full version of the electromagnetic spectrum is shown in Figure 2.

▼ **Figure 2** The electromagnetic spectrum – the spectrum continues above 1×10^{20} Hz and below 1×10^5 Hz).

	Radiofrequency					Microwave		Infrared		Visible	Ultraviolet		X-rays		γ-rays	
Frequency/Hz	10^5	10^6	10^7	10^8	10^9	10^{10}	10^{11}	10^{12}	10^{13}	10^{14}	10^{15}	10^{16}	10^{17}	10^{18}	10^{19}	10^{20}
Wavelength/m		10^3			1			10^{-3}			10^{-6}			10^{-9}		

The particle theory of light

In some situations, the behaviour of light is easier to explain by thinking of it not as waves but as particles. This idea, first proposed by Albert Einstein in 1905, regards light as a stream of tiny 'packets' of energy called **photons**. The energy of the photons is related to the position of the light in the electromagnetic spectrum. For example, photons with energy 3×10^{-19} J correspond to red light.

The two theories of light – the wave and photon models – are linked by a relationship which was initially proposed by Max Planck in 1900:

$$E = h\nu$$

E, the energy of a photon, is equal to the frequency of the light, on the wave model, multiplied by a constant, h. This is known as the **Planck constant** and has a value of $6.63 \times 10^{-34}\,\mathrm{J\,Hz^{-1}}$.

For a photon of red light with an energy of 3.0×10^{-19} J, the frequency is given by $E = h\nu$, or $3.0 \times 10^{-19}\,\mathrm{J} = (6.63 \times 10^{-34}\,\mathrm{J\,Hz^{-1}}) \times \nu$

$$\text{so, } \nu = \frac{3.0 \times 10^{-19}\,\mathrm{J}}{6.63 \times 10^{-34}\,\mathrm{J\,Hz^{-1}}} = 4.5 \times 10^{14}\,\mathrm{Hz}$$

Atomic spectra

Atoms can become **excited** by absorbing energy – for example from hot flames, from an electric discharge or from radiation in the stratosphere or in outer space.

When the excited atoms lose energy and return to their **ground state**, the energy is often emitted as electromagnetic radiation. This radiation is usually in the infrared, visible or ultraviolet regions. The emitted light can be split up into an **atomic spectrum** by passing it through a prism or a diffraction grating (Figure 3).

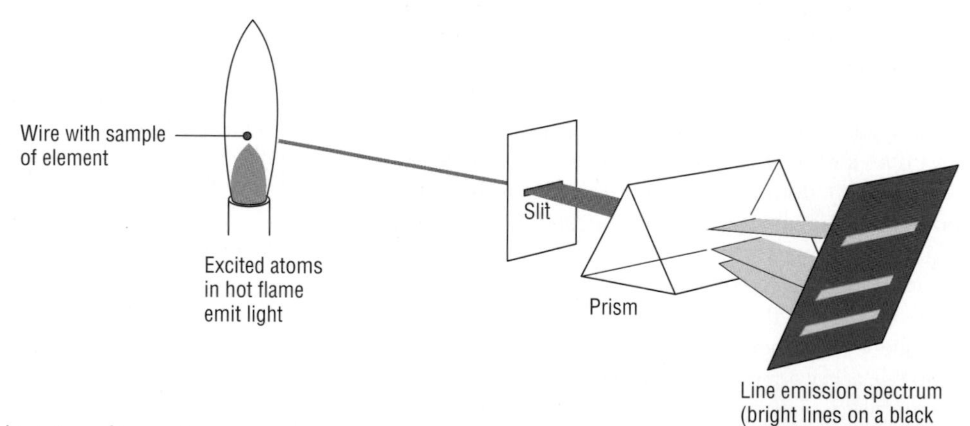

▲ **Figure 3** Obtaining a line emission spectrum.

This type of atomic spectrum is called an **emission spectrum** because it arises from light emitted by excited atoms. In the visible region, the spectrum consists of a series of coloured lines on a black background. A *continuous spectrum* (like one which shows all the colours of the rainbow) is not produced because the atoms can only emit certain precise frequencies.

Suppose that, instead of looking at the light emitted by excited atoms in a hot sample, you looked at the spectrum of white light that has passed through a cooler sample of the element, where most of the atoms are in their ground state.

White light contains all the visible wavelengths and its spectrum is normally continuous, like a rainbow. Figure 4 shows what happens when white light passes through the sample. Black lines appear in the otherwise continuous spectrum. These black lines correspond to light that has been absorbed by the atoms in the sample. They correspond exactly with the coloured lines in the emission spectrum of that element. This type of atomic spectrum is called an **absorption spectrum** because it arises from light absorbed by atoms.

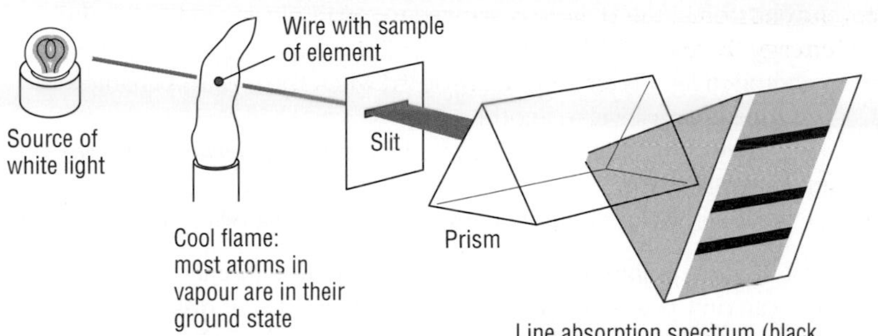

◀ **Figure 4** Obtaining a line absorption spectrum.

The sequence of lines in an atomic spectrum is characteristic of the atoms of the element that gave rise to it. Like a human fingerprint, it can be used to identify the element, even when the element is present in a compound or is part of a mixture. The *intensities* of the lines provide a measure of the element's abundance.

This technique of chemical analysis, called **atomic spectroscopy**, is widely used – for example, to find the composition of a sample of steel or to estimate the sodium content of blood. The composition of stars can be determined from vast distances away using this technique.

The spectrum of hydrogen atoms

The characteristic emission spectrum of hydrogen atoms in the ultraviolet region is shown in Figure 5. This series of lines is called the **Lyman series** after the scientist who first observed it.

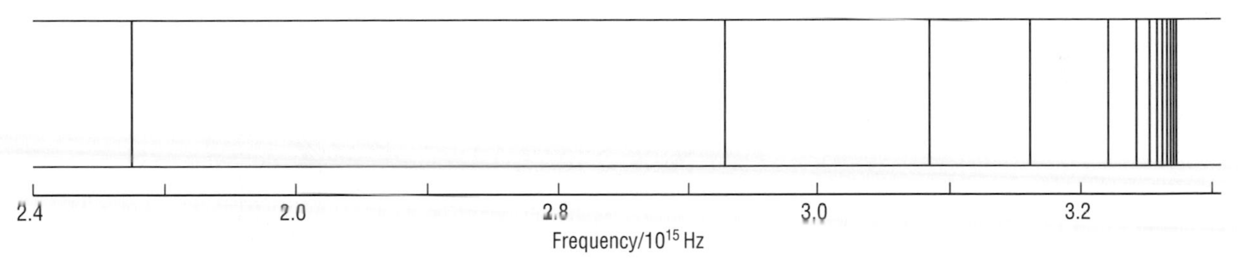

▲ **Figure 5** The Lyman series in the hydrogen atom emission spectrum – remember these are emission lines on a black background. You can see the hydrogen emission spectrum in colour in Figure 13 in the **Elements of Life** module.

The full spectrum contains other series of lines, one in the visible and several in the infrared region. The spectrum was interpreted in 1913 by the Danish scientist Niels Bohr. Bohr's theory explained why the hydrogen atom only emits a limited number of specific frequencies. The frequencies predicted by Bohr's theory match extremely well with the observed lines.

How Bohr's theory explained atomic spectra

The basic idea behind Bohr's theory was this. Atomic spectra are caused by electrons in atoms moving between different energy levels (we now call these energy levels *shells* and *sub-shells* – see **Sections 2.3** and **2.4**). When an atom is excited, electrons jump into higher energy levels. Later, they drop back into lower levels again – and emit the extra energy as electromagnetic radiation, which gives an emission spectrum.

Bohr's theory not only explains how we get absorption and emission spectra, it also gave scientists a model for the electronic structure of atoms.

However, Bohr's theory was controversial. It made use of the idea of **quantisation of energy**. This idea was new at the time and at odds with much of what was thought about energy. But Bohr's theory predicted experimental observations so well that it lent great support to the radical new *quantum theory*.

The main points of Bohr's theory were:

- the electron in the H atom is allowed to exist only in certain definite **energy levels**
- a photon of light is emitted or absorbed when the electron changes from one energy level to another
- the energy of the photon is equal to the difference between the two energy levels (ΔE)
- the frequency of the emitted or absorbed light is related to ΔE by $\Delta E = h\nu$.

The first of these points is the one about quantisation of energy – an electron can only possess definite quantities of energy, or **quanta**. The electron's energy cannot change continuously – it is not able to change to any value, only those values that are allowed.

Figure 6 shows how Bohr's ideas explained the emission lines of the Lyman series.

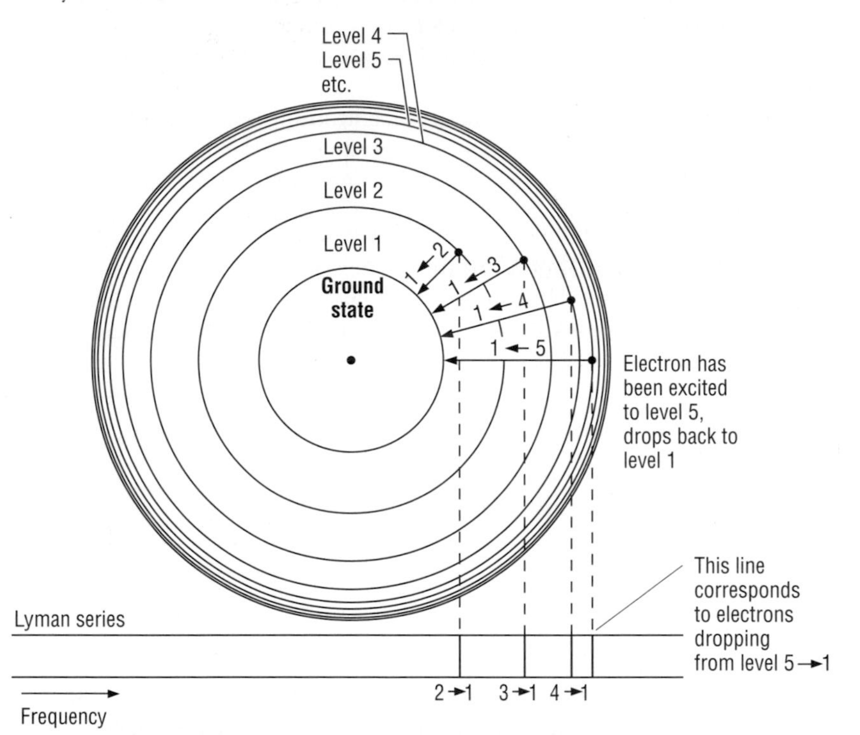

► **Figure 6** How the Lyman series in the emission spectrum is related to energy levels in the H atom.

The rings represent the energy levels of the electron in the hydrogen atom. The further away from the nucleus an electron is, the higher the energy level. Levels are labelled with numbers, starting at 1 for the lowest level – the **ground state**.

The frequencies of the lines of the Lyman series correspond to changes in electronic energy from various upper levels to one common lower level, level 1. Each line corresponds to a particular energy level change, such as level 4 to level 1.

An alternative way of representing emission spectra is by an energy level diagram. Each arrow represents an electron dropping from a *higher* energy level to a *lower* energy level. It can be seen from Figure 7 that the larger the *energy gap* (ΔE) between the two levels, the higher the frequency of electromagnetic radiation emitted.

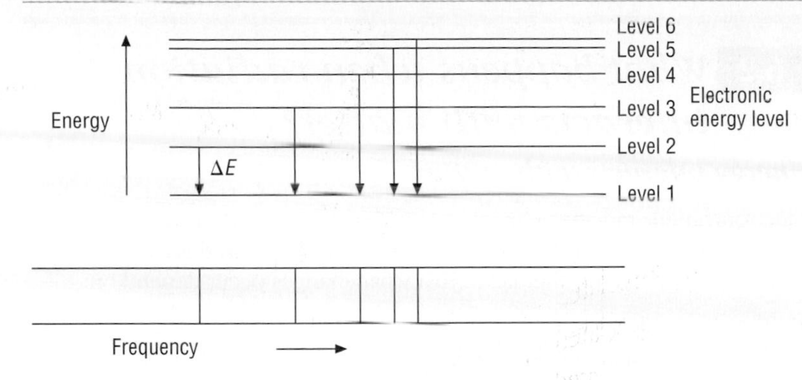

◀ **Figure 7** An energy level diagram and corresponding emission spectrum for the Lyman series – a similar diagram can be used to represent an absorption spectrum – the difference between the diagrams would be that the arrows would point upwards as the electron is raised from a lower energy level to a higher energy level.

You may have looked at the spectrum of a hydrogen lamp in the laboratory. The series of coloured lines that you see in the visible region, the **Balmer series**, arises from electrons falling to level 2 from levels 3, 4, 5 …

Ionisation energy

Notice that as energy increases the levels become more closely spaced, until they converge. After this point, which corresponds to the electron escaping from the atom, the electron is free to move around with any energy. The H atom has lost its electron and become an H^+ ion. This is **ionisation** and the energy difference between this point and the ground state is called the **ionisation energy** (or **ionisation enthalpy**).

Ionisation can be represented by the equation

$$X(g) \rightarrow X^+(g) + e^-$$

where X stands for an atom of any element and e^- for an electron. Notice that X is shown as X(g), indicating that the atoms are separated from one another, in the gaseous state. In the case of hydrogen, we can work out the ionisation energy from the point where the lines of the Lyman series converge together.

Ionisation energies (and the closely related ionisation enthalpies) are useful in deciding which ions a metal will form – there is more about this in **Section 2.5**.

Problems for 6.1

In answering these problems, you will need to use the equation, $E = h\nu$. The Planck constant, $h = 6.63 \times 10^{-34}\,\mathrm{J\,Hz^{-1}}$. This equation gives the energy (in joules) of a single photon for radiation of a particular frequency. If you need to find the energy of a mole of photons, you will need to multiply by the Avogadro constant ($N_A = 6.02 \times 10^{23}\ \mathrm{mol^{-1}}$).

1 **a** Copy the diagram of the energy levels in a hydrogen atom from Figure 6 in this section and draw arrows on it to represent the energy changes which give rise to the four *lowest energy* lines of the Balmer series.

b Underneath your diagram draw a sketch (like the one in Figure 6 for the Lyman series) to show how these first four lines of the Balmer series would be arranged in the emission spectrum of hydrogen.

c Draw *one* line on your energy level diagram to show a transition that would give rise to a line in the *absorption* spectrum of hydrogen. Label this line.

2 **a** The bottom of Figure 7 in this section shows the emission spectrum for the Lyman series.

i If viewed in a laboratory, how would this spectrum appear to the human eye?

ii What would the absorption spectrum from energy level 1 look like to the human eye and how does this differ to your answer to part **i**?

b **i** In problem 1 you looked at the Balmer series related to electron movements to level 2 from higher energy levels. Draw an energy level diagram similar to that in Figure 7 that shows the *absorption* spectrum that would arise from electrons excited *from* energy level 2.

ii How would this differ visibly from the Balmer series energy level diagram?

AS

6.2 *What happens when radiation interacts with matter?*

Energy interacts with matter

Electromagnetic radiation can interact with matter, transferring energy to the chemicals involved. The chemicals absorb energy and this can make changes happen in the chemicals. Just what changes occur depend on:

- the chemical involved
- the amount of energy involved.

In **Section 6.1** you saw what happens when *atoms* absorb radiation. Electrons in the atom jump to higher energy levels – we say the atom moves from its ground state to an excited state.

Let's look now at chemicals with *molecular structures*. Molecules are doing energetic things all the time – they move around, they rotate and the bonds in the molecule vibrate. The electrons in the molecule have energy too, and they can move between the different electronic energy levels.

A molecule has energy associated with several different aspects of its behaviour, including:

- energy associated with **translation** (the molecule moving around as a whole)
- energy associated with **rotation** (of the molecule as a whole)
- energy associated with **vibration** of the bonds
- energy associated with **electrons**.

These different kinds of energetic activities involve different amounts of energy – for example, making the bonds in a molecule vibrate usually involves more energy than making the molecule as a whole rotate. The energy needed increases in the general order shown in Figure 1.

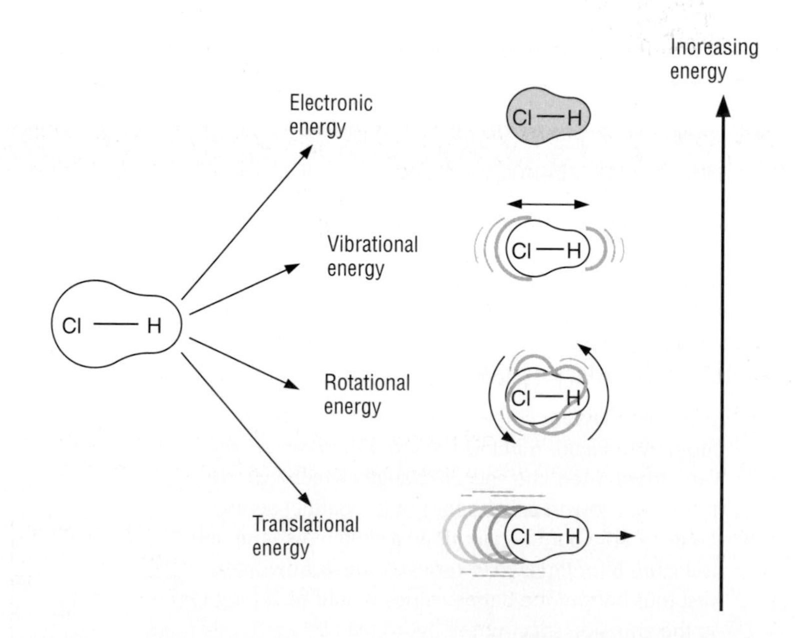

► **Figure 1** An HCl molecule has energy associated with different aspects of its behaviour.

Section 6.1 should have familiarised you with the idea that electrons can occupy definite energy levels. The electronic energy of an atom or molecule changes when an electron moves from one level to another. We say that electronic energy is **quantised**, with fixed levels.

Now here is a crucial point – *all these other types of energy* (translational, rotational and vibrational) *are quantised too*. For example, the HCl molecule can occupy only certain fixed levels of vibrational energy (see Figure 2). When its vibrational energy changes, it moves to a new, fixed energy level, in the same way that changes in electronic energy involve moves to new energy levels.

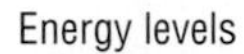

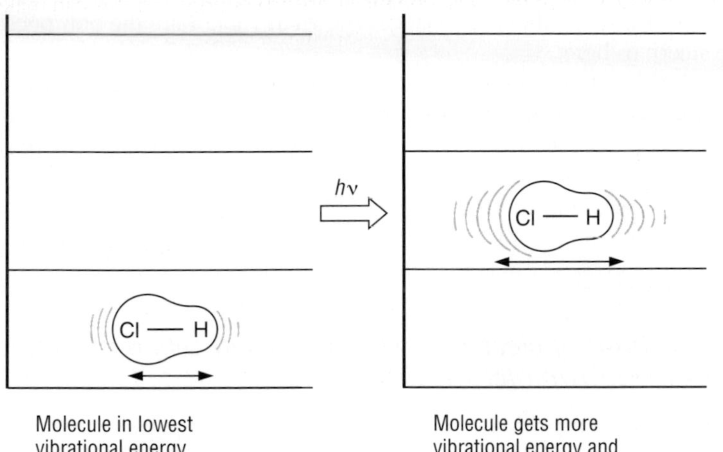

◀ **Figure 2** Vibrational energy changes in an HCl molecule.

It happens that the gap between vibrational energy levels for HCl corresponds to the energy of a photon of infrared radiation. So infrared radiation is able to make HCl molecules – and other molecules – increase their vibrational energy. They vibrate more vigorously and move from one vibrational energy level to a higher one. As a result, we find that HCl absorbs infrared radiation of a certain frequency – the frequency needed to make it vibrate in a particular way. There is more about this in **Section 6.4**.

Different energy changes for different parts of the spectrum

The spacing between **vibrational** energy levels corresponds to the infrared part of the spectrum. We sense infrared radiation as heat – you are sensing infrared radiation when you feel the warmth radiated by a fire. The radiation makes bonds in the chemicals in your skin vibrate more energetically. The energy in the radiation has been converted into kinetic energy and this is why you feel warmer. Making molecules rotate requires less energy than making their bonds vibrate. Therefore, changes in **rotational** energy correspond to a lower energy, a lower frequency part of the electromagnetic spectrum, namely the **microwave** region. Microwaves are the basis of microwave cookery.

The spacings between **translational** energy levels are even smaller – so small, in fact, that we can treat translational energy as being continuous.

Change occurring	Size of energy change/J	Type of radiation absorbed
change of rotational energy level	1×10^{-22} to 1×10^{-20}	microwave
change of vibrational energy level	1×10^{-20} to 1×10^{-19}	infrared
change of electronic energy level	1×10^{-19} to 1×10^{-16}	visible and ultraviolet

◀ **Table 1** Summary of molecular energy changes.

On the other hand, making **electronic** changes occur in a molecule requires *higher* energy than for vibrational changes. Electronic changes involve electrons jumping to higher electronic energy levels within the molecule, and energy corresponding to the **visible** and **ultraviolet** parts of the spectrum is needed to make this happen. Table 1 summarises these changes.

Other changes can occur with energy of different frequencies. For example, gamma rays cause changes in the energy levels within the nuclei of atoms. However, the three changes shown in Table 1 are the only ones that concern us here.

Notice that the table gives *ranges* of energy. The particular value of the energy change depends on the substance involved. Different substances have different chemical structures, and this gives them different energy levels. For example, the C–F bond is stronger than the C–Br bond, so it takes more energy to make a C–F bond vibrate than to make a C–Br bond vibrate. So compounds containing C–F bonds absorb infrared of a higher energy than compounds containing C–Br bonds.

What kind of electronic changes occur when molecules absorb ultraviolet radiation?

Electrons occupy definite energy levels in atoms. The gaps between these electronic energy levels correspond to the energy of visible and ultraviolet light. So electrons jump to higher energy levels when an atom absorbs photons of visible or ultraviolet radiation.

The same kind of thing happens with molecules. The electrons in a molecule, such as Cl_2, occupy definite energy levels. This is true of the electrons that bond the atoms together and also of the non-bonding electrons. The outer shell electrons are in the highest energy levels, and so can move most easily to higher levels.

When a chlorine molecule absorbs radiation, one of three things can happen depending on the amount of energy involved.

- Electrons may be **excited** to a higher energy level. Later the electrons fall back to a lower energy level and nothing permanent has happened to the molecule. Chlorine owes its green colour to this kind of excitation – it so happens that Cl_2 absorbs visible light of such a frequency that the remaining, unabsorbed light looks green.
- If higher energy radiation is used then the molecule may absorb so much energy that the bonding electrons can no longer bond the atoms together – the two atoms in the molecule break apart. This is **dissociation**, or more precisely **photodissociation** because it is caused by light. As a result, radicals are formed (see **Section 6.3**). Radicals are usually very reactive and their formation may lead to further chemical reactions.
- With very high energy photons, the molecules may acquire so much energy that an electron is able to leave it – the molecule has **ionised**.

Figure 3 illustrates these three possibilities:

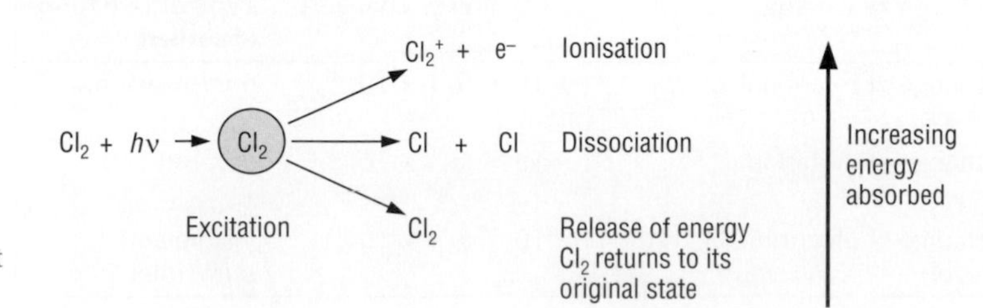

► **Figure 3** When Cl_2 molecules absorb radiation, they become excited – the excited molecules may then ionise, or dissociate or just release the energy.

Problems for 6.2

In answering these problems, you will need to use the equation, $E = h\nu$. This equation gives the energy (in joules) of a single photon for radiation of a particular frequency. The Planck constant, $h = 6.63 \times 10^{-34}\,J\,Hz^{-1}$. If you need to find the energy of a mole of photons, you will need to multiply by the Avogadro constant ($N_A = 6.02 \times 10^{23}\,mol^{-1}$).

1 A beam of infrared radiation has an energy of $3.65 \times 10^{-20}\,J$ per photon. Calculate the frequency of the radiation.

2 Copy and complete the following table. It shows the link between radiation emitted or absorbed and the energy changes that occur in molecules for four different photons (**a–d**).
You will need to use the equation $E = h\nu$ and refer back to Table 1.

	Energy of photon emitted or absorbed/J	Frequency/Hz	Type of radiation	Change between what type of energy levels in molecule
a	4.6×10^{-17}	?	?	?
b	?	3.5×10^{13}	?	?
c	2.1×10^{-22}	?	?	?
d	?	8.3×10^{14}	?	?

3 To change 1 mole of molecular HCl from the lowest vibrational energy level (ground state) to the next vibrational level requires 32.7 kJ.

a How much energy, in joules, would be needed to change *one molecule* of HCl from the ground state to the next vibrational level? (Avogadro constant, $N_A = 6.02 \times 10^{23}\,mol^{-1}$)

b What frequency of radiation would be absorbed when HCl absorbs energy in this way? What *type* of radiation is this? ($h = 6.63 \times 10^{-34}\,J\,Hz^{-1}$)

c Use the equation $c = \lambda\nu$ to calculate the wavelength of this radiation in metres. ($c = 3.00 \times 10^{8}\,m\,s^{-1}$)

4 Ozone in the troposphere is a secondary atmospheric pollutant produced by the reaction of oxygen atoms with dioxygen:

$$O(g) + O_2(g) \rightarrow O_3(g)$$

The oxygen atoms are produced by the photodissociation of nitrogen dioxide in sunlight:

$$NO_2(g) \rightarrow NO(g) + O(g) \quad \Delta H = +214\,kJ\,mol^{-1}$$

The build-up of ozone and other secondary pollutants leads to the formation of photochemical smog.

a Calculate the energy required to break one N–O bond in a *single molecule* of NO_2.

b What is the frequency of radiation (corresponding to this energy) that would break this bond?

c What type of radiation is this?

5 You want to heat up a cup of coffee in a microwave cooker. The cooker uses radiation of frequency $2.45 \times 10^{9}\,Hz$. The cup contains $150\,cm^3$ of coffee, which is mainly water.

a Use the data on the mass, temperature rise and specific heat capacity of the water to calculate the energy needed to raise the temperature of the water by 30 °C.

b How much energy is transferred to the water by each photon of microwave radiation? ($h = 6.63 \times 10^{-34}\,J\,Hz^{-1}$)

c How much energy is transferred to the water by each mole of photons?

d How many moles of photons of this microwave radiation are needed to supply the energy calculated in part **a** and heat up the cup of coffee?

(Specific heat capacity of water = $4.18\,J\,g^{-1}\,K^{-1}$. This is the energy needed to raise the temperature of 1 g of water by 1 °C.)

6 Carbon dioxide is an example of a greenhouse gas. It absorbs some of the infrared radiation given off from the Earth's surface.

a Explain why carbon dioxide molecules absorb only certain frequencies of infrared radiation.

b Explain, in terms of the effect on molecules in the air, why absorbing infrared radiation makes the atmosphere warmer.

6.3 *Radiation and radicals*

Ways of breaking bonds

All reactions involve the breaking and remaking of bonds. Breaking bonds is sometimes called **bond fission**. The way that bonds break has an important influence on reactions.

In a covalent bond, a pair of electrons is shared between two atoms. For example, in the HCl molecule

$$H : Cl$$

When the bond breaks, these electrons get redistributed between the two atoms. There are two ways this can happen – by **heterolytic fission** or by **homolytic fission**.

Heterolytic fission

In this type of fission, both of the shared electrons go to just one of the atoms when the bond breaks. This atom becomes negatively charged, because it has one more electron than it has protons. The other atom becomes positively charged.

In the case of HCl:

$$H : Cl \longrightarrow H^{+} + :Cl^{-}$$

Heterolytic fission is common when a bond is already *polar* (**Section 3.1**). For example, bromomethane contains a polar C–Br bond, and under certain conditions this can break heterolytically:

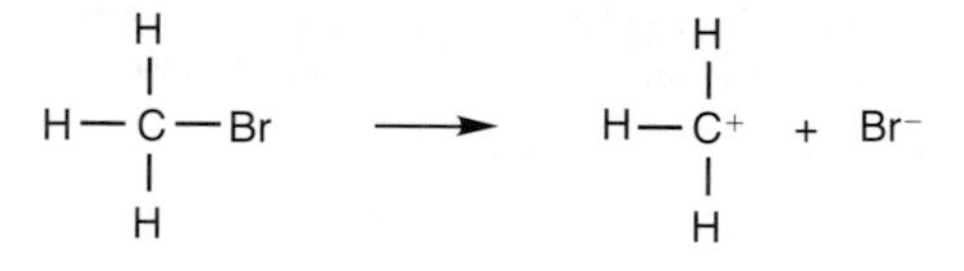

Homolytic fission

In this type of bond fission, one of the two shared electrons goes to each atom. In the case of Br_2

$$Br : Br \longrightarrow Br^{\bullet} + Br^{\bullet}$$

The dot beside each atom shows the unpaired electron that the atom has inherited from the shared pair in the bond. The atoms have no overall electric charge because they have returned to the electronic structure they had before they shared their electrons to form the bond.

The unpaired electron has a strong tendency to pair up again with another electron from another substance. These highly reactive atoms, or groups of atoms, with unpaired electrons are called **radicals**.

Another example of radical formation is when a C–H bond in methane is broken:

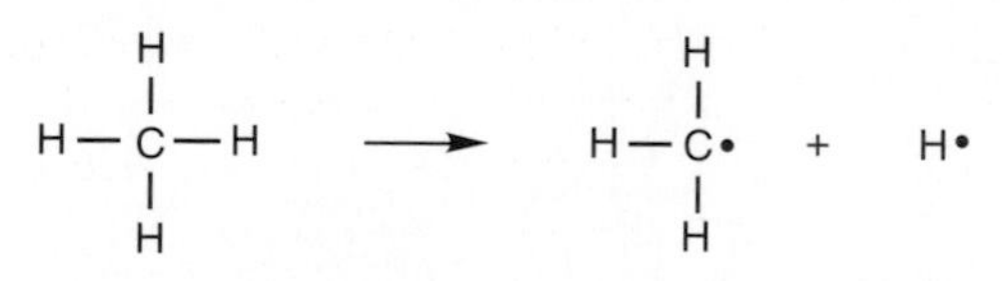

Radicals are most commonly formed when the bond being broken has electrons that are more or less equally shared, but many polar bonds can break this way too – particularly when the reaction is taking place in the gas phase and in the presence of light.

More about radicals

The key feature of a radical is its *unpaired electron*, which makes it particularly reactive. The unpaired electron is often shown as a dot, although of course it is just one of the outer shell electrons.

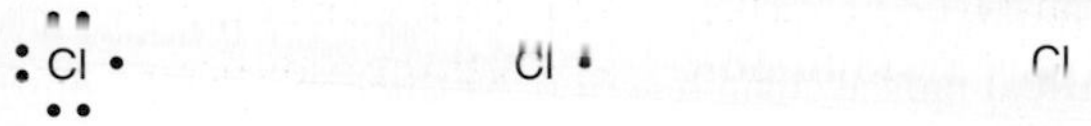

:Cl• (with dot pairs above and below) — Showing all the outer shell electrons

Cl• — Showing the unpaired electron only

Cl — Showing none of the electrons

Sometimes the dot is omitted altogether and a chlorine radical is simply represented by the symbol for the atom – we have used this less cluttered approach in **Chemical Storylines**.

Some radicals are a bit more subdued in their reactivity. This allows them to live long enough to behave as ordinary molecules. Nitrogen monoxide, NO, is an example. One way of showing the arrangement of electrons in NO is

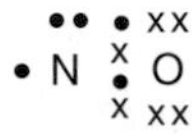

•N:O (dot-cross diagram) *nitrogen monoxide*

Notice that the N atom breaks the 'full outer shell' rule by having only seven electrons in its outer shell. The odd, unpaired electron makes NO a radical.

Some radicals have more than one unpaired electron – for example, the oxygen atom, O, has *two* unpaired electrons and is a **biradical**:

•O• (with dot pairs above and below) — Showing all the outer shell electrons

•O• — Showing the unpaired electrons only

Can you explain why the two electrons in oxygen prefer not to pair with each other? (It may help to look back to **Section 2.4**.)

Another important biradical is the dioxygen molecule, O_2. It has two unpaired electrons, one on each O atom. We can represent it as

•O—O•

The line between the atoms just represents some bonding – it does not imply a single bond in this case.

Like NO, dioxygen is an example of a relatively stable molecular radical. But unlike NO you cannot explain why dioxygen is a biradical by drawing its electron dot–cross diagram. You would need a more sophisticated treatment of bonding than we can give here.

Radicals are reactive

Filled outer electron shells are more stable than unfilled ones. Radicals are reactive because they tend to try to fill their outer shells by grabbing an electron from another atom or molecule.

For example, when a chlorine radical collides with a hydrogen molecule, the chlorine grabs an electron from the pair of electrons in the bond between the H atoms. The effect is to make a new bond between the Cl and H atoms.

Cl• H ˣₓ H ⟶ Cl •ₓ H Hˣ

Cl•

chlorine radical

H ˣₓ H

hydrogen molecule

The curved arrow indicates the movement of an electron. The 'tail' of the arrow shows where the electron starts and the 'head' shows where it finishes. Look carefully at the head of the arrow – it is drawn this way to show the movement of a *single* electron. (You will also meet full-headed arrows which indicate the movement of *pairs* of electrons.)

Notice that a hydrogen radical is also formed in this reaction. This, too, is highly reactive and it will combine with another molecule, once again creating a new radical – and so it goes on. It is a **radical chain reaction**.

The reaction between chlorine and hydrogen is a typical **photochemical** radical chain reaction. Nothing much happens if you mix these two gases in the dark, but they react explosively as soon as you shine ultraviolet light on the mixture.

Like all radical chain reactions, this reaction has three key stages – *initiation, propagation* and *termination*.

Initiation Chlorine radicals are initially formed by the photodissociation of chlorine molecules:

$$Cl_2 + h\nu \rightarrow Cl\cdot + Cl\cdot$$

Only a few chlorine radicals are formed, but they are so reactive that they soon react with something else – they *initiate* the reaction.

Propagation Chlorine radicals react with hydrogen molecules – this produces hydrogen chloride molecules and hydrogen radicals. The hydrogen radicals can then go on to react with chlorine molecules:

$$Cl\cdot + H_2 \rightarrow HCl + H\cdot$$
$$H\cdot + Cl_2 \rightarrow HCl + Cl\cdot$$

These two reactions produce new radicals which keep the reaction going – they *propagate* the reaction.

Termination Every now and then, two radicals collide with each other. This isn't common because very few radicals are present at any one time. When it does happen the reaction chain is *terminated* because the radicals have been taken out of circulation:

$$H\cdot + H\cdot \rightarrow H_2$$
$$Cl\cdot + Cl\cdot \rightarrow Cl_2$$
$$H\cdot + Cl\cdot \rightarrow HCl$$

You can find out more about the manufacture of hydrochloric acid using this mechanism in the **Elements from the Sea** module.

The overall effect of these three stages is to convert hydrogen and chlorine into hydrogen chloride

$$H_2 + Cl_2 \rightarrow 2HCl$$

Methane and chlorine

Apart from their use as fuels (which you studied in the **Developing Fuels** module), alkanes are generally considered to be unreactive, which they are with polar or organic solvents. However, alkanes will react with chlorine, and other halogens, in the presence of light. For example, methane and chlorine do not react at all in the dark, but in sunlight an explosive reaction occurs to form chloromethane and hydrogen chloride. The reaction is another example of a free radical chain reaction, and has the three key stages – initiation, propagation and termination.

Initiation As above, chlorine radicals are initially formed in the presence of light by the photodissociation of chlorine molecules:

$$Cl_2 + h\nu \rightarrow Cl\cdot + Cl\cdot$$

Propagation Chlorine radicals react with methane molecules to produce hydrogen chloride molecules and methyl radicals.

The methyl radicals can then go on to react with chlorine molecules to produce chloromethane and chlorine:

$Cl\cdot + CH_4 \rightarrow HCl + CH_3\cdot$

$CH_3\cdot + Cl_2 \rightarrow CH_3Cl + Cl\cdot$

These two reactions produce new radicals which keep the reaction going – they *propagate* the reaction.

Termination The reaction ends when two free radicals collide and combine:

$Cl\cdot + Cl\cdot \rightarrow Cl_2$

$CH_3\cdot + Cl\cdot \rightarrow CH_3Cl$

$CH_3\cdot + CH_3\cdot \rightarrow C_2H_6$

The overall effect of these three stages is to convert chlorine and methane to hydrogen chloride and chloromethane, although there will also be small amounts of ethane present in the product mixture. In the presence of chlorine, further substitution may occur to form dichloromethane and trichloromethane. This is an example of a **radical substitution** reaction.

Addition polymers such as those found in the **Polymer Revolution** module are manufactured via radical mechanisms.

Summary

Radical chain reactions have particular features:

- they often occur in the gas phase or in a non-polar solvent
- they are often initiated by heating or by light
- they usually go very fast.

Radical chain reactions are very common. Combustion involves radical reactions, and so most explosions. Many of the reactions that occur in the troposphere and stratosphere also involve radicals.

Problems for 6.3

1 Decide whether or not each of the following species is a radical. You may need to draw electron dot–cross diagrams to help you to decide. (They all obey simple bonding rules.)

a F　　c H_2O　　e NO_2

b Ar　　d OH　　f CH_3

2 The hydroxyl radical (HO·) is an important species in atmospheric chemistry. Reaction **A** shows one process in which HO· is produced. The reaction is brought about by radiation with a wavelength below 190 nm.

Reaction A $H_2O + h\nu \rightarrow H\cdot + HO\cdot$

Hydroxyl radicals are very reactive and act as scavengers in the atmosphere. One set of reactions which involve stratospheric ozone is:

Reaction B $HO\cdot + O_3 \rightarrow HO_2\cdot + O_2$

Reaction C $HO_2\cdot + O_3 \rightarrow HO\cdot + 2O_2$

a What term is used to describe a process like reaction **A** in which a molecule is split up by light? Is this splitting a homolytic or a heterolytic process?

b Which of the terms initiation, propagation and termination is most appropriate for each of reactions **A, B** and **C**?

c Which of the chemical species (other than HO·) in the mechanism (for reactions **A, B** and **C**) are radicals?

d i Write an equation which represents the overall result of reactions **B** and **C**.

ii What is the role of HO· in this process?

e Explain why it is important for chemists to have an accurate idea of the concentration of radicals such as HO· when trying to understand the details of atmospheric chemistry.

f One process which removes HO· from the atmosphere is

$HO\cdot + HO_2\cdot \rightarrow H_2O + O_2$

What name is given to a reaction like this, which removes the radicals in a reaction?

3 The creation of significant amounts of nitrogen monoxide from human activities is of concern because it is thought to lead to a loss of ozone from the stratosphere. Reactions **D** and **E** show how this loss can occur.

Reaction D $NO\cdot + O_3 \rightarrow NO_2\cdot + O_2$ $\Delta H = -100\,kJ\,mol^{-1}$

Reaction E $NO_2\cdot + O \rightarrow NO\cdot + O_2$ $\Delta H = -192\,kJ\,mol^{-1}$

a Name one human activity which leads to the production of a significant amount of NO.

b Are reactions **D** and **E** endothermic or exothermic?

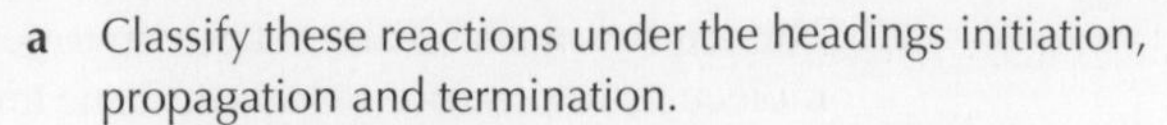

c i What is the overall effect of reactions **D** and **E**?
ii What is the role of NO in this process?
iii Calculate the value of ΔH for the overall process.

d Explain why we should be worried about the effects which might arise from the emission of small amounts of NO into the atmosphere.

4 In the **Developing Fuels** module, you learned about the importance of cracking and reforming reactions to modify a hydrocarbon's performance in a car engine. The details of these reactions are complex, but the effect of strongly heating ethane is well understood. Bringing a reaction about in this way is called **pyrolysis**.
The following reactions occur when ethane is pyrolysed at 620 °C:

F $C_2H_6 \rightarrow 2CH_3\cdot$
G $CH_3\cdot + C_2H_6 \rightarrow CH_4 + C_2H_5\cdot$
H $C_2H_5\cdot \rightarrow C_2H_4 + H\cdot$
I $H\cdot + C_2H_6 \rightarrow H_2 + C_2H_5\cdot$
J $2C_2H_5\cdot \rightarrow C_2H_4 + C_2H_6$
K $2C_2H_5\cdot \rightarrow C_4H_{10}$
(C_2H_4 is ethene, $CH_2{=}CH_2$)

a Classify these reactions under the headings initiation, propagation and termination.

b For most of these reactions, you would need information about bond enthalpies in order to decide whether the process is exothermic or endothermic. But two of the reactions *can* be classified by inspection.
i Which reaction is endothermic?
ii Which reaction is exothermic?

c Which of the chemical species in the reaction sequence are radicals? Give their names and formulae.

5 The reaction of methane (CH_4) with chlorine in the presence of sunlight proceeds via a radical chain reaction. The overall equation for the reaction is

$$CH_4 + Cl_2 \rightarrow CH_3Cl + HCl$$

a Write out a possible mechanism for the reaction showing clearly which reactions correspond to the initiation, propagation and termination stages.

b Suggest why the reaction product also contains some CH_2Cl_2, $CHCl_3$ and CCl_4.

AS

6.4 *Infrared spectroscopy*

The energy possessed by molecules is quantised. In other words, molecules can only have a small number of definite energy values rather than any old energy.

You saw in **Section 6.2** that when a **molecule** interacts with radiation there can be changes in electronic, vibrational or rotational energy, depending on the frequency of the radiation.

Analysis of the energy (or frequency of radiation) needed to produce a change from one energy level to another is the basis of most forms of **molecular spectroscopy**. Note that atomic spectroscopy (**Section 6.1**) is concerned only with changes to electronic energy, because single atoms do not have vibrational or rotational energy.

In **infrared spectroscopy** substances are exposed to radiation in the frequency range 10^{14}–10^{13} Hz, i.e. wavelengths 2.5–15 μm. This makes **vibrational** energy changes occur in the molecules, which absorb infrared radiation of specific frequencies.

Frequency and wavelength are related by the equation: $c = \lambda\nu$ (where c is the speed of light, a constant). This means that the reciprocal of the wavelength ($1/\lambda$) is a direct measure of frequency. It is this reciprocal, called the **wavenumber** of the radiation and usually measured in cm^{-1} units, that is recorded on an infrared spectrum. Figure 1 shows the relationship between wavenumber, wavelength and frequency. Note the directions of the arrows.

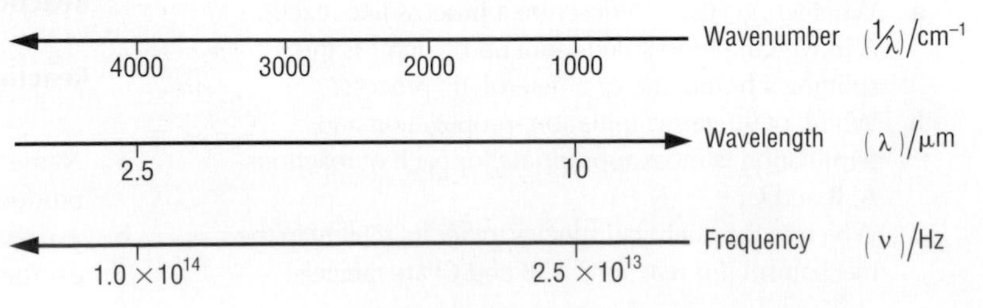

► **Figure 1** The relationship between wavenumber, wavelength and frequency.

Bond deformation

Simple diatomic molecules – such as HCl, HBr and HI – can only vibrate in one way, that is by stretching, where the atoms pull apart and then push together again (Figure 2).

For these molecules there is only one vibrational infrared absorption. This corresponds to the molecules changing from their lowest vibrational energy state to the next higher level, in which the vibration is more vigorous.

The frequencies of the absorptions are different for each molecule. This is because the energy needed to excite a vibration depends on the strength of the bond holding the atoms together; weaker bonds require less energy. It is as if the bonds behave like springs of different strength holding the atoms together. This is illustrated in Table 1.

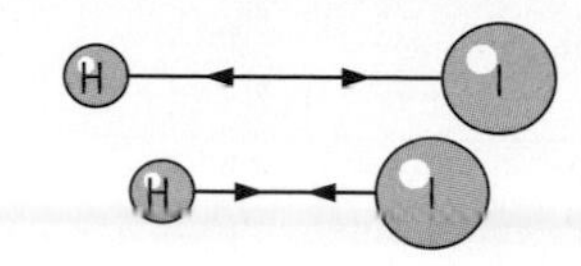

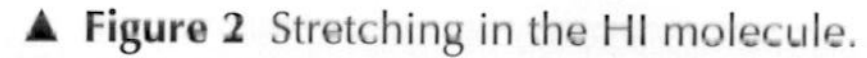

▲ **Figure 2** Stretching in the HI molecule.

◀ **Table 1** Bond enthalpies and infrared absorptions for the hydrogen halides.

Compound	Bond enthalpy/kJ mol^{-1}	Infrared absorption/cm^{-1}
HCl	+432	2886
HBr	+366	2559
HI	+298	2230

In more complex molecules, more bond deformations are possible. Most of these involve more than two atoms. For example, carbon dioxide can vibrate as shown in Figure 3.

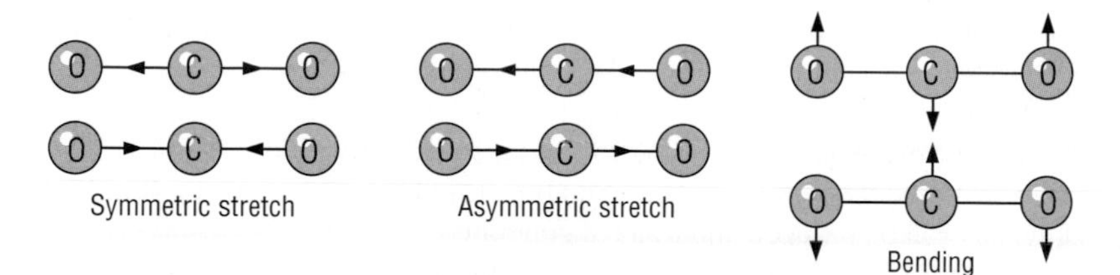

◀ **Figure 3** Vibrations in the CO_2 molecule.

More complex molecules can have very many vibrational modes, with descriptions such as rocking, scissoring, twisting and wagging. Add these to the fact that the molecules also contain bonds with different bond enthalpies and you may have a very complicated spectrum.

The important point to remember about infrared spectroscopy is that you do not try to explain the whole spectrum; you look for one or two signals which are characteristic of particular bonds.

Figure 4 shows the infrared spectrum of ethanol, and Figure 5 shows some of the vibrations which give rise to the signals.

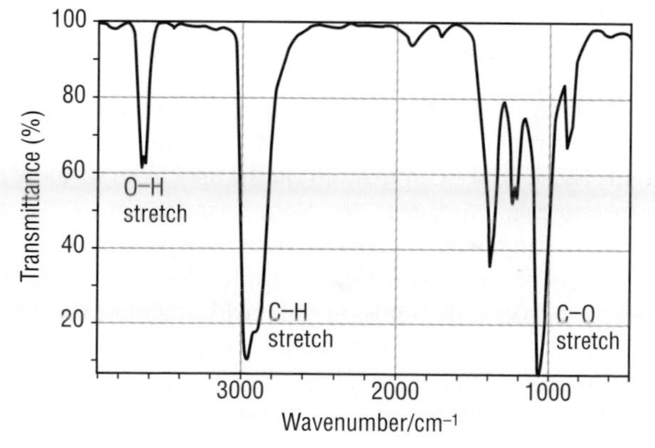

▲ **Figure 4** The infrared spectrum of ethanol in the gas phase.

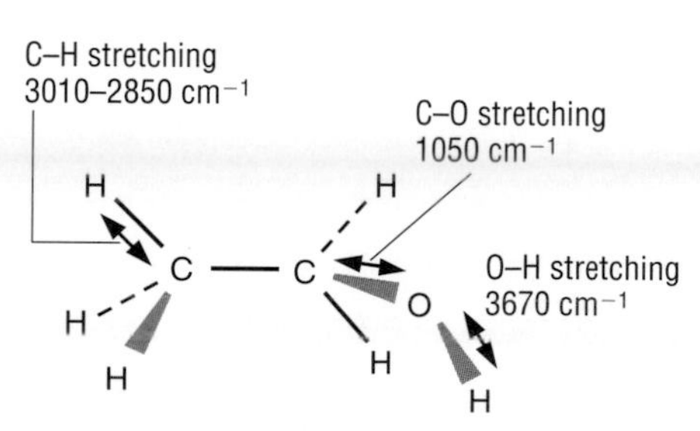

▲ **Figure 5** The ethanol molecule, showing the vibrations which give rise to some characteristic absorptions.

The infrared spectrometer

A typical method of obtaining an infrared spectrum is outlined below. Infrared radiation from a heated filament is split into two parallel beams, one of which passes through the sample, the other through a reference chamber. This ensures that unwanted absorptions from water and carbon dioxide in the air, or from a solvent, are cancelled out. The beams are then directed by mirrors so that they follow parallel paths (Figure 6).

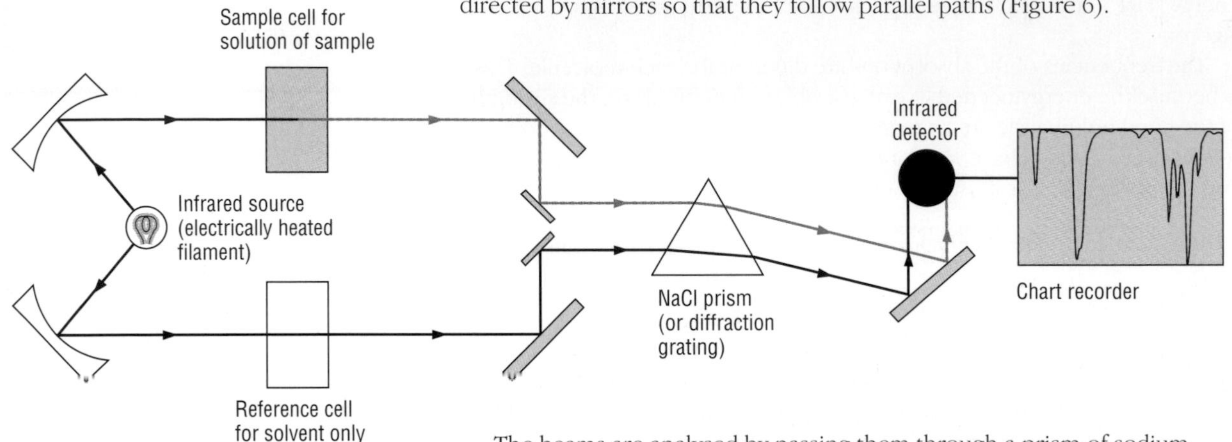

▲ **Figure 6** The basic parts of a double beam infrared spectrometer.

The beams are analysed by passing them through a prism of sodium chloride, which is transparent to infrared radiation, or through a diffraction grating. Light of only one particular frequency will now be focused onto the detector. The spectrum is produced by rotating the prism so that the detector scans the frequencies and records their intensities.

When the sample is not absorbing there will be no difference between the two beams reaching the detector, so no signal is recorded. When a vibration is being excited, the sample absorbs radiation; the sample beam intensity will be reduced and a signal generated.

The record of an infrared spectrum seems to be upside down since the baseline is at the top and the signals (or bands) are recorded as downward movements on the chart recorder. However, as you will see from spectra like that in Figure 4, it is *transmittance* which is being recorded and this is at a maximum when no light is being absorbed.

A newer method called Fourier transform infrared (FTIR) spectroscopy uses a single beam of infrared radiation and the phenomenon of wave interference to produce an infrared spectrum. In effect, the sample is simultaneously subjected to a 'broadband' of energies to produce a complete infrared spectrum. FTIR is cheaper and faster for collecting infrared data.

Interpreting the spectra

The infrared spectrum of but-1-ene (Figure 7) shows most of the characteristic absorptions of an unsaturated hydrocarbon containing a C=C double bond. These have been marked on the spectrum to show the bonds which are responsible.

In general, we can match a particular bond to a particular absorption region. Table 2 (page 136) gives some examples. Note that the precise position of an absorption depends on the environment of the bond in the molecule, so we can only quote wavenumber *regions* in which we can expect absorptions to arise.

Figure 8 shows the infrared spectrum of propanone. The absorption at around $1740\,cm^{-1}$ is characteristic of the C=O group. The absorption is very intense compared with the C=C absorption you saw in a similar region for but-1-ene in Figure 7.

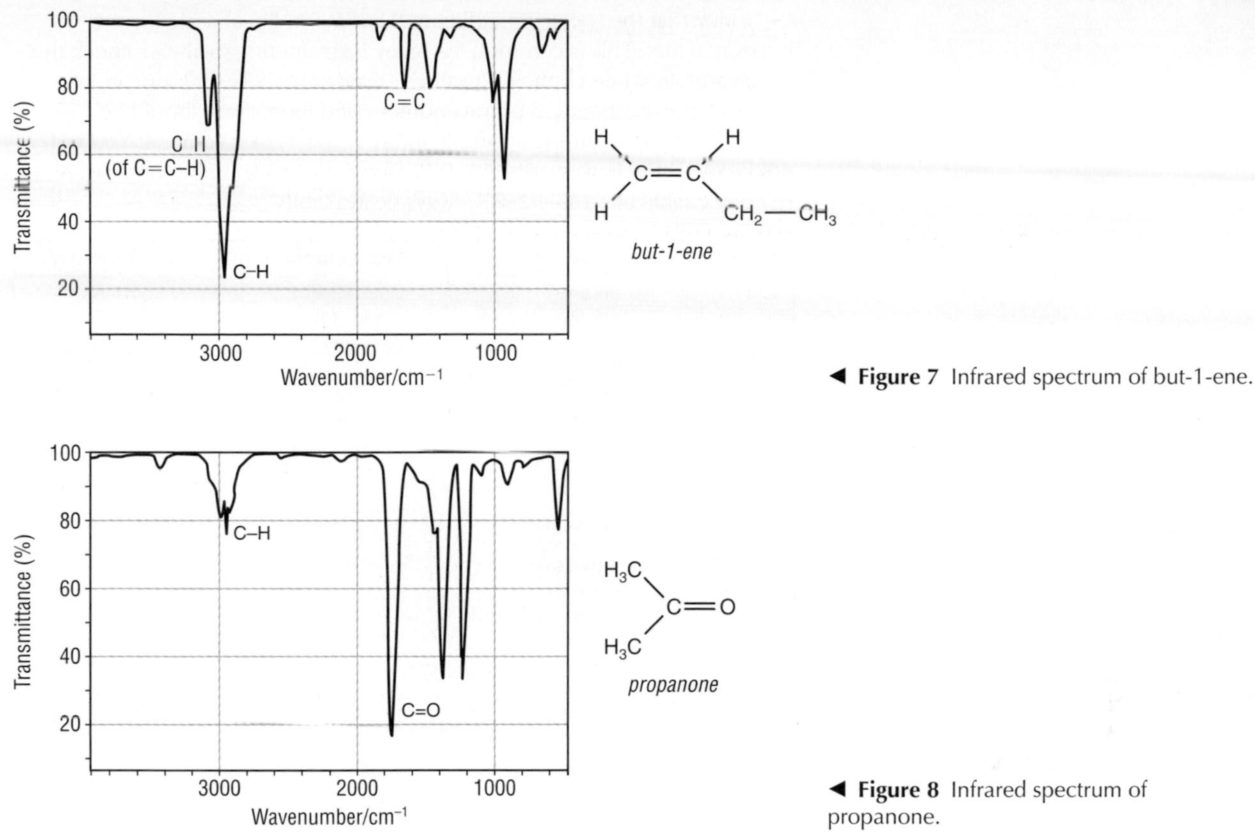

◀ **Figure 7** Infrared spectrum of but-1-ene.

◀ **Figure 8** Infrared spectrum of propanone.

Why are some absorptions intense while others are weaker? The strongest infrared absorptions arise when there is a large change in bond polarity associated with the vibration. Hence O–H, C–O and C=O bonds, which are very polar, give more intense absorptions than the non-polar C–H, C–C and C=C bonds.

Hydrogen bonding affects the absorption due to the O–H stretching vibration. Figure 4 in this section (page 133) shows the infrared spectrum of ethanol in the gas phase. There is little hydrogen bonding and the O–H absorption is a sharp peak at 3670 cm^{-1}.

Now look at Figure 9, which shows the infrared spectrum of a liquid film of ethanol. Hydrogen bonding between the hydroxyl groups changes the O–H vibration and the absorption becomes much broader. It is also shifted to a lower wavenumber and shows maximum absorption at 3340 cm^{-1}.

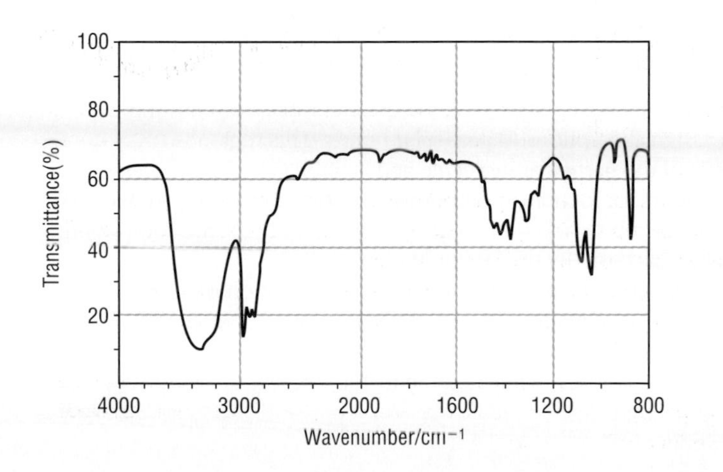

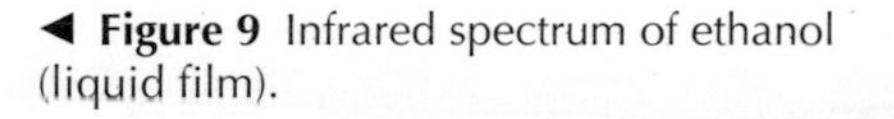
◀ **Figure 9** Infrared spectrum of ethanol (liquid film).

Notice that the spectrum in Figure 9 has a non-linear scale for the horizontal axis. This is common for many instruments, so always check the scale carefully when reading off values of wavenumbers for absorptions.

Using the combination of wavenumber and intensity it should be possible for you to interpret simple infrared spectra for yourself. You will not be expected to remember the characteristic wavenumber regions! A reference table like Table 2 will always be available.

▶ **Table 2** Characteristic infrared absorptions in organic molecules.

Bond	Location	Wavenumber/cm^{-1}	Intensity
C–H	alkanes alkenes, arenes alkynes	2850–2950 3000–3100 *ca* 3300	M–S M–S S
C=C	alkenes	1620–1680	M
(benzene ring)	arenes	several peaks in range 1450–1650	variable
C≡C	alkynes	2100–2260	M
C=O	aldehydes ketones carboxylic acids esters amides	1720–1740 1705–1725 1700–1725 1735–1750 1630–1700	S S S S M
C–O	alcohols, ethers, esters	1050–1300	S
C≡N	nitriles	2200–2260	M
C–F	fluoroalkanes	1000–1400	S
C–Cl	chloroalkanes	600–800	S
C–Br	bromoalkanes	500–600	S
O–H	alcohols, phenols *alcohols, phenols *carboxylic acids	3600–3640 3200–3600 2500–3200	S S (broad) M (broad)
N–H	primary amines amides	3300–3500 *ca* 3500	M–S M
M = medium S = strong * hydrogen-bonded			

Before you try some problems, have a good look at Table 2. It is helpful to divide the infrared spectrum into four regions, as shown in Table 3.

▶ **Table 3** Regions in the infrared spectrum where typical absorptions occur.

Absorption range/cm^{-1}	Bonds responsible	Examples
4000–2500	single bonds to hydrogen	O–H, C–H, N–H
2500–2000	triple bonds	C≡C, C≡N
2000–1500	double bonds	C=C, C=O
below 1500	various	C–O, C–X (halogen)

Below 1500 cm^{-1} the spectrum can be quite complex and it is more difficult to assign absorptions to particular bonds. This region is characteristic of the particular molecule and is often called the *fingerprint region*. It is useful for identification purposes – for example, if you need to compare two spectra to find out if they are spectra of the same compound. It is only rarely used to identify functional groups.

Aromatic compounds often exhibit complex absorption patterns in the fingerprint region (see Figures 12 and 13). Such compounds can often be identified by comparing their infrared spectra with reference spectra.

Figure 10 summarises the information in Table 3.

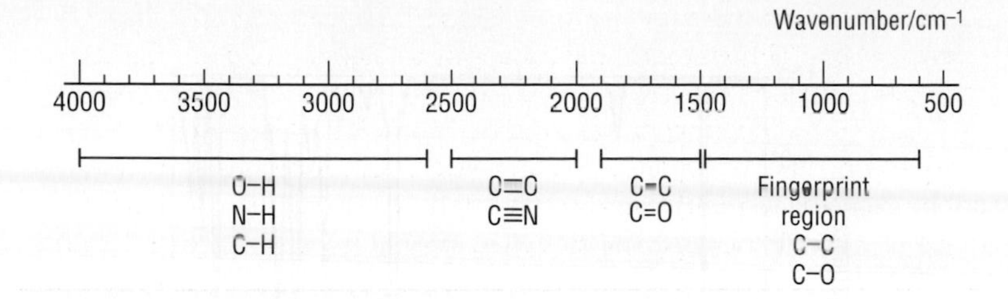

◀ **Figure 10** Typical regions of absorption in the infrared spectrum.

Examples of infrared spectra

The following examples show you how to use infrared spectra to find information about the structure of organic molecules.

Butane

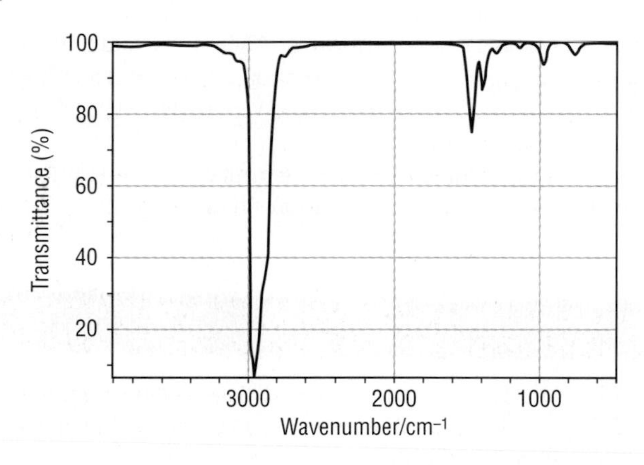

$CH_3-CH_2-CH_2-CH_3$

Absorption/cm^{-1}	Bond
2970	C–H (alkane)

◀ **Figure 11** Infrared spectrum of butane.

The spectrum shows a strong absorption at 2970 cm^{-1} characteristic of C–H stretching in aliphatic compounds. There is no indication of any functional groups.

Methylbenzene

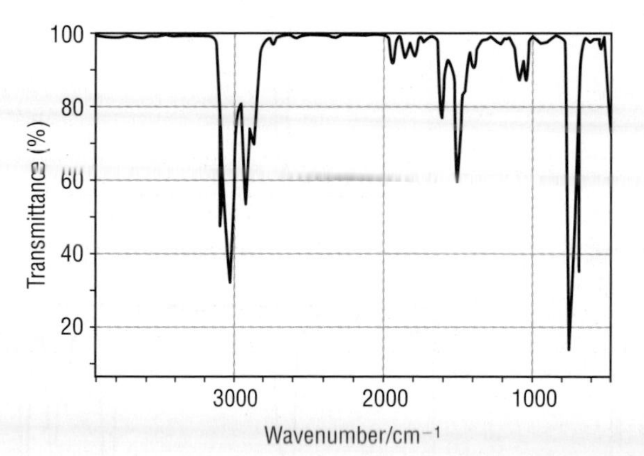

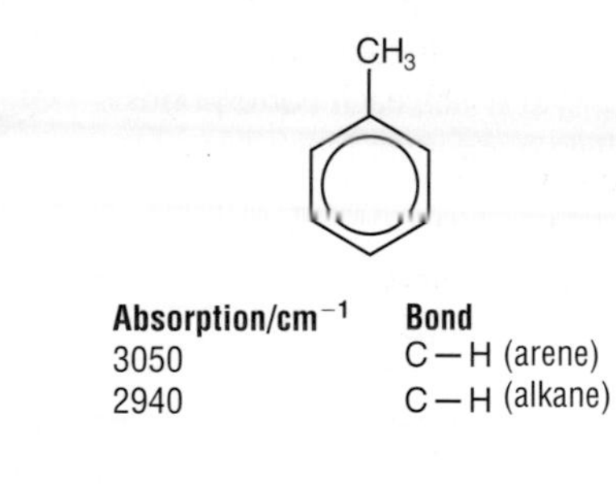

Absorption/cm^{-1}	Bond
3050	C–H (arene)
2940	C–H (alkane)

◀ **Figure 12** Infrared spectrum of methylbenzene.

The spectrum shows two types of C–H absorptions, just above 3000 cm^{-1} for the C–H on the benzene ring and just below 3000 cm^{-1} for C–H on the methyl group. There is no indication of any functional groups. (The absorption pattern around 700 cm^{-1} is typical of a benzene ring with one substituted group.)

Benzoic acid

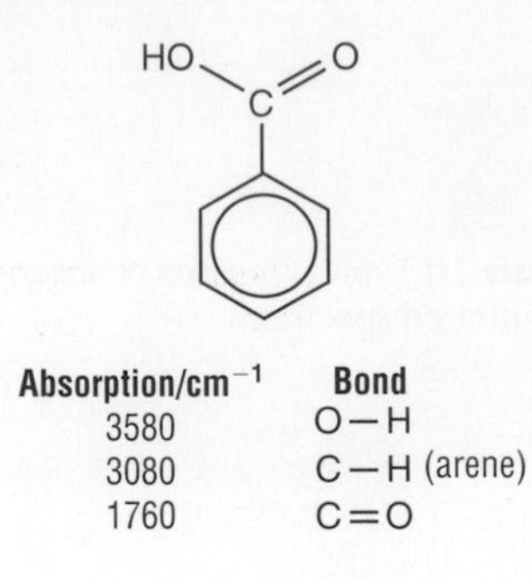

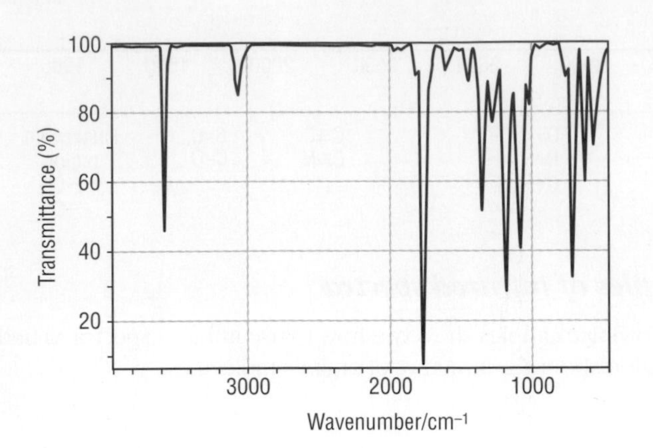

► **Figure 13** Infrared spectrum of benzoic acid.

The spectrum shows a sharp absorption at 3580 cm^{-1} characteristic of an O–H bond (not hydrogen-bonded). The strong absorption at 1760 cm^{-1} shows the presence of a C=O group. The position of the C–H absorption suggests it is an aromatic compound.

The identity of the sample could be confirmed by comparing its spectrum with that of an authentic sample of benzoic acid.

Problems for 6.4

1 The infrared spectrum of carbon dioxide shows a strong absorption at 2360 cm^{-1}.

a Calculate the wavelength of the radiation absorbed. Give your answer in μm.

b Use $c = \lambda \nu$ to calculate the frequency of the radiation absorbed.

($c = 3.00 \times 10^8\,\mathrm{m\,s^{-1}}$; $1\,\mu\mathrm{m} = 1 \times 10^{-6}\,\mathrm{m}$)

2 Figure 14 shows the infrared spectrum of phenol. Identify the key peaks in the spectrum, and the bond to which each corresponds. Give your answer in the form of a table:

OH

phenol

Absorption/cm^{-1}	Bond

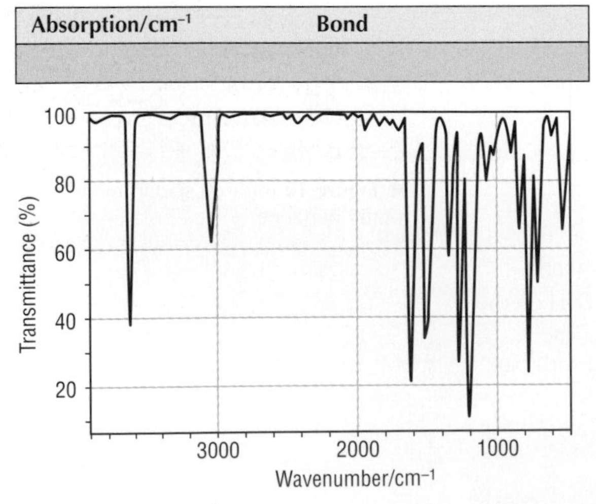

▲ **Figure 14** Infrared spectrum of phenol for problem 2.

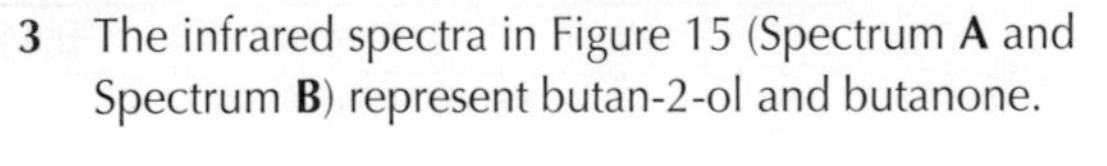

3 The infrared spectra in Figure 15 (Spectrum **A** and Spectrum **B**) represent butan-2-ol and butanone.

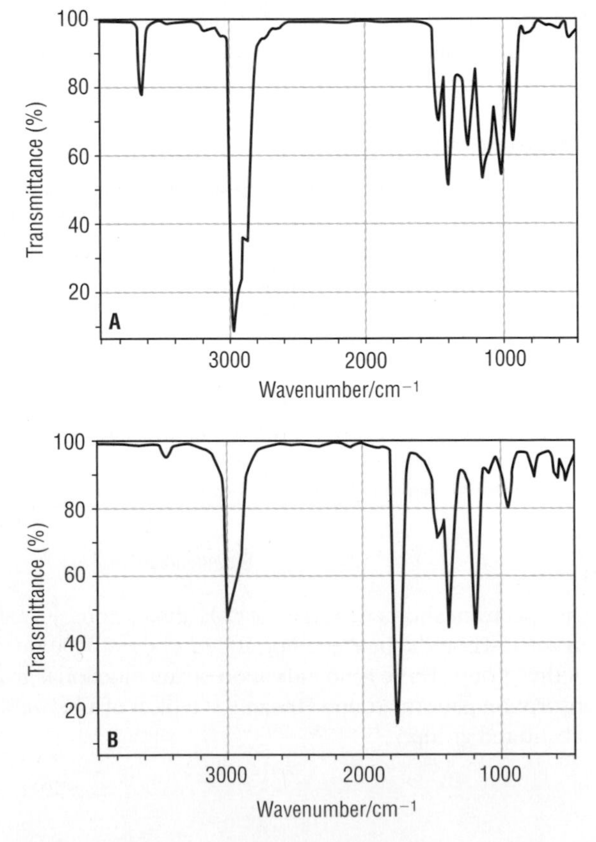

▲ **Figure 15** Spectrum **A** and Spectrum **B**.

- **a** Draw structures for butan-2-ol and butan-2-one.
- **b** Identify the key peaks in each spectrum, and the bond to which each corresponds. Give your answer in the form of a table.
- **c** Decide which spectrum represents butan-2-ol and which represents butan-2-one.

4 The infrared spectra in Figure 16 represent three compounds **C**, **D** and **E**. The compounds are an ester, a carboxylic acid and an alcohol, though not necessarily in that order.

- **a** Identify the key peaks in each spectrum, and the bond to which each corresponds. Give your answer in the form of a table.
- **b** Decide which spectrum represents which type of compound.

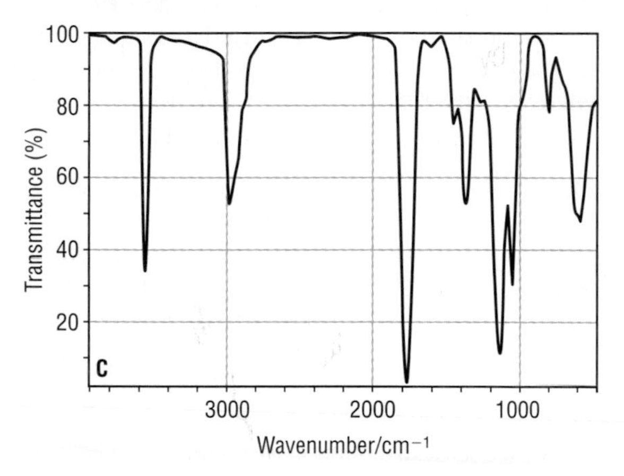

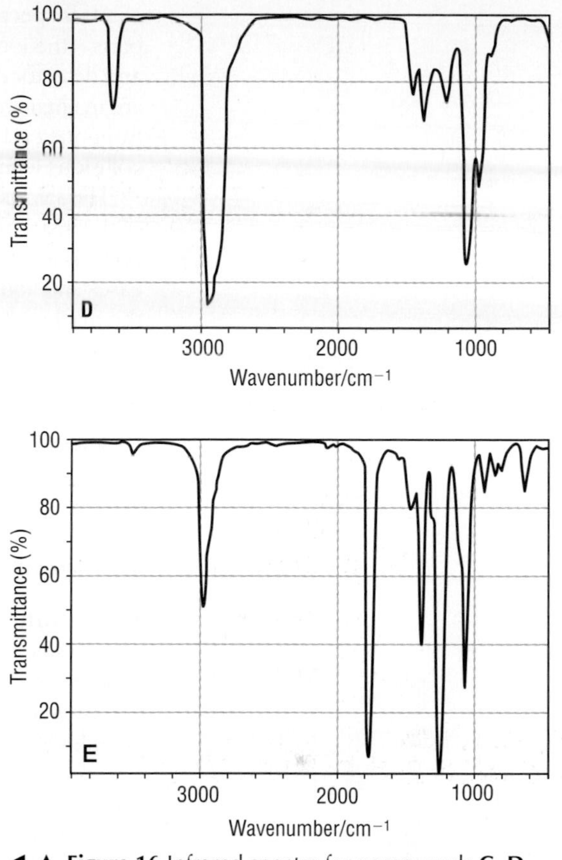

◀ ▲ **Figure 16** Infrared spectra for compounds **C, D** and **E**.

5 Oil of wintergreen has mild pain-killing properties. Its structure is shown below.

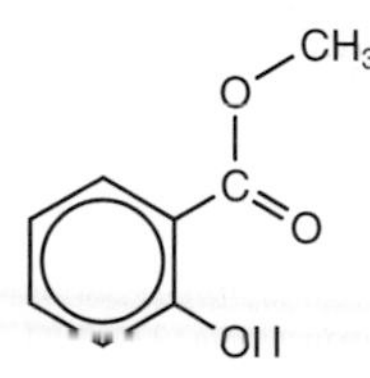

Draw up a table to show the key peaks you would expect to see in the infrared spectrum of oil of wintergreen, and the bond to which each absorption corresponds.

6.5 *Mass spectrometry*

AS

The fundamentals

Mass spectrometry is used to measure the **atomic** or **molecular mass** of different particles (e.g. atoms or molecules) in a sample. The analysis takes place using a **mass spectrometer** – an important instrument for chemists. At a simple level, mass spectrometry can be used to identify the mass and relative abundance of different isotopes in an element, such as the two isotopes of chlorine, ^{37}Cl and ^{35}Cl. At a more complex level, mass spectrometry can be used to investigate the structure of molecular compounds.

Mass spectrometer machines can be divided into three fundamental parts – the ionisation source, the analyser and the detector. The analyser and detector of the mass spectrometer, and often the ionisation source too, are maintained under high vacuum to give the ions a chance of travelling from one end of the instrument to the other without any hindrance by collisions with air molecules.

The sample is introduced via a **sample inlet** into the ionisation source of the instrument. On modern mass spectrometers the sample introduction process is often under complete **computerised** control. Small quantities of gases and liquids are injected. Most liquids vaporise at the low pressure in the machine but solids may need to be heated on special probes.

Once inside the ionisation source, the sample atoms or molecules are ionised to positively charged cations – charged ions are easier to manipulate than neutral molecules.

These ions are extracted into the analyser region of the mass spectrometer where they are separated according to their mass (m) to charge (z) ratios (m/z). The separated ions are detected and signals are sent to a data system where the m/z ratios are stored, together with their relative abundance, for presentation in the format of an m/z spectrum

Ionisation

The sample to be analysed has to be ionised first, forming positively charged cations. There are various techniques for doing this but two of the more common involve electron impact and laser pulsing. Ionisation of solid samples is often done by laser pulse avoiding the need to preheat the solid.

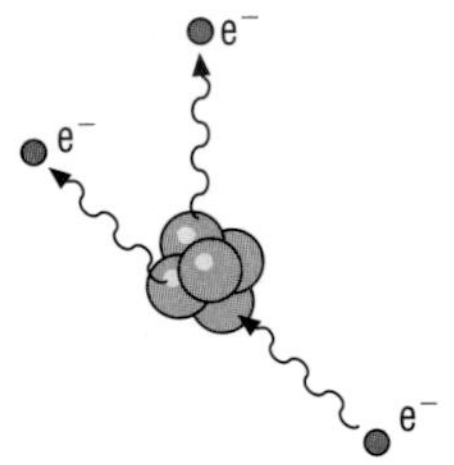

Electron impact

A stream of high-energy electrons from a heated filament bombards the sample of the element. The bombardment knocks electrons from the outside of the atoms, producing positive ions. Occasionally two electrons may be dislodged, but this is unlikely and the general assumption is that all the ions in the equipment carry a single positive charge. Representing the sample to be analysed by 'X', the ionisation process can be summarised by

$$X(g) + e^- \rightarrow X^+(g) + 2e^-$$

▲ **Figure 1** High-energy electron hits a molecule (or atom) knocking out an electron forming a positive ion.

Analyser

The ions produced in the ionisation area are separated according to their mass. There are various techniques available to bring about this separation. Some systems measure the *deflection* of an ionised, accelerated beam of ions in a magnetic or electric field. This deflection depends on mass and, therefore, ions of different mass are separated *spatially*.

However, we will consider one of the more versatile, sensitive and rapid methods of analysis called **'time of flight'** mass spectrometry. In this case, the ions are produced as separate pulses, not continually. As the name implies, a time of flight mass analyser identifies charged sample atoms, or molecules, by measuring their flight time.

Figure 2 shows the working principle of a linear time of flight mass spectrometer. To allow the ions to move along the flight path without hitting anything else, all the air molecules have been pumped out creating an ultra-high vacuum. In a time of flight mass spectrometer, an ion can typically fly on average 600 metres (mean free path) before it hits an air molecule. An electric field accelerates all the ions to the *same kinetic energy* in the acceleration area.

Time of flight mass spectrometers identify ions by measuring the *time* that sample ions, all starting with the same kinetic energy, take to fly a known distance (the flight path) in a constant electric field – this area is called the *drift region*. The time taken by a particular type of ion is a

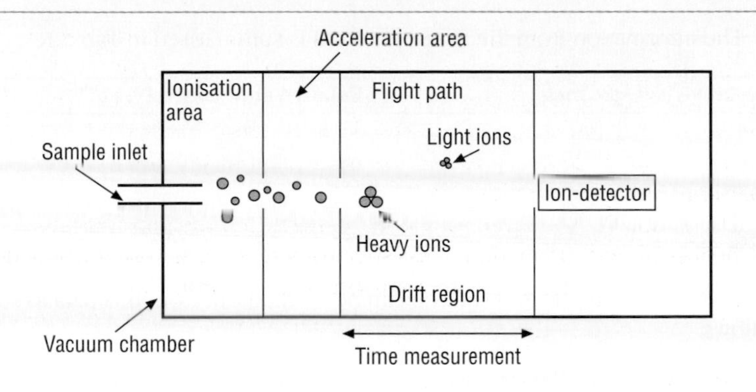

◀ **Figure 2** A schematic diagram showing the main parts of a time of flight mass spectrometer.

measure of its mass (strictly, its mass : charge ratio). Now, kinetic energy = $\frac{1}{2}mv^2$, where m is the mass of the ion and v is its velocity. Clearly, if all the ions of different masses have the same kinetic energy then heavier ions will move more slowly than lighter ions and will arrive at the detector later, having a longer time of flight. These times can be used to differentiate the masses of the different ions arriving at the detector.

Detector

The various ions are detected by an ion detector (detail not needed). The detector produces a varying electric current when it is hit by the ions, and a computer system converts this into a mass spectrum that shows both the masses of the ionised atoms or molecules reaching the detector and their relative abundance (in the case of elements) or intensity (in the case of compounds). See Figure 3 below.

Using mass spectra to calculate relative atomic mass

The mass spectrum for naturally occurring iron is shown in Figure 3. The horizontal axis shows the mass: charge ratio (m/z) of each ion detected. You can assume all the ions are singly charged and, therefore, since $z = 1$, m/z is the same as the mass of the ion detected. The **relative abundance** of each ion can be calculated from the height of each peak.

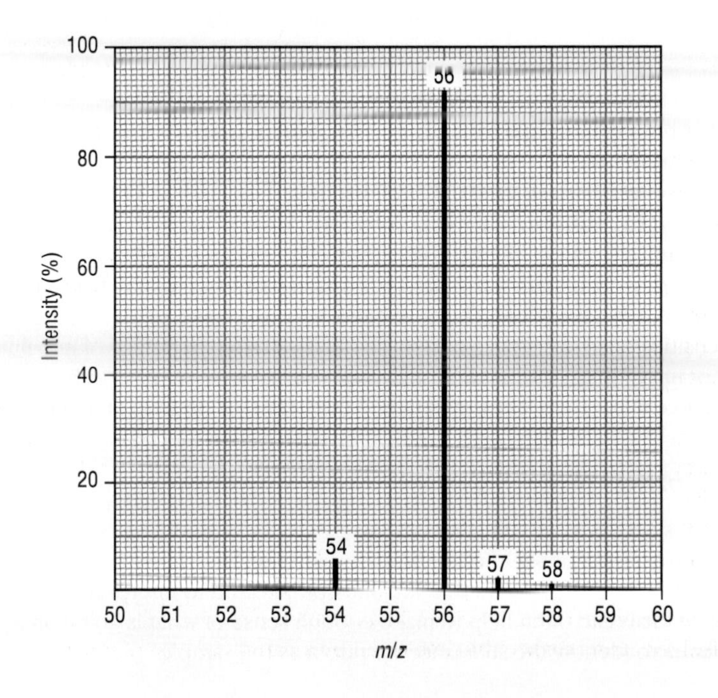

◀ **Figure 3** The mass spectrum for naturally occurring iron.

The information from the mass spectrum is summarised in Table 1.

► **Table 1** Isotopes of naturally occurring iron.

Relative isotopic mass	Relative abundance (%)
54	5.8
56	91.7
57	2.2
58	0.3

Samples of an element may vary slightly in average A_r depending on their source. This is due to slightly different abundances of individual isotopes between samples.

The relative atomic mass of iron can be worked out from this data as follows.

Average mass of 100 atoms
$= (54 \times 5.8) + (56 \times 91.7) + (57 \times 2.2) + (58 \times 0.3)$
$= 313.2 + 5135.2 + 125.4 + 17.4$
$= 5591.2$

So, the average mass of one atom of iron is 55.91, and this is its A_r.

Using mass spectra to investigate the structure of molecules

In this section we are concerned with how the mass spectrometer can be used to investigate the structure of *molecules* – the technique is particularly useful for organic molecules.

$CH_3-CH_2-CH(CH_3)-O-CH_2-CH_3$

2-ethoxybutane

When we look at a typical mass spectrum from an organic compound, such as the one for 2-ethoxybutane shown in Figure 4, we see that it is quite complex.

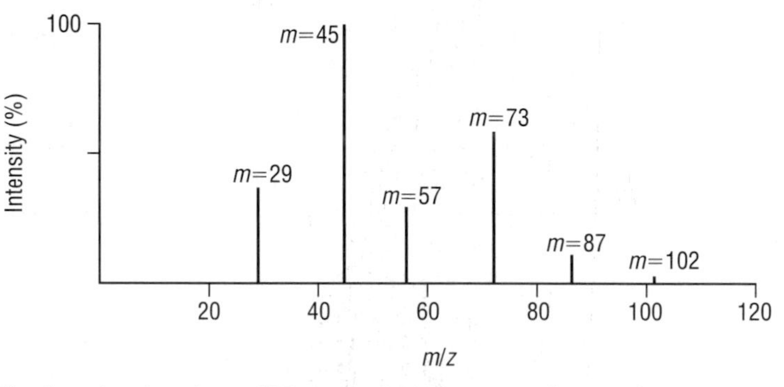

► **Figure 4** Simplified mass spectrum for 2-ethoxybutane, showing the six largest peaks.

The heaviest ion ($m = 102$) is the one corresponding to the ethoxybutane molecule with just one electron removed – this is called the **molecular ion**, X^+.

However, this is not the only ion reaching the detector. Other ions result from **fragmentation** – the breakdown of X^+ into positively charged smaller ions – and, of course, the mass spectrometer detects and analyses these too.

The most abundant ion gives the strongest detector signal, which is set to 100% in the spectrum. This is referred to as the **base peak** and the intensities of all the other peaks are expressed as a percentage of its value. It is not uncommon for the molecular ion peak to be so weak as to be unnoticeable – the spectrum obtained then consists entirely of fragments.

Fragmentation

The positive ions formed in a mass spectrometer can undergo some strange chemistry, quite unlike the reactions you are used to with laboratory chemicals, so mass spectra can be quite complicated to interpret. However, simple ideas can often help to make enough sense of what is going on to allow us to identify the substance we put in as the sample.

If we imagine the molecules of 2-ethoxybutane to be built like 'Lego' models – that can be pulled apart into their constituent 'building blocks' – we can identify nearly all the peaks in its mass spectrum. This is done in Figure 5.

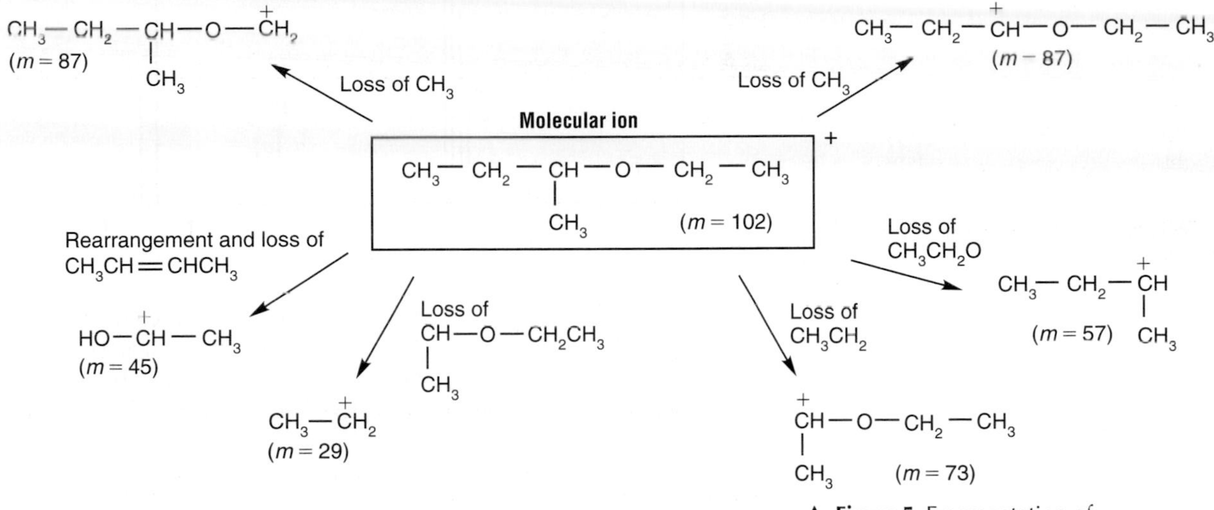

▲ **Figure 5** Fragmentation of 2-ethoxybutane – the fragments may be formed in several steps.

For each fragmentation, one of the products keeps the positive charge. This means that there are always two possibilities:

$$M^+ \rightarrow A^+ + B \quad \text{or} \quad M^+ \rightarrow A + B^+$$

Usually the fragment that gives the most stable positive ion is formed, but you may see products from both routes in a spectrum.

The 'Lego' approach enables you to make use of the mass differences between peaks (the masses of the bits which have fallen off) to help to make sense of the spectrum. For example, in Figure 4, peaks 102 and 87 differ by 15, corresponding to the loss of CH_3. This is a common process in mass spectrometry, so be on the lookout for gaps of 15 in your spectra – they probably mean there is a methyl group in the substance you are investigating. Table 2 gives some other common differences and the groups they suggest.

▼ **Table 2** Some common mass differences, and the groups they suggest.

Mass difference	Group suggested
15	CH_3
17	OH
28	C=O or C_2H_4
29	C_2H_5
43	$COCH_3$
45	COOH
77	C_6H_5

Isotope peaks

Some elements have more than one stable, naturally occurring isotope. Chlorine is an example – natural chlorine is made up of the isotopes ^{35}Cl and ^{37}Cl, in the ratio 75% to 25%.

In the case of a molecule containing one chlorine atom there will be two sorts of ions – one containing a ^{35}Cl atom and giving a signal at mass M, the other containing a ^{37}Cl atom giving a signal at mass $M + 2$. The heights of these two peaks will be in the ratio 75% to 25%, i.e. 3 : 1.

A pair of peaks like this is a tell-tale sign that chlorine is present. For example, Figure 6 (page 144) shows the mass spectrum for chloroethane, C_2H_5Cl. It shows two pairs of peaks with the tell-tale 3 : 1 ratio:

- 64 and 66 due to $[C_2H_5{}^{35}Cl]^+$ and $[C_2H_5{}^{37}Cl]^+$, which are the molecular ions
- 49 and 51 due to $[CH_2{}^{35}Cl]^+$ and $[CH_2{}^{37}Cl]^+$, which are the molecular ions minus CH_3.

One isotope which can be very useful in helping to interpret the mass spectra of more complicated molecules is ^{13}C. This accounts for only 1.1% of

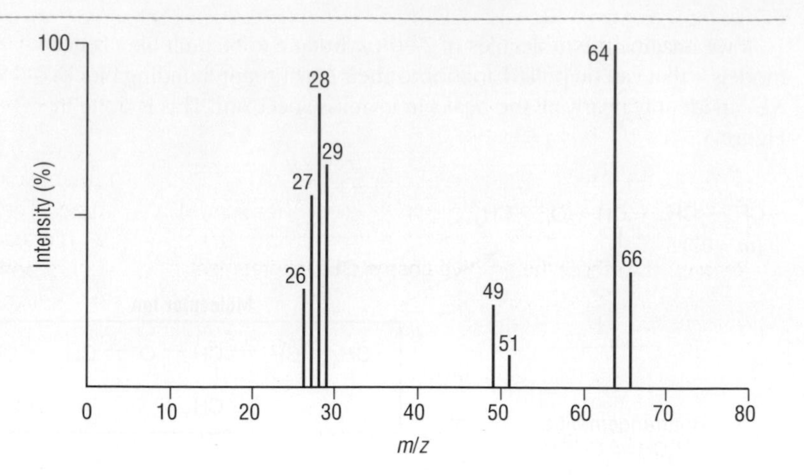

▶ **Figure 6** The mass spectrum of chloroethane.

a sample of naturally occurring carbon atoms. However, when there are several carbon atoms in a molecule of a compound, the abundance of ions containing a ^{13}C becomes significant enough for their signal to be clearly detectable. The more carbon atoms in a molecule, the greater the chance of one of them being ^{13}C.

For example, consider the mass spectrum of decan-1-ol, $C_{10}H_{21}OH$. You would expect a molecular ion peak at 158. However, in a molecule containing 10 carbon atoms there is approximately an 11% (10 × 1.1%) chance of one of them being ^{13}C. This makes the molecular ion heavier by 1 mass unit. So, out of every 100 molecular ions, 11 will have mass 159, and the other 89 will have mass 158. There will, therefore, be two molecular ion peaks, at 158 and 159, with intensities in the ratio 89 : 11.

Problems for 6.5

1 Boron has two isotopes. Use the mass spectrum in Figure 7 to calculate the relative atomic mass of boron.

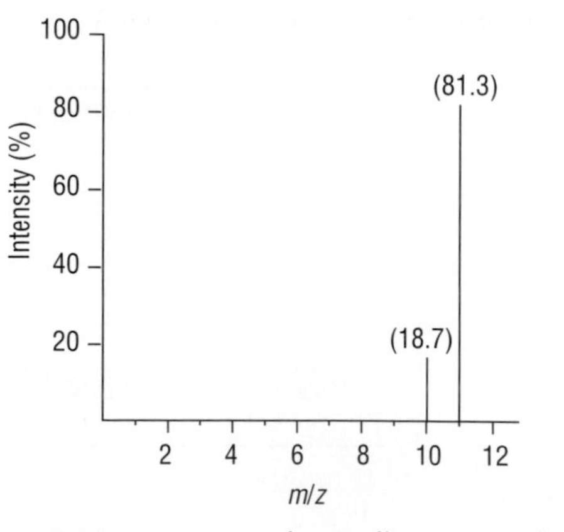

▲ **Figure 7** Mass spectrum of naturally occurring boron.

2 a Figure 8 shows the mass spectrum of a sample of zirconium. How many isotopes is this sample composed of?

b Calculate the relative atomic mass for zirconium using the mass spectrum shown.

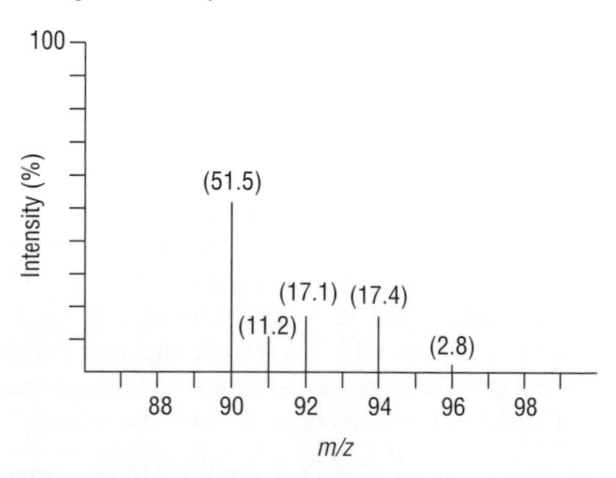

▲ **Figure 8** Mass spectrum of zirconium.

In the remaining problems, you will find that it often helps to draw up a table to show the masses of the peaks and the fragments responsible – Table 2 in this section (page 143) will be helpful too.

Mass of peak	Possible fragment

Remember to show the positive charge on each fragment – all the fragments detected are positive ions.

3 **a** The mass spectra of three compounds are shown in Figure 9. Identify the molecular ion peak in each case.

b Explain why there is a small peak at $M + 2$ in the mass spectrum of 2-chloropropane.

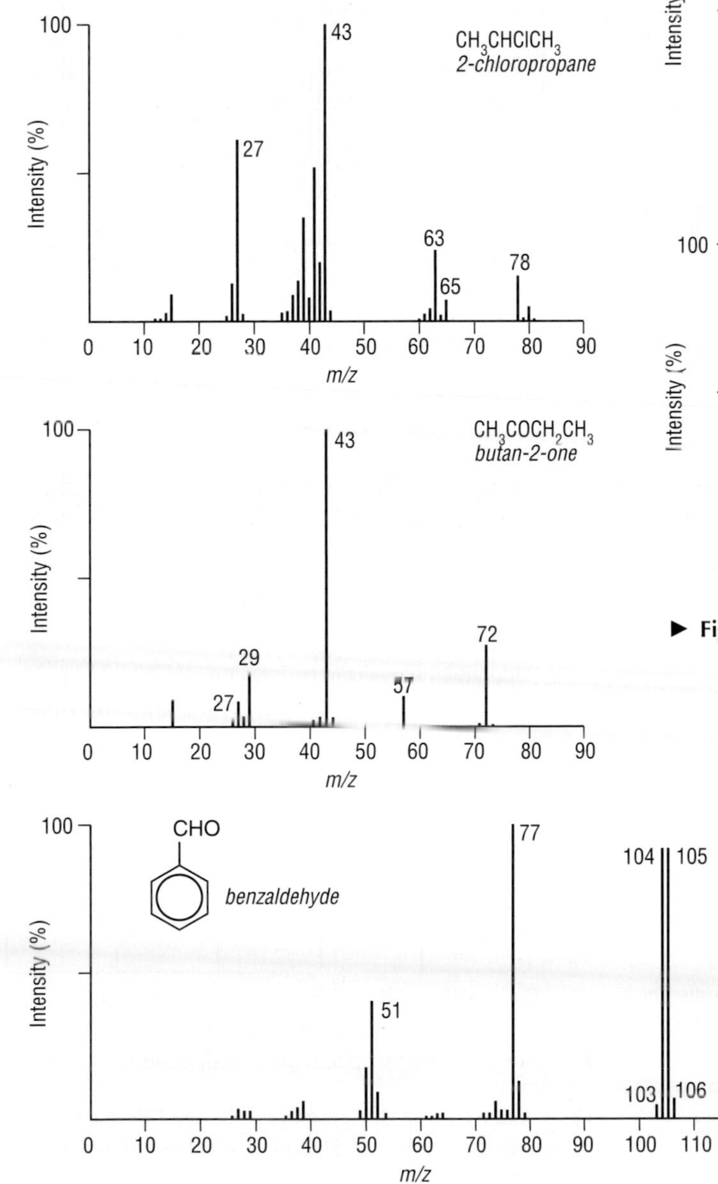

▲ **Figure 9** Mass spectra of 2-chloropropane, butan-2-one and benzaldehyde.

4 For each of the mass spectra in Figure 9, identify the base peak and write down a formula for a possible fragment ion responsible for the peak.

5 Figure 10 shows the mass spectra for ethanol (C_2H_5OH) and ethanoic acid (CH_3COOH), not necessarily in that order. The spectra are labelled **A** and **B**.

a For each spectrum, suggest a possible fragment ion responsible for the peaks marked with an asterisk (*).

b Assign the spectra to the correct compounds.

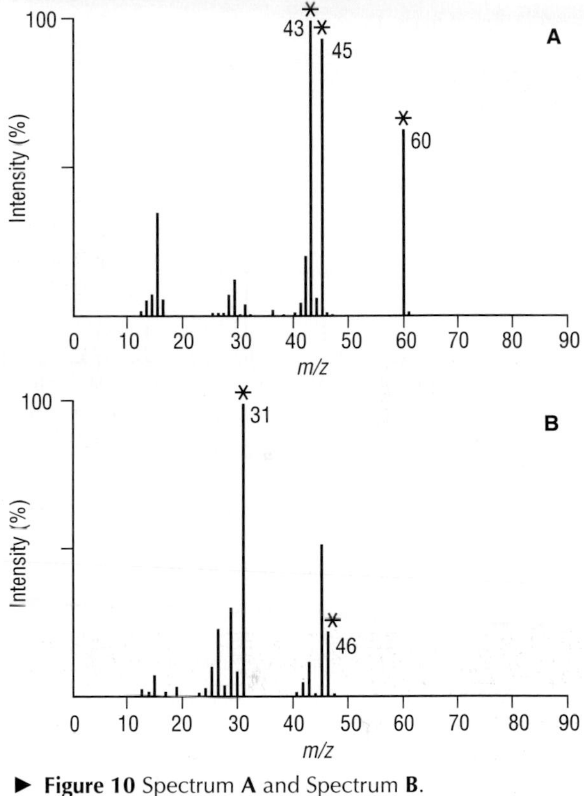

▶ **Figure 10** Spectrum **A** and Spectrum **B**.

6 The mass spectrum of the ester ethyl ethanoate is shown in Figure 11.

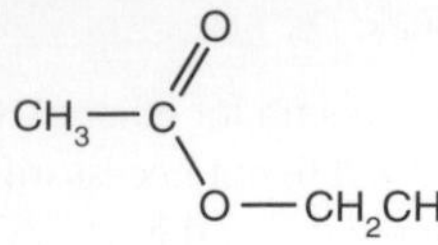

ethyl ethanoate

a Identify the ions responsible for the peaks at masses 88 and 43.

b Write an equation to show the fragmentation that occurs when the ion at mass 88 is converted into the ion at mass 43.

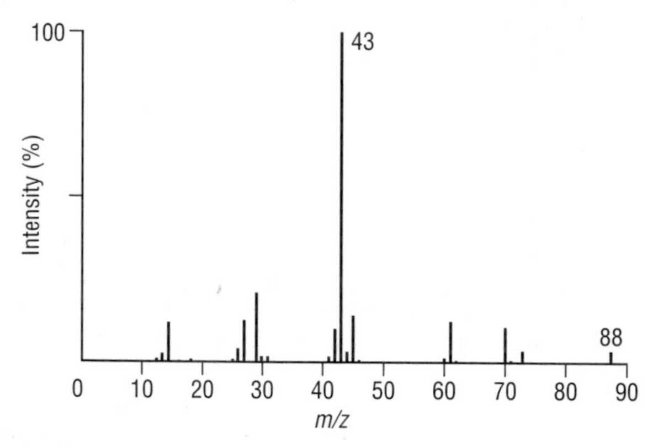

▲ **Figure 11** Mass spectrum of ethyl ethanoate.

7 The mass spectrum of a hydrocarbon with molecular formula C_4H_{10} is shown in Figure 12.

a What ions give rise to the peaks at mass 58, 43, 29, and 15?

b Draw two possible structures for C_4H_{10}.

c Decide which of these structures would give rise to the mass spectrum in Figure 12. Explain your reasoning.

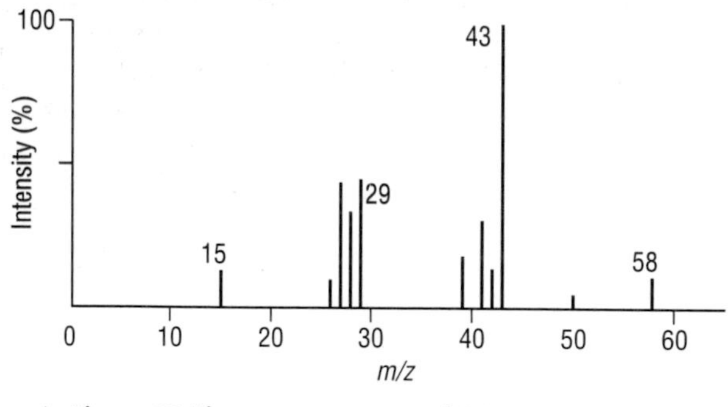

▲ **Figure 12** The mass spectrum of C_4H_{10}.

8 Compounds **C** and **D** both have the same molecular formula, C_3H_6O, but their molecular structures are different – they are isomers. Both contain the carbonyl group, C=O.

a Draw out two possible structures for C_3H_6O.

b For compound **C**, which group of atoms could be lost when the ion of mass 43 forms from the ion of mass 58?

c For compound **D**, what atom or groups of atoms could be lost when

i the molecular ion changes into the ion of mass 57?

ii the ion of mass 57 changes into the ion of mass 29?

d Suggest formulae for the ions of mass

i 43 in compound **C**

ii 28, 29 and 57 in compound **D**.

e Identify **C** and **D**.

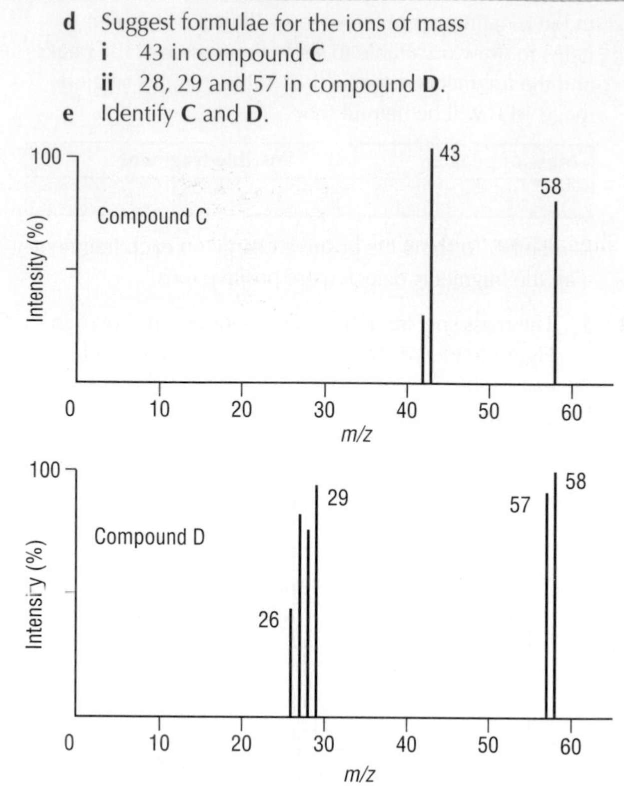

▲ **Figure 13** The mass spectrum of compounds **C** and **D**.

9 Compound **E** contains only carbon, hydrogen and oxygen. Its accurate molecular mass was found to be 72.0573. A database gave four compounds with masses in this region: $C_2H_4N_2O$, $C_3H_8N_2$, $C_3H_4O_2$ and C_4H_8O.

a Use these accurate atomic masses to work out the formula of **E**: H = 1.0078, O = 15.9949, N = 14.0031, C = 12.0000.

b Use the mass spectrum shown in Figure 14 to work out the structure of **E**.

c Identify the peaks in the spectrum at masses 15, 29 and 43 and from these identify the structure.

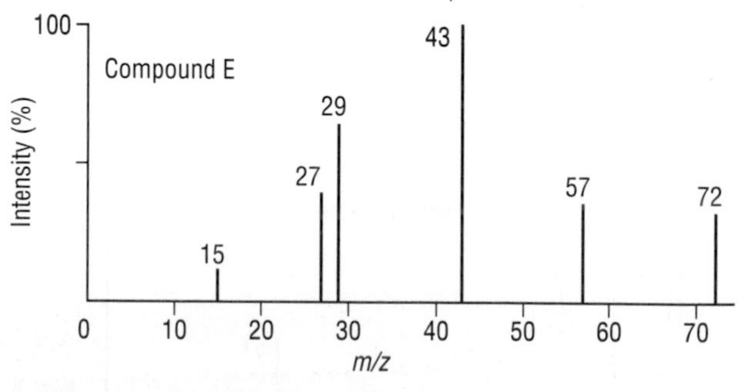

▲ **Figure 14** The mass spectrum of compound **E**.

10 Bromine consists of the isotopes ^{79}Br and ^{81}Br in an approximately 50% to 50% ratio. Use this information, together with the mass spectrum for chloroethane in Figure 6 in this section (page 144), to draw a sketch of what you think the mass spectrum of bromoethane looks like.

6.6 *Nuclear magnetic resonance spectroscopy*

What is n.m.r?

Nuclear magnetic resonance spectroscopy (n.m.r.) is one of the most widely used analytical techniques available to chemists. This might seem surprising – after all, our thinking in chemistry normally stops at the outermost electrons in a substance; the nucleus is just something we accept is there, small and heavy, at the centre of each atom.

However, n.m.r. provides us with very detailed information about the nuclei of certain atoms. For example, **hydrogen nuclei** behave differently in different molecular environments – n.m.r. tells us what the environments are and how many ^{1}H nuclei (usually just called protons) there are in each.

N.m.r can also be used to find out about certain other nuclei, particularly ^{13}C but also ^{19}F and ^{31}P. However, in this section we will be concentrating on ^{1}H nuclei.

Such nuclei behave like tiny magnets and when they are placed in a strong magnetic field they align themselves with or against the magnetic field. If they are aligned *against* the magnetic field, they have a higher energy than if they are aligned *with* it. Slightly more than 50% of nuclei are in the lower energy level. If the correct frequency of radiation is applied, some of these nuclei move up to the higher level, absorbing some energy, ΔE, as they do so (Figure 1). The energy absorbed corresponds to radio frequencies.

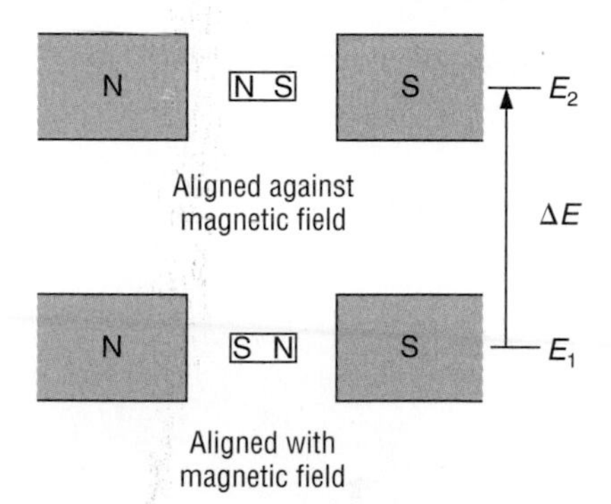

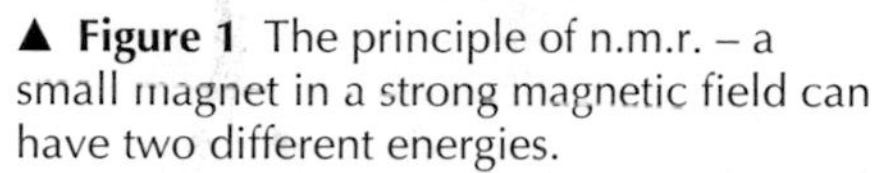
Figure 1 The principle of n.m.r. – a small magnet in a strong magnetic field can have two different energies.

The energy needed for the nuclear 'magnet' to move to a higher energy level depends on the strength of the magnetic field it experiences. This is not quite the same as the field being applied by the instrument because the electrons associated with the neighbouring atoms and groups in the molecule give rise to tiny magnetic fields of their own.

These *local* fields are usually opposed to the *external* field applied by the instrument. The *overall* field experienced by a proton is therefore slightly smaller than the external field, depending on the local field from the surrounding part of the molecule.

So for every type of molecular arrangement (*molecular environment*), there is a very slightly different magnetic field. The ^{1}H nuclei in these different environments have different energy gaps (ΔE) between their high and low energy levels, and so absorb different frequencies of radiation. Therefore, they give different n.m.r. absorption peaks and we can find out how many hydrogen atoms of different types there are in a molecule.

^{1}H n.m.r. spectroscopy is often called **proton n.m.r. spectroscopy** because an ^{1}H nucleus is essentially a proton. This can be confusing because we also use the term *proton* to describe the $H^{+}(aq)$ ion in acid–base reactions (see **Section 8.1**).

Make sure you understand the difference!

The n.m.r. instrument

The main features of a modern n.m.r. spectrometer are illustrated in Figure 2. There is a magnet which produces a strong magnetic field, a radio-frequency (RF) source, a detector and a recorder.

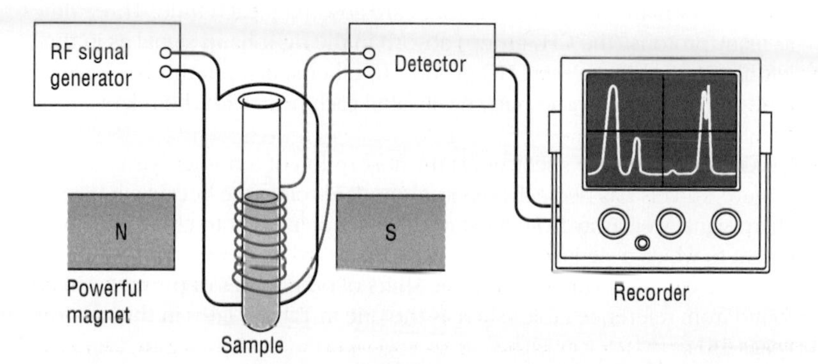

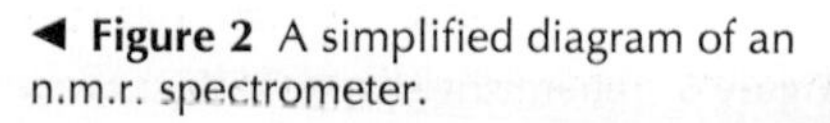
Figure 2 A simplified diagram of an n.m.r. spectrometer.

The sample tube is placed in the magnetic field and is spun to make sure that the field experienced by the sample is as homogeneous as possible. The substances to be investigated are in the form of pure liquids or are dissolved in deuterated solvents such as CD_2Cl_2 or CD_3COCD_3, where ^1H atoms have been replaced by ^2H atoms (deuterium, represented by D). If the solvent contained ^1H atoms, these would give a spectrum of their own.

To obtain the n.m.r. spectrum, the magnetic field is held constant and a band of radio frequencies is applied as a pulse to the sample. Immediately after the radio-frequency pulse, the precise frequency absorbed is re-emitted as ^1H nuclei return to the lower energy state.

The signal detected is weak and the process is over very quickly, so it is repeated many times in rapid succession to build up an accurate record. This is stored and analysed electronically.

Interpreting n.m.r. spectra

Ethanal has two types of ^1H nuclei – one in the CHO group and the other in the CH_3 group. The energy levels for the two types of hydrogens in the same external magnetic field are shown in Figure 3.

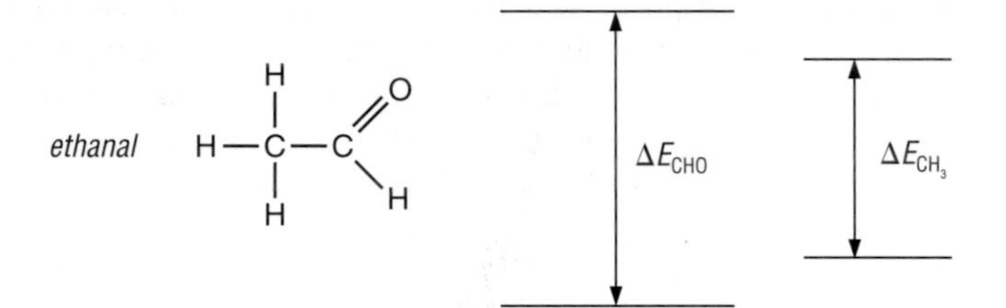

► **Figure 3** Energy levels for the two types of ^1H nuclei in an ethanal molecule.

This molecule will absorb (and emit) two frequencies in an n.m.r. machine. The resulting spectrum is shown in Figure 4.

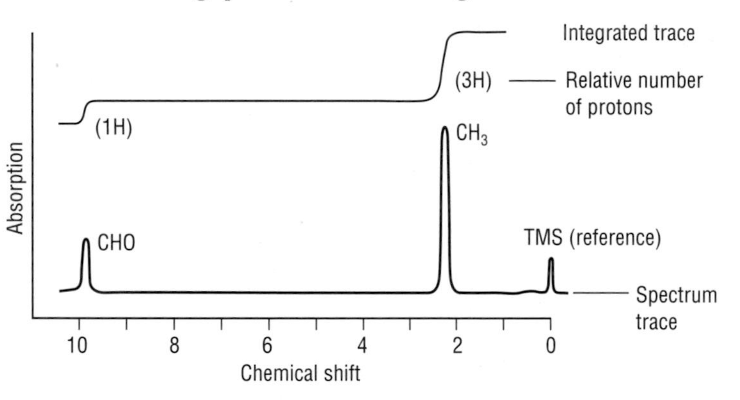

► **Figure 4** The n.m.r. spectrum of ethanal (low resolution).

As well as the **spectrum trace** in Figure 4, the n.m.r. spectrometer has also drawn an **integrated** spectrum trace. This goes upwards in steps, which are proportional to the areas of the absorption signals and therefore tells us how many ^1H nuclei (protons) are absorbing each time. Three times as many protons (the CH_3 group) absorb in the right-hand signal as in the left-hand signal (from the CHO group). In other spectra in this book we will omit the integrated trace, but we will label each peak with the relative numbers of protons.

Also shown on the spectrum is the absorption of tetramethylsilane, TMS (Figure 5). This has been chosen as a standard reference because it gives a sharp signal well away from most of the ones of interest to chemists. The extent to which a signal differs from TMS is called its **chemical shift**. TMS is set at a chemical shift of 0 and the shifts of other types of protons can be found from reference tables such as the one in Table 1 later in this section (page 151).

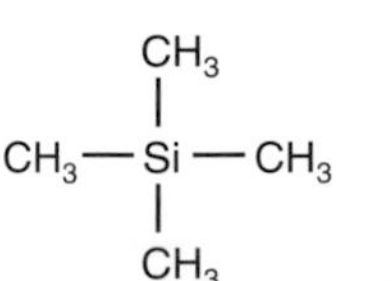

▲ **Figure 5** Tetramethylsilane (TMS).

It isn't quite as simple as that ...

A sensitive n.m.r. machine actually produces a much more detailed spectrum than the one shown in Figure 4. Figure 6 shows what the detailed n.m.r. spectrum of ethanal actually looks like.

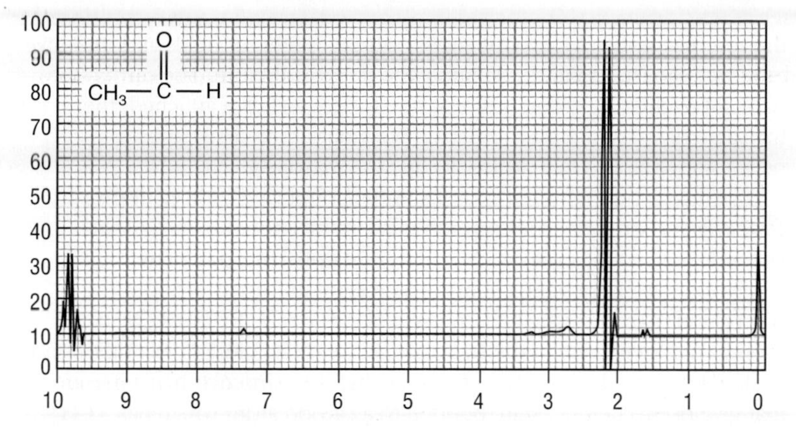

◀ **Figure 6** The n.m.r. spectrum of ethanal at a higher resolution than that in Figure 4.

Figure 7 shows how the less detailed (low resolution) spectrum in Figure 4 is related to the detailed (high resolution) spectrum in Figure 6. How does the extra detail arise?

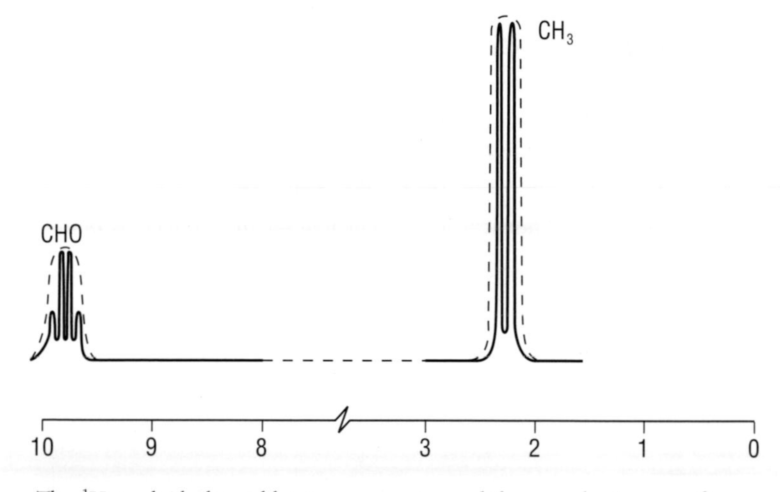

◀ **Figure 7** The dotted lines show how the less detailed spectrum in Figure 4 is related to the detailed spectrum in Figure 6.

The ^{1}H nuclei behave like tiny magnets, and they can be in one of two orientations depending on whether they are in the low or high energy level. In the CH_3 group, there are *four* combinations in which the *three* 'tiny proton magnets' can be arranged:

- all aligned with the external field S–N S–N S–N
- two with and one against the external field S–N S–N N–S
- one with and two against the external field S–N N–S N–S
- all aligned against the external field N–S N–S N–S

The magnetic effect of these arrangements in the CH_3 groups is transmitted to the neighbouring CHO protons so that they sense one of *four* magnetic fields (or $n + 1$ fields, where n is the number of hydrogen atoms bonded to the *neighbouring* carbon atom – this is known as the 'n +1 rule'). They can absorb four different frequencies and give *four* signals. These are centred on the expected chemical shift.

The peak intensities are in the ratio 1 : 3 : 3 : 1 because there are three possible combinations leading to each of the two central peaks:

S–N/S–N/N–S S–N/N–S/S–N N–S/S–N/S–N

and

N–S/N–S/S–N N–S/S–N/N–S S–N/N–S/N–S

Because the CHO protons can only be arranged in one of two ways

S–N or N–S

the protons of the CH_3 groups sense only two different fields and they give only *two* signals. Again we can apply the $n + 1$ rule here. The carbon adjacent to the CH_3 group has only one hydrogen atom attached, therefore the value of $n = 1$ and so $n + 1 = 2$. We would therefore expect the CH_3 peak to be split into *two*, which it is for the reasons given above.

The interaction between protons on neighbouring carbon atoms is called *coupling* and causes the absorption peak in the spectrum to be split. Different interactions will produce different **splitting patterns**. High-resolution spectra provide chemists with extra information about the compound being investigated. In particular, the $n + 1$ rule can be very useful when determining the structure of the molecule. For example, if a peak is split into three then that tells us that the particular hydrogen atoms that give rise to that peak are bonded to a carbon atom which has a CH_2 group attached to it. However, analysing high-resolution spectra and the splitting due to coupling is more tricky. For this course we will not be concerned with the detailed splitting of the peaks.

Examples of n.m.r. spectra

The following examples show you how to use n.m.r. spectra to find information about the structure of organic molecules.

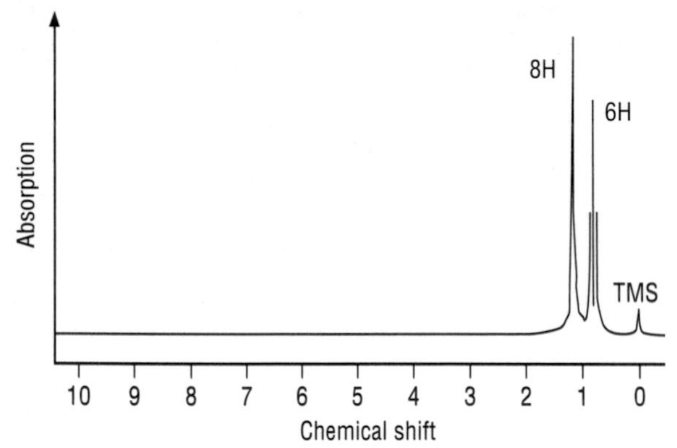

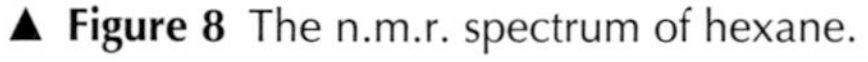

▲ **Figure 8** The n.m.r. spectrum of hexane.

$CH_3–CH_2–CH_2–CH_2–CH_2–CH_3$

Chemical shift	Relative no. of protons	Type of proton
0.9	3	CH_3
1.3	4	CH_2

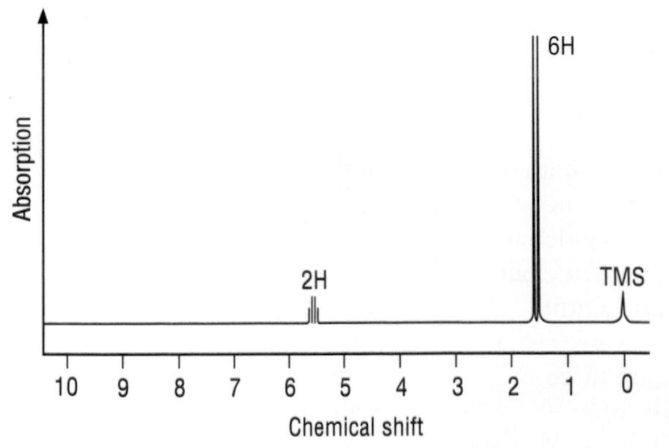

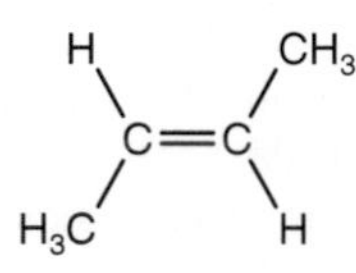

Chemical shift	Relative no. of protons	Type of proton
1.6	3	CH_3
5.6	1	H–C=C–H

▲ **Figure 9** The n.m.r. spectrum of (*E*)-but-2-ene.

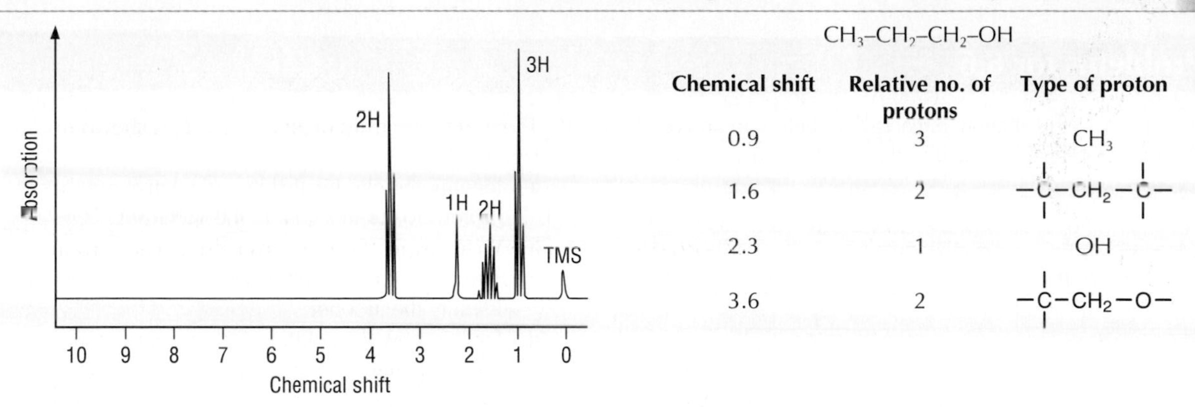

▲ **Figure 10** The n.m.r. spectrum of propan-1-ol.

▼ **Table 1** Chemical shifts for some types of protons (^1H nuclei) – R represents an alkyl group.

Type of proton	Chemical shift (δ) in approximate region of
CH_3—C	0.7–1.6
C—CH_2—C C—CH—C (with H on the central C)	1.4–2.3
CH—C(=O)— carbonyls, esters, amides, acids	2.0–2.7
—CH—N amines, amides	2.3–2.9
benzene ring—CH	2.3–3.0
—O—CH alcohols, esters, ethers	3.3–4.8
—CH—Cl or Br	3.0–4.2
—CH = CH—	4.5–6.0

Type of proton	Chemical shift (δ) in approximate region of
benzene ring—H	6.4–8.2
—C—CHO	9.4–10.0
—C—OH	0.5–4.5*
benzene ring—OH	4.5–10.0*
—C—NH	1.0–5.0*
—CO—NH	5.0–12.0*
—CO—OH	9.0–15.0*

* Signals from hydrogens in –OH and –NH– groups in alcohols, phenols, carboxylic acids, amines and amides are very variable and often broad. The chemical shift is sensitive to temperature, nature of the solvent and the concentration. The stronger the hydrogen bonding the larger the chemical shift.

Problems for 6.6

Use Table 1 in this section (page 151) to help you answer the following problems. You will find it helpful to use a table like the one below.

Chemical shift from spectrum	Relative number of protons	Type of proton

It is important to realise that the value of the chemical shift for a particular type of proton is not always at exactly the same value. You are looking for the best match considering all the evidence given in the problems.

1 A compound with the empirical formula C_4H_8O gave the n.m.r. spectrum shown in Figure 11. The infrared spectrum confirmed the presence of a C=O group.

a How many different types of protons are present in the compound?

b Draw up a table and list the chemical shift and relative number of protons for each signal in the spectrum.

c Which type of proton gives rise to each signal?

d Suggest a structure for the compound.

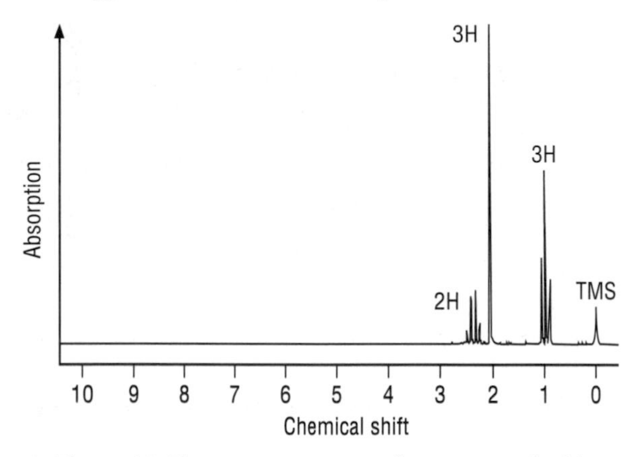

▲ **Figure 11** The n.m.r. spectrum of a compound with empirical formula C_4H_8O.

2 The following acids are among those commonly present in wine.

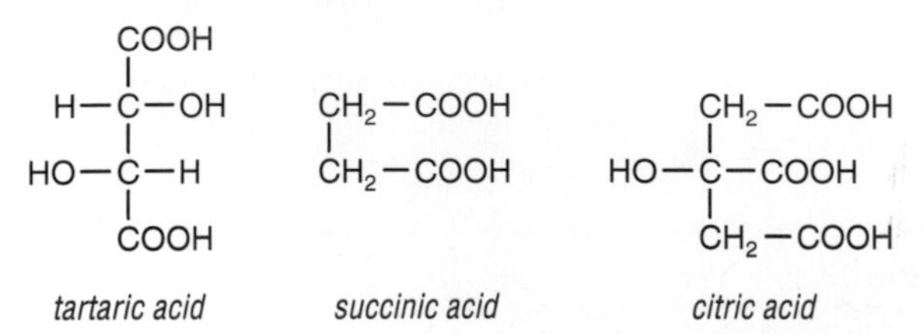

a For each compound, state how many signals you would expect in its n.m.r. spectrum. Explain your reasoning.

b Give the ratios of peak areas in the n.m.r. spectrum of each compound.

3 The n.m.r. spectrum of an alcohol, **E**, is shown in Figure 12.

a Identify the chemical shift and relative number of protons for each signal in the spectrum.

b Which type of proton gives rise to each signal?

c The relative molecular mass of the alcohol is 46. Identify the alcohol.

d Explain how the splitting pattern supports your answer to part **c**.

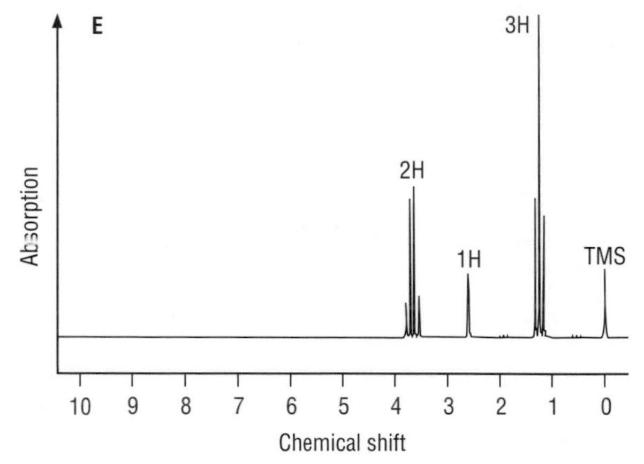

▲ **Figure 12** The n.m.r. spectrum for alcohol **E**.

4 The n.m.r. spectrum of another alcohol, **F**, is shown in Figure 13.

a Identify the chemical shift and relative number of protons for each signal in the spectrum.

b Which type of proton gives rise to each signal?

c Identify the alcohol.

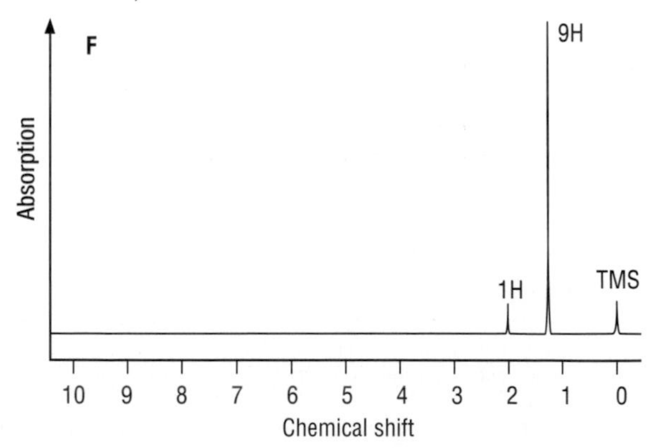

▲ **Figure 13** The n.m.r. spectrum for alcohol **F**.

5 Compound **G** is a hydrocarbon with molecular formula C_8H_{10}. Use the n.m.r. spectrum in Figure 14 to identify **G**.

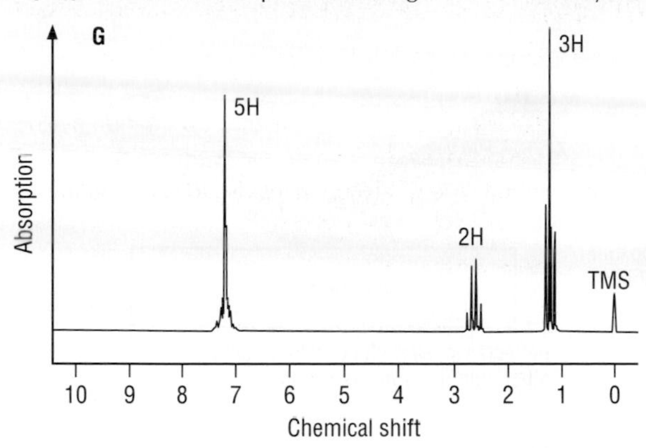

▲ **Figure 14** The n.m.r. spectrum for compound **G**.

6 The n.m.r. spectrum of an ester, **H**, is shown in Figure 15. Work out the structure of the ester.

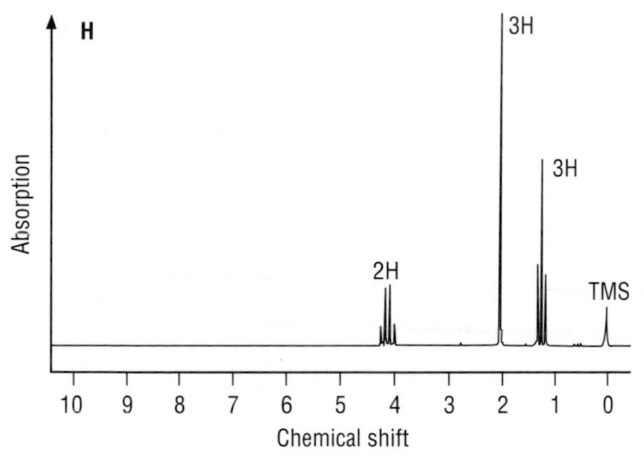

▲ **Figure 15** The n.m.r. spectrum for ester **H**.

7 Vanillin is a widely used flavouring which occurs naturally in the pods of the vanilla orchid. The structure of vanillin is shown here. Figure 16 shows the n.m.r. spectrum of vanillin. Match each of the peaks in the spectrum with protons in the vanillin structure.

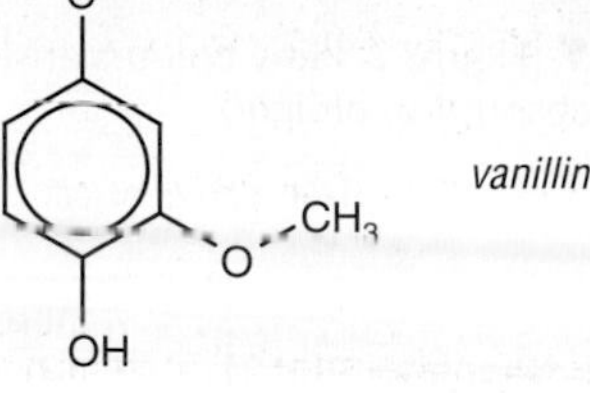

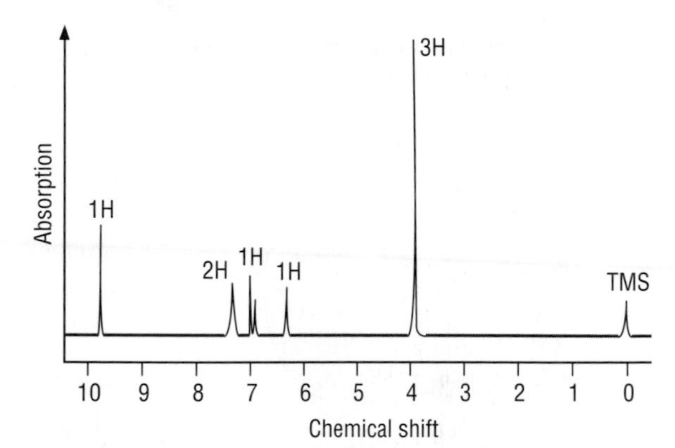

▲ **Figure 16** The n.m.r. spectrum of vanillin.

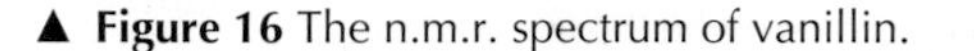

8 DIMP is an insect repellent developed to try to protect forestry workers from insects such as midges. Its structure is shown here.

a Identify the different types of proton in the molecule, the expected chemical shifts and the relative numbers of each type of proton.

b Sketch the n.m.r. spectrum you would expect for DIMP.

6.7 *Where does colour come from?*

You 'see' an object because light reflected from it enters your eyes. If all the incident sunlight is reflected, the object appears white. Some objects are transparent, like glass, and allow light to pass through them. Figure 1 shows what happens if all the wavelengths of visible light are transmitted or reflected.

▼ **Figure 1** How light behaves with transparent and opaque objects.

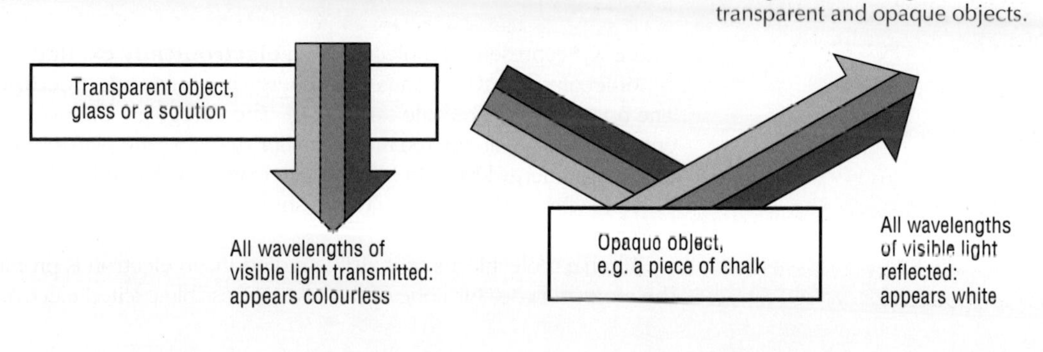

However, most objects appear coloured – this is because wavelengths corresponding to particular colours are being absorbed. Two examples are given in Figure 2.

▼ **Figure 2** How colours arise from absorption of light.

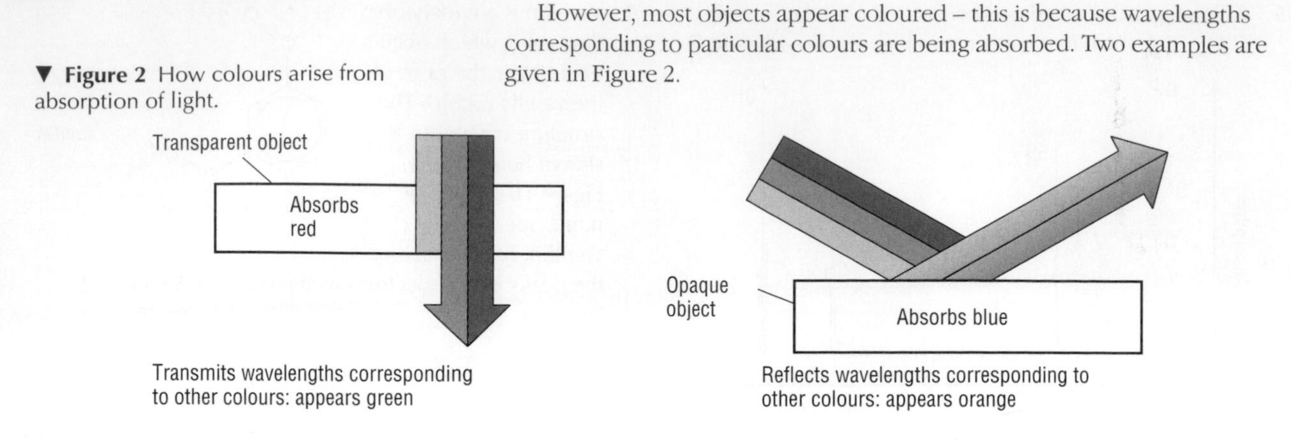

When absorption occurs, wavelengths corresponding to one colour are removed from the white light and you see the **complementary** colour. The relationship is sometimes shown as a colour wheel, in which complementary colours are opposite one another (Figure 3). Thus an object absorbing violet appears yellow-green.

It is common for people working with visible and ultraviolet radiation to use wavelength, rather than frequency, as a unit. The relationship between them is explained in **Section 6.1**.

Figure 4 is a chart showing the frequency, wavelength and colour of light.

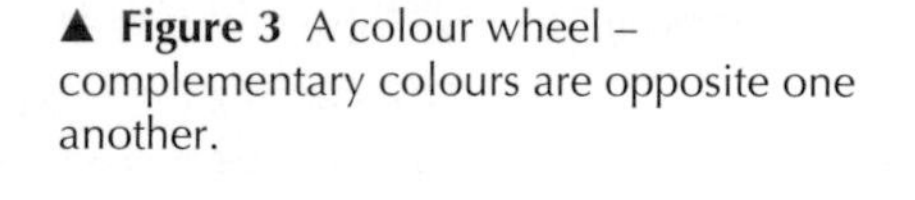

▲ **Figure 3** A colour wheel – complementary colours are opposite one another.

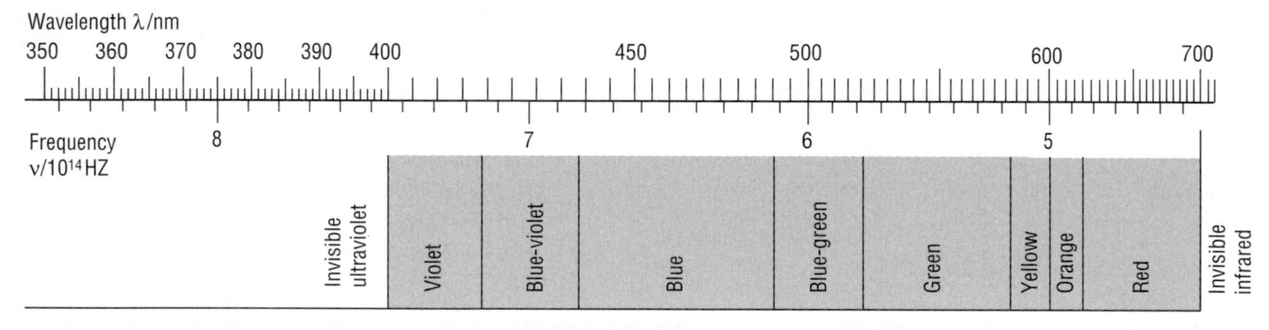

▲ **Figure 4** Wavelength, frequency and colour of light.

What happens when molecules absorb visible radiation?

When an *atom* absorbs visible radiation, an electron moves to a higher energy level. We often represent an atom in an excited (higher energy) state as X*.

The same thing happens when *molecules* absorb visible radiation:

$$X_2 \rightarrow X_2^*$$

where X_2^* represents a molecule in an **electronically excited state**.

Molecules can also change their energy in other ways (**Section 6.2**). The bonds in the molecule can vibrate. The energy gaps between vibrational energy levels are about 100 times smaller than the energy gaps between electronic energy levels. In any one electronic level a molecule has many possible vibrational energies. This is shown for a simple diatomic molecule in Figure 5.

When a molecule absorbs visible radiation, an electron is promoted and the molecule goes into one of the several possible excited electronic levels.

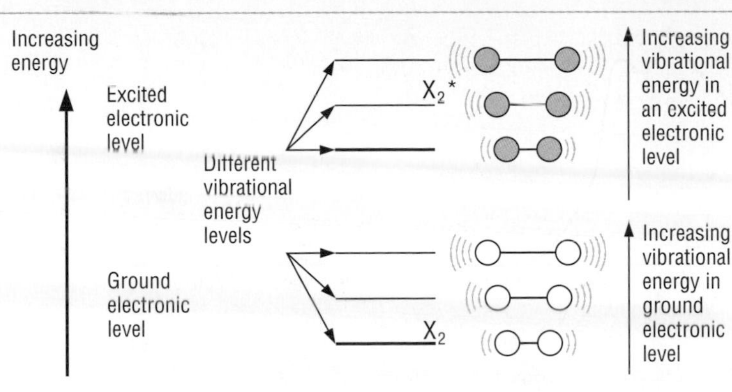

◄ **Figure 5** The electronic and vibrational energy levels of a diatomic molecule X_2. X_2* represents the excited state of X_2.

What happens to the radiation absorbed?

The molecule does not stay in the excited state for very long. The energy absorbed is re-emitted – but not necessarily all at once – as visible radiation. The molecule may emit a smaller quantum of radiation and fall back to an intermediate energy level. Relatively small quantities of vibrational energy are rapidly lost to other molecules during collisions and so the molecule drops to a lower vibrational level in the same electronic level. The energy is converted to kinetic energy of the molecules. This means that the molecules move around more vigorously and the sample becomes warmer.

Problems for 6.7

Refer to Figures 3 and 4 in this section (page 154) when answering these problems.

1 What range of wavelengths do we perceive as
 a green? **b** red?

2 A solution appears orange–red. What range of wavelengths of visible light is being absorbed by the solution?

3 A substance absorbs both red and violet light. What colour is it likely to appear? Sketch a possible absorption spectrum for this substance showing intensity of absorption against wavelength.

4 Calculate the frequency of light corresponding to wavelengths of
 a 430 nm **b** 350 nm **c** 700 nm.

6.8 *Ultraviolet and visible spectroscopy*

Why are carrots orange? If a substance absorbs radiation in the visible region of the spectrum, then the light that reaches your eye from the substance will be lacking in certain colours. It will no longer appear white. For example, carrots contain the pigment *carotene*.

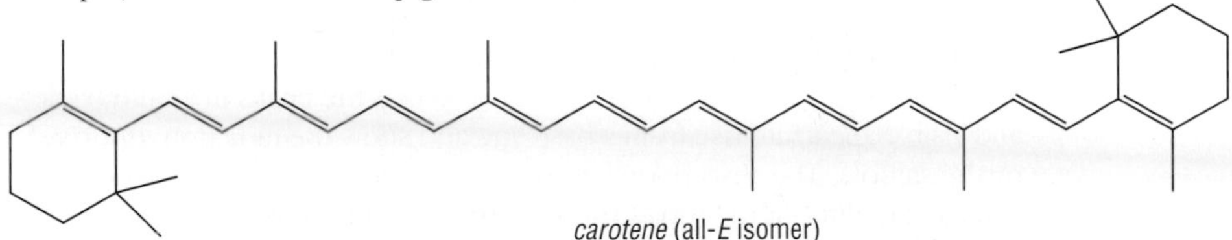

carotene (all-*E* isomer)

Carotene absorbs blue light strongly, so the light reaching your eye is lacking in blue light and the carrot appears orange–red (see **Section 6.7**).

If you make a solution of carotene in a suitable solvent, you can use a **spectrometer** to measure the quantity of light absorbed by the solution at each wavelength. The recorder plots out the intensity of absorption against the wavelength. The result is an **absorption spectrum** of carotene. You can see what this looks like in Figure 1 (page 156).

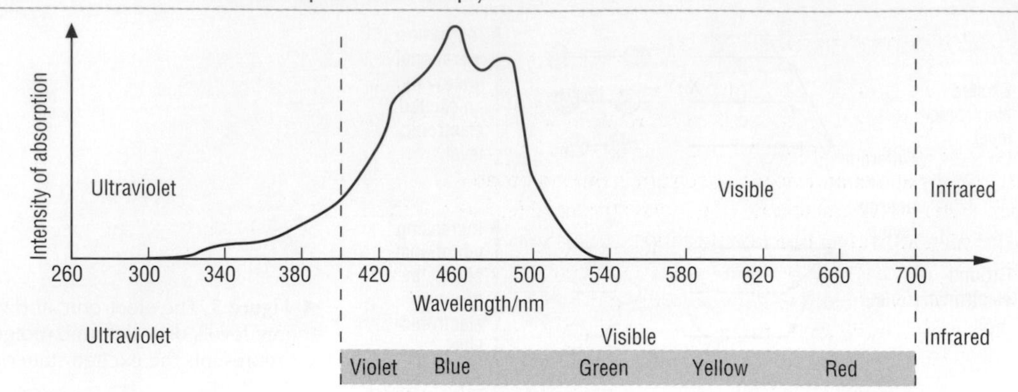

▲ **Figure 1** The absorption spectrum of carotene (in solution in hexane).

Absorption spectra

An **ultraviolet and visible spectrometer** works on much the same principle as the infrared spectrometer (see **Section 6.4**). In this case, the source is ultraviolet and visible radiation rather than infrared. Light from the source is split into two identical beams – one beam passes through the sample solution; the other passes through pure solvent. The light in the two emerging beams is compared to give the absorption spectrum of the sample.

Most spectrometers scan wavelengths in the ultraviolet as well as the visible region to give a continuous spectrum like the one in Figure 1.

Absorption of radiation in the ultraviolet and visible regions is associated with changes in the electronic energy of molecules (**Sections 6.1** and **6.2**). For this reason, ultraviolet and visible spectra are sometimes called **electronic spectra**.

Our eyes cannot detect ultraviolet light, so if a compound absorbs in this region the absorption does not affect its colour. This means that a compound like benzene, which absorbs only in the ultraviolet region (see Figure 2), appears colourless because it transmits all visible radiation.

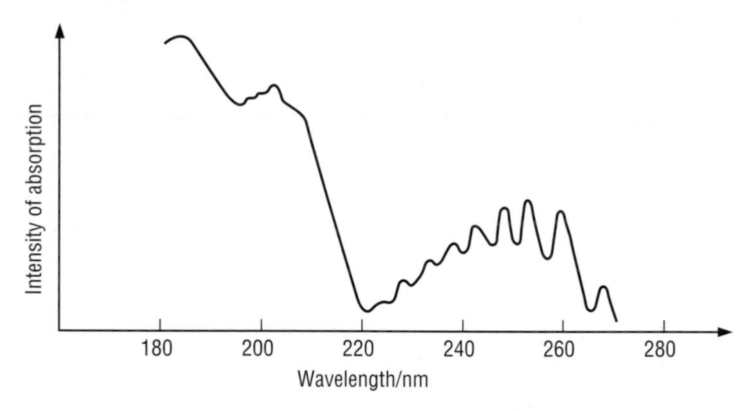

► **Figure 2** The absorption spectrum of benzene.

Ultraviolet and visible spectra look different from the infrared spectra in **Section 6.4**. The most obvious difference is that the peaks in an ultraviolet and visible spectrum rise from a base line and show the radiation *absorbed* by the sample. The absorption bands in an infrared spectrum hang down from a 'base line' at the top of the spectrum, because the spectrometer is plotting the radiation *transmitted* by the sample. It is just two different ways of recording the same effect.

Another difference is in the units used to measure the radiation absorbed. In an infrared spectrum it is the *wavenumber*, $1/\lambda$ (cm^{-1}), which is plotted on the horizontal axis. In an ultraviolet and visible spectrum, it is much more usual to plot the *wavelength*, λ (nm). Don't let this worry you. *Always work in the units of the data you are given, unless you are specifically asked to convert from one set of units to another.*

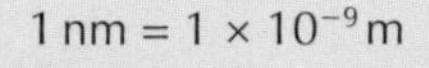

Using ultraviolet and visible spectra

When you interpret an infrared spectrum you can assign specific absorption bands to particular groups in the molecule. For example, a C=O group is usually responsible for an absorption around 1720 cm^{-1} and so on.

In contrast, an ultraviolet and visible spectrum is often a broad absorption band which is characteristic of general structural features of the molecule rather than individual functional groups.

You will find out more about the relationship between the structure of a compound and its ultraviolet and visible spectrum in **Section 6.9**.

A **colorimeter** is a simple type of visible spectrophotometer. Colorimetry is used to measure the intensity of absorption of coloured compounds over a narrow range of frequencies – it provides a useful way of finding the *concentration* of a coloured compound.

You can use a colorimeter to measure the concentration of a coloured compound at different times as a reaction proceeds in a rate experiment to find the order of the reaction (see **Section 10.3**).

Interpreting the spectrum

Colour chemists are interested in three main features of the spectrum:

- the wavelength of the radiation absorbed (remember, for a compound to be coloured, at least part of the absorption band must be in the visible region)
- the intensity of the absorption
- the shape of the absorption band.

When recording the spectrum, chemists often give the wavelength of the maximum absorption (λ_{max}). For carotene, λ_{max} occurs at 453 nm in the blue region of the spectrum (see Figure 1).

For organic molecules with delocalised electron systems, the longer the conjugated carbon chain, the more intense the absorption and the longer the wavelength of λ_{max}.

The intensity of the absorption also depends on the concentration of the solution and on the distance the light travels through the solution. Standard molar values are quoted so that values for different compounds can be compared. (You don't need to know about the units used to measure intensity for this course. It is enough to know that the higher the peak, the more intense the absorption.) The intensity of the absorption is important commercially because it determines the amount of pigment or dye needed to produce a good colour.

The shape and width of the absorption band is important because it governs the shade and purity of the colour seen.

Reflectance spectra

To measure an absorption spectrum, you need to make a solution of the coloured substance. However, it is not always possible to do this – for example, when the coloured substance is a pigment on the surface of a painting. In cases like this, chemists can use a different type of visible spectrum called a **reflectance spectrum**.

Here they shine light onto the paint surface and examine the composition of the reflected light. Since this is the part of the light that was *not absorbed* by the pigment, a reflectance spectrum is a sort of negative of the absorption spectrum. Figure 3 shows the absorption and reflectance spectra for the pigment monastral blue.

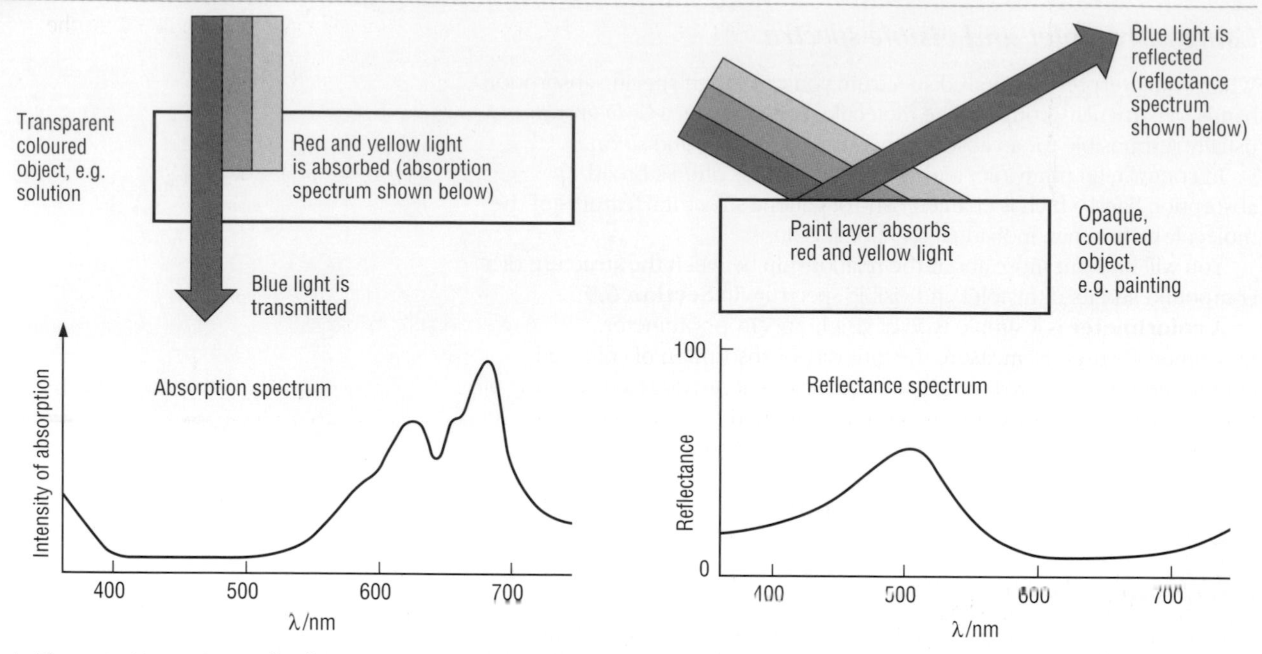

▲ **Figure 3** Absorption and reflectance spectra of monastral blue.

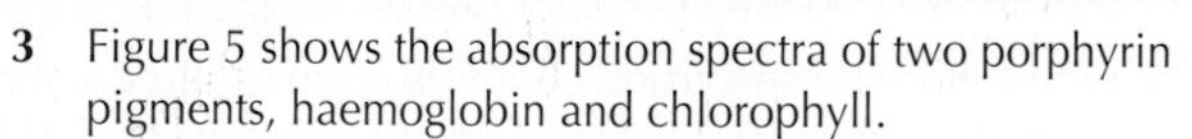

Problems for 6.8

1 Look at the two absorption spectra in Figure 4 – one is the spectrum of a synthetic dye; the other is the spectrum of a colourless compound.

a Which is the spectrum of the dye?

b What colour is the dye? It will help to refer to Figure 4 in **Section 6.7** or to the visible spectrum in the **Data Sheets** (Table 23) to determine which colours correspond to different wavelengths.

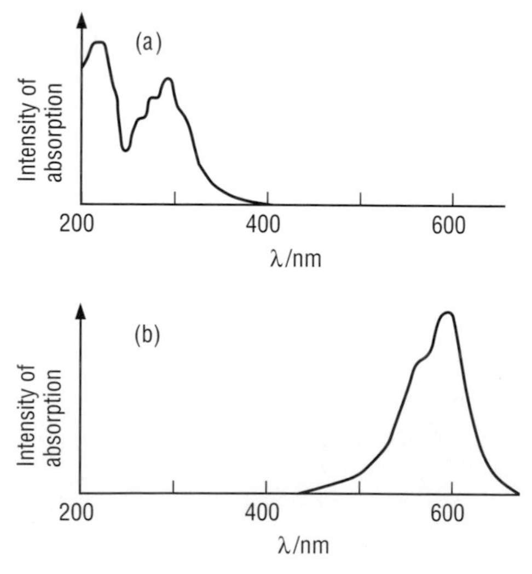

▲ **Figure 4**

2 a Draw a sketch of the absorption spectrum of

i a yellow dye **ii** a black dye.

b What would the reflectance spectrum of a black pigment look like?

3 Figure 5 shows the absorption spectra of two porphyrin pigments, haemoglobin and chlorophyll.

a Give the λ_{max} of the main absorption in each spectrum.

b Which spectrum corresponds to haemoglobin?

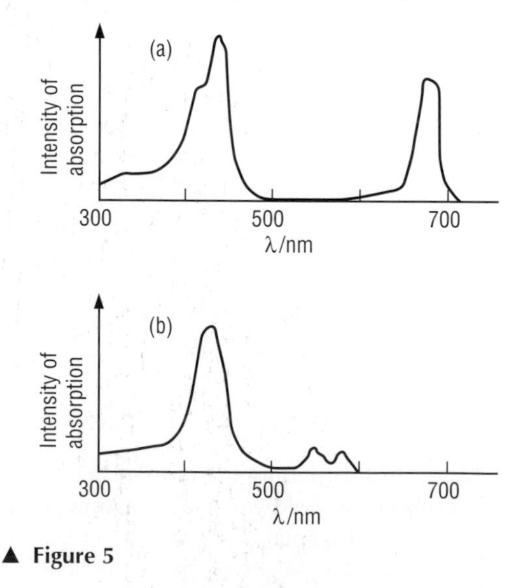

▲ **Figure 5**

4 Chemists at the National Gallery recorded the reflectance spectrum shown in Figure 6 from a pigment on the surface of a painting.

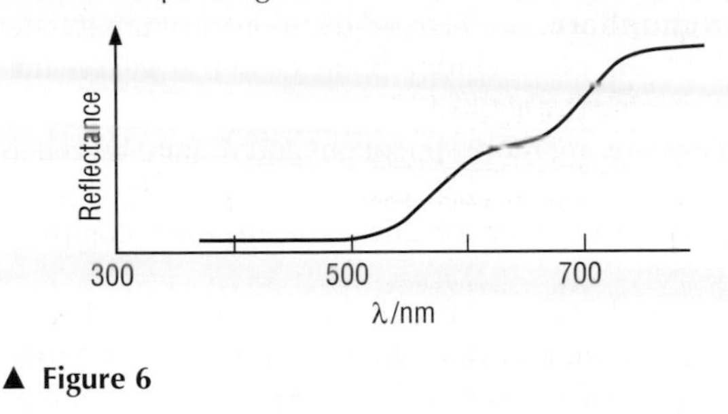

▲ **Figure 6**

a Which of the following pigments do you think the spectrum most likely corresponds to?
i Red ochre
ii Malachite green
iii Prussian blue.

b Give reasons for your choice in part **a**.

6.9 *Chemistry of colour*

Coloured or colourless?

Coloured substances absorb radiation in the visible region of the spectrum. The energy absorbed causes changes in *electronic energy*, and electrons are promoted from the ground state to an *excited state*.

The electrons excited are the outermost ones. These are the electrons involved in bonding or present as lone pairs. The inner electrons are held tightly by the positive nucleus of the atom. The energy needed to excite these electrons is very large.

Not all electronic transitions are brought about by visible light. Many require a greater energy, corresponding to ultraviolet radiation. In this case, the compound absorbs ultraviolet radiation but appears colourless (**Section 6.8**). Figure 1 shows the energy needed to excite an electron in a coloured compound and in a colourless compound – this energy is called the **excitation energy**.

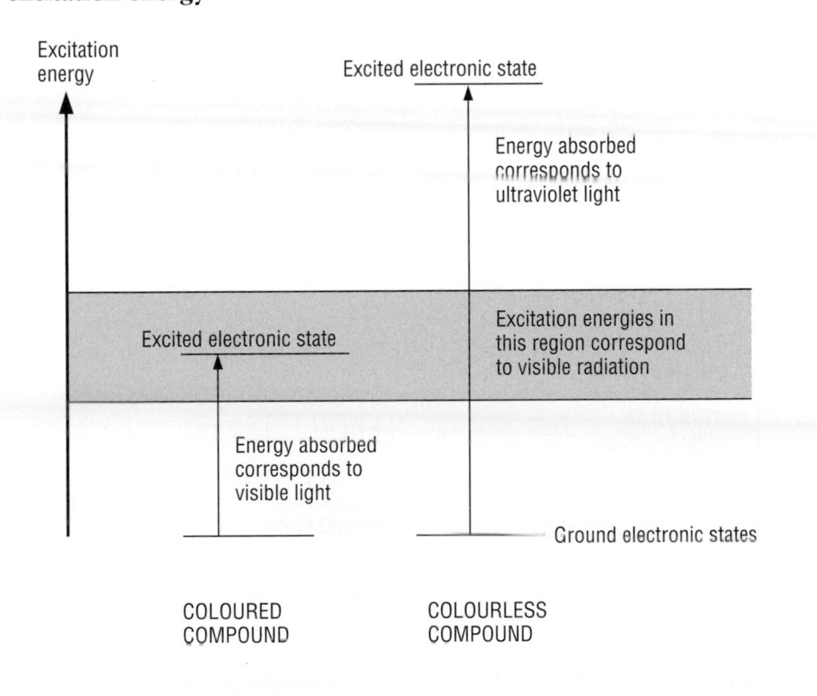

◀ **Figure 1** The energy needed to excite an electron in a coloured compound and in a colourless compound.

Coloured organic compounds

These compounds often contain unsaturated groups such as C=C, C=O or –N=N–. These groups are usually part of an extended delocalised electron system called the **chromophore**. (In the past, the individual unsaturated groups, such as –N=N–, were called chromophores and you will see the term used in this way in some textbooks.)

Electrons in double bonds are more spread out and require less energy to excite than those in single bonds, particularly if the double bond is part of a conjugated system. The lower excitation energy means these compounds absorb in the visible region.

Functional groups such as –OH, $–NH_2$ and $–NR_2$ are often attached to chromophores to enhance or modify the colour of the molecules. These groups all contain lone pairs of electrons which become involved in the delocalised electron system. Small changes to the delocalised system can change the energy of the light absorbed by the molecule, and so change the colour of the compound.

Many dye molecules have different colours in acidic and alkaline solutions. Compounds like this can be used as **acid–base indicators**. For example, methyl orange is red in acidic solutions below pH 3.5. Above this pH, its solutions are yellow. In the red form, one of the N atoms has H^+ bonded to it, and this changes the energy absorbed by the delocalised electron system:

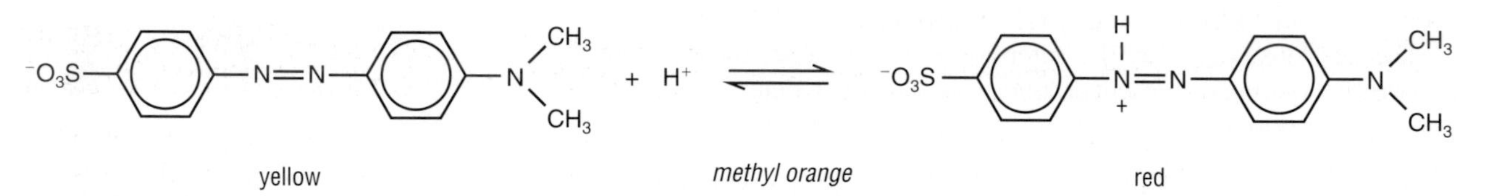

Coloured inorganic compounds

These compounds often contain transition metals. **Section 11.6** introduces the fact that ligands affect the d-orbital electrons of the metal ion they are surrounding. Orbitals close to the ligands are pushed to slightly higher energy levels than those further away. As a result the five d orbitals are split into two groups at different energy levels. Figure 2 shows the relative energy levels of the five 3d orbitals in the hydrated Ti^{3+} ion in the octahedral complex $Ti(H_2O)_6^{3+}$.

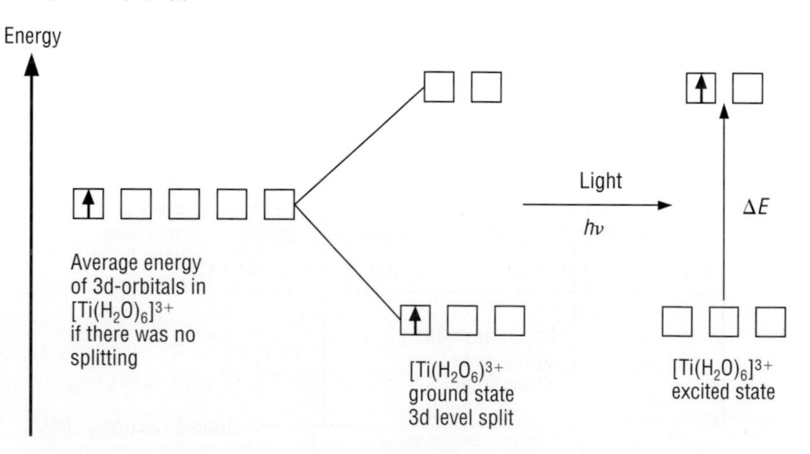

► **Figure 2** Relative energy levels for the five 3d orbitals of the hydrated Ti^{3+} ion.

When light passes through a solution of $[Ti(H_2O)_6]^{3+}$ ions, a photon of light may be absorbed. The energy of this photon corresponds to excitation of an electron from a low energy d orbital to a high energy d orbital.

The frequency (ν) of the light absorbed depends on the energy difference between these two levels, ΔE. ($\Delta E = h\nu$, where h is Planck's constant). For most d-block transition metals, the size of ΔE is such that the light absorbed falls in the visible part of the spectrum. The colour we see is white light minus the frequencies of absorbed light.

The energy needed to excite a d electron to the higher energy level also depends on the oxidation state of the metal. This is why redox reactions of transition metal compounds can be accompanied by spectacular colour changes. For example, vanadium shows a different colour in each of its oxidation states in aqueous solution.

$$V(+5) \rightarrow V(+4) \rightarrow V(+3) \rightarrow V(+2)$$

yellow blue green violet

Some ligands have a more powerful effect than others on the splitting of the d sub-shell. So changing the ligand complexed with a metal ion often results in a colour change.

$$[Ni(H_2O)_6]^{2+}(aq) + 6NH_3(aq) \rightleftharpoons [Ni(NH_3)_6]^{2+}(aq) + 6H_2O$$

light green lilac/blue

Sometimes absorption of visible light can cause the transfer of an electron from the ground state of one atom to an excited state of *another* adjacent atom – this is called **electron transfer**. It is a special sort of electronic transition and is responsible for some very bright pigment colours, such as chrome yellow and Prussian blue.

Problems for 6.9

You may find it helpful to refer to the **Colour by Design** module to help you answer these problems.

1 The structures of three different coloured compounds are given below.

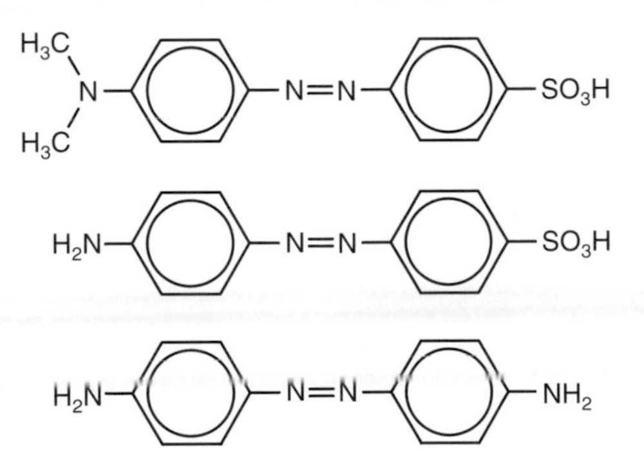

a What is meant by the term *chromophore*?

b Identify the chromophore common to all three molecules.

c β-carotene is a naturally occurring orange-coloured compound with the structure

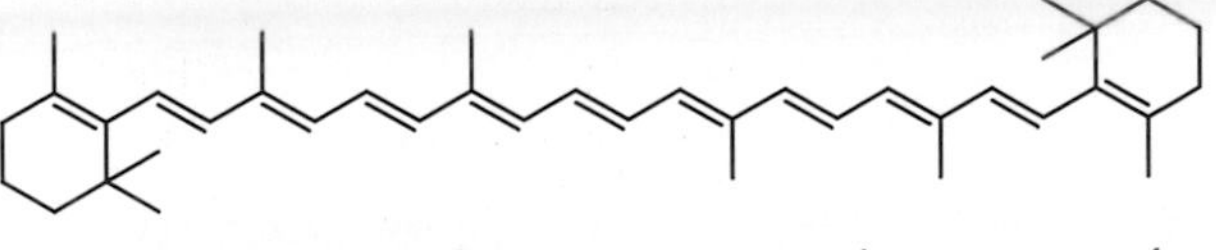

What common feature is present in the structures of β-carotene and the three coloured compounds above?

2 Cyanidin is a pigment whose colour depends on the pH of the solution. It is found in many flowers where it is bound to a sugar molecule and often coordinated to a metal ion.

The neutral compound has the structure shown below and has a purple colour.

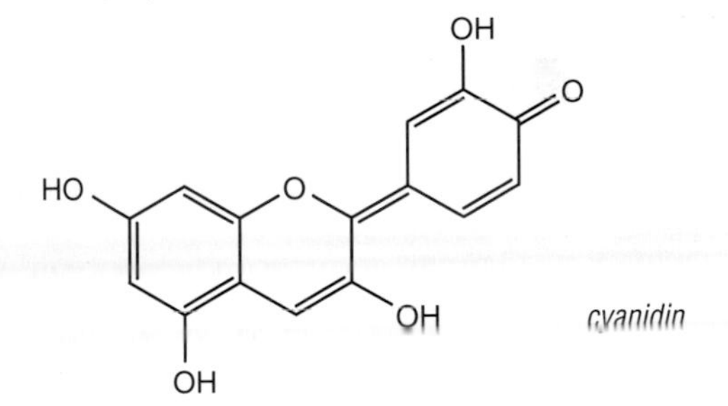

In acid solution, the cyanidin molecule gains H^+ and turns into its red form.

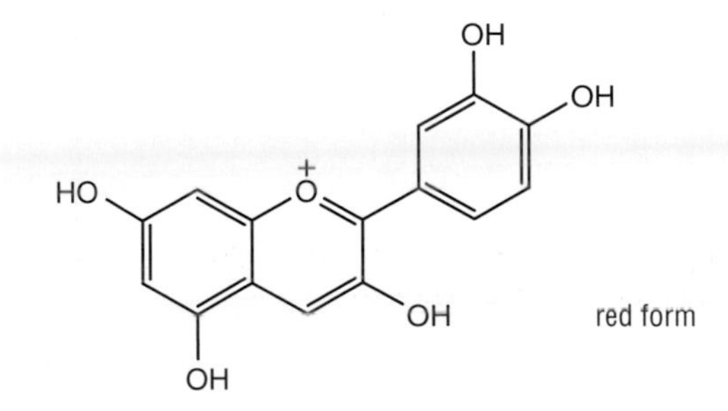

In alkaline solution, the cyanidin molecule loses H^+ and turns into its blue form.

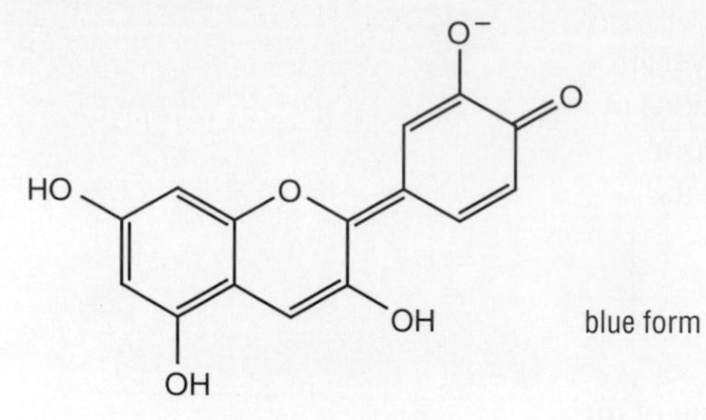

a Explain why molecules of this type absorb energy in the visible region producing coloured compounds.

b Changes in structure lead to changes in electron energy levels, and so to changes in colour. The diagram below shows two energy levels in the red-form molecule and the absorption of light as electrons move from the ground state to the excited state. Copy the diagram, and draw another beside it to represent a transition which corresponds to the absorption of light in the blue form of cyanidin.

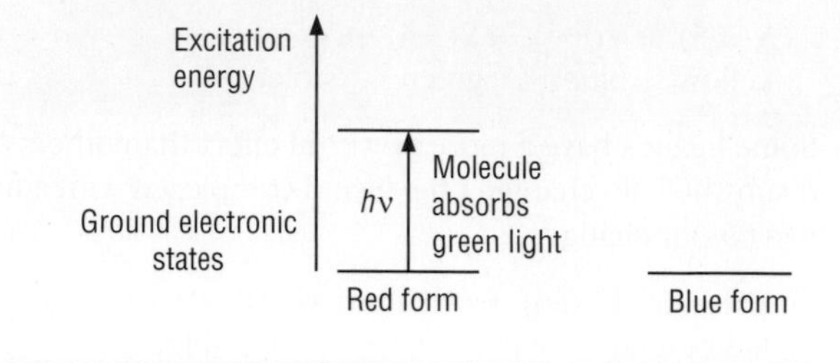

7 EQUILIBRIUM IN CHEMISTRY

7.1 *Chemical equilibrium*

Dynamic equilibrium

The general meaning of the term *equilibrium* is a state of balance in which nothing changes. For example, a see-saw with two people of equal mass, sitting one on each side, is in a state of equilibrium.

In chemistry, a state of equilibrium is also a state of balance, but it has a special feature – chemical equilibrium is **dynamic equilibrium**.

Consider the sealed bottle of soda water shown in Figure 1. The bottle and its contents make up a **closed system**. Nothing can enter or leave the bottle. In this system, carbon dioxide is present in two states – as a gas above the liquid, $CO_2(g)$, and as an aqueous solution in water, $CO_2(aq)$.

Suppose the bottle has been standing at a steady temperature for some time. If you measure the pressure of carbon dioxide gas above the water, you find it remains constant. If you measure the concentration of carbon dioxide dissolved in the water, you find that remains constant too. The system is at equilibrium and nothing appears to be changing, at least on the **macroscopic scale** – the scale which we humans are used to.

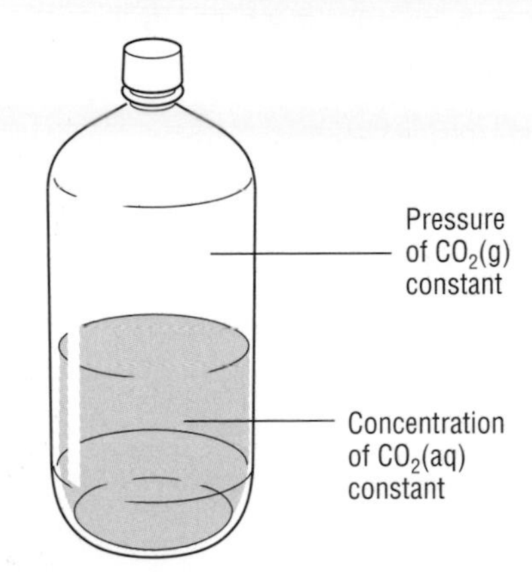

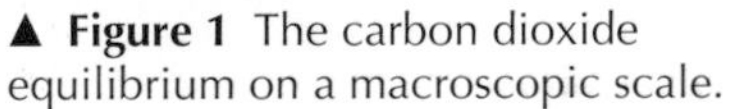

▲ **Figure 1** The carbon dioxide equilibrium on a macroscopic scale.

If you were able to see how the individual molecules are behaving – that is on a **microscopic scale** – the picture would be rather different. In any gas, the molecules are constantly moving. They move rapidly in random directions, and inevitably they collide with the molecules in the surface of the water (see Figure 2). Some bounce back into the gas phase, but some dissolve and enter the aqueous phase.

At the same time, the molecules of carbon dioxide in the water are also constantly moving around, colliding with water molecules and with each other. Near the surface of the water, some of these molecules escape into the gaseous phase (see Figure 2).

So there are molecules entering the aqueous phase, and molecules leaving it – it is a **reversible change**. *When the system is at equilibrium, the molecules enter and leave at the same rate*. On the macroscopic scale it *seems* as though nothing is changing, but on the molecular scale molecules are constantly moving from one phase to the other. That is why the situation is described as *dynamic* equilibrium. It can be represented by the equation

$$CO_2(g) \rightleftharpoons CO_2(aq)$$

The $\rightleftharpoons$ sign represents dynamic equilibrium.

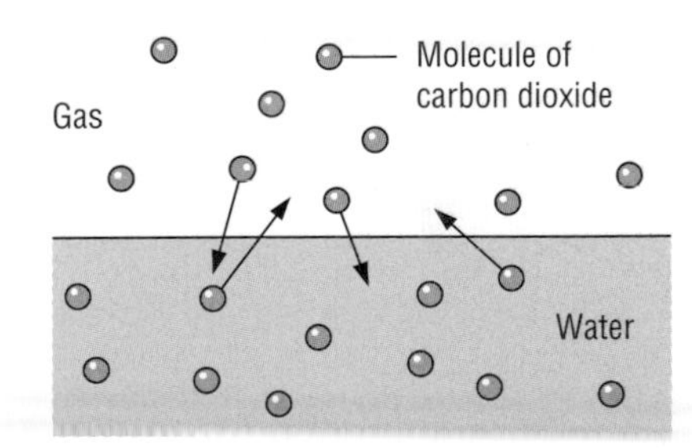

▲ **Figure 2** The carbon dioxide equilibrium on a microscopic scale.

Other examples of dynamic equilibrium

Dynamic equilibrium is a basic principle of chemistry. You can find examples everywhere – and you can find two more without going any further than the bottle of soda water shown in Figure 1.

Water in equilibrium

The water itself is present in two phases – liquid and gas. Water molecules from the liquid are constantly escaping into the gaseous phase. At the same time, molecules from the gas are constantly colliding with the surface of the liquid and condensing. In a closed system (the bottle with the top on), after a while the molecules reach the situation where they are escaping and condensing at the same rate. The system is then in dynamic equilibrium, and it is represented by

$$H_2O(l) \rightleftharpoons H_2O(g)$$

Carbon dioxide reacts with water

Neither of the two examples just mentioned involve chemical reactions – they are physical processes involving the liquid/gas phase change. But when carbon dioxide dissolves in water, a small proportion of it actually *reacts* with the water instead of just dissolving. Hydrogencarbonate ions and hydrogen ions are formed

$$CO_2(aq) + H_2O(l) \rightleftharpoons \underset{\text{hydrogencarbonate ion}}{HCO_3^-(aq)} + H^+(aq)$$

Only a small proportion of the carbon dioxide reacts in this way. The majority of it remains as $CO_2(aq)$, with only a small amount converted to $HCO_3^-(aq)$.

On the molecular scale, $CO_2(aq)$ molecules are constantly reacting with $H_2O(l)$ molecules to form $HCO_3^-(aq)$ and $H^+(aq)$. At the same time, $HCO_3^-(aq)$ and $H^+(aq)$ are constantly reacting together to reform $CO_2(aq)$ and $H_2O(l)$. It is a **reversible reaction**.

Many chemical reactions are reversible reactions. When such reactions are represented by equations, like the one above, the reaction going from left to right is known as the **forward reaction** – the reaction going in the opposite direction is known as the **reverse reaction**. Even though the reaction is reversible, it is the practice to refer to substances on the left of the equilibrium sign as the **reactants** and to those on the right as the **products**.

When the reaction between carbon dioxide and water in the soda water bottle has had a chance to settle down, the forward and reverse reactions go at the same rate. Nothing *seems* to change on the overall, macroscopic level. If you measure the pH of the soda water, you find it remains constant – the concentration of $H^+(aq)$ ions is constant, and so are the concentrations of all the other substances involved. It is an example of dynamic equilibrium.

Another example of chemical equilibrium

If a mixture of nitrogen dioxide, NO_2, and carbon monoxide, CO, is heated in a sealed container, the following reversible reaction takes place:

$$NO_2(g) + CO(g) \rightleftharpoons NO(g) + CO_2(g)$$

At the start, no molecules of NO(g) or $CO_2(g)$ are present so there will be no reverse reaction to begin with. As soon as some NO(g) and $CO_2(g)$ are produced, the reverse reaction will start to occur – slowly at first but with increasing rate as the concentrations of NO(g) and $CO_2(g)$ increase. At the same time, the concentrations of $NO_2(g)$ and CO(g) will be decreasing and so the rate of the forward reaction will start to decrease.

Eventually a point is reached when the rate of the forward reaction equals the rate of the reverse reaction. At this point, reactants are turning into products as quickly as products are turning into reactants. The reversible reaction is now at equilibrium (see Figure 3). It is a dynamic equilibrium because both forward and reverse reactions are still occurring, although the concentrations of both products and reactants now remain constant.

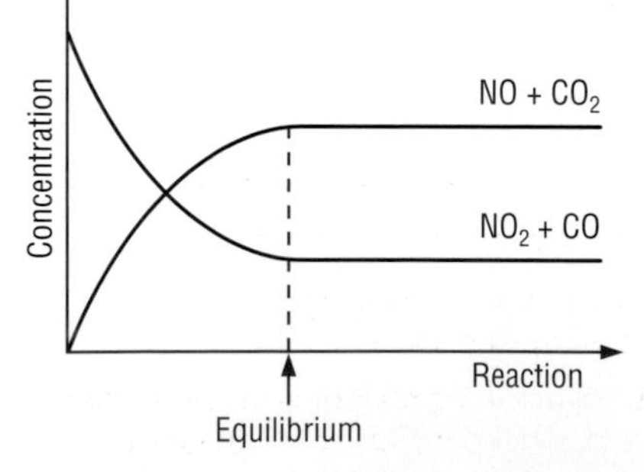

▲ **Figure 3** Reaction of $NO_2(g)$ and CO(g).

Position of equilibrium

There are many equilibrium mixtures possible for a given reaction system, depending on the concentrations of the substances you mix and the conditions, such as temperature and pressure. We often use the term **position of equilibrium** to describe one particular set of equilibrium concentrations for a reaction. If one of the concentrations is changed then the system is no longer in equilibrium, and the concentrations of all the substances will change until a new position of equilibrium is reached.

Although the rates of the forward and reverse reactions are the same when equilibrium is reached, this does not mean that the concentrations of the reactants and the concentrations of the products are the same at equilibrium – in fact, they rarely are. Usually, at the point of equilibrium there are more products than reactants or vice versa.

If the forward reaction is nearly complete before the rate of the reverse reaction has increased sufficiently to establish equilibrium, we say that the position of equilibrium 'lies to the products' or to the right-hand side of the reaction. In other cases, little of the reactants will have changed to products when the rate of the reverse reaction becomes equal to the rate of the forward reaction, and we say that the position of equilibrium 'lies to the reactants' or to the left-hand side of the reaction.

Once a reaction system is at equilibrium, it is impossible to tell whether the equilibrium mixture was arrived at starting with reactants or products. Under the same conditions, the same equilibrium position is arrived at whether you start with the reactants or the products. The reversible reaction of hydrogen and iodine to form hydrogen iodide can be used to illustrate this:

$$H_2(g) + I_2(g) \rightleftharpoons 2HI(g)$$

Suppose we heat a mixture of equal amounts of hydrogen and iodine (1 mole of each) in a sealed reaction vessel at 500 °C and measure the amounts of H_2, I_2 and HI during the course of the reaction. If we then plot these results as graphs we obtain Figure 4a. Equilibrium is reached at the point on the graphs where the lines become horizontal.

If we now repeat the experiment with hydrogen iodide (2 moles) we obtain the graphs shown in Figure 4b. Notice that we have ended up with identical amounts of HI and identical amounts of H_2 and I_2 in both graphs. This shows that we reach the same equilibrium position irrespective of whether we start with the reactants or the products.

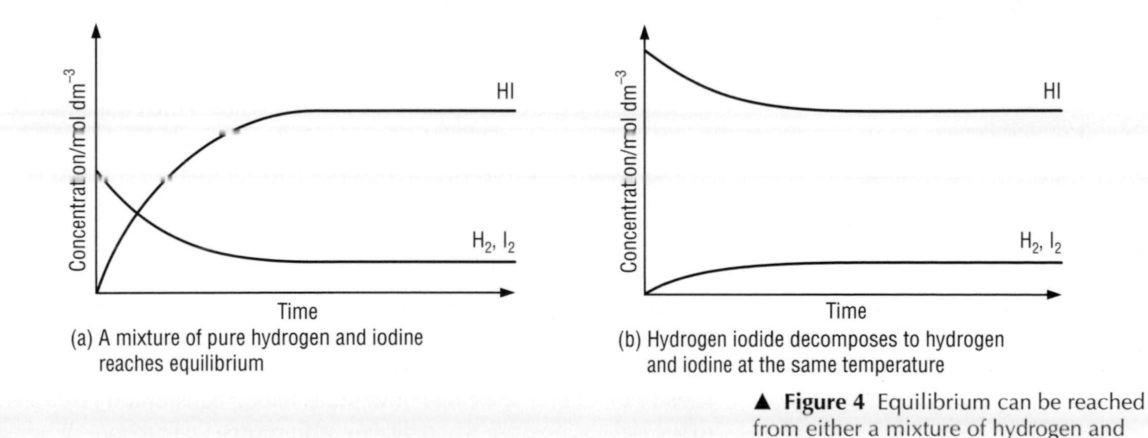

(a) A mixture of pure hydrogen and iodine reaches equilibrium

(b) Hydrogen iodide decomposes to hydrogen and iodine at the same temperature

▲ **Figure 4** Equilibrium can be reached from either a mixture of hydrogen and iodine, or from pure hydrogen iodide.

Shifting the position of equilibrium

The position of an equilibrium can be altered by changing:

- the concentrations of reacting substances if dealing with solutions
- the pressures of reacting gases
- the temperature.

By studying data from many reactions, Henri le Chatelier was able (in 1888) to propose rules which enable us to make qualitative predictions about the effect of each of these changes on an equilibrium. Each of these rules is an

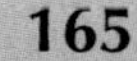

example of the simple general rule that if a system is at equilibrium and a change is made in any of the conditions, then the system responds to counteract the change as much as possible.

This general rule is known as **le Chatelier's principle**. As you work through the following sections you will be able to see how each of the separate rules is an example of this general rule.

Using a catalyst does not shift the position of equilibrium. A catalyst lowers the activation enthalpy of a reaction. As the reaction is reversible, the activation enthalpy of both the forward and reverse reactions will be reduced by the same amount, so both will speed up by the same proportion (see **Section 10.6**). This means that the same equilibrium position is reached. The catalyst causes the equilibrium position to be reached much more quickly than it would be if no catalyst were used.

Le Chatelier's principle

If a system is at equilibrium and a change is made in any of the conditions, then the system responds to counteract the change as much as possible.

Changing the concentration

Yellow iron(III) ions, $Fe^{3+}(aq)$, and colourless thiocyanate ions, $SCN^{-}(aq)$, react in solution to give the deep red ion, $[Fe(SCN)]^{2+}(aq)$. The reaction is reversible and the equilibrium set up is

$$\underset{\text{yellow}}{Fe^{3+}(aq)} + \underset{\text{colourless}}{SCN^{-}(aq)} \rightleftharpoons \underset{\text{deep red}}{[Fe(SCN)]^{2+}(aq)}$$

The intensity of the red colour of the solution can be used as a measure of how much $[Fe(SCN)]^{2+}(aq)$ is present. If the solution gets darker it means that the equilibrium is shifting to the right; if lighter, it is shifting to the left.

If we increase the concentration of either $Fe^{3+}(aq)$ or $SCN^{-}(aq)$ then the solution gets darker, showing that the position of equilibrium has moved to the right.

The concentration of $Fe^{3+}(aq)$ ions in the solution can be reduced by adding ammonium chloride – chloride ions from the ammonium chloride react to form $[FeCl_4]^{-}(aq)$. If we do this then the red colour becomes paler, indicating that the equilibrium is being moved to the left.

From these observations, and others involving a variety of reversible reactions, we can arrive at some rules for deciding how concentration changes will affect the position of an equilibrium. These are:

- increasing the concentration of reactants causes the equilibrium to move to the product side
- increasing the concentration of products causes the equilibrium to move to the reactant side
- decreasing the concentration of reactants causes the equilibrium to move to the reactant side
- decreasing the concentration of products causes the equilibrium to move to the product side.

In each case, the system at equilibrium is responding to counteract the change as much as possible and so is consistent with le Chatelier's principle.

Since much chemistry is about turning reactants into products, chemists spend a lot of time working out ways of shifting equilibrium positions as far to the right as possible when dealing with reversible reactions. When a gas is one of the products of a reversible reaction, the simplest way to achieve this is to keep removing the gas from the reaction chamber. This ensures that the reaction is being constantly shifted to the product side.

For example, when limestone ($CaCO_3$) is heated in an open container, carbon dioxide is allowed to escape so the reaction constantly shifts to the product side. See page 168 for more details of this.

Changing the pressure

Many important industrial processes involve reversible reactions that take place in the gas phase. For these processes it is essential that conditions are identified which ensure that the equilibrium is shifted as far to the right as possible.

From the study of equilibria in gas-phase reactions, a simple rule has emerged:

- increasing the pressure moves the equilibrium to the side of the equation with fewer gas molecules as this tends to reduce the pressure
- decreasing the pressure moves the equilibrium to the side of the equation with more gas molecules as this tends to increase the pressure.

In each case, the position of equilibrium shifts so as to counteract the change in pressure as much as possible, which again is consistent with le Chatelier's principle.

The steam reforming of methane provides a good illustration of these rules. This process is used to make methanol, an important industrial chemical. In the first stage, methane reacts with steam to form CO and H_2:

$$\underbrace{CH_4(g) + H_2O(g)}_{\text{2 molecules}} \rightleftharpoons \underbrace{CO(g) + 3H_2(g)}_{\text{4 molecules}}$$

There are more gas molecules on the product side of the equation – so, as the rules predict, a reduction in the pressure results in the position of equilibrium shifting to the right. The *lower* the pressure, the more product will be obtained.

The second stage of methanol manufacture involves the reaction

$$\underbrace{CO(g) + 2H_2(g)}_{\text{3 molecules}} \rightleftharpoons \underbrace{CH_3OH(g)}_{\text{1 molecule}}$$

For this reaction, increasing the pressure will move the equilibrium to the product side – the *higher* the pressure, the more product will be obtained.

Changing the temperature

Heating a reaction always makes it go faster. How does heating (or cooling) affect the position of an equilibrium? For a reversible reaction, if the forward reaction is exothermic then the reverse reaction will be endothermic to the same extent, and vice versa.

Nitrogen dioxide, NO_2, is a dark brown gas which exists in equilibrium with its colourless dimer, dinitrogen tetraoxide, N_2O_4.

$$\underset{\text{brown}}{2NO_2(g)} \underset{\text{endothermic}}{\overset{\text{exothermic}}{\rightleftharpoons}} \underset{\text{colourless}}{N_2O_4(g)}$$

The reaction that forms $N_2O_4(g)$, the forward reaction, is exothermic and releases thermal energy to the surroundings. The reverse reaction, forming $NO_2(g)$, is endothermic and thermal energy is taken in from the surroundings.

If a sealed container of the brown equilibrium mixture is heated in boiling water, it becomes darker. On cooling in ice it turns almost colourless. From this evidence we can conclude that changing the temperature shifts the position of this equilibrium. It also illustrates the rule for deciding how a change of temperature affects the position of an equilibrium. This is:

- heating a reversible reaction at equilibrium shifts the reaction in the direction of the endothermic reaction
- cooling a reversible reaction at equilibrium shifts the reaction in the direction of the exothermic reaction.

Again, these are examples of le Chatelier's general principle.

Note that when a description of a reversible reaction includes a ΔH value, it is accepted practice that the ΔH value given is always for the forward reaction.

Linking equilibria together

In many systems, particularly naturally occurring systems, two or more equilibria are linked together so that the product of one equilibrium is the reactant in a second equilibrium. This is the case for the two reversible reactions involving carbon dioxide discussed earlier in this section.

$$CO_2(g) \rightleftharpoons CO_2(aq)$$

and

$$CO_2(aq) + H_2O(l) \rightleftharpoons HCO_3^-(aq) + H^+(aq)$$

The product of the first equilibrium is a reactant in the second, so that the dissolved carbon dioxide is involved in both equilibria.

Le Chatelier's principle can be used to predict the effect of imposing a change on this equilibrium system. For example, if more gaseous carbon dioxide is added, the position of the first equilibrium will move to the right and more $CO_2(aq)$ will be formed. This increase in the concentration of $CO_2(aq)$ will then cause the position of the second equilibrium to move to the right to produce more $HCO_3^-(aq)$ and $H^+(aq)$.

Clearly, le Chatelier's principle helps in the development of theories to explain naturally occurring systems that involve reversible reactions, such as the dissolving of atmospheric carbon dioxide in the oceans.

Chemical equilibria and steady state systems

Strictly speaking, a chemical equilibrium can only be established in a **closed system** which is sealed off from its surroundings. For example, if you heat calcium carbonate in a closed container it decomposes and an equilibrium mixture is established

$$CaCO_3(s) \rightleftharpoons CaO(s) + CO_2(g)$$

If you heat calcium carbonate in an *open* tube, the carbon dioxide is lost into the air. The reaction constantly shifts to the right to try to replace the lost carbon dioxide until eventually all the calcium carbonate is converted into calcium oxide. It is an open system, so equilibrium is never established.

Sometimes in an open system, a series of reactions can reach a **steady state**, where the concentrations of reactants and products remain constant. A blue Bunsen burner flame is in a steady state. It doesn't *seem* to change but reactants are being used up as fast as they arrive, so it certainly isn't in equilibrium.

Another example of a steady state is the series of reactions which produce and destroy ozone in the stratosphere.

$$O + O_2 \rightarrow O_3 \qquad \text{ozone production}$$

$$\left.\begin{array}{l} O_3 + h\nu \rightarrow O_2 + O \\ O + O_3 \rightarrow O_2 + O_2 \end{array}\right\} \text{ozone destruction}$$

None of these reactions comes to equilibrium, but left to themselves they will reach a point at which ozone is being produced as fast as it is being used up – the series of reactions has reached a steady state, but is not in equilibrium.

Problems for 7.1

1 You will be very familiar with the equilibrium

$H_2O(l) \rightleftharpoons H_2O(g)$

a Why is this described as a *dynamic* equilibrium?

b One place where this equilibrium is established is in a plastic bag full of wet socks. Use the idea of dynamic equilibrium to explain why the socks in the bag never get dry.

c You can dry the socks by taking them out of the bag. Once they are out of the bag the system is no longer in equilibrium. Explain why.

d In which direction does this equilibrium move

i when dew forms in the evening?

ii when you use the car's fan to demist a car window?

2 In a reversible reaction the concentrations of reactants and products were measured during the reaction and the following graph was drawn.

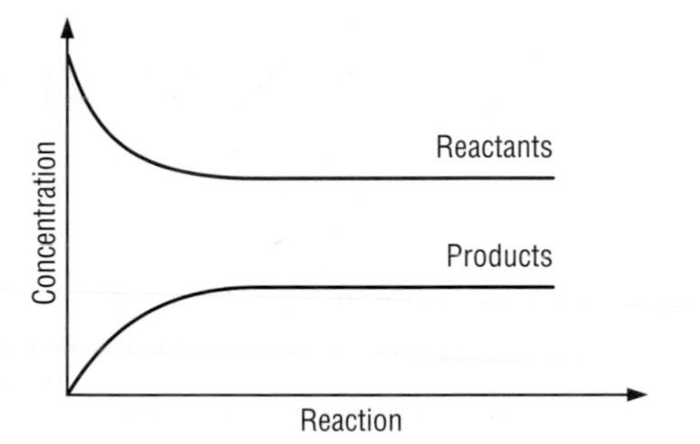

a To which side, reactants or products, does the position of equilibrium lie in this reaction? Explain how you arrived at your decision.

b At which point on the graph is equilibrium reached?

3 Ethanol is produced industrially at about 70 atmospheres pressure and 300°C by the following reaction. The reaction needs a catalyst.

$$C_2H_4(g) + H_2O(g) \rightleftharpoons C_2H_5OH(g) \quad \Delta H = -46\,kJ\,mol^{-1}$$

Which of the following would move the position of equilibrium to the right?

A increasing the temperature

B increasing the concentration of steam

C decreasing the pressure

D adding more catalyst.

4 State the direction in which the position of equilibrium of each system would move (if at all) if the pressure was increased by compressing the reaction mixture. Give your answer as 'left → right', 'right → left', or 'no change'.

a $2NO(g) + O_2(g) \rightleftharpoons 2NO_2(g)$

b $C_2H_6(g) \rightleftharpoons C_2H_4(g) + H_2(g)$

c $2HI(g) \rightleftharpoons H_2(g) + I_2(g)$

d $2NO_2(g) \rightleftharpoons N_2O_4(g)$

e $2CO(g) + O_2(g) \rightleftharpoons 2CO_2(g)$

5 If solid bismuth trichloride, $BiCl_3$, is added to water, a reaction occurs producing a white precipitate of BiOCl:

$$BiCl_3(aq) + H_2O(l) \rightleftharpoons BiOCl(s) + 2HCl(aq)$$

A colourless aqueous solution of $BiCl_3$ may be prepared by dissolving the solid in concentrated hydrochloric acid.

a Explain, in terms of the equilibrium, why concentrated hydrochloric acid enables an aqueous solution of bismuth trichloride, $BiCl_3(aq)$, to be formed.

b Predict what you would see if $BiCl_3(aq)$ was diluted with a large volume of water. Explain your prediction in terms of the equilibrium.

6 a In a half-empty, but tightly stoppered, bottle of cola, the following equilibrium is established:

$$CO_2(g) \rightleftharpoons CO_2(aq)$$

If you squeezed the bottle to expel most of the air and then stoppered the bottle tightly, would the cola keep its fizziness better or not? Explain your answer.

b The $CO_2(aq)$ is also involved in the following equilibrium:

$$CO_2(aq) + H_2O(l) \rightleftharpoons HCO_3^-(aq) + H^+(aq)$$

If, after the system had been allowed to reach equilibrium, you added some additional carbon dioxide to the space above the cola, what effect would this have on the concentration of $H^+(aq)$ in the cola? Explain your answer.

c Explain what the effect would be on the equilibrium pressure of carbon dioxide above the cola if you added some dilute alkali to the cola.

7.2 *Equilibria and concentrations*

Starting to put numbers to equilibria

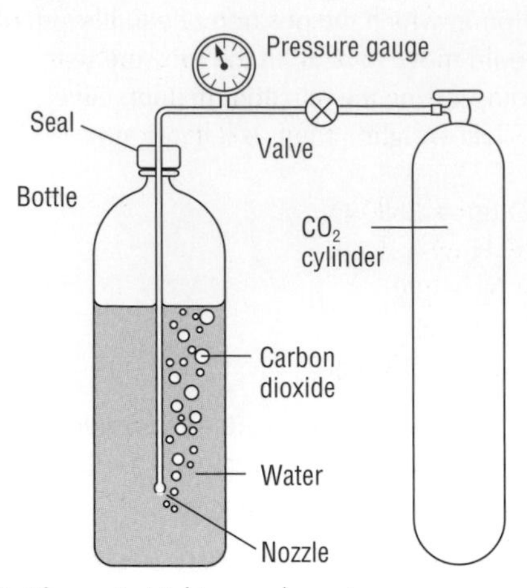

▲ **Figure 1** Making soda water.

Let's start by looking again at the equilibria involving carbon dioxide and water that were discussed in **Section 7.1**. To investigate chemical equilibrium further, we need to ask questions about the *quantities* of substances involved.

Figure 1 shows a device for making soda water. It forces carbon dioxide under pressure into water. The same principle is used to make most fizzy drinks.

Normally, carbon dioxide gas is injected into the water until it reaches a certain pressure, and then escapes from under the seal. But if you control the valve you can *vary* the pressure of carbon dioxide in the space above the water. Some of this carbon dioxide gas will then dissolve in the water, and after a while it will reach equilibrium. If you then measure the pH of the water, you can find out the concentration of H^+ ions in the aqueous phase and use this to work out the concentration of carbon dioxide in the aqueous phase. If you introduce a pressure gauge, you can measure the pressure of carbon dioxide in the gas phase.

Table 1 shows the figures that were obtained when measurements of this kind were carried out at 292 K.

► **Table 1** Measurements of the pressure of carbon dioxide in the gas phase and the concentration of carbon dioxide in the aqueous phase when the system is in equilibrium at 292 K.

Experiment	Pressure of CO_2(g)/atm	Concentration of CO_2(aq)/mol dm^{-3}
1	1.0	0.035
2	1.2	0.043
3	1.5	0.053
4	1.9	0.065
5	2.3	0.080

If, for each of these experiments, you work out the ratio

$$\frac{\text{concentration of } CO_2\text{(aq) (mol dm}^{-3}\text{)}}{\text{pressure of } CO_2\text{(g) (atm)}}$$

you will find the value is always the same.

[ ] represents concentration in mol dm^{-3}.

p represents the pressure of a gas in atmosphere (atm).

The ratio $\frac{\text{concentration of } CO_2\text{(aq)}}{\text{pressure of } CO_2\text{(g)}}$ or $\frac{[CO_2\text{(aq)}]}{p_{CO_2\text{(g)}}}$ is always

found to be constant at a particular temperature. This makes sense if you think about it – if you have a greater pressure of carbon dioxide gas, you would expect proportionately more of it to dissolve in the water. (Note the use of square brackets to represent the *concentration* of CO_2(aq) in mol dm^{-3}.)

The ratio above is an example of an **equilibrium constant**, and it is given the symbol ***K***. If you take an average of the five ratios you worked out from the data in Table 1, you will have a value for the equilibrium constant for the reaction

$$CO_2\text{(g)} \rightleftharpoons CO_2\text{(aq)}$$

Similarly, there is an equilibrium constant for the reaction

$$CO_2\text{(aq)} + H_2O\text{(l)} \rightleftharpoons HCO_3^-\text{(aq)} + H^+\text{(aq)}$$

It has a value of 4.3×10^{-7} mol dm^{-3} at 292 K.

Now think what happens if we try to force more carbon dioxide to dissolve in water by using a higher pressure for the gas. Table 1 shows you that the concentration of dissolved gas will indeed rise, so that the equilibrium constant remains at its fixed value. The position of equilibrium shifts to the right. This will have an effect on the reaction of carbon dioxide with water to form hydrogencarbonate ion. If it didn't, the increased amount of dissolved carbon dioxide would mean that you had raised the concentration of reactants but left the products alone, and then the reactants and products would no longer be in equilibrium.

To keep K at $4.3 \times 10^{-7}\,\text{mol dm}^{-3}$, the reaction must create some more products – make more $H^+(aq)$ ions – and so the pH will fall. Hence we can use the pH of the solution to measure the pressure of carbon dioxide in the gas phase.

These observations are consistent with the qualitative predictions made using le Chatelier's principle in **Section 7.1**.

Equilibrium constants

You can write equilibrium constants for *all* equilibrium processes. Table 2 shows data obtained for the hydrolysis of an ester, ethyl ethanoate:

$$CH_3COOC_2H_5(l) + H_2O(l) \rightleftharpoons CH_3COOH(l) + C_2H_5OH(l)$$

▼ **Table 2** Equilibrium concentrations for the hydrolysis of ethyl ethanoate at 293 K.

Equilibrium concentrations/mol dm^{-3}	$[CH_3COOC_2H_5(l)]_{eq}$	$[H_2O(l)]_{eq}$	$[CH_3COOH(l)]_{eq}$	$[C_2H_5OH(l)]_{eq}$
Experiment 1	0.090	0.531	0.114	0.114
Experiment 2	0.204	0.118	0.082	0.082
Experiment 3	0.151	0.261	0.105	0.105

For a reaction that occurs in solution, the quantities that matter are *concentrations*. The table shows the equilibrium concentrations of different reaction mixtures, all at 293 K. 'Equilibrium concentration' means the concentration when the reaction has reached equilibrium. It is indicated by the subscript 'eq', though this is often omitted.

If you look at the data in the table, you will find that the expression

$$K_c = \frac{[CH_3COOH(l)]\,[C_2H_5OH(l)]}{[CH_3COOC_2H_5(l)]\,[H_2O(l)]}$$

is constant for the three experiments carried out at the same temperature. The constant $\boldsymbol{K_c}$ is the **equilibrium constant** for this reaction.

The concentrations in the expression for K_c *must* be those at equilibrium. From the data given in Table 2 you can check that the value of K_c is constant at about 0.28 for the hydrolysis of ethyl ethanoate at 293 K. This value, which is less than 1, tells you that, at equilibrium, a substantial proportion of the reactants is left unreacted – the reaction is incomplete.

Table 3 (page 172) shows the results obtained for another reaction, between hydrogen and iodine to form hydrogen iodide

$$H_2(g) + I_2(g) \rightleftharpoons 2HI(g)$$

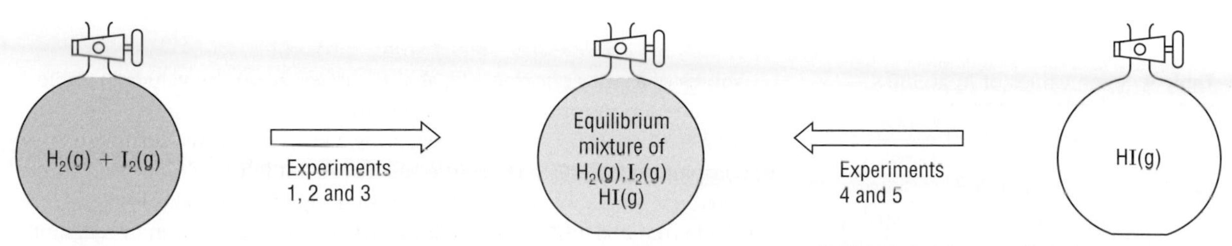

▲ **Figure 2** The equilibrium $H_2(g) + I_2(g) \rightleftharpoons 2HI(g)$ can be approached from two opposite directions.

In the first three experiments, mixtures of hydrogen and iodine were put into sealed reaction vessels. In the final two experiments hydrogen iodide alone was sealed into the vessel (Figure 2). The mixtures were held at a

Experiment	Initial concentrations/mol dm^{-3}			Equilibrium concentrations/mol dm^{-3}			K_c
	$[H_2(g)]$	$[I_2(g)]$	$[HI(g)]$	$[H_2(g)]$	$[I_2(g)]$	$[HI(g)]$	
1	2.40×10^{-2}	1.38×10^{-2}	0	1.14×10^{-2}	0.12×10^{-2}	2.52×10^{-2}	46.4
2	2.44×10^{-2}	1.98×10^{-2}	0	0.77×10^{-2}	0.31×10^{-2}	3.34×10^{-2}	46.7
3	2.46×10^{-2}	1.76×10^{-2}	0	0.92×10^{-2}	0.22×10^{-2}	3.08×10^{-2}	46.9
4	0	0	3.04×10^{-2}	0.345×10^{-2}	0.345×10^{-2}	2.35×10^{-2}	46.9
5	0	0	7.58×10^{-2}	0.86×10^{-2}	0.86×10^{-2}	5.86×10^{-2}	46.4

▲ **Table 3** Initial and equilibrium concentrations for the reaction $H_2(g) + I_2(g) \rightleftharpoons 2HI(g)$.

constant temperature of 731 K until equilibrium was reached. The concentrations of all three substances were then recorded.

For this reaction, the expression that is constant at equilibrium is

$$K_c = \frac{[HI(g)]^2}{[H_2(g)]\,[I_2(g)]}$$

The values of K_c are shown in Table 3. You can see that the equilibrium constant is the same whether we start from a mixture which is all $H_2 + I_2$ (Experiments 1 to 3) or one which is all HI (Experiments 4 and 5). The graphs in Figure 4 in **Section 7.1** show how the concentrations of reactants and products change as the reaction approaches equilibrium.

The mean value of K_c for this reaction at 731 K is about 46.7. This means that, at equilibrium, most of the H_2 and I_2 has been converted to HI, with a bit left unreacted.

Writing the expression for K_c

The rules for writing K_c expressions have been discovered by using the results of many experiments. This makes it possible to write an expression for K_c for any reaction, without having to examine data.

Start by writing a balanced equation for the reaction. For example, take the reaction in which insulin dimers, In_2, break down to form insulin monomers, In.

$$In_2(aq) \rightleftharpoons 2In(aq)$$

In the expression for the K_c, the products of the forward reaction appear on the top line and the reactants on the bottom line. The power to which you raise the concentration of a substance is the same as the number which appears in front of it in the balanced equation.

$$K_c = \frac{[In(aq)]^2}{[In_2(aq)]}$$

In general, if an equilibrium mixture contains substances A, B, C, and D which react according to the equation.

$$aA + bB \rightleftharpoons cC + dD$$

$$\text{then } K_c = \frac{[C]^c[D]^d}{[A]^a[B]^b}$$

▼ **Table 4** Some values of K_c at 550 K.

Reaction	K_c
$H_2(g) + I_2(g) \rightleftharpoons 2HI(g)$	~2
$H_2(g) + Br_2(g) \rightleftharpoons 2HBr(g)$	~10^{10}
$H_2(g) + Cl_2(g) \rightleftharpoons 2HCl(g)$	~10^{18}

Values of K_c vary enormously, as you can see from the values in Table 4 for three reactions at the same temperature.

All reactions are equilibrium reactions – even reactions that seem to go to completion actually have a little bit of reactant left in equilibrium with the product. In some cases, the equilibrium position is so far towards products that it is impossible to detect the tiny concentration of reactants left in the equilibrium mixture. Such reactions have very large equilibrium constants.

Similarly, a reaction that is observed not to go at all may be regarded as having a vanishingly small equilibrium constant.

For any value of K_c you must quote the units and the temperature for which it applies.

What are the units of K_c?

The units of K_c vary – it depends on the expression for K_c for the particular reaction you are studying.

Example 1

$$H_2(g) + Br_2(g) \rightleftharpoons 2HBr(g)$$

$$K_c = \frac{[HBr(g)]^2}{[H_2(g)]\ [Br_2(g)]}$$

The units of K_c are given by $\dfrac{(mol\,dm^{-3})^2}{(mol\,dm^{-3})\ (mol\,dm^{-3})}$

So K_c for this reaction has *no units*, since they cancel out on the top and bottom of the expression.

Example 2

$$In_2(aq) \rightleftharpoons 2In(aq)$$

$$K_c = \frac{[In(aq)]^2}{[In_2(aq)]}$$

The units of K_c are given by $\dfrac{(mol\,dm^{-3})^2}{(mol\,dm^{-3})} = \mathbf{mol\,dm^{-3}}$.

You can see that the units of K_c vary from reaction to reaction, and need to be worked out from the equilibrium expression. When you quote a value for K_c you must always show a balanced equation for the reaction.

What happens to the equilibrium constant if the conditions are changed?

From the sound of it, nothing should happen to an equilibrium *constant* if anything is changed. This turns out to be true for changing the concentration, the pressure and for the use of a catalyst – but not for changing the temperature.

Changing concentrations

Suppose a system is at equilibrium and you suddenly disturb it, say by adding more of a reagent. The composition of the system will change until equilibrium is reached again. The composition of the mixture will always adjust to keep the value of K_c constant, *provided the temperature stays constant*.

Let's look at an example. In an experiment involving the formation of ethyl ethanoate

$$CH_3COOH(l) + C_2H_5OH(l) \rightleftharpoons CH_3COOC_2H_5(l) + H_2O(l)$$

the system was allowed to reach equilibrium (Experiment 1). The equilibrium concentrations are shown in Figure 3 (page 174).

Using the equilibrium concentrations from Experiment 1,

$$K_c = \frac{[CH_3COOC_2H_5]\ [H_2O]}{[CH_3COOH]\ [C_2H_5OH]}$$

$$= \frac{(0.67\,mol\,dm^{-3})(0.67\,mol\,dm^{-3})}{(0.33\,mol\,dm^{-3})(0.33\,mol\,dm^{-3})}$$

$$= 4.1 \text{ at } 298\,K.$$

In Experiment 2, one of the concentrations was deliberately changed by adding more C_2H_5OH to give a new concentration of $0.67\,mol\,dm^{-3}$.

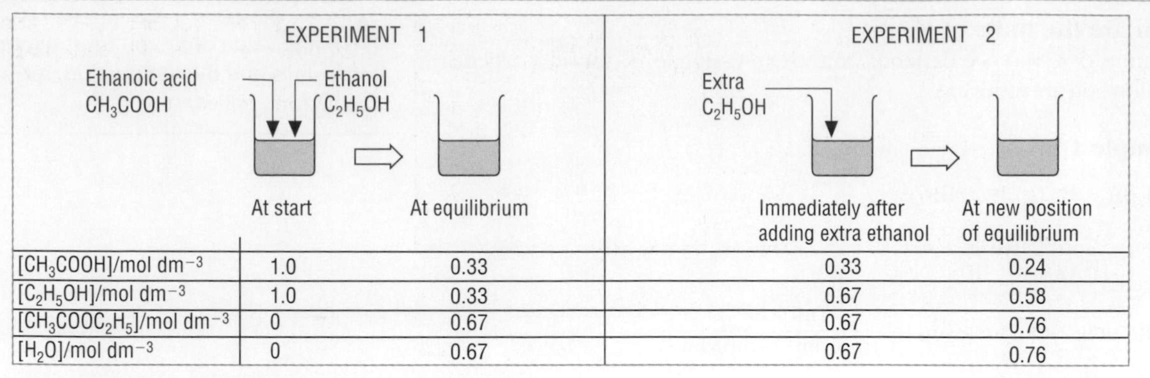

	At start	At equilibrium	Immediately after adding extra ethanol	At new position of equilibrium
$[CH_3COOH]$/mol dm^{-3}	1.0	0.33	0.33	0.24
$[C_2H_5OH]$/mol dm^{-3}	1.0	0.33	0.67	0.58
$[CH_3COOC_2H_5]$/mol dm^{-3}	0	0.67	0.67	0.76
$[H_2O]$/mol dm^{-3}	0	0.67	0.67	0.76

▲ **Figure 3** Equilibrium is set up in Experiment 1, starting with equal concentrations of ethanoic acid and ethanol. In Experiment 2, the equilibrium is disturbed by adding extra ethanol. Both experiments are carried out at 298 K.

Immediately after adding the extra C_2H_5OH, before any changes occur, the new concentration ratio is

$$\frac{(0.67\,\text{mol}\,\text{dm}^{-3})(0.67\,\text{mol}\,\text{dm}^{-3})}{(0.33\,\text{mol}\,\text{dm}^{-3})(0.67\,\text{mol}\,\text{dm}^{-3})} = 2.0\text{, which is much smaller than } K_c.$$

In order to restore the value of K_c to 4.1, some C_2H_5OH and CH_3COOH must react (making the bottom line smaller) to produce $CH_3COOC_2H_5$ and H_2O (making the top line bigger).

The system was left to reach equilibrium again, and the new equilibrium concentrations were measured. The values are shown in Figure 3.

$$K_c = \frac{(0.76\,\text{mol}\,\text{dm}^{-3})(0.76\,\text{mol}\,\text{dm}^{-3})}{(0.24\,\text{mol}\,\text{dm}^{-3})(0.58\,\text{mol}\,\text{dm}^{-3})}$$

$$= 4.1$$

So when the equilibrium was disturbed it moved in such a way that K_c remained constant.

Changing the total pressure

You saw in **Section 7.1** that for equilibria in the *gas phase* a change in total pressure affects the position of equilibrium if there is a change in the number of gas molecules in the reaction.

Le Chatelier's principle can be used to predict the *qualitative* effect of changing the total pressure – the position of equilibrium shifts so as to counteract the change in pressure. However, as with changes of concentration, the equilibrium will move in such a way that K_c remains constant. This involves calculating the pressures exerted by each gas in a mixture. (This is used in preference to concentrations in mol dm^{-3}.) We can then calculate a value called K_p. In this course, we do not deal quantitatively with K_p.

Changing the temperature

Changes in concentration and total pressure may alter the composition of equilibrium mixtures, but *they do not alter the value of the equilibrium constant itself*, provided the temperature does not change. The proportions of reactants and products alter in such a way as to keep the ratio in the K_c expression unchanged.

However, changes in temperature actually alter the value of the equilibrium constant itself.

Table 5 shows some experimental measurements of how temperature changes affect K_c for two reactions.

	$N_2(g) + 3H_2(g) \rightleftharpoons 2NH_3(g)$ $\Delta H^\ominus = -92\,kJ\,mol^{-1}$		$N_2O_4 \rightleftharpoons 2NO_2(g)$ $\Delta H^\ominus = +57\,kJ\,mol^{-1}$
T/K	$K_c/mol^{-2}\,dm^6$	*T*/K	$K_c/mol\,dm^{-3}$
400	4.39×10^1	200	5.51×10^{-8}
600	4.03	400	1.46
800	3.00×10^{-2}	600	3.62×10^2
	$K_c = \frac{[NH_3(g)^2]}{[N_2(g)]\,[H_2(g)]^3}$		$K_c = \frac{[NO_2(g)^2]}{[N_2O_4(g)]}$

◀ **Table 5** The effect of temperature on the equilibrium constants of two reacting systems – the value of $\Delta H^\ominus$ is for the forward reaction in each case.

You can see from the data that a rise in temperature favours the products and increases K_c for endothermic reactions. For exothermic reactions a rise in temperature favours the reactants and decreases K_c.

Using a catalyst

Catalysts *do not* affect the position of equilibrium or the equilibrium constant. They alter the *rate* at which equilibrium is attained but not the composition of the equilibrium mixture.

Summary

The effect of changing conditions on equilibrium mixtures is summarised in Table 6. The important thing to remember is that K_c is constant unless the temperature is changed.

Change in:	Composition	K_c
concentration	changed	unchanged
total pressure	may change	unchanged
temperature	changed	*changed*
catalyst	unchanged	unchanged

◀ **Table 6** The effect of changing conditions on equilibrium mixtures.

Problems for 7.2

1 a Use the values in Table 1 in this section (page 170) to plot a graph of concentration of carbon dioxide in the aqueous phase, $[CO_2(aq)]$ (*y*-axis), against its pressure in the gas phase, $p_{CO_2(g)}$ (*x*-axis).

b Use your graph to work out an accurate value for the equilibrium constant, at 292 K, of the reaction
$CO_2(g) \rightleftharpoons CO_2(aq)$

c Why is this better than taking an average of the five ratios you worked out from the data in Table 1?

2 Write expressions for K_c for the following reactions – in each case, give the units of K_c.

a $2NO(g) + O_2(g) \rightleftharpoons 2NO_2(g)$

b $C_2H_6(g) \rightleftharpoons C_2H_4(g) + H_2(g)$

c $2HI(g) \rightleftharpoons H_2(g) + I_2(g)$

d $CO_2(aq) + H_2O(l) \rightleftharpoons HCO_3^-(aq) + H^+(aq)$

e $In_6(aq) \rightleftharpoons 3In_2(aq)$

f $CH_3COOH(l) + C_3H_7OH(l) \rightleftharpoons CH_3COOC_3H_7(l) + H_2O(l)$

3 The equilibrium constant, K_c, for a reaction is given by the expression

$$K_c = \frac{[SO_3(g)]^2}{[SO_2(g)]^2[O_2(g)]}$$

Write the balanced chemical equation for the reaction.

4 A mixture of nitrogen and hydrogen was sealed in a steel vessel and held at 1000 K until equilibrium was reached. The contents were then analysed – the results are given in the following table.

Substance	Equilibrium concentration/$mol\,dm^{-3}$
$N_2(g)$	0.142
$H_2(g)$	1.84
$NH_3(g)$	1.36

a Write an expression for K_c for the reaction
$N_2(g) + 3H_2(g) \rightleftharpoons 2NH_3(g)$

b Calculate a value for K_c – remember to give the units.

5 Consider the reaction between hydrogen and oxygen to produce steam.
  a Write an equation for the reaction with state symbols.
  b Write an expression for K_c.
  c State how the equilibrium position is affected by
    i an increase in temperature
    ii an increase in the total pressure.
  d State how the value of K_c is affected by
    i an increase in temperature
    ii an increase in the total pressure.

6 When PCl_5 is heated in a sealed container and maintained at a constant temperature, an equilibrium is established. At 523 K, the following equilibrium concentrations were determined.

Substance	Equilibrium concentration/mol dm^{-3}
PCl_5	0.077
PCl_3	0.123
Cl_2	0.123

  a Write an expression for K_c for the reaction
    $PCl_5(g) \rightleftharpoons PCl_3(g) + Cl_2(g)$
  b Calculate a value for K_c.

7 For the reaction

$$2H_2(g) + S_2(g) \rightleftharpoons 2H_2S(g)$$

K_c was found to be $9.4 \times 10^5\,mol^{-1}\,dm^3$ at 1020 K. Equilibrium concentrations were measured as

$[H_2(g)] = 0.234\,mol\,dm^{-3}$
$[H_2S(g)] = 0.442\,mol\,dm^{-3}$

  a Write an expression for K_c for the reaction.
  b What is the equilibrium concentration of $S_2(g)$?

8 The equilibrium constant K_c for the reaction

$$2NO_2(g) \rightleftharpoons 2NO(g) + O_2(g)$$

is 9.0 mol dm^{-3} at 683 K.
  a Do you expect reactant or products to predominate at equilibrium?
  b Write an expression for K_c for the reaction.
  c Equilibrium concentrations were measured as
    $[NO(g)] = 0.50\,mol\,dm^{-3}$
    $[O_2(g)] = 0.25\,mol\,dm^{-3}$
    What is the equilibrium concentration of $NO_2(g)$ in the above mixture?
  d Does this agree with your prediction in part **a**?

9 Ethanol and ethanal react together according to the equation

$$2C_2H_5OH(l) + CH_3CHO(l) \rightleftharpoons CH_3CH(OC_2H_5)_2(l) + H_2O(l)$$

ethanol *ethanal*

An excess of ethanol was mixed with ethanal, and the system allowed to reach equilibrium at 298 K. The equilibrium concentrations were measured and are shown in the table below.

Substance	Equilibrium concentration/ mol dm^{-3}
$C_2H_5OH(l)$	13.24
$CH_3CHO(l)$	0.133
$CH_3CH(OC_2H_5)_2(l)$	1.311
$H_2O(l)$	1.311

  a Write an expression for K_c for this equilibrium.
  b Examine the equilibrium concentrations. What do you think predominates at equilibrium – reactants or products?
  c What does this lead you to expect for the magnitude of K_c?
  d Calculate a value for K_c.

10 For the reaction of aqueous chloromethane with alkali

$$OH^-(aq) + CH_3Cl(aq) \rightleftharpoons CH_3OH(aq) + Cl^-(aq)$$

the equilibrium constant has a value of 1×10^{16} at room temperature. What does this tell you about the concentration of chloromethane at equilibrium?

7.3 *Chromatography*

The general principle

Chromatography is an important analytical technique for separating and identifying the components of a mixture.

There are a number of different types of chromatography – they all depend on the equilibrium set up when a compound distributes itself between two phases. One phase stands still (the **stationary phase**) while the other moves over it (the **mobile phase**). Different compounds distribute themselves between the two phases to different extents, and so move along with the mobile phase at different speeds.

You may already be familiar with paper chromatography through your work on the activities, and you may also have used the **thin-layer chromatography** (t.l.c.) technique shown in Figure 1.

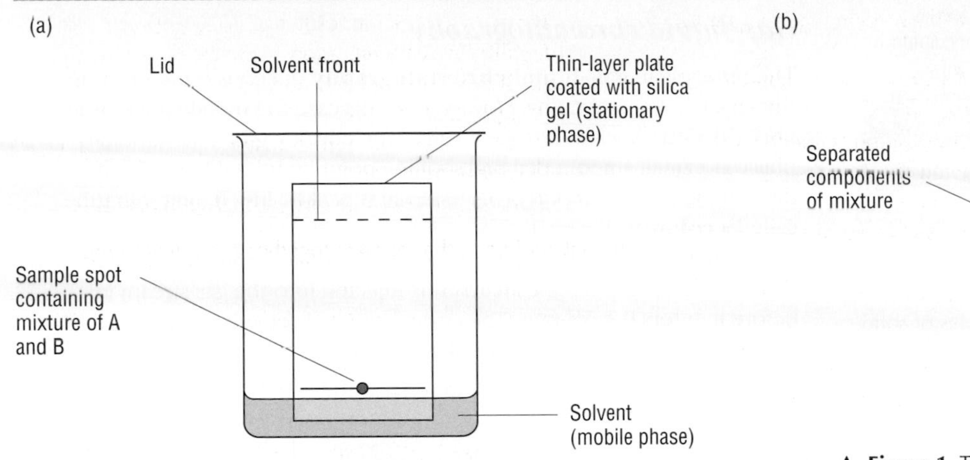

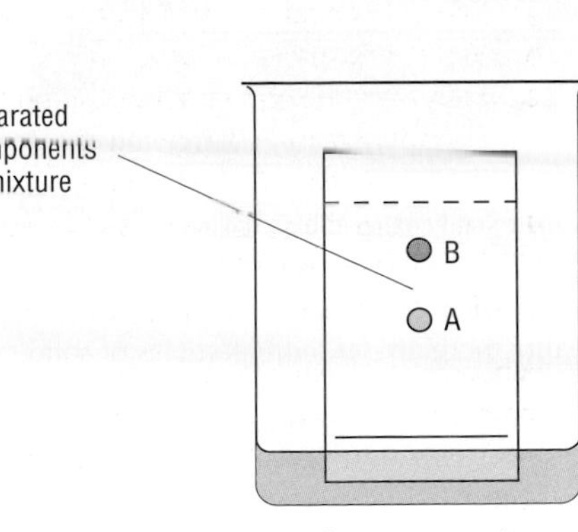

▲ **Figure 1** Thin-layer chromatography separating a mixture of substances A and B.
a The start of the process.
b The end of the process.

Here a small sample of a mixture is spotted onto a solid support material (the stationary phase). In the case of t.l.c., this might be silica gel spread in a thin uniform layer on a plastic plate. In the case of paper chromatography, this might be a rectangular piece of filter paper. A suitable solvent now rises up the plate. This is the mobile phase.

What you see when you examine the developed plate is a series of spots – one for each compound in the mixture. If the compounds are not coloured then you will need to 'develop' the plate to show the spots. In the case of amino acids, ninhydrin is used to develop the spots. Amino acids then show up in shades of purple. Ninhydrin is said to be the 'locating agent'. Iodine and uv light are often used as locating agents for other materials. For more details, see **Chemical Ideas Appendix 1: Experimental Techniques** (Technique 4).

Figure 3 shows, in a simplified way, how two components get separated in thin-layer chromatography.

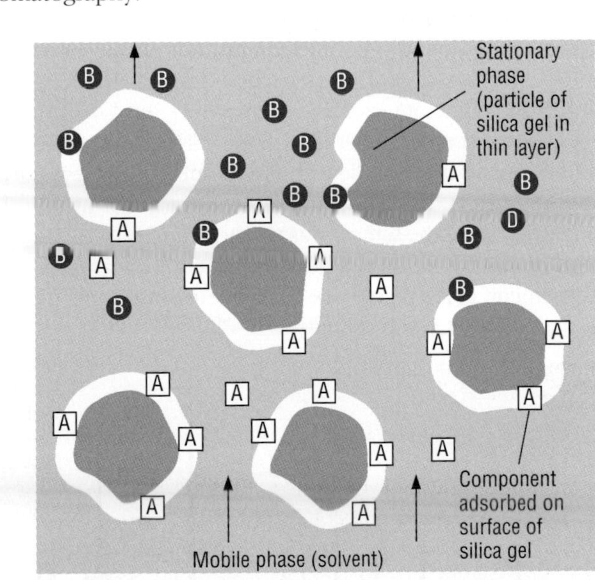

▲ **Figure 3** A simplified explanation of thin-layer chromatography. This magnified view shows the behaviour of components A and B as the solvent passes between the particles of silica gel in the thin layer.

R_f values

The distance a particular substance travels in paper or thin-layer chromatography depends on

- the nature of the substance
- the total distance travelled by the solvent front
- the conditions under which chromatography is carried out – such as temperature, type of paper or thin-layer plate, and the type of solvent used.

The **R_f value** for a substance is the distance that substance travels relative to the solvent front (Figure 2). R_f values are constant for a particular set of conditions, so we can use them to identify the different spots on a chromatogram.

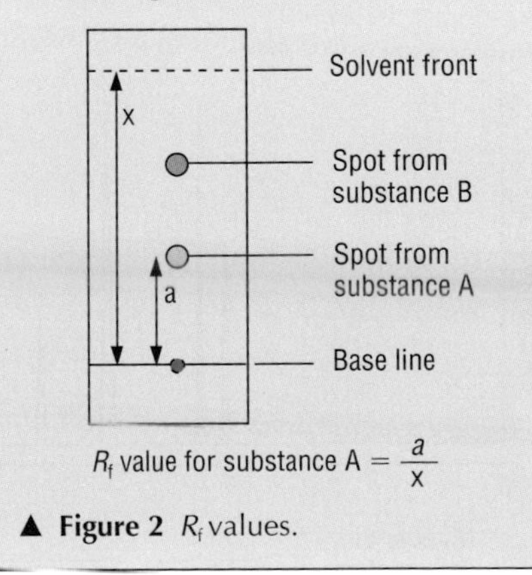

▲ **Figure 2** R_f values.

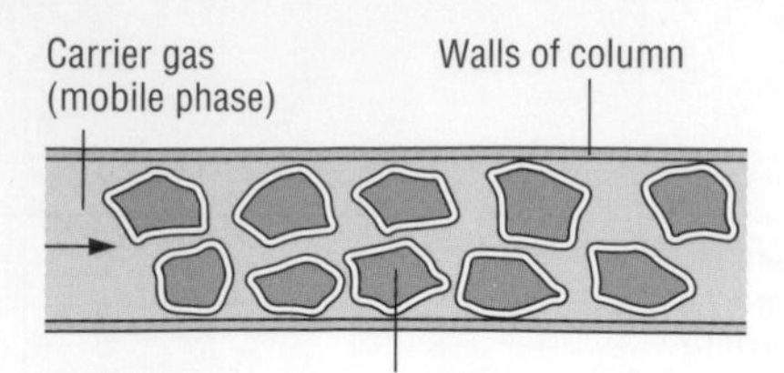

▲ **Figure 4** Inside a g.l.c. column – in reality there are far more particles of solid than this.

Gas–liquid chromatography

The principle in **gas–liquid chromatography** (g.l.c.) is the same as in thin-layer chromatography. However, in this case the mobile phase is an unreactive gas, such as nitrogen, called the **carrier gas**. The stationary phase is a small amount of a high boiling point liquid held on a finely divided inert solid support. This material is packed into a long thin tube called a **column** (Figure 4). The column is coiled inside an oven.

The main parts of a simple gas–liquid chromatograph are shown in Figure 5. The sample to be analysed is injected into the gas stream just before it enters the column. The components of the mixture are carried through the column in a stream of gas.

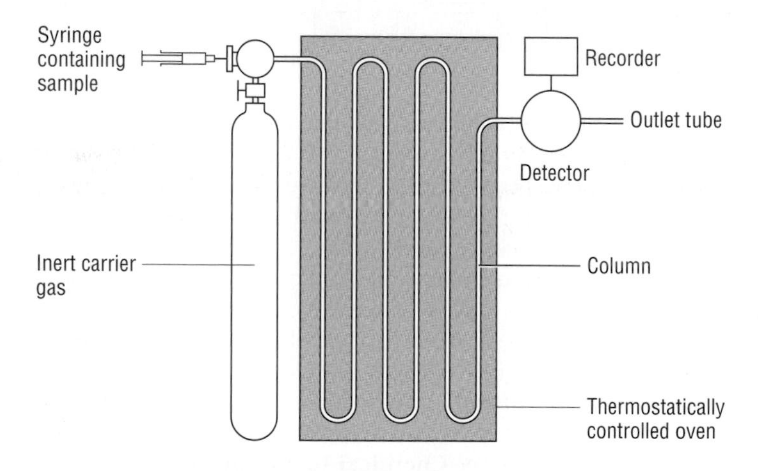

► **Figure 5** A gas–liquid chromatograph.

Each component has a different affinity for the stationary phase compared with the mobile phase. Each distributes itself to different extents between the two phases, and so emerges from the column at different times – those compounds that favour the mobile phase (the carrier gas) are carried along more quickly. The most volatile compounds usually emerge first. The compounds that favour the stationary phase (the liquid) get held up in the column and come out last.

A **detector** on the outlet tube monitors the compounds coming out of the column. Signals from the detector are plotted out by a **recorder** as a **chromatogram**. This shows the recorder response against the time which has elapsed since the sample was injected onto the column. Each component of the mixture gives rise to a peak. Figure 6 shows a typical gas chromatogram of a premium grade petrol (see the **Developing Fuels** module).

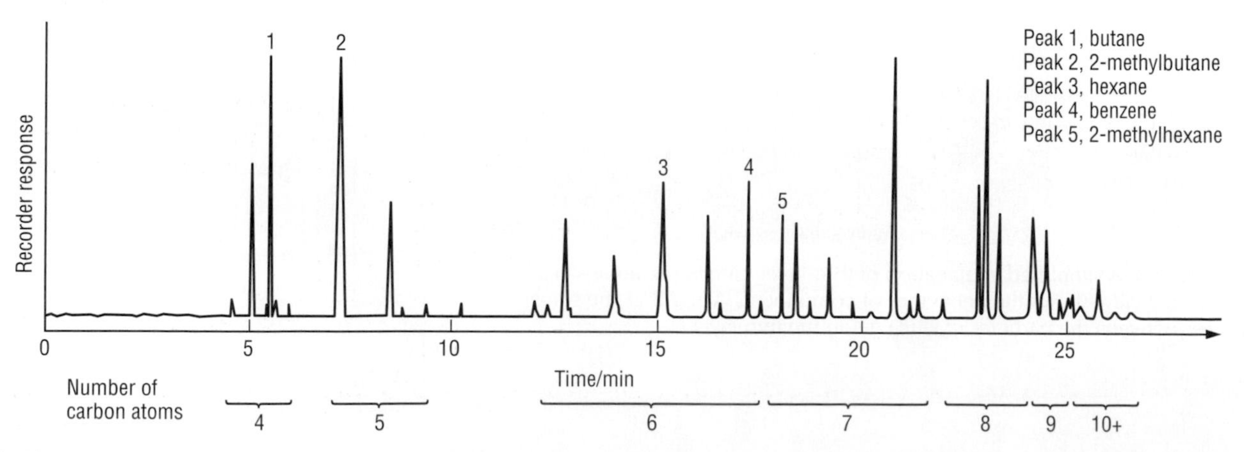

▲ **Figure 6** Gas chromatogram of a premium grade petrol. 2-methylhexane took about 18 minutes to travel along the column.

The time that a compound is held on a column under given conditions is characteristic of the compound and is called its **retention time**.

Lots of factors can affect the retention time including the length and packing of the column, the nature and flow rate of the carrier gas, and the temperature of the column. So you have to calibrate an instrument with known compounds and keep the conditions constant throughout the analysis.

The area under each peak depends on the amount of compound present, so you can use a gas chromatogram to work out the *relative amount* of each component in the mixture. If the peaks are very sharp, their relative *heights* can be used.

The technique is very sensitive and very small quantities can be detected, such as traces of explosives or drugs in forensic tests. With larger instruments, a pure sample of each compound can be collected as it emerges from the outlet tube. In more sophisticated instruments, the outlet tube is connected to a **mass spectrometer** so that each compound can be identified directly.

Problems for 7.3

1 Figure 7 shows the result of separating two compounds, X and Y, using t.l.c.

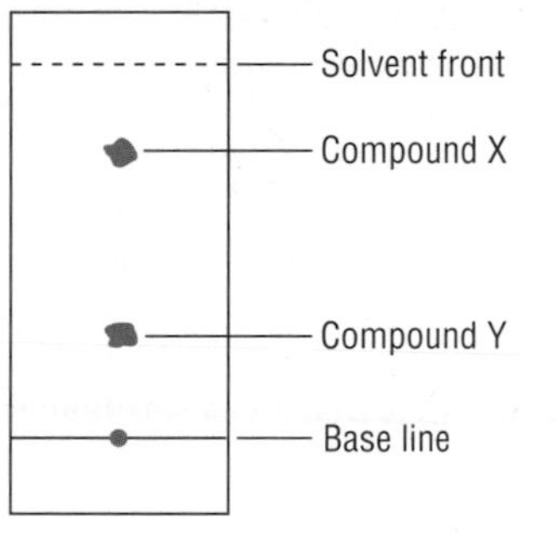

▲ **Figure 7** Developed t.l.c. plate for the mixture of X and Y.

a Calculate the R_f values for X and Y under these conditions.

b Refer to Figure 3 in this section (page 177) and explain why compound Y moves more slowly up the plate than compound X.

c Why is it necessary to cover the container in which you carry out thin-layer chromatography if you want to get reliable results?

2 **a** List the factors that can affect the retention time of a compound on a gas chromatography column.

b Explain why it is important to enclose the column of a gas–liquid chromatograph inside an oven kept at a constant temperature.

c What types of substances do you think would not be suitable for analysis using a gas–liquid chromatograph? Explain your answer.

3 The chromatogram in Figure 8 was obtained from a mixture containing a pair of *Z* and *E* isomers.

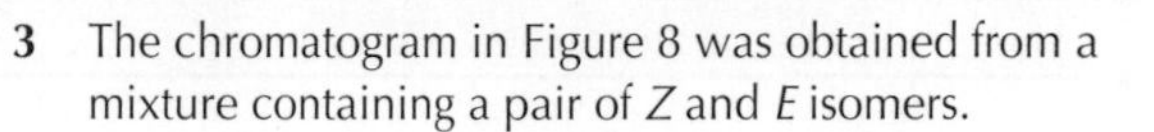

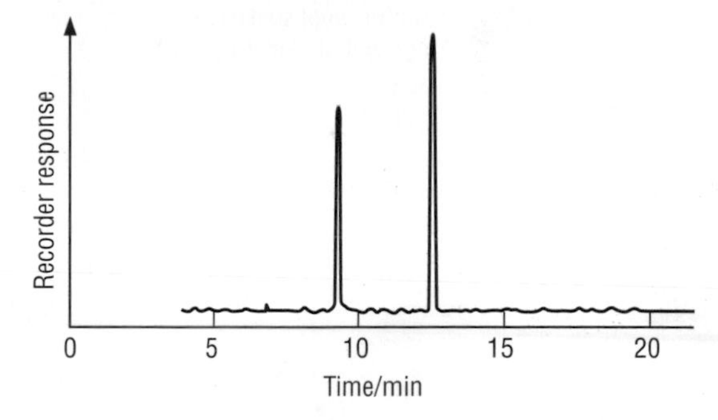

▲ **Figure 8** Gas chromatogram for a mixture containing a pair of *Z* and *E* isomers.

a The pure *Z* isomer had a retention time of 9.0 min under identical conditions. The pure *E* isomer had a retention time of 12.5 min. What were the proportions of the *Z* and *E* isomers in the mixture?

b The gas chromatograph used was linked to a mass spectrometer. A mass spectrum was recorded for each substance as it emerged from the column. What can you say about the molecular ion peaks in the two spectra? Explain your answer.

4 Look at the gas chromatogram in Figure 6 in this section.

a There are two peaks corresponding to four carbon atoms. One is due to butane – what do you think the other is due to?

b How does the number of carbon atoms affect how long it takes a compound to travel along the column? Suggest a reason for your answer.

c How would the chromatogram change if the column were longer?

ACIDS AND BASES

8.1 Acid–base reactions

What do we mean by acid and base?

Properties of acidic solutions

- turn litmus red
- neutralised by bases
- pH < 7
- liberate CO_2 from carbonates

We say that hydrochloric acid is an 'acid' because of the things it does. It turns litmus red, reacts with carbonates to give CO_2 and is neutralised by bases – all properties that we expect of acids.

Chemists try to *explain* properties in terms of what goes on at the level of atoms, molecules and ions. They explain the characteristic properties of acids by saying that acids have *the ability to transfer H^+ ions to something else*. The substance which accepts the H^+ ion is called a **base**.

For example, take the reaction of hydrogen chloride with ammonia. You may have seen this reaction demonstrated in apparatus like that shown in Figure 1. The reaction forms a white salt, ammonium chloride:

$$HCl(g) + NH_3(g) \rightarrow NH_4^+Cl^-(s)$$

HCl(g) and NH_3(g) meet here and react

Cotton wool soaked in concentrated hydrochloric acid

Cotton wool soaked in concentrated ammonia solution

▲ **Figure 1** The reaction of hydrogen chloride with ammonia – an acid–base reaction.

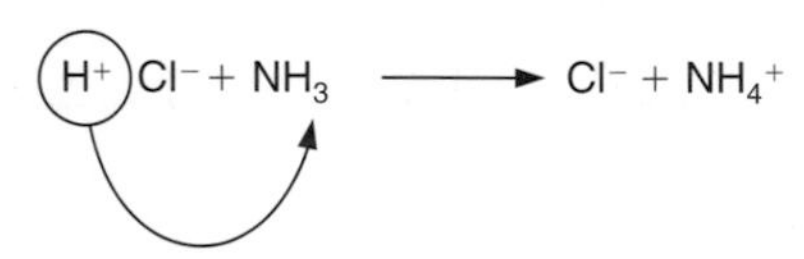

▲ **Figure 2** HCl transfers an H^+ ion to NH_3.

In this reaction, the HCl transfers H^+ to NH_3 (Figure 2). The HCl is behaving as an acid, and the NH_3 is behaving as a base.

Notice that this is not a redox reaction. The oxidation states of H, Cl and N are unchanged during the reaction – check for yourself.

The general definition of an acid is that it is a substance which donates H^+ in a chemical reaction. The substance that accepts the H^+ is a base. The reaction in which this happens is called an **acid–base reaction**.

Since a hydrogen atom consists of only a proton and an electron, an H^+ ion corresponds to just one proton. Sometimes we refer to acids as *proton donors* and bases as *proton acceptors*.

This theory of H^+ transfer is known as the *Brönsted–Lowry theory* of acids and bases.

The Brönsted–Lowry theory

An acid is an H^+ donor.

A base is an H^+ acceptor.

Solutions of acids and bases

Hydrogen chloride is a gas and contains HCl molecules. Water is almost totally made up of H_2O molecules. Yet hydrochloric acid, a solution of hydrogen chloride in water, readily conducts electricity so it must contain ions. There must be a reaction between the hydrogen chloride molecules and the water molecules which produces these ions. The reaction is

$$\underset{\text{acid}}{HCl(aq)} + \underset{\text{base}}{H_2O(l)} \rightarrow H_3O^+(aq) + Cl^-(aq)$$

In this reaction, H_2O is behaving as a base.

You may not previously have come across the ion with the formula H_3O^+. It is called the **oxonium ion**. It is a very common ion – it is present in every solution of an acid in water. In other words it occurs in every *acidic* solution. The acid donates H^+ to H_2O to form H_3O^+. Figure 3 shows how the H_2O bonds to H^+.

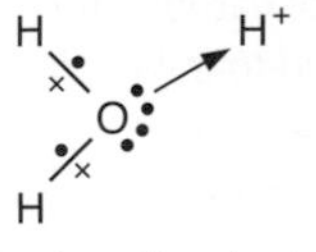

▲ **Figure 3** The bonding in the H_3O^+ ion. A lone pair on the O atom forms a **dative** covalent bond to H^+. (Once the bond has formed, however, it is indistinguishable from the two other O–H bonds).

The H_3O^+ ion can itself act as an acid – it can donate H^+ and turn into an H_2O molecule. The familiar properties of acidic solutions are all properties of the H_3O^+ ion.

You will often see the formula $H_3O^+(aq)$ shortened to $H^+(aq)$ and the dissociation of HCl(aq) into ions represented by

$$HCl(aq) \rightarrow H^+(aq) + Cl^-(aq)$$

When an acid dissolves in water, $H^+(aq)$ ions form in solution.

An **alkali** is a base that dissolves in water to produce hydroxide ions, $OH^-(aq)$. This is illustrated in Figure 4.

Some alkalis, such as sodium hydroxide and potassium hydroxide, already contain hydroxide ions, whereas others, such as sodium carbonate and ammonia, form $OH^-(aq)$ ions when they react with water:

$$CO_3^{2-}(aq) + H_2O(l) \rightleftharpoons HCO_3^-(aq) + OH^-(aq)$$
$$NH_3(aq) + H_2O(l) \rightleftharpoons NH_4^+(aq) + OH^-(aq)$$

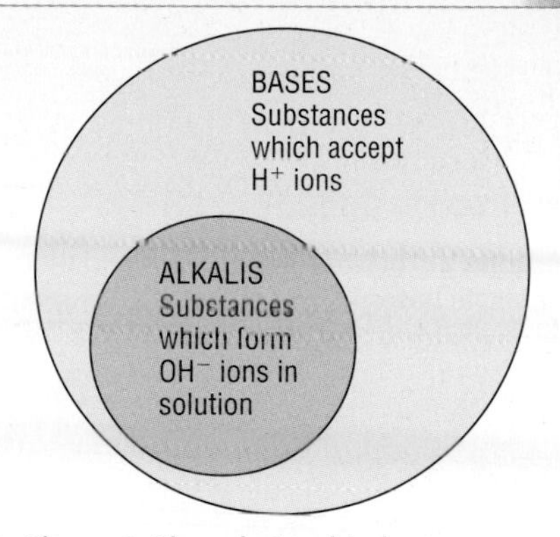

▲ **Figure 4** The relationship between alkalis and bases.

Acid–base pairs

Once an acid has donated an H^+ ion, there is always the possibility that it will take it back again. For example, consider ethanoic acid (acetic acid), the acid in vinegar. When ethanoic acid behaves as an acid, this is the change that happens:

$$CH_3COOH(aq) \rightarrow CH_3COO^-(aq) + H^+(aq)$$

But if you add a strong acid to a solution containing ethanoate ions, CH_3COO^-, the ethanoate ions accept H^+ from the stronger acid and go back to ethanoic acid:

$$CH_3COO^-(aq) + H^+(aq) \rightarrow CH_3COOH(aq)$$

In this reaction, the ethanoate ion is behaving as a *base* – it is called the **conjugate base** of ethanoic acid.

Every acid has a conjugate base, and every base has a conjugate acid. They are called a **conjugate acid–base pair.** If we represent a general acid as HA, then we have

$$\underset{\text{conjugate acid}}{HA(aq)} \rightarrow H^+(aq) + \underset{\text{conjugate base}}{A^-(aq)}$$

For example, the conjugate base of HCl(aq) is $Cl^-(aq)$. The conjugate acid of $NH_3(aq)$ is $NH_4^+(aq)$. Table 1 shows more examples – the state symbols have been omitted for clarity.

◀ **Table 1** Conjugate acid–base pairs.

Acid			Base
chloric(VII) acid	$HClO_4$	→	$H^+ + ClO_4^-$
hydrochloric acid	HCl	→	$H^+ + Cl^-$
sulfuric acid	H_2SO_4	→	$H^+ + HSO_4^-$
oxonium ion	H_3O^+	→	$H^+ + H_2O$
ethanoic acid	CH_3COOH	→	$H^+ + CH_3COO^-$
hydrogen sulfide	H_2S	→	$H^+ + SH^-$
ammonium ion	NH_4^+	→	$H^+ + NH_3$
water	H_2O	→	$H^+ + OH^-$
ethanol	C_2H_5OH	→	$H^+ + C_2H_5O^-$

Notice in Table 1 that water, H_2O, can be both an acid and a base – it all depends on what the water is reacting with. If it is reacting with a strong acid, such as HCl(aq), water acts as a base accepting $H^+(aq)$ and forming $H_3O^+(aq)$. If it is reacting with a strong base, such as CaO, water acts as an acid donating H^+ to form $OH^-(aq)$ ions.

Many other substances can behave as both acid and base – they are described as **amphoteric**.

Strength of acids and bases

Not all acids have the same strength. Some are powerful H^+ donors and are described as *strong acids*. Others are *weak acids* – they are moderate or weak H^+ donors. Table 2 shows the range of strengths found in acids.

▶ **Table 2** The strength of acids and their conjugate bases.

	Acid	Base	
STRONGEST ACID ↑	$HClO_4 \longrightarrow$	$H^+ + ClO_4^-$	WEAKEST BASE ↓
	$HCl \longrightarrow$	$H^+ + Cl^-$	
	$H_2SO_4 \longrightarrow$	$H^+ + HSO_4^-$	
increasing acid strength	$H_3O^+ \longrightarrow$	$H^+ + H_2O$	increasing base strength
	$CH_3COOH \rightarrow$	$H^+ + CH_3COO^-$	
	$H_2S \longrightarrow$	$H^+ + SH^-$	
	$NH_4^+ \longrightarrow$	$H^+ + NH_3$	
	$H_2O \longrightarrow$	$H^+ + OH^-$	
WEAKEST ACID	$C_2H_5OH \longrightarrow$	$H^+ + C_2H_5O^-$	STRONGEST BASE

Notice that a strong acid has a weak conjugate base, and vice versa. This makes sense – if an acid has a strong tendency to donate H^+ ions, its conjugate base will have a weak tendency to accept them back.

Table 2 includes some substances, such as ethanol, that we do not normally think of as acids. Ethanol has a very small tendency to donate protons – it *can* behave as an acid, but only in the presence of a very strong base.

Indicators

Acid–base indicators, such as litmus, are coloured organic substances which are themselves weak acids. The special thing about indicators is that the conjugate acid and base forms have *different colours*.

For example, the acid form of litmus is red and its conjugate base is blue. If we represent litmus as HIn, we have

$$\underset{\textit{red}}{\underset{\text{acid}}{HIn(aq)}} \rightleftharpoons H^+(aq) + \underset{\textit{blue}}{\underset{\text{base}}{In^-(aq)}}$$

When you add $H_3O^+(aq)$ from an acid to blue litmus, this reaction occurs:

$$\underset{\textit{blue}}{In^-(aq)} + H_3O^+(aq) \rightarrow \underset{\textit{red}}{HIn(aq)} + H_2O(l)$$

The blue litmus turns red.

When you add an alkali containing OH^- to red litmus, this reaction happens:

$$\underset{\textit{red}}{HIn(aq)} + OH^-(aq) \rightarrow \underset{\textit{blue}}{In^-(aq)} + H_2O(l)$$

The red litmus turns blue.

Problems for 8.1

1 Use the Brönsted–Lowry theory to explain what happens when an acid reacts with a base.

2 Write out each of the following equations, and indicate which of the reactants is the acid and which is the base.
 a $HNO_3 + H_2O \rightarrow H_3O^+ + NO_3^-$
 b $NH_3 + H_2O \rightarrow NH_4^+ + OH^-$
 c $NH_4^+ + OH^- \rightarrow NH_3 + H_2O$
 d $SO_4^{2-} + H_3O^+ \rightarrow HSO_4^- + H_2O$
 e $H_2O + H^- \rightarrow H_2 + OH^-$
 f $H_3O^+ + OH^- \rightarrow 2H_2O$
 g $NH_3 + HBr \rightarrow NH_4^+ + Br^-$
 h $H_2SO_4 + HNO_3 \rightarrow HSO_4^- + H_2NO_3^+$
 i $CH_3COOH + H_2O \rightarrow CH_3COO^- + H_3O^+$

3 Classify each of these reactions as either acid–base or redox.
 a $NH_4^+ + CO_3^{2-} \rightarrow NH_3 + HCO_3^-$
 b $H_2S + 2OH^- \rightarrow S^{2-} + 2H_2O$
 c $I_2 + 2OH^- \rightarrow I^- + IO^- + H_2O$
 d $Mg + 2H^+ \rightarrow Mg^{2+} + H_2$

4 Write out the following equations and in each case identify the two conjugate acid–base pairs. For example, for hydrochloric acid and the hydroxide ion the answer would be:
 Equation: $HCl + OH^- \rightarrow Cl^- + H_2O$
 Conjugate pairs: acid/base
 HCl/Cl^-
 H_2O/OH^-
 a $NH_4^+ + H_2O \rightarrow NH_3 + H_3O^+$
 b $H_2SO_4 + HNO_3 \rightarrow HSO_4^- + H_2NO_3^+$
 c $CH_3COOH + HClO_4 \rightarrow CH_3COOH_2^+ + ClO_4^-$

5 Write balanced equations for each of the following reactions between acids and bases. In each case identify the two conjugate acid–base pairs (see problem 4).
 a Ethanoic acid and hydroxide ion.
 b Hydrogencarbonate ion, HCO_3^-, and hydrochloric acid to produce carbonic acid, H_2CO_3.
 c Water reacting with the hydrogensulfate ion, HSO_4^-.

8.2 *Strong and weak acids and pH*

Strong and weak acids

The Brönsted–Lowry theory of acids and bases was introduced in **Section 8.1**. This theory describes an **acid** as an **H^+ donor**. An important consequence of this is that acids donate H^+ to water molecules in aqueous solution to produce **oxonium ions, H_3O^+**. The presence of oxonium ions gives rise to the familiar acidic properties of all aqueous solutions of acids.

Acids vary in **strength**. Different acids donate H^+ to differing extents. **Strong acids** have a strong tendency to donate H^+ – the donation of H^+ is essentially complete. The reaction with water can be regarded as going to completion and can be described by the following equation, where HA represents the strong acid (we can assume that *no* unreacted HA remains in solution):

$$HA(aq) + H_2O(l) \rightarrow H_3O^+(aq) + A^-(aq)$$

We can simplify this to

$$HA(aq) \rightarrow H^+(aq) + A^-(aq)$$

by leaving out the water, which is present in excess.

Hydrochloric acid, HCl, and sulfuric acid, H_2SO_4, are examples of strong acids.

$$HCl(aq) \rightarrow H^+(aq) + Cl^-(aq)$$
$$H_2SO_4(aq) \rightarrow 2H^+(aq) + SO_4^{2-}(aq)$$

If a substance is a **weak acid**, its tendency to donate H^+ is weaker and the reaction with water is incomplete. Some $H^+(aq)$ ions are formed but there is still some unreacted acid in solution. If we represent the weak acid as HA, its reaction with water can be represented as

$$HA(aq) \rightleftharpoons H^+(aq) + A^-(aq)$$

The equilibrium sign shows that a significant concentration of HA(aq) is present along with the A^-(aq) and H^+(aq) ions formed from it.

There are many more weak acids than strong acids and they vary in strength depending on the extent to which the equilibrium lies to the right-hand side. The more it lies to the right-hand side, the stronger is the acid. Most carboxylic acids are weak acids – ethanoic acid, for example, reacts with water and establishes this dynamic equilibrium:

$$CH_3COOH(aq) \rightleftharpoons H^+(aq) + CH_3COO^-(aq)$$

The position of this equilibrium is well to the left, so ethanoic acid is a weak acid. You can read about this in **Section 13.4**.

Carbon dioxide and sulfur dioxide both react with water to form weakly acidic solutions. For example:

$$CO_2(aq) + H_2O(l) \rightleftharpoons H^+(aq) + HCO_3^-(aq)$$
$$HCO_3^-(aq) \rightleftharpoons H^+(aq) + CO_3^{2-}(aq)$$

This equilibrium system, where HCO_3^-(aq) is the product of the first equilibrium and the reactant of the second, is sometimes simplified by combining the two equilibria:

$$H_2CO_3(aq) \rightleftharpoons 2H^+(aq) + CO_3^{2-}(aq)$$

In this equation, the solution of carbon dioxide in water is called carbonic acid and is represented by H_2CO_3(aq).

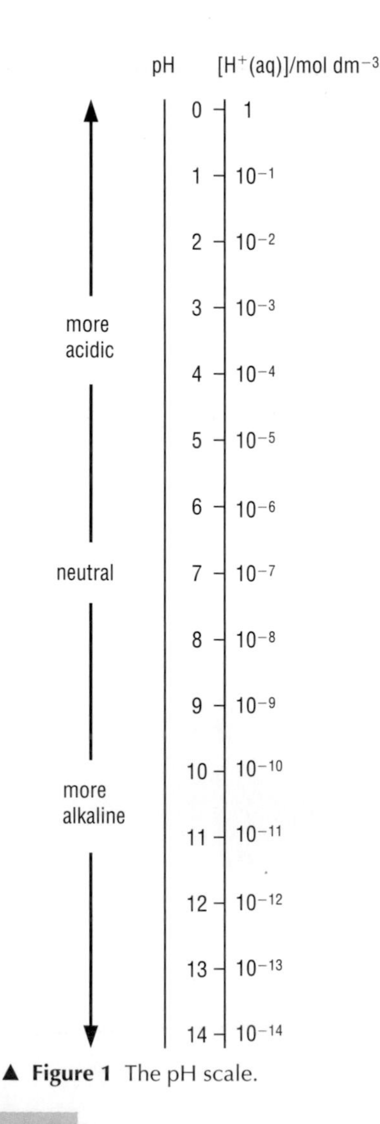

▲ **Figure 1** The pH scale.

The pH scale

Acids and alkalis are involved in many chemical processes, including the biochemical reactions that take place inside our bodies. Chemists often find it useful to know the concentration of H^+(aq) or OH^-(aq) present in a solution. Because it is not easy to work with figures such as 1×10^{-8} and 1×10^{-4} all the time, they use a measuring system called the **pH scale**.

The pH scale was devised at the beginning of the twentieth century by a Danish chemist called Soren Sorensen. He wanted a simple way to indicate how much acid or alkali was present in a solution. The 'p' in pH stands for *potens*, which is Latin for power – so pH is measuring the power of H^+(aq) in a solution, i.e. its concentration.

pH is defined as

$$pH = -lg[H^+(aq)]$$

If you are unfamiliar with the lg function, you can think of it as a technique for bringing numbers of very different magnitudes onto the same scale. Try entering the following numbers into your calculator and finding their lg values:

$$1 \times 10^{-14}; \quad 1 \times 10^{-7}; \quad 1 \times 10^{-3}; \quad 1$$

Figure 1 shows how the pH of a solution is related to $[H^+(aq)]$ in a solution. The pH value is the same as the negative power of 10 that relates to $[H^+(aq)]$. When $[H^+(aq)]$ changes by a factor of 10, the pH changes by 1. This is known as a *logarithmic scale*. Solutions with a pH of 7 are neutral. Acids have a pH of less than 7 and alkalis have a pH of more than 7.

Notice that the pH scale runs in the opposite direction to the scale of $[H^+(aq)]$ values – a *low* pH corresponds to a *high* concentration of H^+(aq). This is because pH is equal to *minus* lg $[H^+(aq)]$.

Many solutions of acids or alkalis have a concentration that is less than 1 mol dm^{-3} and so the pH values of these solutions lie between 0 and 14.

An acid where $[H^+(aq)]$ is more than $1\,mol\,dm^{-3}$ has a pH of less than 0, i.e. a negative pH. Similarly, an alkaline solution where $[OH^-(aq)]$ is greater than $1\,mol\,dm^{-3}$ has a pH greater than 14.

You can use indicator paper to measure pH by comparing the colour of the paper with a colour chart. You will get a more accurate answer, and you can record a numerical value directly, if you use a special **glass electrode** connected to a specially calibrated voltmeter called a **pH meter**.

Calculating pH values

Strong acids

You can find the pH of a solution of a strong acid by a straightforward calculation. Since the reaction with water goes to completion effectively, the amount in moles of $H^+(aq)$ ions is equal to the amount in moles of acid (HA) put into solution.

For a $0.01\,mol\,dm^{-3}$ solution of a strong acid HA

$[H^+(aq)] = 0.01\,mol\,dm^{-3}$
$pH = -lg(0.01) = 2$

For a $1\,mol\,dm^{-3}$ solution of a strong acid HA

$[H^+(aq)] = 1\,mol\,dm^{-3}$
$pH = -lg(1) = 0$

Weak acids

For a weak acid,

$$HA(aq) \rightleftharpoons H^+(aq) + A^-(aq)$$

We can write an equilibrium constant in the usual way as

$$K_a = \frac{[H^+(aq)][A^-(aq)]}{[HA(aq)]}$$

This equilibrium constant is called the **acidity constant** (or often the *acid dissociation constant*) and so is given the symbol K_a. Table 1 gives values for some weak acids.

Acid		K_a/mol dm^{-3}
methanoic	HCOOH	1.6×10^{-4}
benzoic	C_6H_5COOH	6.3×10^{-5}
ethanoic	CH_3COOH	1.7×10^{-5}
chloric(I)	HClO	3.7×10^{-8}
hydrocyanic	HCN	4.9×10^{-10}
nitrous (nitric(III))	HNO_2	4.7×10^{-4}

◀ **Table 1** Acidity constants for some weak acids at 298 K.

To work out the pH, we need first to find $[H^+(aq)]$. In doing so, we can make two assumptions about weak acids that simplify the calculation considerably.

Assumption 1: *$[H^+(aq)] = [A^-(aq)]$*

At first sight this might seem obvious because equal amounts of $H^+(aq)$ and $A^-(aq)$ are formed from HA. But there is another source of $H^+(aq)$ – from water itself. This is through the equilibrium

$$H_2O(l) \rightleftharpoons H^+(aq) + OH^-(aq)$$

Water produces far fewer $H^+(aq)$ ions than most weak acids, so we do not introduce a significant inaccuracy into the calculation by neglecting the ionisation of water.

Assumption 2: *The amount of HA at equilibrium is equal to the amount of HA put into the solution.*
In other words, when we calculate the concentration of HA at equilibrium, we can neglect the fraction of HA which has lost H^+. We can do this because we are dealing with weak acids and this fraction is very small. For example, in a solution of HA of concentration 0.1 mol dm^{-3} we can assume that when equilibrium has been established

$$[HA(aq)] = 0.1\,mol\,dm^{-3}$$

Let us use these ideas to calculate the value of $[H^+(aq)]$ and the pH for ethanoic acid solutions at 298 K with concentrations of

a 1 mol dm^{-3}
b 0.01 mol dm^{-3}

(K_a for ethanoic acid is 1.7×10^{-5} mol dm^{-3} at 298 K.)

We can represent the reaction between ethanoic acid and water by the general equation

$$HA(aq) \rightleftharpoons H^+(aq) + A^-(aq)$$

and the acidity constant is therefore given by

$$K_a = \frac{[H^+(aq)]\,[A^-(aq)]}{[HA(aq)]} = 1.7 \times 10^{-5}\,mol\,dm^{-3}\ \text{(at 298 K)}$$

Assumptions 1 and **2** then allow us to write

a $$1.7 \times 10^{-5}\,mol\,dm^{-3} = \frac{[H^+(aq)]^2}{1\,mol\,dm^{-3}}$$

Therefore $[H^+(aq)]^2 = 1.7 \times 10^{-5}\,mol^2\,dm^{-6}$
So $[H^+(aq)] = 4.12 \times 10^{-3}\,mol\,dm^{-3}$ and pH = 2.39 (at 298 K)

b $$1.7 \times 10^{-5}\,mol\,dm^{-3} = \frac{[H^+(aq)]^2}{0.01\,mol\,dm^{-3}}$$

Therefore $[H^+(aq)]^2 = 1.7 \times 10^{-7}\,mol^2\,dm^{-6}$
So $[H^+(aq)] = 4.12 \times 10^{-4}\,mol\,dm^{-3}$ and pH = 3.4 (at 298 K)

Notice that when $[HA(aq)]$ is decreased by a factor of 100, $[H^+(aq)]$ falls by a factor of 10 and pH increases by 1 unit. The calculations on strong and weak acids are compared in Table 2.

► **Table 2** A comparison of strong and weak acids.

	Strong acid	Typical weak acid
acid is diluted by a factor of 100	$[H^+(aq)]$ becomes 100× smaller	$[H^+(aq)]$ becomes 10× smaller
	pH increases by 2	pH increases by 1

Strong or concentrated?

Measurement of $[H^+(aq)]$ or pH alone does not allow us to distinguish strong and weak acids. For example, solutions of both could have $[H^+(aq)] = 1 \times 10^{-4}$ mol dm^{-3}, but the solutions would have different concentrations of acid. The strong acid would have a concentration of 1×10^{-4} mol dm^{-3}; the weak acid would have to be much more concentrated to have a pH as low as 4.

It is important to distinguish carefully between *concentration* and *strength*. Concentration is a measure of the amount of substance in a given volume of solution – typically measured in mol dm^{-3}. Strength is a measure of the extent to which an acid can donate H^+. The two terms have very different meanings in chemistry but are often interchangeable in everyday language.

A table of K_a values, such as Table 3, provides an easy way of indicating the strength of an acid – the greater the value of K_a, the stronger the acid.

Like $[H^+(aq)]$ values, K_a values are spread over a wide range and it is convenient to define a term, pK_a, as a measure of acid strength

$$pK_a = -\lg K_a$$

(like $pH = -\lg[H^+(aq)]$). Just as a lower pH means a greater $[H^+(aq)]$, so a smaller pK_a corresponds to a greater acid strength. Table 3 gives the pK_a values for the acids in Table 1.

Acid		K_a/mol dm^{-3}	pK_a
methanoic	HCOOH	1.6×10^{-4}	3.8
benzoic	C_6H_5COOH	6.3×10^{-5}	4.2
ethanoic	CH_3COOH	1.7×10^{-5}	4.8
chloric(I)	HClO	3.7×10^{-8}	7.4
hydrocyanic	HCN	4.9×10^{-10}	9.3
nitrous (nitric(III))	HNO_2	4.7×10^{-4}	3.3

◀ **Table 3** K_a and pK_a values for some weak acids.

Ionisation of water

We do not think of water as an ionic substance, but water does in fact ionise slightly. When water ionises, it behaves as both an acid and a base. In the equilibrium

$$H_2O(l) \rightleftharpoons H^+(aq) + OH^-(aq)$$

water is acting like other weak acids because H–OH can be thought of as HA, and OH^- as A^-. We can write an expression for K_a for water as

$$K_a = \frac{[H^+(aq)]\,[OH^-(aq)]}{[H_2O(l)]}$$

However, we can leave out the $[H_2O(l)]$ term, which is effectively constant because water is present in excess. The expression becomes

$$K_w = [H^+(aq)]\,[OH^-(aq)]$$

Notice the special symbol, $\boldsymbol{K_w}$, to denote the **ionic product** of water.

At 298 K, $K_w = 1 \times 10^{-14}\,mol^2\,dm^{-6}$.

In pure water $[H^+(aq)] = [OH^-(aq)]$, so
$K_w = 1 \times 10^{-14}\,mol^2\,dm^{-6} = [H^+(aq)]^2$
Therefore, $[H^+(aq)] = 1 \times 10^{-7}\,mol\,dm^{-3}$
and pH = 7

An acidic solution is defined as one in which $[H^+(aq)] > [OH^-(aq)]$.
An alkaline solution is defined as one in which $[H^+(aq)] < [OH^-(aq)]$.

The idea that pure water, or a neutral solution, has pH = 7 should be familiar to you. But it is only true at 298 K. Ionisation of water is endothermic. Le Chatelier's principle predicts that K_w will increase with increasing temperature. $[H^+(aq)]$ will therefore increase in a similar way and the pH of water will fall as it gets hotter.

K_w also allows us to calculate the pH of a solution of a strong base – a solution in which the production of hydroxide ions is complete. For example, sodium hydroxide is completely ionised in solution – it is a **strong base** – so in 0.1 mol dm^{-3} NaOH(aq), $[OH^-(aq)] = 0.1\,mol\,dm^{-3}$. We can neglect the small amount of OH^- formed from water, so

$K_w = 1 \times 10^{-14}\,mol^2\,dm^{-6} = [H^+(aq)] \times 0.1\,mol\,dm^{-3}$

Therefore, $[H^+(aq)] = \dfrac{1 \times 10^{-14}\,mol^2\,dm^{-6}}{0.1\,mol\,dm^{-3}}$

or $[H^+(aq)] = 1 \times 10^{-13}\,mol\,dm^{-3}$
and pH = 13

Many bases, however, are weak bases in that in an aqueous solution the production of hydroxide ions is not complete. An example is ammonia solution:

$$NH_3(aq) + H_2O(l) \rightleftharpoons NH_4^+(aq) + OH^-(aq)$$

The position of this equilibrium is well to the left and so a solution of ammonia is less alkaline (lower pH) than a solution of a strong base of identical concentration.

Comparing strong and weak acids

Strong acids may be distinguished from weak acids in a number of ways.

pH

A 0.1 mol dm^{-3} solution of a strong acid such as hydrochloric acid has a pH value of 1. A weak acid of equivalent concentration has a pH value greater than 1, due to a lower concentration of $H^+(aq)$ ions in the solution. For example, a 0.1 mol dm^{-3} solution of sulfur dioxide has a pH of about 1.5, while a 0.1 mol dm^{-3} solution of ethanoic acid has a pH of about 3 and a 0.1 mol dm^{-3} solution of carbon dioxide has a pH of about 4.

Conductivity

A strong acid has a much higher electrical conductivity than a weak acid of equivalent concentration. This is because there are fewer ions present in the solution of the weak acid because the reaction with water is incomplete, and the position of the equilibrium is well to the left.

Reaction rate

Acids can react with metals, alkalis, metal oxides and metal carbonates, and all of these reactions involve the $H^+(aq)$ ion. The rates of these reactions decrease as $[H^+(aq)]$ decreases. No matter which reaction you choose, a weak acid of a certain concentration reacts more slowly than a strong acid of the same concentration. This is because the position of the equilibrium of a weak acid means that, at any given time, there are relatively few $H^+(aq)$ ions available for reaction.

For example, in the reaction

$$2HX(aq) + Mg(s) \rightarrow MgX_2(aq) + H_2(g)$$

where HX is HCl or ethanoic acid (CH_3COOH), the reaction with the strong hydrochloric acid is much faster than the reaction with the weak ethanoic acid. However, if we add excess magnesium to identical volumes of the two acids and if the two acids are of identical concentration then we get identical volumes of hydrogen produced by the time the reaction has finished.

Why don't we get less hydrogen with the weak acid? This is because when the reaction starts the concentration of $H^+(aq)$ ions is reduced as they react with the magnesium. In the weak acid this means that the equilibrium is unbalanced, so some weak acid molecules react with water to form ions, producing more $H^+(aq)$ ions, which then react. This process continues until eventually all of the acid molecules have reacted.

Although a weak acid reacts more slowly than a strong acid with metals, alkalis and other substances, it will eventually react with the stoichiometric quantities indicated by the overall equation for the reaction. Both acids will neutralise equal amounts of alkali.

Problems for 8.2

1 Calculate the pH values of the following solutions of strong acids at 298 K. An example has been done for you.
Example Calculate the pH of 0.1 mol dm^{-3} hydrochloric acid.

$$pH = -lg[H^+(aq)]$$

For strong acids, $[H^+(aq)]$ = concentration of the original acid solution, so pH = −lg 0.1 = 1

- **a** 0.01 mol dm^{-3} hydrochloric acid
- **b** 0.2 mol dm^{-3} nitric acid
- **c** 0.2 mol dm^{-3} sulfuric acid
- **d** 0.1 mol of $HClO_4(l)$ in 250 cm^3 of aqueous solution.

2 Calculate the pH values of the following solutions of weak acids at 298 K. You will need to use the K_a values in Table 1 (page 185) and follow the worked example on page 186.

- **a** 0.1 mol dm^{-3} ethanoic acid
- **b** 0.05 mol dm^{-3} ethanoic acid
- **c** 0.001 mol dm^{-3} benzoic acid
- **d** 0.25 mol of methanoic acid in 100 cm^3 of solution.

3 For each of the following types of pH calculations state what assumptions are made:

- **a** pH of a strong acid
- **b** pH of a weak acid.

4 A 0.01 mol dm^{-3} hydrochloric acid solution and a 0.2 mol dm^{-3} nitric(III) acid (HNO_2) solution have a similar pH value of 2.

- **a** Classify each of these acids as strong or weak.
- **b** Explain the difference between a strong acid and a weak acid using HCl and HNO_2 as examples.

5 Calculate the pH of the following solutions of strong bases at 298 K.

- **a** 1 mol dm^{-3} KOH(aq)
- **b** 0.01 mol dm^{-3} NaOH(aq)
- **c** 0.1 mol dm^{-3} $Ba(OH)_2(aq)$.

6 Indicators are weak acids. The acidic form, HIn, of an indicator is one colour, with its conjugate base, In^-, being a different colour. At the end point for a titration, $[HIn(aq)] = [In^-(aq)]$.
For the indicator phenolphthalein, HIn is colourless and In^- is pink.

$$HIn(aq) \rightleftharpoons H^+(aq) + In^-(aq)$$

- **a** What is the colour of phenolphthalein in alkaline solution? Explain your answer.
- **b** Write an expression for the acidity constant, K_a, for phenolphthalein.
- **c** Use this expression to calculate $[H^+(aq)]$, and hence the pH at the end point of a titration when phenolphthalein is used as an indicator. The pK_a for phenolphthalein at 298 K is 9.3.

7 **a** Use the pH values of the following acidic solutions to calculate the acidity constant, K_a, and the value of pK_a in each case.

- **i** 0.10 mol dm^{-3} HCN, pH = 5.15
- **ii** 0.005 mol dm^{-3} phenol, pH = 6.10
- **iii** 1.00 mol dm^{-3} HF, pH = 1.66

b Place the acids in order of increasing acid strength.

8.3 *Buffer solutions*

Where do we find buffer solutions?

Many processes in living systems must take place under fairly precise pH conditions. If the pH changes to a value outside a narrow range, the process will not occur at the correct rate, or it may not take place at all and the organism will die. The pH ranges for some fluids in our bodies are shown in Table 1.

Fluid	pH range
stomach juices	1.6–1.8
saliva	6.4–6.8
blood	7.35–7.45

◀ **Table 1** pH ranges of some body fluids.

Small organisms must also be surrounded by liquid at the correct pH. This is true, for example, for the bacteria and moulds used for fermentation in biotechnology processes, as well as for many life forms in the sea.

Controlling pH may not be a matter of life or death in a chemical manufacturing process, but it is nevertheless often very important. For example, using reactive dyes to colour fabrics can be ineffective at the wrong pH.

So, during evolution and in the practice of chemistry, solutions have been developed which can *resist changes in pH despite the addition of acid or alkali* – such solutions are called **buffer solutions**. Their pH stays approximately constant even if small amounts of acid or alkali are added.

Buffer solutions are usually made from:

- either a weak acid and one of its salts
- or a weak base and one of its salts.

An example of the first type is ethanoic acid plus sodium ethanoate. The salt has an ionic structure and acts as a source of ethanoate ions. The presence of ethanoic acid leads to an acidic pH.

Ammonia solution plus ammonium chloride is an example of a buffer made from a weak base and one of its salts. The salt provides NH_4^+ ions and the ammonia leads to an alkaline pH. The action of both types of buffer depends on the presence of an H^+ donor (CH_3COOH or NH_4^+) and an H^+ acceptor (CH_3COO^- or NH_3).

By choosing suitable pairs of conjugate acids and bases, we can make buffer solutions with almost any desired pH. Living systems often use H_2CO_3 and HCO_3^-, or $H_2PO_4^-$ and HPO_4^{2-}, in combination with proteins.

How do buffers work?

The action of a buffer solution (**buffer**) depends on the weak acid equilibrium discussed in **Section 8.2**, which we can represent as

$$HA(aq) \rightleftharpoons H^+(aq) + A^-(aq)$$

We can make two very good approximations about the species present in this equilibrium in the case of a buffer solution.

Assumption 1: *All the A^- ions come from the salt*

The weak acid, HA, supplies very few A^- ions in comparison with the fully ionised salt.

Assumption 2: *Almost all the HA molecules put into the buffer remain unchanged*

We make the same assumption when calculating the pH of a weak acid, but it is a better approximation for a buffer solution because the high concentration of A^-(aq) from the salt pushes the equilibrium even further to the left.

What happens if an acidic substance is added to a buffer? Any rise in $[H^+(aq)]$ disturbs the equilibrium. Some A^-(aq) ions from the salt react with the extra H^+(aq) ions to form HA(aq) and water. A significant fall in pH is prevented.

If alkali is added, H^+(aq) ions are removed from the solution. The buffer solution counteracts this because H^+(aq) can be regenerated from the acid HA. A significant pH rise is avoided.

The presence of *both* a weak acid *and* its salt are necessary for a buffer to work. There must be plenty of HA to act as a *source* of extra H^+(aq) ions when they are needed, and plenty of A^- to act as a *sink* for any extra H^+(aq) ions which have been added (Figure 1).

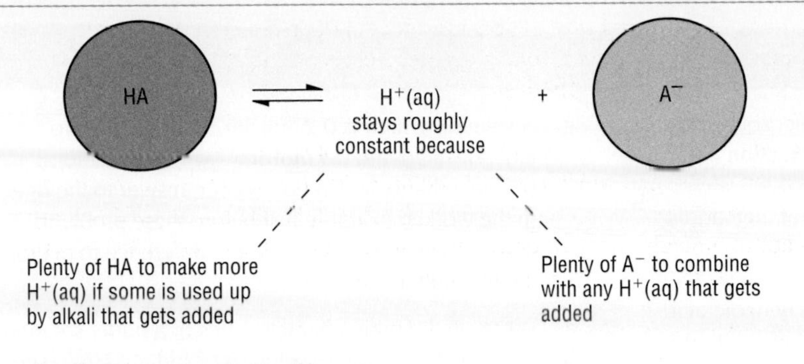

◀ **Figure 1** How a buffer solution keeps the pH constant.

Calculations with buffers

All we need to do calculations on buffer solutions is the K_a expression for the relevant weak acid.

$$K_a = \frac{[H^+(aq)]\,[A^-(aq)]}{[HA(aq)]} = [H^+(aq)] \times \frac{[A^-(aq)]}{[HA(aq)]}$$

If we make use of **assumptions 1** and **2** above, we get

$$K_a = [H^+(aq)] \times \frac{[\text{salt}]}{[\text{acid}]}$$

The value of $[H^+(aq)]$, and therefore the pH of the buffer solution, depends on *two* factors.

- The value of K_a

This provides a 'coarse tuning' of the buffer's pH. K_a values normally lie in the range 1×10^{-4} mol dm^{-3} to 1×10^{-10} mol dm^{-3}. Choice of a particular weak acid determines which *region* of the pH range the buffer is in, from about pH = 4 to pH = 10.

- Ratio of [salt] : [acid]

This provides a 'fine tuning' of the buffer pH. Changing the ratio from about 3 : 1 to about 1 : 3 changes $[H^+(aq)]$ by a factor of 9, and alters pH by approximately 1 unit. The ratio should not be too far outside this range, otherwise there will be insufficient HA or A^- for the buffer to be effective.

The expression

$$K_a = [H^+(aq)] \times \frac{[\text{salt}]}{[\text{acid}]}$$

shows that the pH of a buffer is not affected by dilution. When you add water, the concentrations of both the salt and the acid are reduced equally. Therefore the ratio of their concentrations remains the same and the pH is unchanged.

Finally, let's use this expression to calculate the pH of a buffer solution which contains 0.1 mol dm^{-3} ethanoic acid and 0.2 mol dm^{-3} sodium ethanoate. K_a for ethanoic acid is 1.7×10^{-5} mol dm^{-3} at 298 K.

$$K_a = [H^+(aq)] \times \frac{[CH_3COO^-(aq)]}{[CH_3COOH(aq)]}$$

Using **assumptions 1** and **2** we can write

$$1.7 \times 10^{-5}\,\text{mol dm}^{-3} = [H^+(aq)] \times \frac{0.2\,\text{mol dm}^{-3}}{0.1\,\text{mol dm}^{-3}}$$

Therefore $[H^+(aq)] = 1.7 \times 10^{-5}\,\text{mol dm}^{-3} \times \dfrac{0.1\,\text{mol dm}^{-3}}{0.2\,\text{mol dm}^{-3}}$

$[H^+(aq)] = 8.5 \times 10^{-6}$ mol dm^{-3}
and pH = 5.07 (at 298 K)

Problems for 8.3

1 Calculate the pH values of the following buffer solutions at 298 K. (You will find K_a values in Table 1 in **Section 8.2**, page 185)

a A solution in which the concentrations of methanoic acid and potassium methanoate are both 0.1 mol dm^{-3}.

b A solution made by dissolving 0.01 mol benzoic acid and 0.03 mol sodium benzoate in 1 dm^3 of solution.

c A solution made by mixing equal volumes of 0.1 mol dm^{-3} methanoic acid and 0.1 mol dm^{-3} potassium methanoate.

d A solution which is 0.1 mol dm^{-3} with respect to propanoic acid and 0.005 mol dm^{-3} with respect to sodium propanoate (K_a of propanoic acid = 1.3 × 10^{-5} mol dm^{-3}).

e A solution made by dissolving 0.005 mol of methanoic acid and 0.015 mol of sodium methanoate in 500 cm^3 of solution.

f A solution made by mixing 250 cm^3 of 0.1 mol dm^{-3} ethanoic acid and 500 cm^3 of 0.1 mol dm^{-3} sodium ethanoate.

g A solution which is 0.2 mol dm^{-3} with respect to ethanoic acid and 0.1 mol dm^{-3} with respect to sodium ethanoate. Compare your answer to the one in the worked example just before these problems (page 191) and comment on the difference in pH for the two buffer solutions.

2 a What are the main characteristics of a buffer solution?

b Describe how a buffer solution of sodium ethanoate and ethanoic acid reacts in order to maintain a constant pH when:

i a small amount of hydrochloric acid is added

ii a small amount of sodium hydroxide is added

iii a small amount of water is added.

3 Which of the acids listed in Table 1 in **Section 8.2** (page 185) would be the most suitable choice for the preparation of a buffer with pH 5.2? Give a reason for your answer.

9.1 *Oxidation and reduction*

When we heat a piece of copper in oxygen, the surface of the copper becomes coated with a black solid. The copper's surface is oxidised – it gains oxygen. The product, copper(II) oxide, is an ionic compound:

$$Cu(s) + \frac{1}{2}O_2(g) \rightarrow Cu^{2+}O^{2-}(s)$$

If you think carefully about what is happening, you will see that the reaction is composed of two **half-reactions** which can be described by two **half-equations**:

$$Cu \rightarrow Cu^{2+} + 2e^- \text{ and } \frac{1}{2}O_2 + 2e^- \rightarrow O^{2-}$$

The copper is losing electrons in one half-reaction, and the oxygen is gaining them in the other. Another way of looking at *oxidation* is to say that it is the *loss of electrons*.

Reduction is the opposite of oxidation, so *reduction* must occur when *electrons are gained*. The oxygen is being reduced in this example.

Reduction and oxidation occur together when copper reacts with oxygen – it is a reduction/oxidation reaction, or a **redox** reaction (Figure 1).

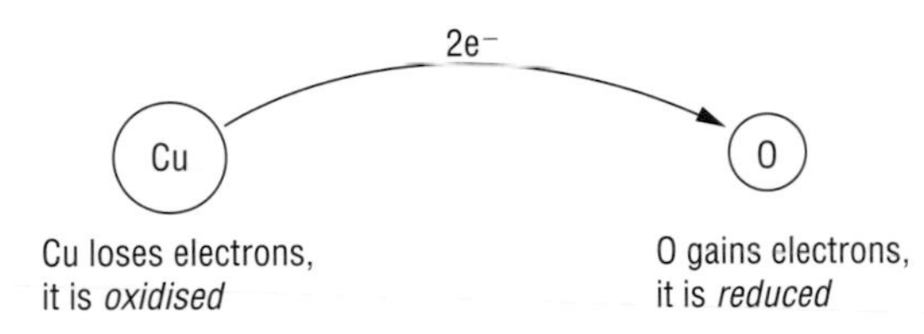

◀ **Figure 1** A redox reaction occurs when electrons are transferred from copper to oxygen.

If you had reacted copper with chlorine, you would have performed another redox reaction. The half-equations show this:

$$Cu \rightarrow Cu^{2+} + 2e^- \text{ and } Cl_2 + 2e^- \rightarrow 2Cl^-$$

Copper loses electrons and is oxidised; chlorine gains electrons and is reduced. In this reaction, the copper is said to be the **reducing agent** – this is because the copper atoms have given electrons to the chlorine atoms to reduce them to chloride ions. The chlorine is the **oxidising agent** because it caused the copper to lose electrons.

Redox is a very important type of chemical reaction. Many reactions can be classified as redox reactions, using the ideas in the box on the right.

Oxidation is loss of electrons.
An **oxidising agent** removes electrons from something else.

Reduction is gain of electrons.
A **reducing agent** gives electrons to something else.

Displacement reactions

When chlorine is bubbled through potassium iodide solution, a **displacement reaction** takes place. The colourless solution turns brown as iodine is produced. We say that the chlorine (Cl_2) has displaced the iodine (I_2) from the solution.

The overall equation for the reaction is:

$$Cl_2(g) + 2KI(aq) \rightarrow 2KCl(aq) + I_2(aq)$$

but this is not the most helpful way of showing what is happening. Look at Figure 2 (page 194), which shows the reaction in diagrammatic form.

One reactant, potassium iodide solution, contains a mixture of potassium ions and iodide ions that move around independently of each other. The same is true for the potassium ions and chloride ions in the product, potassium chloride solution.

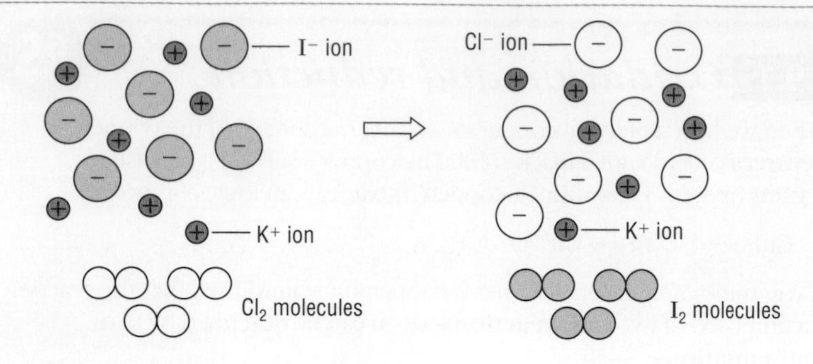

▶ **Figure 2** The reaction of chlorine with potassium iodide solution to give potassium chloride solution and iodine.

The potassium ions take no part in the chemical reaction and are **spectator ions**. On removing them from the overall equation we obtain the ionic equation:

$Cl_2(g) + 2I^-(aq) \rightarrow 2Cl^-(aq) + I_2(aq)$ redox

This can be written as two half-equations:

$2I^-(aq) \rightarrow I_2(aq) + 2e^-$ oxidation
$Cl_2(aq) + 2e^- \rightarrow 2Cl^-(aq)$ reduction

By the end of a redox reaction:

- the oxidising agent has been reduced;
- the reducing agent has been oxidised.

The electrons lost by the iodide ions are accepted by the chlorine molecules, which makes the chlorine molecules the **oxidising agent** in this reaction. The iodide ions must therefore be the **reducing agent**, and this is confirmed by the fact that they donate the electrons needed to reduce the chlorine molecules.

You will meet a similar redox reaction in the **Elements from the Sea** module – between chlorine and bromide ions, which is used to displace bromine from sea water.

Oxidation state

How would you describe the reaction that takes place when hydrogen burns in oxygen?

$H_2(g) + \frac{1}{2}O_2(g) \rightarrow H_2O(l)$

The equation is so similar to the one for the magnesium/oxygen reaction that it must be a redox reaction – but the product, water, is not ionic but molecular. This means that we cannot break the overall equation down into half-equations involving electron transfer, as we did for magnesium and oxygen and for chlorine and potassium iodide.

An element is oxidised when its oxidation state increases.	An element is reduced when its oxidation state decreases.

We need to extend the idea of redox to include reactions like this – and there are many of them. We can do this by using the idea of **oxidation state** (also known as **oxidation number**).

Each atom in a molecule or ion is assigned an oxidation state to show how much it is oxidised or reduced. Two very useful rules about oxidation states are:

- atoms in *elements* are in oxidation state *zero*
- in *simple ions* the oxidation state is the same as the *charge* on the ion.

Thus bromine has oxidation state 0 in Br_2, but oxidation state −1 in Br^-. Magnesium has oxidation state 0 in Mg, but oxidation state +2 in Mg^{2+}.

Since *compounds* have no overall charge, the oxidation states of all the constituent elements must add up to zero.

Some elements have oxidation states that rarely vary, whatever the compound they are in. For example, fluorine's oxidation state is always −1 in its compounds. Table 1 shows some more elements whose oxidation states are invariable, with a few exceptions.

Element	Oxidation state of elements
F	−1
O	−2 (except in O_2^{2-} and OF_2)
H	+1 (except in H^-)
Cl	−1 (except when combined with O or F)

◀ **Table 1** Oxidation states of some elements in compounds.

Here is an example of how to use these rules to work out the oxidation states of the elements in compounds.

Example 1: CO_2

Oxygen

In CO_2, the oxidation state of the oxygen must be −2 (it is one of the rules).

There are two oxygens, so the total contribution of oxygen to the oxidation states in the compound comes to $(-2) \times 2 = -4$.

Carbon

The oxidation state of the carbon must be +4 to balance the total contribution from the oxygens.

This makes the oxidation states for CO_2 add up to zero.

Table 2 shows some further examples. Make sure you understand how the oxidation states in Table 2 have been worked out.

Compound	Oxidation states of elements	
H_2O	H +1	O −2
CH_4	H +1	C −4
BrF_3	F −1	Br +3
SO_2	O −2	S +4
PCl_3	Cl −1	P +3

◀ **Table 2** Oxidation states of elements in some compounds.

The same rules can be used for the elements which make up *ions*, such as PO_4^{3-}, but this time the oxidation state of the constituent elements add up to the overall charge on the ion.

Example 2: PO_4^{3-}

Oxygen

In PO_4^{3-}, the oxidation state of the oxygen is −2.

There are four oxygens, so the total contribution of oxygen to the oxidation states of the ion comes to $(-2) \times 4 = -8$.

Phosphorus

The oxidation state of the phosphorus must be +5.

This makes the oxidation states of the ion add up to −3, which is the total charge on the ion.

Table 3 gives some other examples.

Ion	Oxidation states of elements	
NH_4^+	H +1	N −3
ClO_3^-	O −2	Cl +5
VO^{2+}	O −2	V +4

◀ **Table 3** Oxidation states of elements in some ions.

Summary of rules for assigning oxidation states

- The oxidation state of atoms in elements is 0.
- In compounds, the sum of all the oxidation states is 0.
- In ions, the sum of all the oxidation states is equal to the charge on the ion.
- In compounds and ions, oxidation states are assigned as shown in Table 4 (page 196).

▶ **Table 4** Assigning oxidation states.

F	−1
O	−2 (except in O_2^{2-} and OF_2)
H	+1 (except in H^-)
Cl	−1 (except when combined with O or F)

Oxidation states in names

Oxidation states can help us to give systematic names to compounds and ions that contain elements capable of existing in more than one oxidation state. For example:

FeO is called iron(II) oxide
Fe_2O_3 is called iron(III) oxide.

Notice that:

- Roman numerals are used
- the number shows the oxidation state of the preceding element
- the number is placed close up to the element it refers to – there is no space between the name and the number.

Here are some other examples:

CuS is copper(II) sulfide
$Fe(NO_3)_2$ is lead(II) nitrate(V)
$KMnO_4$ is potassium manganate(VII).

$Fe(NO_3)_2$
iron(II) nitrate(V)
Oxidation state of Fe
Oxidation state of N

▲ **Figure 3** Oxidation states in the naming of $Fe(NO_3)_2$.

With ions, oxidation states are used to help clarify the names of **oxyanions** (sometimes called **oxoanions**) – negative ions that contain oxygen (see Figure 3). For example, there are several different chlorate ions (ions containing chlorine and oxygen), with formulae ClO^-, ClO_2^-, ClO_3^- and ClO_4^-. The oxidation states of Cl are given in Table 5, together with the names of the oxyanions.

▶ **Table 5** Oxidation state of chlorine in different chlorate ions.

Ion	Oxidation state	Name of ion
ClO^-	Cl +1	chlorate(I)
ClO_2^-	Cl +3	chlorate(III)
ClO_3^-	Cl +5	chlorate(V)
ClO_4^-	Cl +7	chlorate(VII)

Using oxidation states

We can use oxidation states to find what has been oxidised and what has been reduced in a redox reaction. For example, look again at the reaction

$$Cl_2(aq) + 2I^-(aq) \rightarrow 2Cl^-(aq) + I_2(aq)$$

In terms of electron transfer, we know that chlorine is reduced and iodine oxidised. Now look at the oxidation states.

Oxidation states of chlorine:
in Cl_2 0
in Cl^- −1

The oxidation state of chlorine decreases; chlorine has been reduced.

Oxidation states of iodine:
in I^- −1
in I_2 0

The oxidation state of iodine increases; iodine has been oxidised.

Summary

Chemists use the following ideas for oxidation and reduction:

Something is oxidised if
- it gains oxygen
- it loses electrons
- its oxidation state increases

Something is reduced if
- it loses oxygen
- it gains electrons
- its oxidation state decreases

Remember:
Oxidation
Is
Loss of electrons
Reduction:
Is
Gain of electrons

Problems for 9.1

1 a Insert electrons (e^-) on the appropriate side of the following half-equations in order to balance and complete them, so that the electrical charges on both sides are equal.

i $K \rightarrow K^+$
ii $H_2 \rightarrow 2H^+$
iii $O \rightarrow O^{2-}$
iv $Cu^+ \rightarrow Cu^{2+}$
v $Cr^{3+} \rightarrow Cr^{2+}$

b For each completed half-equation, describe the process as oxidation or reduction.

2 Write down the oxidation states of the *elements* in the following examples.

a Ag^+ **h** CO_2
b Br_2 **i** PCl_5
c P_4 **j** Al_2O_3
d H^+ **k** SF_6
e H^- **l** SO_4^{2-}
f N^{3-} **m** NO_3^-
g $MgCl_2$ **n** PO_4^{3-}

3 Some reactions of the halogens are shown below – they are all examples of redox reactions. In each case state which element is oxidised and which is reduced, and give the oxidation states of each atom or ion before and after the reaction.
For example:

$$Cl_2 + 2Br^- \rightarrow 2Cl^- + Br_2$$

Cl: 2 × (0) 2 × (−1)
chlorine reduced

Br: 2 × (−1) 2 × (0)
bromine oxidised

a $2Fe + 3Cl_2 \rightarrow 2FeCl_3$
b $H_2 + Cl_2 \rightarrow 2HCl$
c $2FeCl_2 + Cl_2 \rightarrow 2FeCl_3$
d $2H_2O + 2F_2 \rightarrow 4HF + O_2$

4 For each of the redox reactions in parts **a** to **d** of problem 3, identify by formula the
i oxidising agent
ii reducing agent.

5 Some further reactions of halogens and halogen ions are shown below – again they are all redox reactions. In each case, state which element is oxidised and which is reduced, and give the oxidation states before and after the reaction.

a $2ClO_3^- \rightarrow 2Cl^- + 3O_2$
b $2Br^- + 2H^+ + H_2SO_4 \rightarrow Br_2 + SO_2 + 2H_2O$
c $8I^- + 8H^+ + H_2SO_4 \rightarrow 4I_2 + H_2S + 4H_2O$
d $I_2 + SO_3^{2-} + H_2O \rightarrow 2I^- + SO_4^{2-} + 2H^+$

6 In some redox reactions the same element undergoes both oxidation and reduction. For example:

$$Cl_2 + 2OH^- \rightarrow Cl^- + ClO^- + H_2O$$

Cl: 2 × (0) −1 +1
oxidised and reduced

O: 2 × (−2) −2 −2
no change

H: 2 × (+1) 2 × (+1)
no change

Using the same format as in this example, for each of the following:
- state the oxidation states of the elements before and after the reaction
- indicate whether each element has been oxidised, reduced, both oxidised and reduced, or has remained unchanged.

a $Cu_2O + 2H^+ \rightarrow Cu^{2+} + Cu + H_2O$
b $3Br_2 + 6OH^- \rightarrow BrO_3^- + 5Br^- + 3H_2O$
c $4IO_3^- \rightarrow 3IO_4^- + I^-$

7 Use oxidation states to name the ions or compounds with the formulae:

a SnO **f** Cu_2O **k** SO_4^{2-}
b SnO_2 **g** $Mn(OH)_2$ **l** MnO_4^-
c $FeCl_2$ **h** NO_2^- **m** CrO_4^{2-}
d $FeCl_3$ **i** NO_3^- **n** VO_3^-
e $PbCl_4$ **j** SO_3^{2-}

8 Write formulae for the following compounds – in each case, the negative ion has a charge of −1.
a potassium chlorate(III)
b sodium chlorate(V)
c iron(III) hydroxide
d copper(II) nitrate(V)

9.2 Redox reactions and electrode potentials

Redox reactions

Redox reactions involve electron transfer (see **Section 9.1**). Redox reactions can be split into two *half-reactions* – one producing electrons and one accepting them.

For example, when zinc is added to copper(II) sulfate solution, a redox reaction takes place. The blue colour of the solution becomes paler, and copper metal deposits on the zinc. The temperature rises – it is an exothermic reaction. The overall equation is

$$Zn(s) + CuSO_4(aq) \rightarrow ZnSO_4(aq) + Cu(s)$$

This is an example of a *displacement reaction* (see **Section 9.1**). Figure 1 shows the reaction in diagrammatic form.

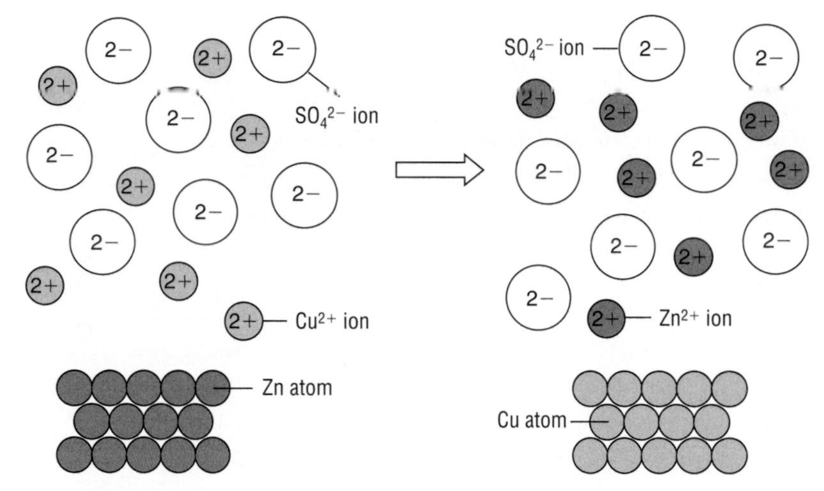

▶ **Figure 1** The reaction of zinc with copper(II) sulfate solution.

The sulfate ions play no part in the reaction and are *spectator ions*. On removing them from the overall equation we obtain the ionic equation:

$Zn(s)$ +	$Cu^{2+}(aq)$ →	$Cu(s)$ +	$Zn^{2+}(aq)$
grey	blue	orange	colourless
solid	solution	solid	solution

What the reaction amounts to is Zn atoms *transferring* electrons to Cu^{2+} ions.

The ionic equation can be written as two half-equations:

$Zn(s) \rightarrow Zn^{2+}(aq) + 2e^-$ oxidation, electrons produced in this half-reaction

$Cu^{2+}(aq) + 2e^- \rightarrow Cu(s)$ reduction, electrons accepted in this half-reaction

Zinc provides the electrons which reduce Cu^{2+} to Cu, so we say that zinc is the *reducing agent*, while Cu^{2+} is the *oxidising agent*.

If copper is added to zinc sulfate solution, no change is observed – reaction does not occur in the reverse direction. Zinc reacts with copper ions, but copper ions do not react with zinc.

However, if copper is added to silver nitrate(V) solution the copper *does* react. A grey precipitate forms, and the solution turns from colourless to blue. The overall reaction is

$Cu(s)$ +	$2Ag^+(aq)$ →	$Cu^{2+}(aq)$ +	$2Ag(s)$
orange	colourless	blue	grey
solid	solution	solution	solid

and the half-reactions are

$$Cu(s) \rightarrow Cu^{2+}(aq) + 2e^- \quad \text{oxidation}$$
$$2Ag^+(aq) + 2e^- \rightarrow 2Ag(s) \quad \text{reduction}$$

No reaction is observed when silver is added to copper(II) sulfate solution.

Individual half-reactions are reversible – they can go either way

$$Cu^{2+}(aq) + 2e^- \rightleftharpoons Cu(s)$$

The actual direction they take depends on what they are reacting with. For example, *zinc atoms* can supply electrons to *copper ions*, so that the copper half-reaction is

$$Cu^{2+}(aq) + 2e^- \rightarrow Cu(s)$$

But *copper atoms* supply electrons to *silver ions*, so in this case the copper half-reaction is

$$Cu(s) \rightarrow Cu^{2+}(aq) + 2e^-$$

Combining half-equations

Once we know the direction in which each half-reaction will go, we can add the half-equations together to get an equation for the overall reaction. For example, if you add zinc to silver ions, the zinc atoms supply electrons to the silver ions. The half-equations are

$$Zn(s) \rightarrow Zn^{2+}(aq) + 2e^-$$
$$Ag^+(aq) + e^- \rightarrow Ag(s)$$

To combine the two half-equations together, we need to make sure the number of electrons is the same in each half-equation – because every electron released by a zinc atom must be accepted by a silver ion.

This means we have to multiply the silver half-equation by two so there are $2e^-$ in each half-equation:

$$Zn(s) \rightarrow Zn^{2+}(aq) + 2e^-$$
$$2Ag^+(aq) + 2e^- \rightarrow 2Ag(s)$$

Now we can add the two half-equations together to give the overall equation:

$$Zn(s) + 2Ag^+(aq) \rightarrow 2Ag(s) + Zn^{2+}(aq)$$

The $2e^-$ disappear because they are on both sides of the equation.

Electrochemical cells

Something must control the direction of electron transfer in a redox reaction. To find out more about redox reactions, and what makes them go in a particular direction, we need to be able to study the half-reactions.

We can arrange for the two half-reactions to occur separately with electrons flowing through an external wire from one half-reaction to the other.

A system like this is used in all batteries and 'dry' cells. In one part of the cell an oxidation reaction occurs. Electrons are produced and transferred through an external circuit to the other part of the cell, where a reduction reaction takes place accepting the electrons. The two parts are called **half-cells** which, when combined, make an **electrochemical cell**. Figure 2 shows the general arrangement, and Figure 3 (page 200) shows a familiar example.

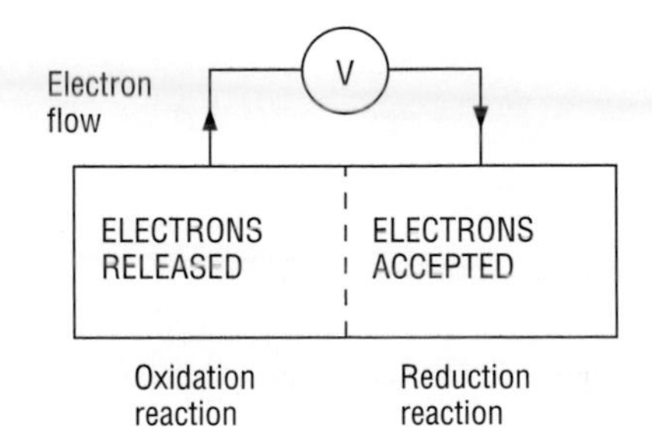

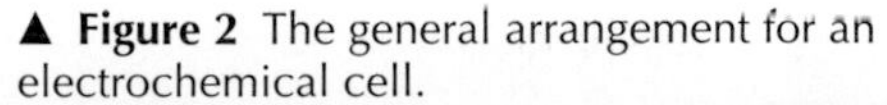
▲ **Figure 2** The general arrangement for an electrochemical cell.

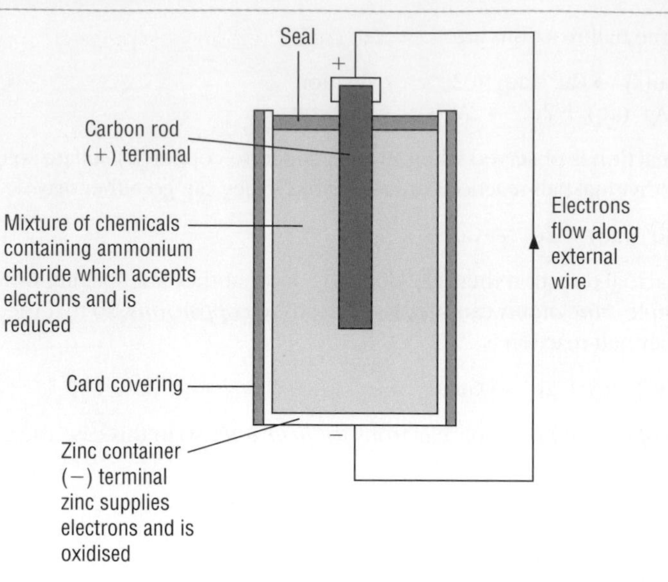

▶ **Figure 3** An ordinary dry cell – the kind you use in a torch.

Electrical units

Electric charge is measured in *coulombs* (C).

Electric current is a flow of charge and is measured in *amps* (A).

One amp is a flow of charge of one coulomb per second.

The *potential difference* between the terminals of the cell is measured in *volts* (V). The voltage of the cell tells you the number of joules of energy transferred whenever one coulomb of charge flows round the circuit.

$1\,V = 1\,J\,C^{-1}$

For example, if one coulomb of charge flows through a potential difference of 3 V, then 3 J are transferred.

The energy given out, instead of heating the surroundings, becomes available as electrical energy which we use to do work for us.

Cells are labelled with positive and negative terminals and a voltage. The voltage measures the potential difference between the two terminals.

As current flows in a circuit, the voltage can drop – the higher the current drawn, the lower the voltage the cell may give.

If we want to compare cells, and half-cells, by measuring voltages, we need to be careful to compare like with like. We can do this if we measure the potential difference between the terminals of the cell when *no* current flows. This potential difference is given the symbol $\boldsymbol{E_{cell}}$. (It is sometimes called the electromotive force or e.m.f. of the cell, although this is not a very good term because it is not a force.)

To measure E_{cell} we use a high-resistance voltmeter so that almost zero current flows. We record the maximum potential difference between the electrodes of the two half-cells. The potential difference is a measure of how much each electrode is tending to release or accept electrons.

Metal ion–metal half-cells

You can set up a simple half-cell by using a strip of metal dipping into a solution of metal ions. For example, the copper–zinc cell consists of the two half-cells shown in Figure 4.

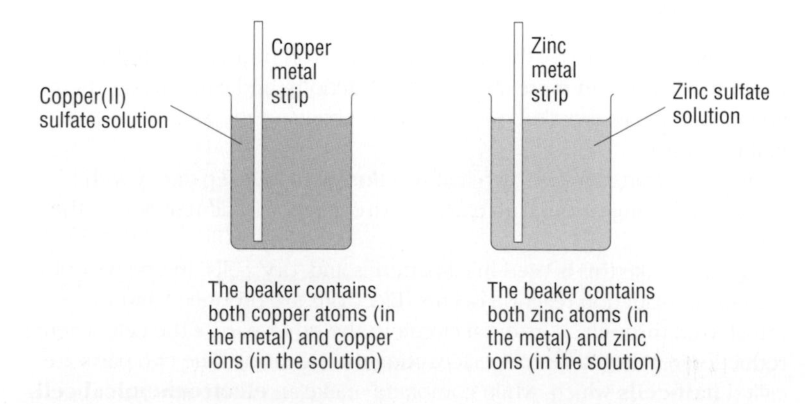

▶ **Figure 4** The copper and zinc half-cells.

Each of these half-cells has its own electrode potential. Take the zinc half-cell, for example, in Figure 5.

The Zn atoms in the zinc strip form Zn^{2+} ions by releasing electrons:

$$Zn(s) \rightarrow Zn^{2+}(aq) + 2e^-$$

The electrons released make the Zn strip negatively charged relative to the solution, so there is a potential difference between the zinc strip and the solution. The Zn^{2+} ions in the solution accept electrons, reforming Zn atoms:

$$Zn^{2+}(aq) + 2e^- \rightarrow Zn(s)$$

When Zn^{2+} ions are turning back to Zn as fast as they are being formed, an equilibrium is set up:

$$Zn^{2+}(aq) + 2e^- \rightleftharpoons Zn(s)$$

For a general metal, M:

$$M^{2+}(aq) + 2e^- \rightleftharpoons M(s)$$

The position of this equilibrium determines the size of the potential difference (the electrode potential) between the metal strip and the solution of metal ions. The further to the right the equilibrium lies, the greater the tendency of the electrode to accept electrons, and the more positive the electrode potential.

When we put two half-cells together, the one with the more positive potential will become the positive terminal of the cell and the other one will become the negative terminal.

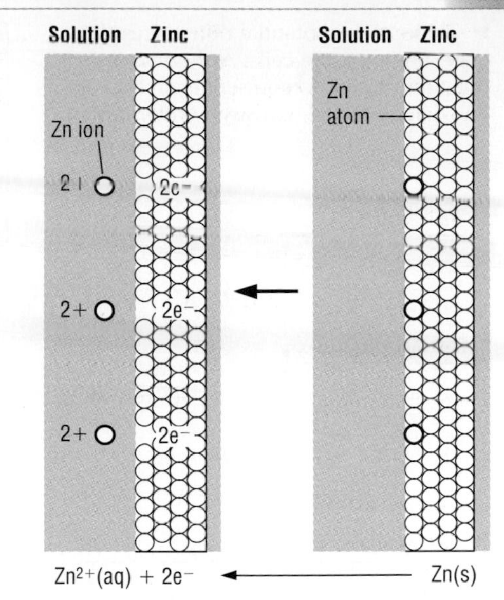

▲ **Figure 5** Zinc atoms form zinc ions, releasing electrons and setting up a potential difference between the metal strip and the solution of metal ions.

Making a cell from two half-cells

A connection is needed between the two solutions, but the solutions should not mix together. A strip of filter paper soaked in saturated potassium nitrate(V) solution can be used as a junction, or **salt bridge**, between the half-cells. Sometimes this is called an **ion bridge** because the current is carried by the movement of ions, not electrons. The potassium ions and nitrate(V) ions carry the current in the salt bridge so that there is electrical contact between the solutions, but no mixing. The complete set up is shown in Figure 6.

An **electrochemical cell** consists of two half-cells connected by a salt bridge. We measure the maximum potential difference between the two electrodes of the half-cells with a high-resistance voltmeter, so that negligible current flows.

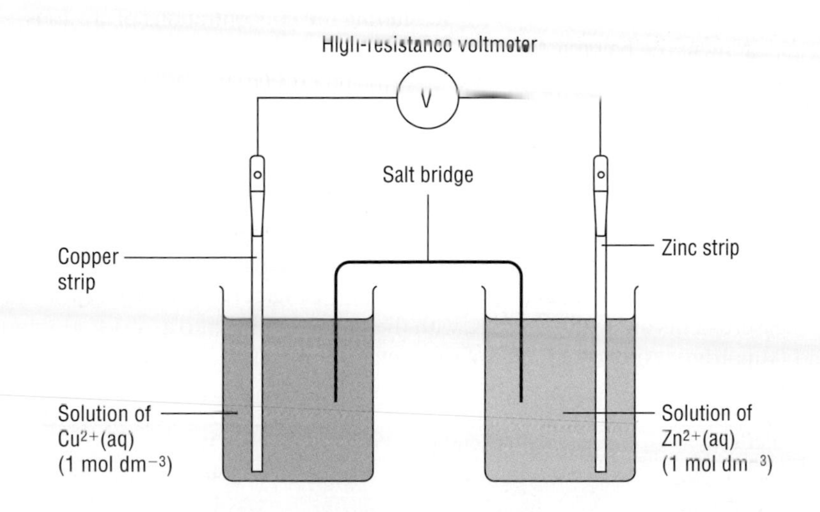

◀ **Figure 6** A copper–zinc cell.

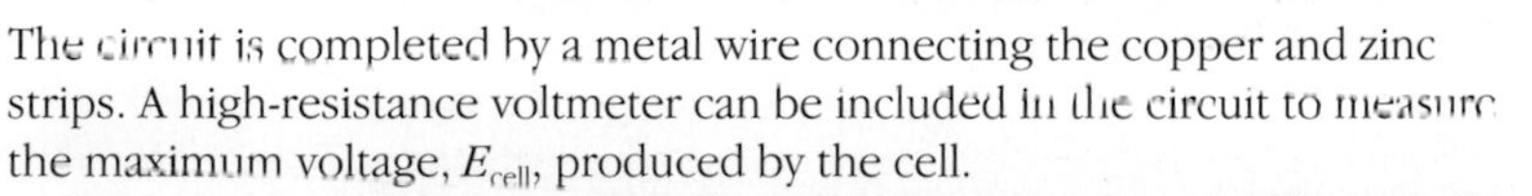
The circuit is completed by a metal wire connecting the copper and zinc strips. A high-resistance voltmeter can be included in the circuit to measure the maximum voltage, E_{cell}, produced by the cell.

Table 1 (page 202) shows data obtained from a number of different cells.

► **Table 1** The potential differences, E_{cell}, generated by some cells. A shorthand notation is used to represent each half-cell. The oxidised form is always written first.

Positive half-cell	Negative half-cell	E_{cell}/V
$Cu^{2+}(aq)/Cu(s)$	$Zn^{2+}(aq)/Zn(s)$	1.10
$Cu^{2+}(aq)/Cu(s)$	$Fe^{2+}(aq)/Fe(s)$	0.78
$Zn^{2+}(aq)/Zn(s)$	$Mg^{2+}(aq)/Mg(s)$	1.60
$Au^{3+}(aq)/Au(s)$	$Zn^{2+}(aq)/Zn(s)$	2.26
$Cu^{2+}(aq)/Cu(s)$	$Mg^{2+}(aq)/Mg(s)$	2.70
$Ag^{+}(aq)/Ag(s)$	$Zn^{2+}(aq)/Zn(s)$	1.56
$Fe^{2+}(aq)/Fe(s)$	$Zn^{2+}(aq)/Zn(s)$	0.32

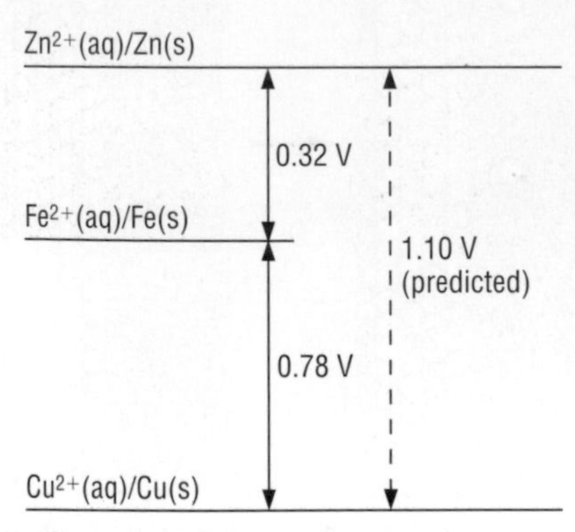

▲ **Figure 7** A diagram showing the potential differences between three half-cells.

Can you see any patterns in the voltages in Table 1? It is difficult – this is because we are measuring *differences* between the electrode potentials of the various half-cells.

Figure 7 shows one way of sorting out some of the data. It shows the potential differences between three half-cells – copper, iron and zinc. The potential difference measured between the electrodes of the copper and iron half-cells is 0.78 V, with the copper positive. The potential difference measured between the iron and zinc half-cells is 0.32 V with the iron positive.

What potential difference would you expect between the electrodes of the copper and zinc half-cells? Using Figure 7, we can predict it will be 1.10 V. The value we actually get when we measure the voltage is indeed 1.10 V.

Standard electrode potentials

It would help to sort things out if we selected one reference half-cell and measured all the others against it. We would then have a common reference point, and we could construct a list of electrode potentials relative to it. A **standard hydrogen half-cell** is chosen as the reference (Figure 8). The half-reaction in this half-cell is

$$2H^+(aq) + 2e^- \rightarrow H_2(g)$$

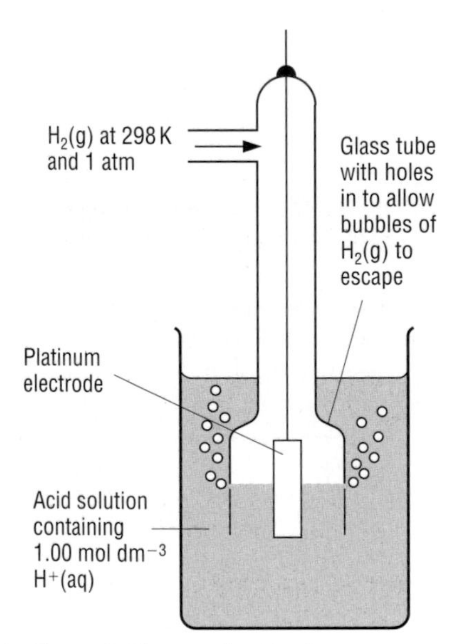

▲ **Figure 8** The standard hydrogen half-cell – sometimes called a standard hydrogen electrode.

Standard conditions

Electrode potentials vary with temperature and so for all cells a standard temperature is defined. This is 298 K. Altering the concentrations of any ions appearing in the half-reactions also affects the voltages and so a standard concentration of 1.00 mol dm^{-3} is chosen. Standard pressure is 1 atmosphere (101.3 kPa).

The potential of the standard hydrogen half-cell is *defined* as 0.00 V, a value chosen for convenience.

The **standard electrode potential** of a half cell, $E^{\ominus}$, is defined as the potential difference between it and a standard hydrogen half-cell.

$E^{\ominus}$ values have a sign depending on whether the half-cell is at a higher or lower positive potential than the standard hydrogen half-cell. Measurements are made at 298 K, with the metal dipping into a 1.00 mol dm^{-3} solution of a salt of the metal. Some values are shown in Table 2.

Standard conditions for the hydrogen half-cell

$[H^+(aq)] = 1.00$ mol dm^{-3}
pressure of H_2 gas = 1 atm
$T = 298$ K

► **Table 2** The standard electrode potentials of some half-cells.

Half-cell	Half-reaction	$E^{\ominus}$/V
$Mg^{2+}(aq)/Mg(s)$	$Mg^{2+}(aq) + 2e^- \rightarrow Mg(s)$	−2.36
$Zn^{2+}(aq)/Zn(s)$	$Zn^{2+}(aq) + 2e^- \rightarrow Zn(s)$	−0.76
$2H^+(aq)/H_2(g)$	$2H^+(aq) + 2e^- \rightarrow H_2(g)$	0 (by definition)
$Cu^{2+}(aq)/Cu(s)$	$Cu^{2+}(aq) + 2e^- \rightarrow Cu(s)$	+0.34
$Ag^+(aq)/Ag(s)$	$Ag^+(aq) + e^- \rightarrow Ag(s)$	+0.80

Listed like this, with the most positive potential at the bottom, the series is called the **electrochemical series**.

The half-cell at the bottom of the series has the greatest tendency to *accept* electrons. The half-cell at the top has the least tendency to accept electrons and the most negative $E^\ominus$ values. In fact, the half-cell at the top has the greatest tendency to go in the reverse direction and *release* electrons. For this reason, the most reactive metals are at the top of the series. You will notice that the order is very similar to the 'reactivity series' of metals you have met in earlier studies.

Other half-cell reactions

Metal ion–metal reactions are only one type of redox reaction. There are many others – for example, between ions:

$$Fe^{3+}(aq) + e^- \rightarrow Fe^{2+}(aq)$$
$$Cr_2O_7^{2-}(aq) + 14H^+(aq) + 6e^- \rightarrow 2Cr^{3+}(aq) + 7H_2O(l)$$
$$MnO_4^-(aq) + 8H^+(aq) + 5e^- \rightarrow Mn^{2+}(aq) + 4H_2O(l)$$

and between molecules and ions:

$$Cl_2(aq) + 2e^- \rightarrow 2Cl^-(aq)$$

All these half-reactions can be set up as half-cells. However, there is no metal in the half-reaction to make electrical contact, so an electrode made of an unreactive metal, such as platinum, is used. It dips into a solution containing all the ions and molecules involved in the half-reaction. For example, Figure 9 shows the set-up for a $Fe^{3+}(aq)/Fe^{2+}(aq)$ half-cell.

Table 3 shows values of standard electrode potentials for a selection of half-cells of this type. There is a fuller table in the **Data Sheets** (Table 20).

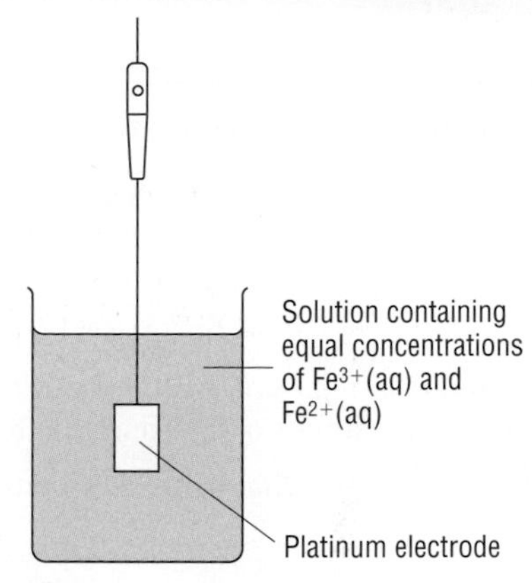

▲ **Figure 9** A standard half-cell for the $Fe^{3+}(aq)/Fe^{2+}(aq)$ half-reaction.

◀ **Table 3** Some standard electrode potentials.

Half-reaction	$E^\ominus$/V
$I_2(aq) + 2e^- \rightarrow 2I^-(aq)$	+0.54
$Br_2(aq) + 2e^- \rightarrow 2Br^-(aq)$	+1.07
$Cl_2(g) + 2e^- \rightarrow 2Cl^-(aq)$	+1.36
$MnO_4^-(aq) + 8H^+(aq) + 5e^- \rightarrow Mn^{2+}(aq) + 4H_2O(l)$	+1.51

Working out E_{cell} from standard electrode potentials

An electrochemical cell consists of two half-cells. If you know the electrode potential of each half-cell then you can work out the potential difference, E_{cell}, of the cell as a whole.

For example, suppose we set up a cell using the two half-reactions:

$$Zn^{2+}(aq) + 2e^- \rightarrow Zn(s) \quad E^\ominus = -0.76\,V$$
$$Cu^{2+}(aq) + 2e^- \rightarrow Cu(s) \quad E^\ominus = +0.34\,V$$

(This is called a Daniell cell – it was invented in 1836.)

E_{cell} is simply the voltage difference between the standard electrode potentials of the two half-cells. It is easiest to see this if you draw up an electrode potential chart, as shown in Figure 10.

From Figure 10,

$$E_{cell} = (+0.34\,V) - (-0.76\,V) = +1.10\,V$$

E_{cell} is an experimentally measured potential difference and is always positive, so always subtract the less positive potential from the more positive one.

To find the overall reaction occurring in the cell as a whole, we can add together the two half-equations.

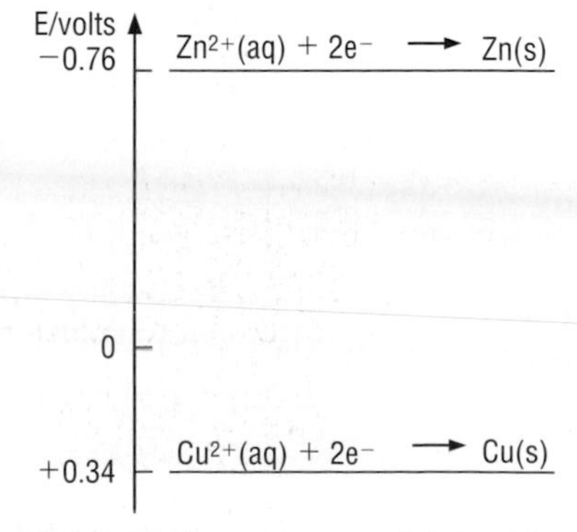

▲ **Figure 10** Electrode potential chart for the Daniell cell.

$Zn(s) \rightarrow Zn^{2+}(aq) + 2e^-$
$Cu^{2+}(aq) + 2e^- \rightarrow Cu(s)$
$Zn(s) + Cu^{2+}(aq) \rightarrow Cu(s) + Zn^{2+}(aq)$

This is the reaction we discussed at the very beginning of this section.

A table of standard electrode potentials allows us to calculate the maximum voltage obtainable from any cell under standard conditions. But electrode potentials are also useful in other ways. We can use them to *predict* the reactions which can take place when two half-cells are connected so that electrons can flow. In fact, we can use them to make predictions about any redox reaction, whether or not it occurs in a cell. **Section 9.3** explains how.

Rusting is a redox reaction

Rusting is an electrochemical process. Electrochemical cells are set up in the metal surface, where different areas act as sites of oxidation and reduction. The two half-reactions involved in rusting are:

$Fe^{2+}(aq) + 2e^- \rightarrow Fe(s) \qquad E^\ominus = -0.44\,V$
$\frac{1}{2}O_2(g) + H_2O(l) + 2e^- \rightarrow 2OH^-(aq) \qquad E^\ominus = +0.40\,V$

The reduction of oxygen to hydroxide ions occurs at the more positive potential, and so electrons flow to the half-cell in which the iron is oxidised to iron(II) ions.

Figure 11 shows what happens when a drop of water is left in contact with iron or steel.

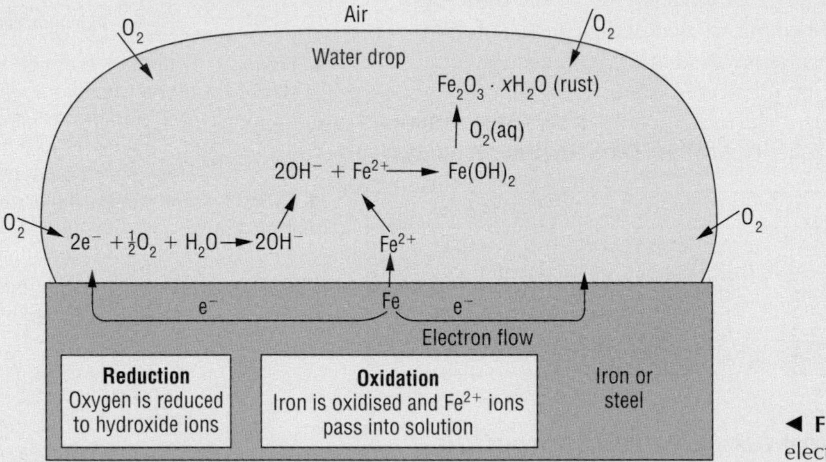

◀ **Figure 11** Rusting is an electrochemical process.

The concentration of dissolved oxygen in the water droplet determines which regions of the metal surface are sites of reduction and which are sites of oxidation. At the edges of the droplet, where the concentration of dissolved oxygen is higher, oxygen is reduced to hydroxide ions.

$\frac{1}{2}O_2(g) + H_2O(l) + 2e^- \rightarrow 2OH^-(aq)$

The electrons needed to reduce the oxygen come from the oxidation of iron at the centre of the water droplet, where the concentration of dissolved oxygen is low. The $Fe^{2+}(aq)$ pass into solution:

$Fe(s) \rightarrow Fe^{2+}(aq) + 2e^-$

The electrons released flow in the metal surface to the edges of the droplet.

This explains why corrosion is always greatest at the centre of a droplet of water or under a layer of paint – these are the regions where the oxygen supply is limited. Pits are formed here where the iron has dissolved away.

Rust forms in a series of secondary processes within the solution as Fe^{2+} and OH^- ions diffuse away from the metal surface. It does not form as a protective layer in contact with the iron surface.

$Fe^{2+}(aq) + 2OH^-(aq) \rightarrow Fe(OH)_2(s)$

$Fe(OH)_2(s) \xrightarrow{O_2(aq)} Fe_2O_3{\cdot}xH_2O(s)$

Some ionic impurities, such as sodium chloride from salt spray near the sea, promote rusting by increasing the conductivity of the water.

Summary

When two half-cells are connected to form an electrochemical cell:

- the half-cell with the more positive standard electrode potential becomes the positive terminal of the cell
- the half-cell with the more negative standard electrode potential becomes the negative terminal of the cell
- electrons flow in the external circuit from the negative to the positive terminal.

An oxidation reaction takes place at the negative terminal and the electrons produced are supplied to the reduction reaction at the positive terminal. E_{cell} is the difference between the standard electrode potentials of the two half-cells.

Problems for 9.2

Note: You will need to consult the table of standard electrode potentials in the **Data Sheets** (Table 20) to answer some of these problems.

1 For each of the following redox reactions, write equations for the two half-reactions.

a $Zn(s) + Pb^{2+}(aq) \rightarrow Zn^{2+}(aq) + Pb(s)$
b $2Al(s) + 6H^{+}(aq) \rightarrow 2Al^{3+}(aq) + 3H_2(g)$
c $2Ag^{+}(aq) + Cu(s) \rightarrow Cu^{2+}(aq) + 2Ag(s)$
d $Cl_2(g) + 2I^{-}(aq) \rightarrow I_2(aq) + 2Cl^{-}(aq)$
e $Zn(s) + S(s) \rightarrow ZnS(s)$

2 Combine each of the following pairs of half-reactions in order to write the overall balanced equation for the reaction occurring in the cell. (**Hint**: make sure that you have the same number of electrons in each half-reaction before you combine the two.)

	Positive half-cell	*Negative half-cell*
a	$Cu^{2+}(aq) + 2e^{-} \rightarrow Cu(s)$	$Mg(s) \rightarrow Mg^{2+}(aq) + 2e^{-}$
b	$Ag^{+}(aq) + e^{-} \rightarrow Ag(s)$	$Zn(s) \rightarrow Zn^{2+}(aq) + 2e^{-}$
c	$Au^{3+}(aq) + 3e^{-} \rightarrow Au(s)$	$Mg(s) \rightarrow Mg^{2+}(aq) + 2e^{-}$
d	$Cl_2(g) + 2e^{-} \rightarrow 2Cl^{-}(aq)$	$Fe^{2+}(aq) \rightarrow Fe^{3+}(aq) + e^{-}$

3 a Calculate the $E^{\ominus}_{cell}$ values for the cells made from the following pairs of standard half-cells.

i $Ag^{+}(aq)/Ag(s)$ and $Mg^{2+}(aq)/Mg(s)$
ii $Zn^{2+}(aq)/Zn(s)$ and $Cu^{2+}(aq)/Cu(s)$
iii $Fe^{2+}(aq)/Fe(s)$ and $Ni^{2+}(aq)/Ni(s)$
iv $Zn^{2+}(aq)/Zn(s)$ and $Fe^{2+}(aq)/Fe(s)$
v $Sn^{4+}(aq)/Sn^{2+}(aq)$ and $MnO_4^{-}(aq)/Mn^{2+}(aq)$

b For each cell in part **a** identify which electrode will be the positive terminal of the cell.

c Write balanced equations for the overall reactions occurring in each cell in part **a**.

4 When a cell is made from a standard cadmium half-cell and a standard copper half-cell, the reading on a high-resistance voltmeter is 0.74 V. The copper electrode forms the positive terminal of the cell. What is the electrode potential of a standard cadmium half-cell? (You will need to look up the standard electrode potential of the copper half-cell.)

5 A cell is made from a $Co^{2+}(aq)/Co(s)$ half-cell and a $Cu^{2+}(aq)/Cu(s)$ half-cell. $E^{\ominus}_{cell}$ is 0.62 V with the copper half-cell positive. Calculate the standard electrode potential of the cobalt half-cell.

6 A cell is made from a $Pb^{2+}(aq)/Pb(s)$ half-cell and a $Mg^{2+}(aq)/Mg(s)$ half-cell. $E^{\ominus}_{cell}$ is 2.23 V, with lead as the positive terminal. Calculate the standard electrode potential of the lead half-cell.

7 A cell is made from an $Al^{3+}(aq)/Al(s)$ half-cell and a $Fe^{3+}(aq)/Fe^{2+}(aq)$ half-cell. The aluminium is the negative terminal and $E^{\ominus}_{cell}$ is 2.45 V. Calculate the standard electrode potential of the $Fe^{3+}(aq)/Fe^{2+}(aq)$ half-cell.

8 A cell is made from the half-cells chlorine, $Cl_2(g)/2Cl^{-}(aq)$, and fluorine, $F_2(g)/2F^{-}(aq)$. $E^{\ominus}_{cell}$ is 1.49 V, with chlorine as the negative terminal. Calculate the standard electrode potential of the fluorine half-cell.

9 If a metal is placed high in the electrochemical series, its ability to release electrons and form hydrated positive ions is high – the metal is a strong reducing agent. Consult a table of electrode potentials and arrange the following metals in order of their strength as reducing agents:

Ag Ce Sn Ni Cd K

10 A series of electrochemical cells was set up. The table shows the half-cells used and readings obtained using a high-resistance voltmeter. Use the results in the table to work out a value for the standard electrode potential of each half-cell. (Constructing an electrode potential chart will help you to do this.)

Positive half-cell	**Negative half-cell**	$E^{\ominus}_{cell}$/V
$2H^{+}(aq), H_2(g)/Pt$	$Pb^{2+}(aq)/Pb(s)$	0.13
$Cd^{2+}(aq)/Cd(s)$	$Cr^{3+}(aq)/Cr(s)$	0.34
$Pb^{2+}(aq)/Pb(s)$	$Cd^{2+}(aq)/Cd(s)$	0.27
$Ag^{+}(aq)/Ag(s)$	$Cd^{2+}(aq)/Cd(s)$	1.20

9.3 Predicting the direction of redox reactions

Electrode potentials measure the tendency of a half-reaction to accept electrons. The more positive the electrode potential, the greater is this tendency. Figure 1 illustrates this for three half-reactions. If a half-reaction has a large tendency to *accept* electrons, it will have a small tendency to *supply* them – and vice versa.

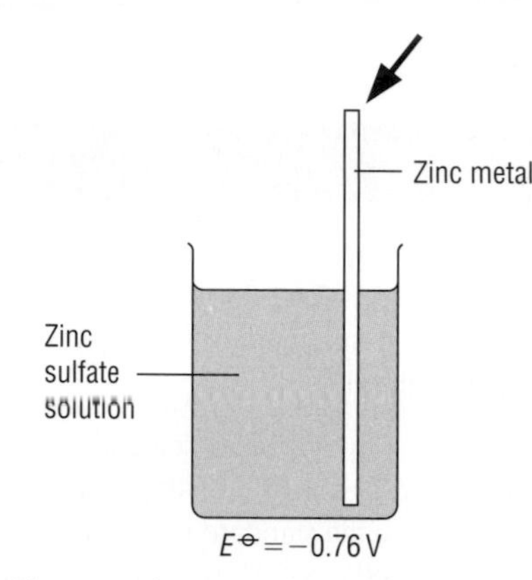

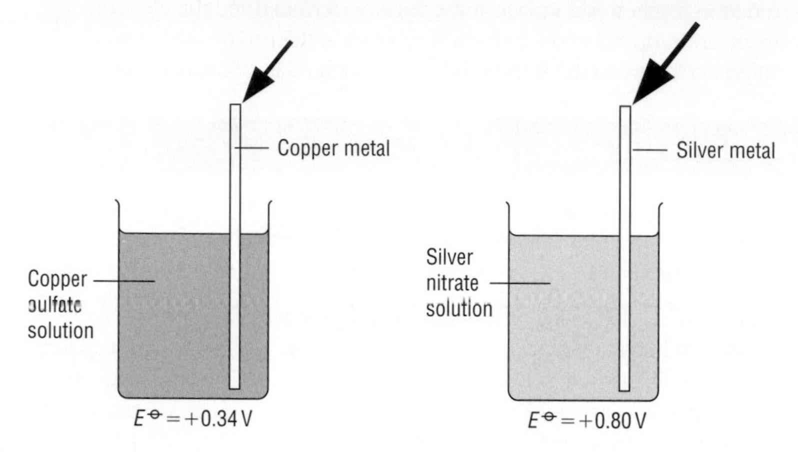

▲ **Figure 1** The sizes of the arrows indicate the tendency of the half-cells to accept electrons.

Copper ions will accept electrons supplied by zinc. Silver ions will accept electrons supplied by copper.

For zinc to supply electrons the reaction must be

$$Zn(s) \rightarrow Zn^{2+}(aq) + 2e^-$$

For the copper ions to accept electrons, the reaction must be

$$Cu^{2+}(aq) + 2e^- \rightarrow Cu(s)$$

The prediction for the overall reaction agrees with the observed changes.

$$Zn(s) + Cu^{2+}(aq) \rightarrow Zn^{2+}(aq) + Cu(s)$$

When the copper and silver half-cells are connected, the predicted changes are:

$$2Ag^+(aq) + 2e^- \rightarrow 2Ag(s)$$

$$Cu(s) \rightarrow Cu^{2+}(aq) + 2e^-$$

The overall reaction predicted is

$$Cu(s) + 2Ag^+(aq) \rightarrow Cu^{2+}(aq) + 2Ag(s)$$

This again agrees with the observed changes.

Electrode potential charts

Electrode potential charts provide a useful way of displaying and using the data. We can use them to make predictions about the direction a particular redox reaction will take. Figure 2 shows an electrode potential chart for the three half-reactions that we have just discussed.

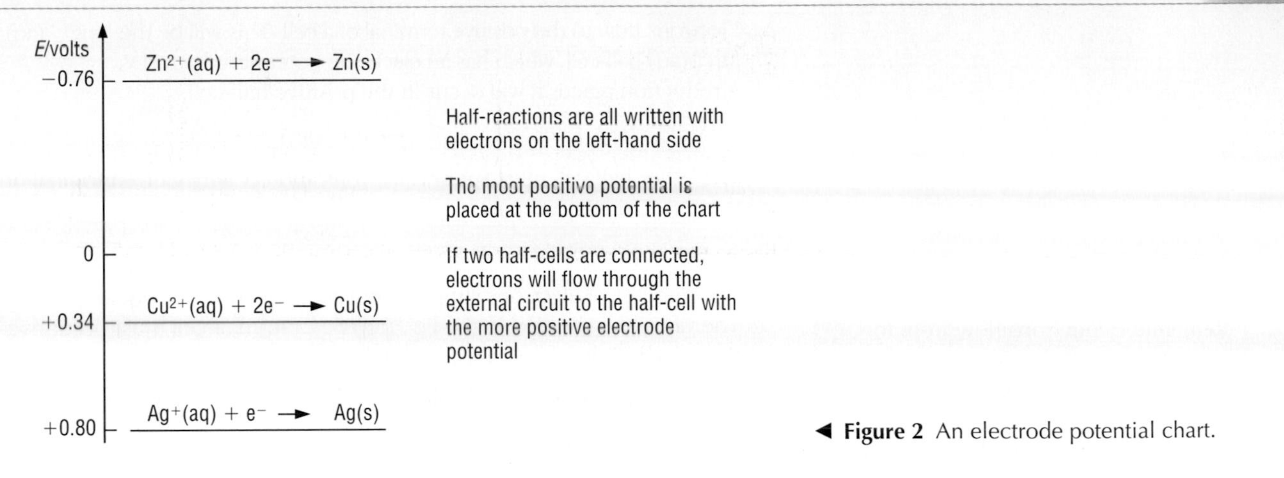

◀ **Figure 2** An electrode potential chart.

Example

Table 1 gives standard electrode potentials of some half-reactions.

◀ **Table 1** Standard electrode potentials for a number of half-cells.

Half-cell	Half-reaction	$E^{\ominus}$/V
$I_2(aq)/2I^-(aq)$	$I_2(aq) + 2e^- \rightarrow 2I^-(aq)$	+0.54
$Fe^{3+}(aq)/Fe^{2+}(aq)$	$Fe^{3+}(aq) + e^- \rightarrow Fe^{2+}(aq)$	+0.77
$Br_2(aq)/2Br^-(aq)$	$Br_2(aq) + 2e^- \rightarrow 2Br^-(aq)$	+1.07
$Cl_2(g)/2Cl^-(aq)$	$Cl_2(aq) + 2e^- \rightarrow 2Cl^-(aq)$	+1.36
$MnO_4^-(aq)/Mn^{2+}(aq)$	$MnO_4^-(aq) + 8H^+(aq) + 5e^- \rightarrow Mn^{2+}(aq) + 4H_2O(l)$	+1.51

What reaction could occur if we connect the $MnO_4^-(aq)/Mn^{2+}(aq)$ and $Fe^{3+}(aq)/Fe^{2+}(aq)$ half-cells so that electrons can flow? Write an equation for the overall reaction.

Steps

1 Construct an electrode potential chart.
2 Use it to make predictions about whether a reaction could occur.
3 Use the half-equations to give an overall equation.

Solution

1 The electrode potential chart for the half-reactions in Table 1 is shown in Figure 3. To answer the problem, we need only look at the half-reactions for $MnO_4^-(aq)/Mn^{2+}(aq)$ and $Fe^{3+}(aq)/Fe^{2+}(aq)$.

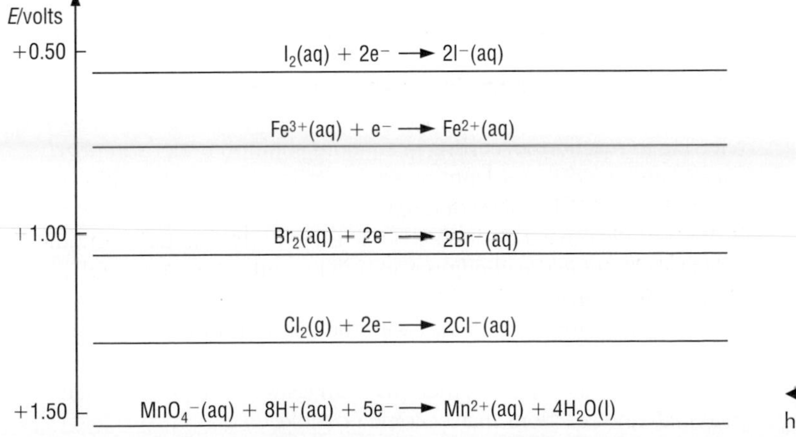

◀ **Figure 3** Electrode potential chart for the half-reactions in Table 1.

2 Electrons flow to the positive terminal of a cell. This will be the $MnO_4^-(aq)/Mn^{2+}(aq)$ half-cell, which has an electrode potential of +1.51 V.
A reduction reaction will occur in the positive half-cell.

$$MnO_4^-(aq) + 8H^+(aq) + 5e^- \rightarrow Mn^{2+}(aq) + 4H_2O(l)$$

The other half-cell must supply electrons and is the negative terminal of the cell. Oxidation occurs.

$$Fe^{2+}(aq) \rightarrow Fe^{3+}(aq) + e^-$$

3 The number of electrons supplied and accepted must be equal. Therefore the reaction in the $Fe^{3+}(aq)/Fe^{2+}(aq)$ half-cell must occur five times each time one $MnO_4^-(aq)$ ion is reduced.

$$5Fe^{2+}(aq) \rightarrow 5Fe^{3+}(aq) + 5e^-$$

The overall equation is

$$MnO_4^-(aq) + 8H^+(aq) + 5Fe^{2+}(aq) \rightarrow Mn^{2+}(aq) + 4H_2O(l) + 5Fe^{3+}(aq)$$

Try some of the problems at the end of this section.

Remember: Electrons flow from the more negative half-cell to the more positive half-cell. Once you have decided which half-cell supplies electrons, and which half-cell receives them, it is easy to predict the direction of the half-reactions.

What exactly can we predict?

Are we limited to making predictions about redox reactions occurring in half-cells?

No – we can use the electrode potentials to make predictions about the feasibility of redox reactions whether or not we have physically arranged the reagents into two half-cells with electrodes. This makes electrode potentials useful for making predictions about *any* redox reactions.

Are we able to predict for certain whether a particular reaction will happen or not?

No – we can use electrode potentials to say whether it is *possible* for a reaction to happen. But we know nothing about the *rate* of the reaction. So a reaction which we have predicted as possible may not actually occur in practice because it is too slow.

Can we *make* reactions happen?

We may predict that a change is feasible, mix the reagents and find that nothing in fact happens. The rate of the reaction is so slow that no change is observable. But we can sometimes change reaction rates.

If a reaction is slow it means that the activation enthalpy for the reaction must be very high. If we want the reaction to happen faster, we could look for a catalyst to provide an alternative route with a lower activation enthalpy, and so increase the rate.

However, if we predict from the electrode potentials that the reaction is *not* possible, no catalyst in the world is going to make it happen. Electrons will not flow spontaneously from a positive potential to a less positive one.

What else could we try to make reactions happen?

We must remember that if we use standard electrode potentials, we are referring to reactions occurring in aqueous solution under *standard conditions*, at 298 K and 1 atm pressure. Under different conditions, the electrode potentials will be different.

If we predict that a reaction is not feasible under standard conditions, we could try changing the conditions in order to alter the values of the electrode potentials.

Electrode potentials may vary with the concentration of the ions and molecules involved in the cell reaction. For example, if hydrogen or hydroxide ions are involved then pH changes will change electrode potentials and reactions may become possible.

Problems for 9.3

1 Use the chart in Figure 3 in this section (page 207) to help you answer this problem.
  a What reactions can occur if the following half-cells are connected so that electrons can flow?
    i $I_2(aq)/2I^-(aq)$ and $Cl_2(g)/2Cl^-(aq)$
    ii $Br_2(aq)/2Br^-(aq)$ and $MnO_4^-(aq)/Mn^{2+}(aq)$
    iii $Br_2(aq)/2Br^-(aq)$ and $I_2(aq)/2I^-(aq)$
  b Write a balanced equation for the overall reaction in each case.

2 The standard electrode potentials for some half-reactions are:

$Sn^{2+}(aq) + 2e^- \rightarrow Sn(s)$ $-0.14\,V$
$Fe^{3+}(aq) + e^- \rightarrow Fe^{2+}(aq)$ $+0.77\,V$
$2Hg^{2+}(aq) + 2e^- \rightarrow Hg_2^{2+}(aq)$ $+0.92\,V$
$Cl_2 + 2e^- \rightarrow 2Cl^-(aq)$ $+1.36\,V$

Construct an electrode potential chart and use it to predict which of the following reactions can occur.
  a $2Fe^{2+}(aq) + 2Hg^{2+}(aq) \rightarrow 2Fe^{3+}(aq) + Hg_2^{2+}(aq)$
  b $Sn(s) + 2Hg^{2+}(aq) \rightarrow Sn^{2+}(aq) + Hg_2^{2+}(aq)$
  c $2Cl^-(aq) + Sn^{2+}(aq) \rightarrow Cl_2(g) + Sn(s)$
  d $2Fe^{2+}(aq) + Cl_2(aq) \rightarrow 2Fe^{3+}(aq) + 2Cl^-(aq)$

3 Here are two half-cells:

Half-cell	$E^\ominus$/V
$I_2(aq)/2I^-(aq)$	+0.54
$Cr_2O_7^{2-}(aq)/Cr^{3+}(aq)$	+1.36

  a Write equations for the two half-reactions which can occur if the half-cells are connected.
  b Write a balanced equation for the overall reaction.

4 a Construct an electrode potential chart for the following half-reactions:
    i $Fe^{3+}(aq) + e^- \rightarrow Fe^{2+}(aq)$
    ii $I_2(aq) + 2e^- \rightarrow 2I^-(aq)$
    iii $Sn^{4+}(aq) + 2e^- \rightarrow Sn^{2+}(aq)$
  b Predict whether a reaction is possible between the following pairs:
    i Fe^{2+} and I_2
    ii Sn^{4+} and I^-
    iii I^- and Fe^{3+}

5 The following table gives data for a series of half-reactions.

Half-reaction	$E^\ominus$/V
$2H^+(aq) + 2e^- \rightarrow H_2(g)$	0
$I_2(aq) + 2e^- \rightarrow 2I^-(aq)$	+0.54
$Fe^{3+}(aq) + e^- \rightarrow Fe^{2+}(aq)$	+0.77
$Br_2(aq) + 2e^- \rightarrow 2Br^-(aq)$	+1.07
$IO_3^-(aq) + 6H^+ + 5e^- \rightarrow \frac{1}{2}I_2(aq) + 3H_2O(l)$	+1.19
$Cl_2(g) + 2e^- \rightarrow 2Cl^-(aq)$	+1.36

Use the data to decide which of the following can be oxidised by $IO_3^-(aq)$ in acidic solution:
  a $Fe^{2+}(aq)$
  b $Cl^-(aq)$
  c $I^-(aq)$
  d $H_2(g)$
  e $Br^-(aq)$

6 Electrode potentials can be useful when considering redox reactions of organic compounds.
  a Use the data given below to decide which reactions are feasible, under standard conditions, between oxygen and
    i methane
    ii methanol
    iii methanal (HCHO)
  b Give equations for those reactions which are feasible.

Half-reaction	$E^\ominus$/V
$HCOOH(aq) + 2H^+(aq) + 2e^- \rightarrow HCHO(aq) + H_2O(l)$	+0.06
$O_2(g) + 4H^+(aq) + 4e^- \rightarrow 2H_2O(l)$	+1.23
$CO_2(g) + 8H^+(aq) + 8e^- \rightarrow CH_4(g) + 2H_2O(l)$	+0.17
$HCHO(aq) + 2H^+(aq) + 2e^- \rightarrow CH_3OH(aq)$	+0.23
$CH_3OH(aq) + 2H^+(aq) + 2e^- \rightarrow CH_4(g) + H_2O(l)$	+0.59

10 RATES OF REACTIONS

10.1 *Factors affecting reaction rates* AS

Different chemical reactions go at different rates. Some reactions, such as burning fuel in a cylinder of a car engine or precipitating silver chloride from solution, go very fast. Others, such as the souring of milk or the rusting of iron, are much slower. The study of reaction rates, or **reaction kinetics**, is an important branch of chemistry. Studying reaction kinetics helps chemists to find ways of speeding up or slowing down chemical processes in industry. It also helps them to make predictions about important reactions, such as those that occur between gases in the atmosphere, and to understand the mechanisms of chemical reactions.

What affects the rate of a reaction?

The rate of a chemical reaction may be affected by:

- the **concentration** of the reactants. For example, the rate of reaction of chlorine atoms with ozone in the stratosphere increases as the concentration of chlorine atoms increases. In the case of solutions, concentration is measured in mol dm^{-3}; in the case of gases, the concentration is proportional to the **pressure**.
- the **temperature**. Nearly all reactions go faster at higher temperatures.
- the **intensity of radiation**, if the reaction involves radiation. For example, ultraviolet radiation of a certain frequency causes O_2 molecules to split into O atoms, and the reaction goes faster when the intensity of the ultraviolet radiation increases.
- the **particle size** of a solid. A solid such as magnesium reacts much faster when it is finely powdered than when it is in a large lump because there is a much larger **surface area** of solid exposed for reaction to take place on.
- the presence of a **catalyst**.

Low concentration

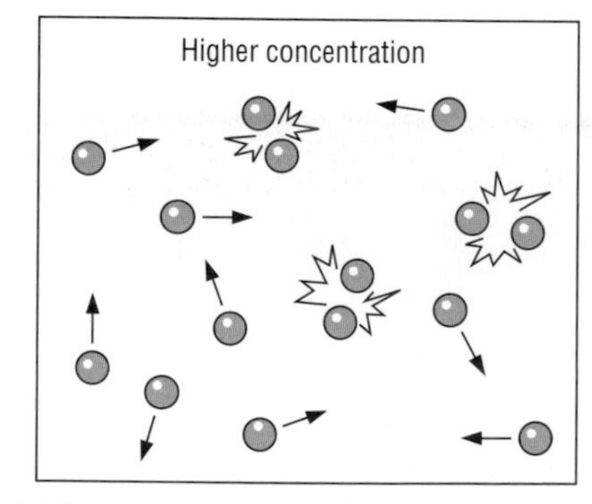

▲ **Figure 1** When particles are closer together they have a greater chance of reacting.

The collision theory of reactions

We can explain the effect of these factors using a simple **collision theory**. The basic idea is that reactions occur when the particles of reactants collide, provided they collide with a certain minimum kinetic energy.

For example, imagine two particles, say an ozone molecule and a chlorine atom, moving around in the stratosphere. For a reaction to occur, the two particles must first collide so that they come into contact with each other. This will happen more often if there are more particles in a given volume – so it is easy to see why increasing the concentration of the reactants speeds up the reaction (see Figure 1).

Indeed, any factor which increases the number of collisions will increase the rate of the reaction.

But, for most reactions, simply colliding isn't enough – not every collision causes a reaction. As the particles approach and collide, kinetic energy is converted into potential energy and the potential energy of the reactants rises. When a plot is drawn of reaction progress against enthalpy, this is known as an **enthalpy profile** (Figure 2).

Existing bonds start to stretch and break and new bonds start to form. Only those pairs with enough combined kinetic energy on collision to overcome the energy barrier, or **activation enthalpy**, for the reaction will go on to produce products.

The proportion of colliding pairs with sufficient kinetic energy to overcome the energy barrier depends very much on the temperature of

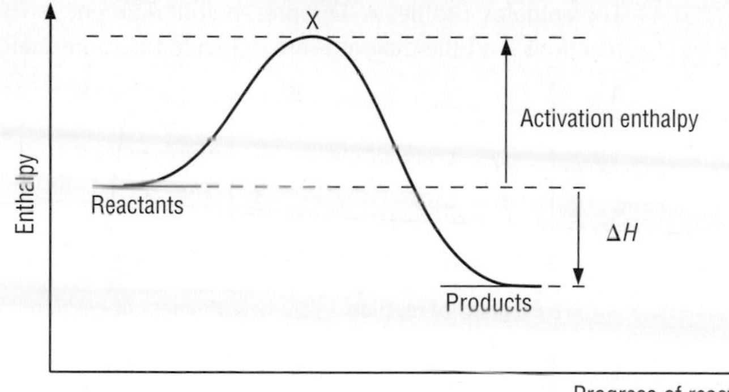

◀ **Figure 2** Enthalpy profile for an exothermic reaction.

the gas. At higher temperatures, a much larger proportion of colliding pairs have enough energy to react.

Temperature often has a dramatic effect on reaction rates – for example, methane and oxygen do not react at room temperature, but react explosively when heated.

You will study the effect of temperature on the rate of a reaction in more detail in **Section 10.2**.

Activation enthalpy

Activation enthalpy is the minimum kinetic energy required by a pair of colliding particles before reaction will occur. It is the energy that must be supplied to enable the bonds in the reactants to stretch and break as new bonds form in the product.

Enthalpy profiles

Plots like the one in Figure 2 are a useful way of picturing the energy changes that take place as a reaction proceeds. The curved line is the energy pathway for a pair of colliding molecules – it is called the **energy** (or **enthalpy**) **profile** for the reaction.

In going from reactants to products, the highest point (X) on the pathway corresponds to an arrangement of atoms where old bonds are stretched and new bonds are starting to form.

For the reaction of chlorine atoms and ozone, this would be an arrangement in which one of the O–O bonds in ozone is breaking and a new bond from Cl to O is forming.

$$Cl + O_3 \rightarrow \underset{\text{at X}}{Cl \cdots\cdots O \cdots\cdots O{-}O} \rightarrow ClO + O_2$$

Such an arrangement is very unstable and exists for only a very short time.

The curve in Figure 2 applies to a simple one-step reaction. Many reactions, such as the combustion of methane, actually take place in a series of steps and the equation simply represents the overall reaction. In cases like this, it is meaningless to draw a single curve like the one in Figure 2 for the reaction pathway, because there will be many curves – one for each step in the reaction!

Problems for 10.1

1 For each of the following reactions, say how, if at all, you would expect its rate to be affected by the following factors:

A temperature
B total pressure of gas
C concentration of solution
D surface area of solid.

a The reaction of magnesium with hydrochloric acid

$$Mg(s) + 2HCl(aq) \rightarrow MgCl_2(aq) + H_2(g)$$

b The reaction of nitrogen with hydrogen in the presence of an iron catalyst

$$N_2(g) + 3H_2(g) \xrightarrow{Fe} 2NH_3(g)$$

c The decomposition of aqueous hydrogen peroxide

$$2H_2O_2(aq) \rightarrow 2H_2O(l) + O_2(g)$$

2 The rate of hydrolysis of proteins increases in the presence of an acid or an enzyme. What is the function of the acid or enzyme?

3 The reaction of chlorine atoms with ozone in the upper atmosphere (stratosphere) is thought to be responsible for the destruction of the ozone layer:

$Cl + O_3 \rightarrow ClO + O_2$

The chlorine atoms are produced by the action of high energy solar radiation on CFCs.

a Use collision theory to explain how an increase in the concentration of chlorine atoms increases the rate of reaction with ozone.

b The activation enthalpy for the reaction of chlorine atoms with ozone is relatively small. What effect would you expect a change of temperature to have on the rate of this reaction? Explain your reasoning.

4 The enthalpy profiles **A–D** represent four different reactions – all the diagrams are drawn to the same scale.

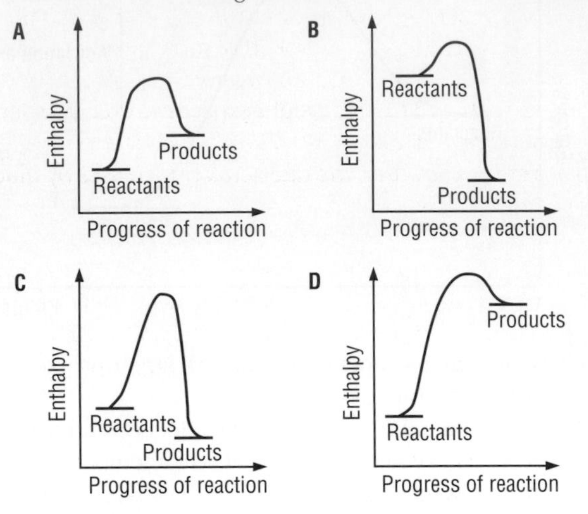

Which of the enthalpy profiles **A–D**:

a represent exothermic reactions?
b represent endothermic reactions?
c shows the largest activation enthalpy?
d shows the smallest activation enthalpy?
e represents the most exothermic reaction?
f represents the most endothermic reaction?

10.2 *The effect of temperature on rate*

How can I dissolve this sugar more quickly?
How can I get this cake to cook more quickly?
How can I get this glue to set more quickly?

The answer to all these questions is likely to involve raising the temperature. Temperature has an important effect on the rate of chemical reactions, and we make use of it constantly in our everyday lives – the chemical industry also depends heavily on it. Without it, there would be no Haber process for making ammonia, for example.

If you've measured the rates of reactions at different temperatures, you'll know that for many reactions the rate is roughly *doubled* by a temperature rise of just 10°C.

Extending the collision theory of reactions

You met the basic ideas of the collision theory in **Section 10.1**. In this section we will take those ideas a little further.

Think of the reaction between nitrogen and hydrogen to make ammonia in the Haber process:

$$N_2(g) + 3H_2(g) \rightleftharpoons 2NH_3(g)$$

The collision theory says that reaction can only occur when N_2 and H_2 molecules collide. The more frequently they collide, the faster the reaction. If you increase the pressure, for example, the N_2 and H_2 molecules are forced closer together, so they collide and react more often. That's one of the reasons the Haber Process uses high pressures – to make ammonia more quickly.

What about the effect of temperature? Think about what happens to the molecules when you raise the temperature. They move faster, so they collide more frequently. We can work out *how much* more frequently they collide because the average speed of molecules is proportional to the square root of the absolute temperature. So if we increase the temperature from, say, 300 to 310 K we would expect the average speed of the molecules to increase by a factor of $(310/300)^{1/2}$, which is about 1.016, i.e. an increase of 1.6%. Yet we know that the rate actually increases by much more than this, sometimes by 200 to 300%.

Clearly there is more to the effect of temperature than simply making the particles collide more frequently. What matters is not just *how frequently* they collide but also *with how much energy*. Unless the molecules collide with a certain minimum kinetic energy, they just bounce off one another and stay unreacted. In fact, this is what happens to most of the molecules most of the time. In the reaction between N_2 and H_2 at 300 K, only 1 in 10^{11} collisions results in a reaction! Even at 800 K (a temperature used in the Haber process) only 1 in 10^4 collision results in a reaction.

So the collision theory says that reactions occur when molecules collide **with a certain minimum kinetic energy**. The more frequent these collisions, the faster the reaction.

This energy is needed to overcome the energy barrier, the **activation enthalpy**, for the reaction – this is the energy needed to start breaking the bonds in the colliding molecules so that the collision can lead to a reaction (see **Section 10.1**).

Temperature

The temperature of a substance is related to the kinetic energy of its particles.

A substance feels hotter when its particles are moving more energetically. This is because movement energy can easily be passed on to the particles in you, the thermometer or anything else.

The distribution of energies

At any temperature, the speeds – and therefore the kinetic energies – of the molecules in a substance are spread over a wide range. It is like the walking speeds of people in a street – at any one moment some are moving slowly and some quickly, but the majority are moving at moderate speeds. Molecules are similar – some have high kinetic energies, many have medium energies and some have low energies.

This distribution of kinetic energies in a gas (say a mixture of N_2 and H_2 molecules) at a given temperature is shown in Figure 1. It is called the Maxwell–Boltzmann distribution.

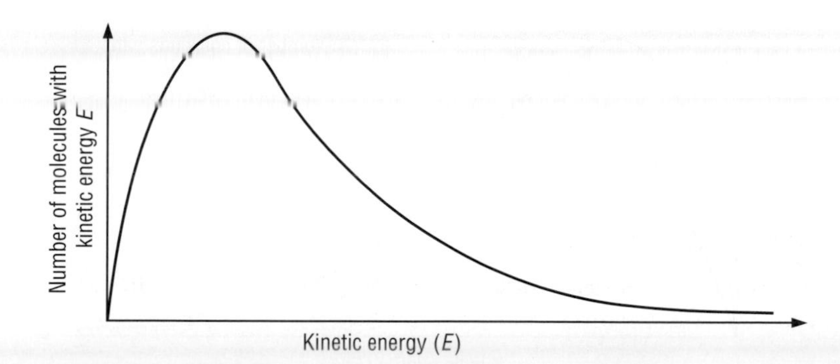

◀ **Figure 1** Distribution curve for molecular kinetic energies in a gas at a particular temperature.

As the temperature increases, more molecules move at higher speeds and have higher kinetic energies. Figure 2 (page 214) shows how the distribution of energies changes when you increase the temperature by 10 °C, from 300 to 310 K. You can see that there is still a spread of energies, but now a greater proportion of molecules have higher energies.

Now let's look at the significance of this for reaction rates. We shall take as our example a reaction whose activation enthalpy, E_a, is $+50\,kJ\,mol^{-1}$, a value typical of many reactions.

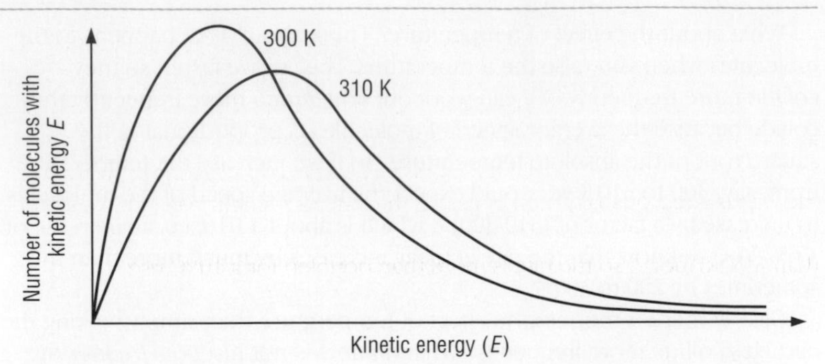

▶ **Figure 2** Distribution curves for molecular kinetic energies in a gas at 300 K and 310 K.

We need to think about how many *collisions* have energy greater than 50 kJ mol^{-1}, because these are the collisions that can lead to a reaction. Of course, each colliding pair of molecules will have energy far less than 50 kJ – this is the energy possessed by 6×10^{23} (1 mole) of them.

Figure 3 shows the number of collisions with energy greater than 50 kJ mol^{-1} for the reaction at 300 K – it's given by the shaded area underneath the curve. Only those collisions with energies in the shaded area can lead to a reaction. (The shape of the energy distribution curve for collisions is the same as for individual molecules.)

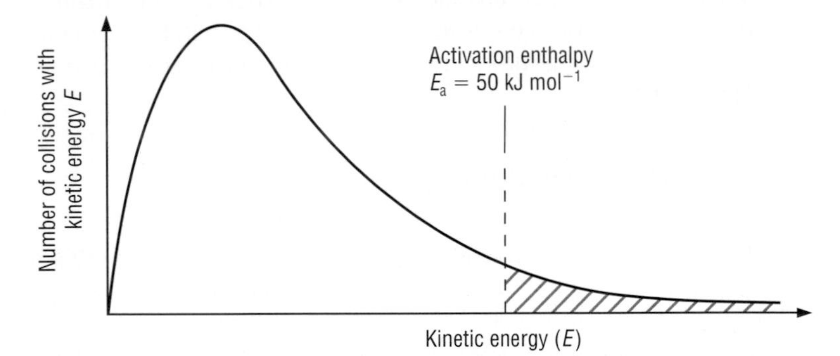

▶ **Figure 3** Distribution curve showing collisions with energy 50 kJ mol^{-1} and above.

Now look at the graph in Figure 4, which shows the curves for both 300 K and 310 K. You can see that, at the higher temperature, a significantly higher proportion of molecules have energies above 50 kJ mol^{-1} – about twice as many in fact. This means that twice as many molecules have enough energy to react – so the reaction goes twice as fast.

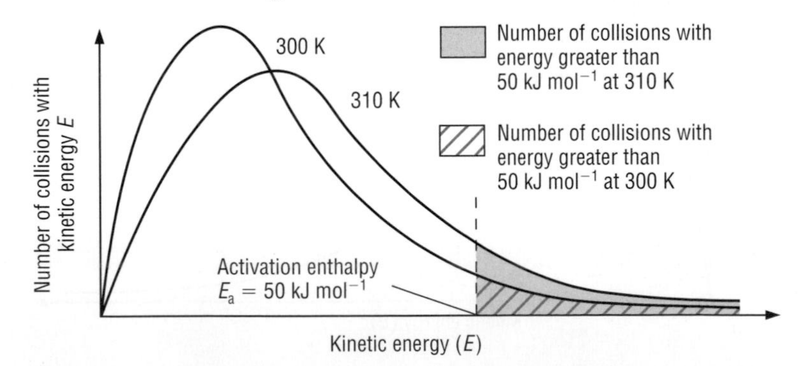

▶ **Figure 4** Distribution curves showing the effect of changing the temperature, from 300 K to 310 K, on the proportion of collisions with energy 50 kJ mol^{-1} and above.

We can summarise all this by saying that **reactions go faster at higher temperatures** because a larger proportion of the colliding molecules have the minimum activation enthalpy needed to react. Increasing the temperature may not make much difference to the energy of individual molecules, but it makes a big difference to the proportion of molecules with enough energy to react. In the example above, just a 10 °C increase is enough to place twice as many molecules above the minimum activation enthalpy of +50 kJ mol^{-1}.

For many reactions, the rate is roughly doubled by a 10 °C temperature rise – even though everything else stays the same. It is only a rough rule – in fact, it only works exactly for reactions with an activation enthalpy of $+50\,kJ\,mol^{-1}$ and for a temperature rise from 300 to 310 K. Many reactions have an activation enthalpy of about this value and a lot of chemistry is done at about 300 K, so the rule is a reasonable rough guide. In the case of the reaction of nitrogen with hydrogen in the Haber process, the activation enthalpy is rather greater than $+50\,kJ\,mol^{-1}$, so the rate is more than doubled for a 10 °C rise.

The greater the activation enthalpy, the greater is the effect of increasing the temperature on the rate of the reaction.

Problems for 10.2

1 The collision theory assumes that the rate of a reaction depends on:
  A the rate at which reactant molecules collide with one another
  B the proportion of reactant molecules that have enough energy to react once they have collided.
  Which, out of **A** and **B**, explains each of the following observations?
  a Reactions in solution go faster at higher concentration.
  b Solids react faster with liquids or gases when their surface area is greater.
  c Catalysts increase the rate of reactions.
  d Increasing the temperature increases the rate of a reaction.

2 A mixture of hydrogen and oxygen doesn't react until it is ignited by a spark – then it explodes. The mixture also explodes if you add some powdered platinum.
  a The energy of a spark is tiny, yet it is enough to ignite any quantity of a hydrogen/oxygen mixture, large or small. Suggest an explanation for this.
  b Explain why platinum makes the hydrogen/oxygen reaction occur at room temperature.

3 Explain why, above a certain temperature, enzyme-catalysed reactions actually go *more slowly* if the temperature is raised.

4 Use the collision theory to explain:
  a why coal burns faster when it is finely powdered than when it is in a lump
  b why nitrogen and oxygen in the atmosphere do not normally react to form nitrogen oxides
  c why reactions between two solids take place very slowly
  d why flour dust in the air can ignite with explosive violence.

5 Catalytic converters in car exhausts are designed to remove pollutant gases such as CO and NO_x yet the converters do not work effectively until the car engine has warmed up.
  Use the ideas in this section to suggest a reason.

6 Excess zinc was added to $1\,mol\,dm^{-3}$ sulfuric acid at room temperature and the reaction followed by plotting the volume of hydrogen given off against time, as shown below.

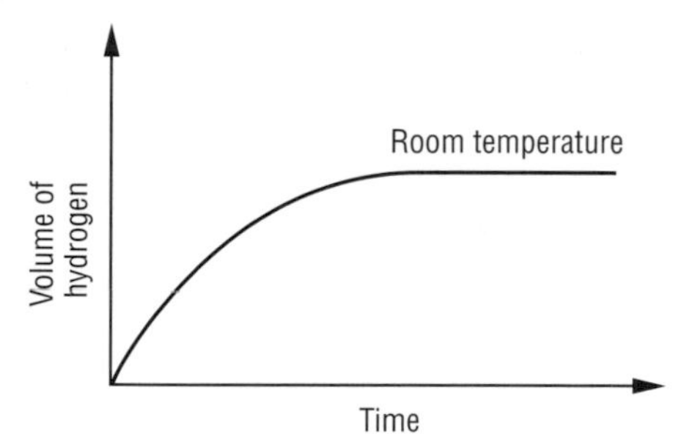

Copy the graph and add the curve that would be obtained if the reaction were carried out at a higher temperature.

7 The diagram below shows the energy distribution for collisions between Cl atoms and O_3 molecules at temperature T_1 K. E_a is the activation enthalpy for the reaction of Cl atoms and O_3 molecules:

$$Cl + O_3 \rightarrow ClO + O_2$$

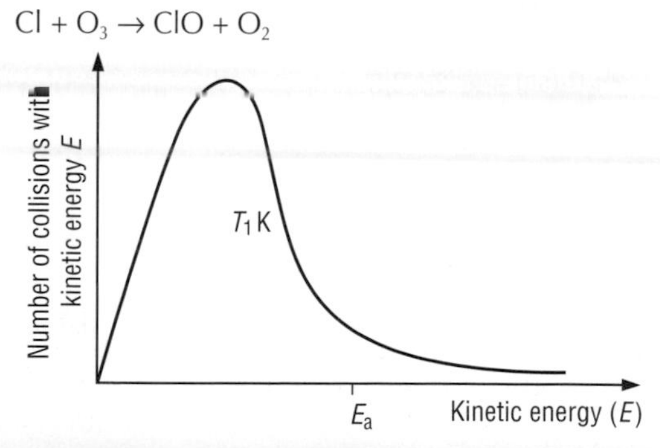

  a Copy the diagram and shade the area of the graph representing the number of collisions with sufficient energy to lead to a reaction.
  b On your graph, show the energy distribution for collisions between Cl atoms and O_3 molecules when the temperature of the reactants is increased by 10 K to T_2 K. Shade (using a different colour) the area under the graph representing the number of collisions with sufficient energy to lead to a reaction at T_2 K.

10.3 *The effect of concentration on rate*

In this section we look at how the rates of reactions depend on concentration. Normally this means the concentration of a *solution* but we can also apply the idea to gases – for a gas we normally measure its concentration in terms of its *pressure*.

It is useful to know how concentration affects rates for all sorts of reasons. You might want to use changes of concentration to alter the rate of a reaction in an industrial process. Knowing how concentration affects rate can also tell us a great deal about the way that reactions occur – their **mechanisms**.

Before looking closely at the way that concentration affects rate, we need to know a little about how reaction rate is measured.

The meaning of 'rate of reaction'

When we talk about the rate of something, we mean the rate at which some quantity changes. Speed is rate of change of distance, the inflation rate is the rate of change of prices. Whenever you are measuring a rate you need to be clear about the units you are using. Speed is often measured in metres per second, inflation rate is often measured as percentage change in prices per year.

When we talk about the **rate of a reaction** we are talking about the rate at which reactants are converted into products. In the decomposition of hydrogen peroxide to form water and oxygen,

$$2H_2O_2(aq) \rightarrow 2H_2O(l) + O_2(g)$$
hydrogen peroxide

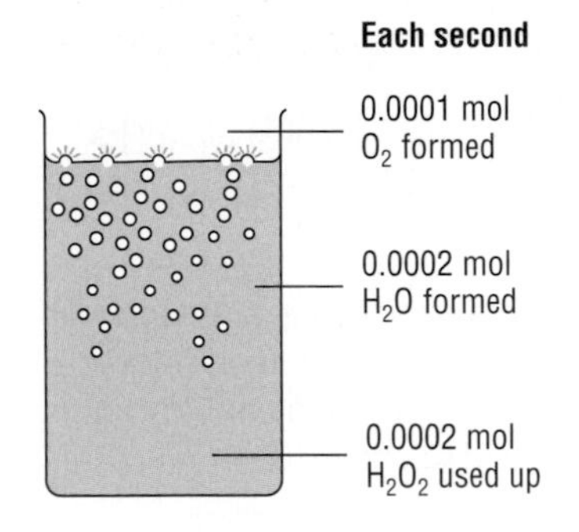

▲ **Figure 1** The relationship between the rates of reaction in terms of the reactant, and each of the products, for the decomposition of hydrogen peroxide.

the rate of the reaction means the rate at which $H_2O(l)$ and $O_2(g)$ are formed, which is the same as the rate at which $H_2O_2(aq)$ is used up. We could measure the rate of this reaction in moles of product (water or oxygen) formed per second, or moles of hydrogen peroxide used up per second. Suppose it turns out that 0.0001 mol of oxygen are being formed per second (Figure 1). The rate of the reaction is

$$0.0001\,\text{mol}\,(O_2)\,s^{-1} \text{ or } 0.0002\,\text{mol}\,(H_2O)\,s^{-1} \text{ or } -0.0002\,\text{mol}\,(H_2O_2)\,s^{-1}$$

Notice that the rate in terms of moles of H_2O is twice the rate in terms of moles of O_2 – this is because 2 moles of H_2O are formed for 1 mole of O_2. Notice also that the rate in terms of H_2O_2 has a *minus* sign – this is because H_2O_2 is getting used up instead of being produced.

In this case, the units for the rate of reaction are $\text{mol}\,s^{-1}$, though they can also be $\text{mol}\,\text{min}^{-1}$, or even $\text{mol}\,h^{-1}$.

Measuring rate of reaction

In order to determine the rate of a reaction you need to be able to measure the rate at which one of the reactants is used up, or the rate at which one of the products is formed. In practical terms what this means is that you need to measure a property that changes during the reaction and that is proportional to the concentration of a particular reactant or product. There are many experimental methods that can be used in the school laboratory to follow the rate of a reaction.

Measuring volumes of gases evolved

The reaction between calcium carbonate and hydrochloric acid produces carbon dioxide as one of the products:

$$CaCO_3(s) + 2HCl(aq) \rightarrow CaCl_2(aq) + H_2O(l) + CO_2(g)$$

The carbon dioxide can be collected in a gas syringe. The volume produced can be used to follow the reaction rate. The apparatus you would use for this is shown in Figure 2.

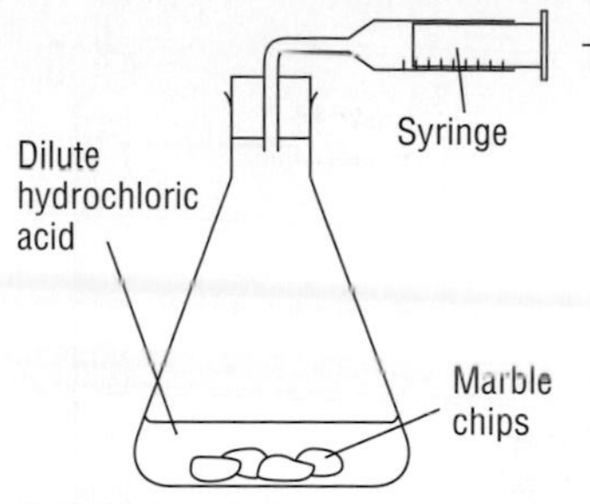

▲ **Figure 2** The apparatus needed to measure the rate of a gas produced using a gas syringe.

Measuring mass changes

Figure 3 shows a different way of monitoring the same reaction. Instead of measuring the volume of gas produced, the rate is being monitored by recording the mass lost (in the form of carbon dioxide gas) from the reaction.

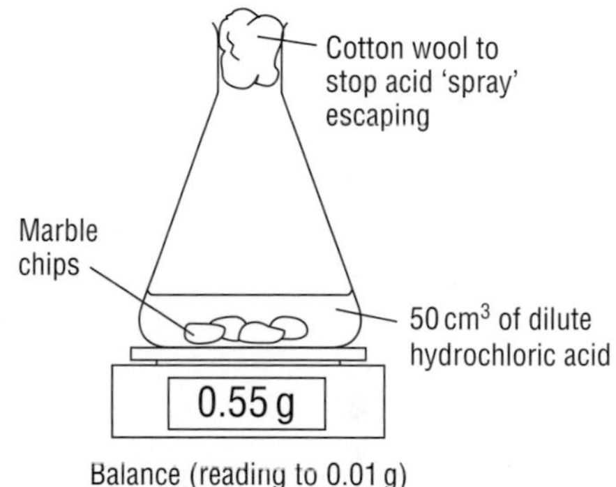

▲ **Figure 3** The apparatus needed to measure the rate of reaction by recording the loss in mass.

pH measurement

There is a third way that the above reaction can be monitored. As the reaction proceeds, the hydrochloric acid concentration will fall, as it reacts with the calcium carbonate. Monitoring the acid concentration by measuring the pH of the reaction mixture is another way of following the reaction.

Colorimetry

A colorimeter can be used to measure the change in colour of a reaction. (See **Chemical Ideas Appendix 1: Experimental Techniques** (Technique 12) for information on how a colorimeter works.) For example, in the reaction between zinc and aqueous copper(II) sulfate

$$Zn(s) + CuSO_4(aq) \rightarrow ZnSO_4(aq) + Cu(s)$$

the blue coloration of the copper sulfate solution decreases as the reaction proceeds, and the reaction can be followed using a colorimeter.

Chemical analysis and titration

All the techniques described so far have not interfered with the progress of the reaction. Chemical analysis, however, involves taking samples of the reaction mixture at regular intervals, and stopping the reaction in the sample, by a process known as 'quenching', before analysis.

For example, in the acid-catalysed reaction between iodine and propanone, a sample can be extracted from the reaction mixture and quenched by the addition of sodium hydrogencarbonate. This neutralises the acid catalyst, effectively stopping the reaction. Any unreacted iodine can then be analysed by titration with a thiosulfate solution.

The hydrolysis of methyl methanoate using sodium hydroxide is as follows:

$$CH_3COOCH_3 + NaOH \rightarrow CH_3COO^-Na^+ + CH_3OH$$

The rate of the reaction can be followed by monitoring the concentration of sodium hydroxide as it is used up during the course of the reaction. Samples taken from the reaction mixture as it proceeds are quenched by dilution with a known volume of ice-cold water. The concentration of the sodium hydroxide remaining in each reaction sample is determined by titrating it with an acid such as dilute hydrochloric acid.

Summary

The procedure for measuring rate of reaction is:

Step 1 Decide on a property of a reactant or product which you can measure, such as volume of gas produced or total mass of the reaction mixture.

Step 2 Measure the change in the property in a certain time.

Step 3 Find the rate in terms of $\frac{\text{change in property}}{\text{time taken}}$.

Notice that the units of this will not be $mol\,s^{-1}$ but, say, $cm^3\,s^{-1}$ if you are following the course of the reaction by measuring volume of gas produced, $g\,s^{-1}$ if you are measuring mass, or $mol\,dm^{-3}\,s^{-1}$ if you are measuring concentration. However, you can convert these to $mol\,s^{-1}$.

Investigating how rate depends on concentration

Now let's look more closely at the decomposition of hydrogen peroxide in solution. This reaction proceeds slowly under normal conditions but it is greatly speeded up by catalysts. A particularly effective catalyst is the enzyme *catalase*.

Figure 4 shows the apparatus used to carry out an experiment on the enzyme-catalysed decomposition of hydrogen peroxide solution. The volume of oxygen produced is measured in the inverted burette.

Let's look at the kind of results you might expect when you investigate how the rate of oxygen formation depends on the concentration of hydrogen peroxide solution.

You measure the total volume of oxygen given off at different times from the start of the experiment and plot a graph of this against time. This allows you to work out the rate of the reaction in $cm^3 s^{-1}$. We could convert this to $mol s^{-1}$ because we know that 1 mol of oxygen occupies about 24 000 cm^3 at room temperature. But what we are interested in is *comparing* rates, and for these purposes we can use cm^3 (of O_2) without bothering to convert to moles.

Figure 5 shows a graph of the results that were obtained by starting with hydrogen peroxide of concentration 0.4 mol dm^{-3}.

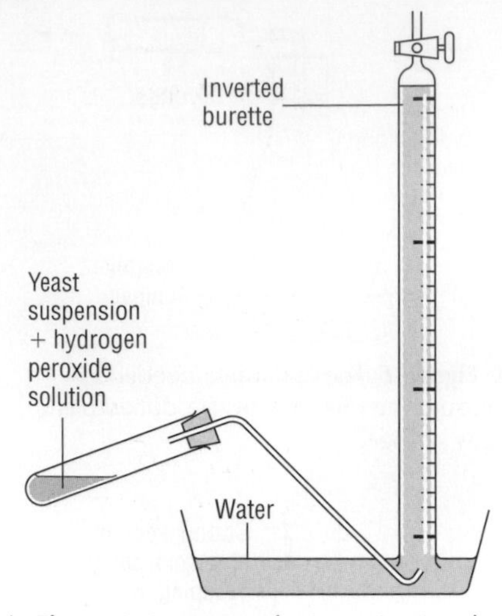

▲ **Figure 4** Apparatus for investigating the rate of decomposition of hydrogen peroxide – the yeast provides the enzyme catalase.

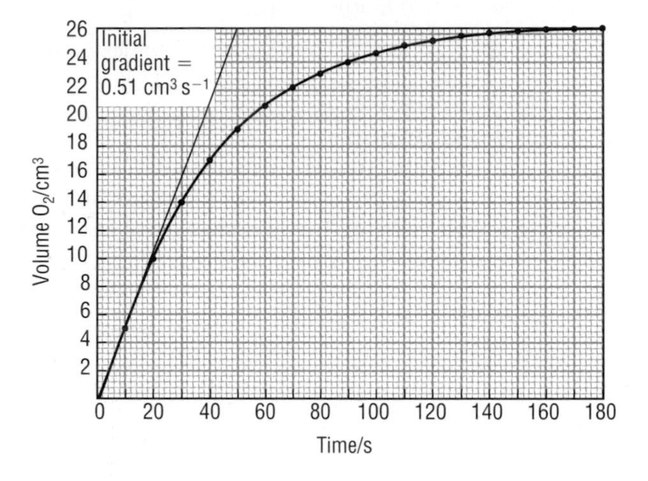

► **Figure 5** The decomposition of hydrogen peroxide using a solution of concentration 0.4 mol dm^{-3}. The rate is proportional to the volume of oxygen given off per second.

Notice these points about the graph in Figure 5:

- *The graph is steep at first.* The gradient of the graph gives us the rate of the reaction – the steeper the gradient, the faster the reaction. The reaction is at its fastest at the start, when the concentration of hydrogen peroxide in solution is high, before any has been used up.
- *The graph gradually flattens out.* This is because the hydrogen peroxide is used up and its concentration falls. The lower the concentration, the slower the reaction. Eventually, the graph is horizontal – the gradient is zero, and the reaction has come to a stop.

The rate of the reaction at the start is called the **initial rate**. We can find the initial rate by drawing a tangent to the curve at the point $t = 0$ and measuring the gradient of this tangent. In the example in Figure 5, the gradient is 0.51, so the initial rate = 0.51 $cm^3 s^{-1}$.

Figure 6 shows some results that were obtained when the same experiment was done using hydrogen peroxide solutions of different concentrations. In each case, the concentration of the catalase enzyme was kept constant, as were all other conditions such as temperature. As you would expect, the graphs start off with differing gradients, depending on the initial concentration of hydrogen peroxide. Table 1 shows the initial rates of the experiments in Figure 6.

▼ **Table 1** Initial rates of decomposition of hydrogen peroxide.

Concentration of hydrogen peroxide at start/mol dm^{-3}	Initial rate/$cm^3 s^{-1}$
0.40	0.51
0.32	0.41
0.24	0.32
0.16	0.21
0.08	0.10

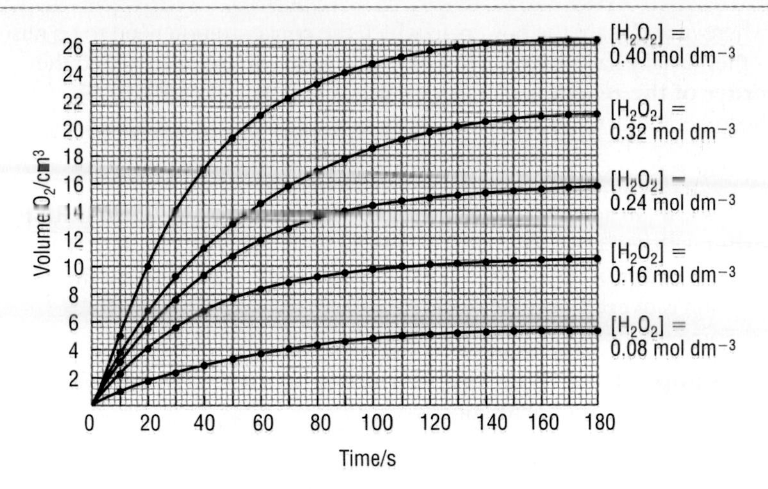

◀ **Figure 6** The decomposition of hydrogen peroxide solutions of differing concentrations.

Now we are in a position to answer the question 'How does the rate of the reaction depend on the concentration of hydrogen peroxide?' Figure 7 shows the initial rates plotted against concentration of hydrogen peroxide and you can see that it is a straight line.

This means that the rate is directly proportional to the concentration of hydrogen peroxide. In other words,

$$\text{rate} \propto [H_2O_2(aq)]$$

or

$$\text{rate} = \text{constant} \times [H_2O_2(aq)]$$

Hydrogen peroxide is not the only substance whose concentration affects the rate of this reaction. It is also affected by the concentration of the catalase enzyme, but in the series of experiments shown in Figure 6 we kept the concentration of catalase constant. However, we could do another set of experiments to find the effect of changing the concentration of catalase – this time we need to keep the concentration of hydrogen peroxide constant.

When we vary the concentration of catalase in this way, we find that the rate of the reaction is also proportional to the concentration of catalase. In other words,

$$\text{rate} = \text{constant} \times [\text{catalase}]$$

If we combine this with the equation involving H_2O_2, we get

$$\text{rate} = \text{constant} \times [H_2O_2(aq)] \times [\text{catalase}] \text{ or}$$
$$\text{rate} = k\,[H_2O_2(aq)]\,[\text{catalase}]$$

This is called the **rate equation** for the reaction – the constant k is called the **rate constant**. The value of k varies with temperature, so you must always say at what temperature the measurements were made when you give the rate, or the rate constant, of a reaction.

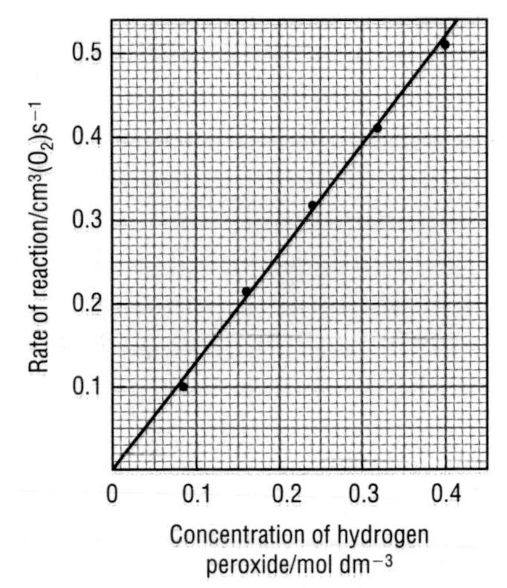

▲ **Figure 7** The initial rate of decomposition of hydrogen peroxide plotted against concentration of hydrogen peroxide.

Order of reaction

We can write a rate equation for *any* chemical reaction – provided we can do an experiment first to find out how the rate depends on the concentration of the reactants. For a general reaction in which A and B are the reactants

$$A + B \rightarrow \text{products}$$

the general rate equation is

$$\text{rate} = k\,[A]^m[B]^n$$

where m and n are the powers to which the concentration need to be raised – they usually have values of 0, 1 or 2. The terms $\boldsymbol{m}$ and $\boldsymbol{n}$ are called the **order of the reaction** with respect to A and B. For example, in the hydrogen peroxide example which we met earlier,

$$\text{rate} = k\,[H_2O_2]\,[\text{catalase}]$$

In this case, m and n are both equal to 1. We say that the reaction is **first order** with respect to H_2O_2 and first order with respect to catalase. The **overall order** of the reaction is given by $(m + n)$, so in this case the reaction is **overall second order**.

Some examples will help to explain the idea of reaction order.

Example 1: *the reaction of Br radicals to form Br_2 molecules*
The equation for the reaction is

$$2Br(g) \rightarrow Br_2(g)$$

Experiments show that the rate equation for the formation of Br_2 is

$$\text{rate} = k\,[Br]^2$$

So this reaction is second order with respect to Br. Since Br is the only reactant involved, the reaction is also second order overall.

Example 2: *the reaction of iodide ions, I^-, with peroxodisulfate(VI) ions, $S_2O_8^{2-}$*
The equation for the reaction is

$$S_2O_8^{2-}(aq) + 2I^-(aq) \rightarrow 2SO_4^{2-}(aq) + I_2(aq)$$

Experiments show that the rate equation for the reaction is

$$\text{rate} = k\,[S_2O_8^{2-}(aq)]\,[I^-(aq)]$$

So this reaction is first order with respect to $S_2O_8^{2-}$, first order with respect to I^- and second order overall.

Notice that in Example 2, the order of the reaction with respect to I^- is *one*, even though there are *two* I^- ions in the balanced equation for the reaction. This raises an important point – *you cannot predict the rate equation for a reaction from its balanced equation*. It *might* work (as in Example 1, above), but it often won't. They *only* way to find the rate equation for a reaction is by doing experiments to find the effect of varying the concentrations of reactants. This point becomes very clear if you look at the reactions in Examples 3 and 4.

Example 3: *the reaction between bromide ions, Br^-, and bromate(V) ions, BrO_3^-*
This reaction takes place in acidic solution:

$$BrO_3^-(aq) + 5Br^-(aq) + 6H^+(aq) \rightarrow 3Br_2(aq) + 3H_2O(l)$$

Experiments show that the rate equation for this reaction is

$$\text{rate} = k\,[BrO_3^-]\,[Br^-]\,[H^+]^2$$

So the reaction is first order with respect to both BrO_3^- and Br^-, and second order with respect to H^+. Notice that the orders do *not* correspond to the numbers in the balanced equation.

Example 4: *the reaction of propanone with iodine*
This reaction is catalysed by acid.

$$CH_3COCH_3(aq) + I_2(aq) \xrightarrow{\text{acid catalyst}} CH_3COCH_2I(aq) + H^+(aq) + I^-(aq)$$

The rate equation found by experiments is rate $= k\,[CH_3COCH_3]\,[H^+]$

(Note that the acid catalyst, H^+, appears in the rate equation even though it is not used up.) So the reaction is first order with respect to both CH_3COCH_3 and H^+. Notice that $[I_2]$ does not appear in the rate equation even though iodine is one of the reactants. The reaction is **zero order** with respect to I_2.

You may find it surprising that the reaction can be zero order with respect to one of the reactants – the reason is explained later in this section under the 'rate-determining steps' heading (page 226).

Temperature and rate constants

You saw in **Section 10.2** that the rate of many reactions is roughly doubled for a 10 °C temperature rise – even though everything else stays the same. Since the concentrations are not changed, the rate equation

$$\text{rate} = k[\text{A}]^m[\text{B}]^n$$

shows that it must be the value of k, the rate constant, which has roughly doubled. Changing the temperature almost always changes the value of k – though the 10 °C rule is only a rough one.

The rate constant, k, increases with increasing temperature.

Half-life

To explain the important idea of half-life, let's go back to the decomposition of hydrogen peroxide:

$$2H_2O_2(aq) \rightarrow 2H_2O(l) + O_2(g)$$

The graph in Figure 5 earlier in this section (page 218) shows the volume of oxygen produced in the decomposition of hydrogen peroxide. We can convert this graph to show *the amount of H_2O_2 remaining* at different times in this experiment. We can do this because we know how much H_2O_2 we started with, and we know that for every 1 mole of O_2 produced, 2 moles of H_2O_2 get used up.

These calculations have been used to produce the graph in Figure 8. Notice that the graph shows that the amount of H_2O_2 *decreases* with time – this is the opposite of Figure 5, in which the amount of O_2 *increases* with time.

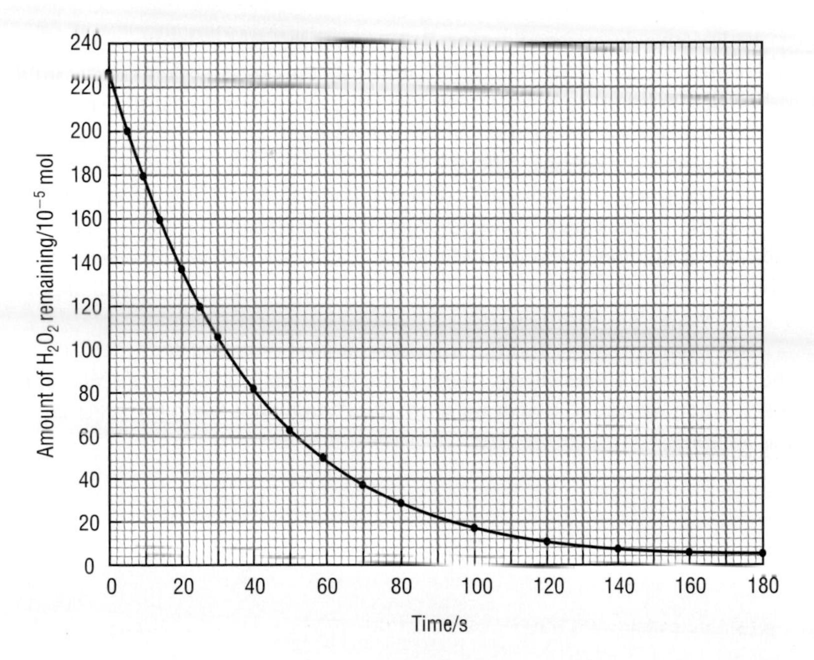

◀ **Figure 8** The same experiment as shown in Figure 5, but this time the progress of the reaction is followed in terms of the amount of H_2O_2 remaining.

We can use Figure 8 to find the **half-life**, $t_{½}$, of H_2O_2 in this experiment. The half-life means *the time taken for half the H_2O_2 to get used up.*

In Figure 9 this has been done in three cases, using exactly the same graph as in Figure 8.

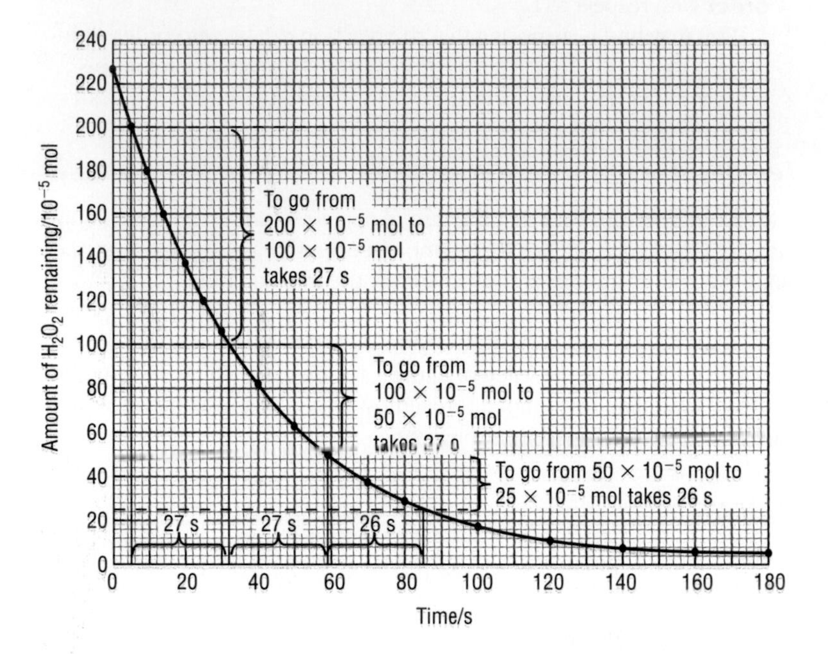

► **Figure 9** Finding half-lives, $t_{½}$, for the decomposition of hydrogen peroxide.

Looking at Figure 9, you can see that to go from 200×10^{-5} mol of H_2O_2 to 100×10^{-5} mol takes 27 s. In other words, starting with 200×10^{-5} mol of H_2O_2, the half-life = 27 s. Starting with 100×10^{-5} mol, $t_{½}$ is again 27 s. Starting with 50×10^{-5} mol, $t_{½}$ is 26 s. In fact, allowing for experimental error, we find that $t_{½}$ is always 27 s *whatever* the starting amount.

The decomposition of hydrogen peroxide is a first-order reaction with respect to H_2O_2, and we find the same rule for the half-lives of *all* first-order reactions. *For a first-order reaction, the half-life is constant, whatever the starting amount.* This characteristic gives us a useful way of deciding whether a reaction is first order or not – zero-order and second-order reactions do *not* have constant half-lives.

Radioactive decay (see **Section 2.2**) is an important example of a first-order process. The time taken for a sample of any one isotope to decay until only half of it is left has always been the same ever since the Earth was formed (in fact, ever since the isotope was formed).

Different isotopes have different half-lives. For some, such as radium-226 ($t_{½}$ = 1622 years), it is thousands of years. For others it is only a fraction of a second. Both times are really quite short compared with the Earth's 4.6 billion year lifetime, but some radioactive isotopes which were present when the Earth was formed are still around. Uranium-235, with a half-life of 7×10^8 years, is an example.

This situation shows another important aspect of first-order processes – they never end. There will always be some reactants left. We say a reaction is 'over' when we can no longer measure the change, not when the reactant has all gone.

Finding the order of reaction

You must do experiments to find the order of a reaction. Most reactions involve more than one reactant, and you have to do several experiments to find the order with respect to each reactant separately. You have to control the variables, so that the concentration of only one substance is changing at a time, and you must make all your measurements at the same temperature.

Let's look again at our trusty example of the decomposition of hydrogen peroxide in the presence of the enzyme catalase.

$$2H_2O_2(aq) \xrightarrow{\text{catalase}} 2H_2O(l) + O_2(g)$$

If we are looking at the effect of changing the concentration of hydrogen peroxide, there is no need to worry about the catalase – it is an enzyme so it doesn't get used up, and its concentration doesn't change. But if we want to look at the effect of varying the concentration of *catalase*, we must control the concentration of hydrogen peroxide to keep it constant – otherwise we will have two variables changing at the same time. One way of doing this is to do several experiments to measure the initial rate of the reaction, keeping the concentration of hydrogen peroxide the same each time but varying the concentration of catalase.

Another way of controlling the concentration of a reactant is to have a large excess of it, so that over the course of the experiment the concentration does not change significantly.

Once we have collected a set of data for the effect of changing the concentration of a particular reactant, there are several methods we can use to find the order with respect to that reactant.

The progress curve method

A **progress curve** shows how the concentration of a reactant (or product) changes as the reaction proceeds. A progress curve for the decomposition of hydrogen peroxide is shown in Figure 8. Figure 10 shows how you can use a progress curve to find the rate of the reaction at different concentrations.

The gradient of the tangent gives the rate of reaction for a particular concentration of hydrogen peroxide. You can then find the order with respect to hydrogen peroxide as in the initial rate method (see below).

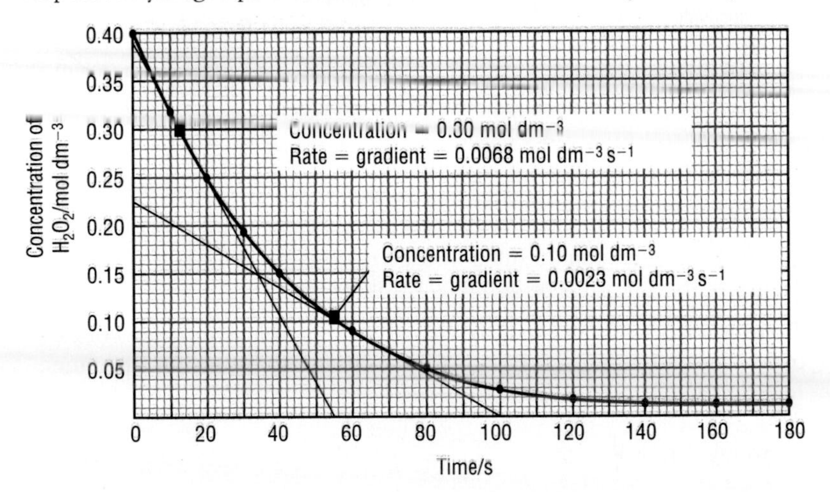

◄ **Figure 10** Using a progress curve to find the rate of a reaction at different concentrations, by drawing tangents.

The initial rate method

By drawing tangents at the origin of different progress curves

This is the method used in the hydrogen peroxide investigation. We do several experimental 'runs' at different concentrations. For each run, we can find the initial rate graphically, as shown in Figure 5 earlier in this section (page 218).

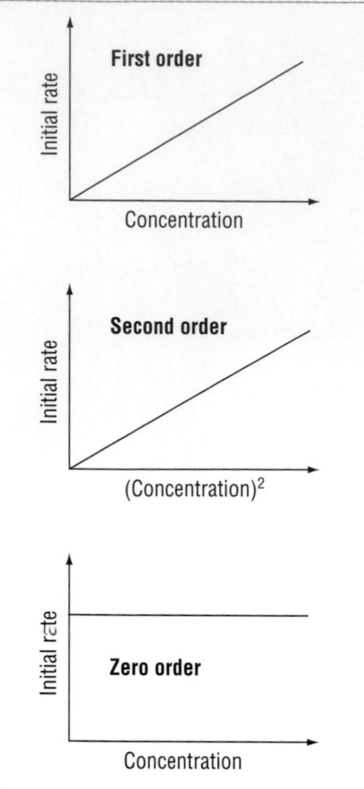

▲ **Figure 11** Plots of initial rate against concentration can tell you the order with respect to a reactant.

Once we know the initial rates for different concentrations, we can find the order. If a graph of initial rate against concentration is a straight line (as in Figure 7, page 219) then the reaction is first order. If a graph of initial rate against (concentration)^2 is a straight line then the reaction is second order. If the rate doesn't change with changing concentration at all then the reaction is zero order. Figure 11 shows these graphs.

Using the reciprocal of the reaction time as a measure of the rate

One way of measuring the **initial rate** of a reaction is to measure how long the reaction takes to produce a small, fixed amount of one of the products – the time taken is called the **reaction time**. If the rate is high (the reaction proceeds quickly) then the reaction time will be small. If the rate is low (the reaction proceeds slowly) then the reaction time will be large.

A good example of this method is the reaction between sodium thiosulfate solution and hydrochloric acid.

$$Na_2S_2O_3(aq) + 2HCl(aq) \rightarrow 2NaCl(aq) + SO_2(g) + S(s) + H_2O(l)$$

sodium thiosulfate *sulfur*

As the reaction proceeds, solid sulfur forms as a colloidal suspension of fine particles and the mixture becomes cloudy. The reaction flask is placed on a cross drawn on a piece of white paper and viewed from above as shown in Figure 12. The cross becomes less visible and is obscured when a certain amount of sulfur has formed in the reaction mixture. The reaction time to reach this point can be measured using different starting concentrations of sodium thiosulfate solution. The volume of each solution and the concentration of the hydrochloric acid must be kept constant in each experiment.

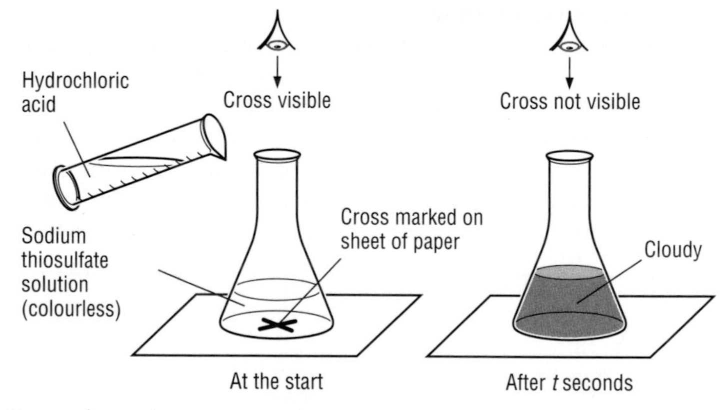

► **Figure 12** Following the reaction of sodium thiosulfate solution and hydrochloric acid.

For each starting concentration,

$$\text{average rate for this stage of the reaction} = \frac{\text{amount of sulfur needed to obscure the cross}}{\text{reaction time for sulfur to form}}$$

The amount of sulfur needed to obscure the cross will be the same for each experiment, so the rate of reaction is proportional to the reciprocal of the reaction time for the cross to be obscured, $1/t$.

$$\text{average rate} \propto \frac{1}{t}$$

The shorter the time taken, the faster the reaction; the longer the time taken, the slower the reaction.

If a graph of $1/t$ is plotted against the concentrations of the thiosulfate solutions used, a straight line plot is obtained, showing that the rate of the reaction is proportional to the concentration of sodium thiosulfate – i.e. it is first order with respect to $[Na_2S_2O_3]$.

The half-lives method

Alternatively, you can use the progress curve to find half-lives for the reaction. If the half-lives are constant then the reaction is first order.

Rate equations and reaction mechanisms

Once we know the rate equation, we can then link it to the reaction mechanism. Most reaction mechanisms involve several individual steps. The rate equation gives us information about the slowest step in the mechanism – the **rate-determining step**. We have already seen that the rate equation for a reaction cannot be predicted from the balanced chemical equation.

One way of using the rate equation to find information about the rate-determining step and the reaction mechanism is described below.

- What substances from the chemical equation are not involved in the rate equation?

These substances are *not* involved in the rate-determining step, they are part of a *faster* step than that of the rate-determining step and therefore nothing can be found *directly* from this knowledge.

- What substances from the chemical equation are involved in the rate equation?

These substances *are* the ones involved in the rate-determining step – the slowest step in the mechanism.

- What order are each of the substances?

This will tell you the relative number of moles of each substance involved in the rate-determining step. For example, if a reaction is first order with respect to a substance there will be one molecule, atom or ion of this substance involved in the rate-determining step.

It is now possible to make predictions about the mechanism of the rate-determining step because you now know what substances are involved, and in what ratio, in this particular part of the chemical reaction.

Once you know the rate-determining step, it is often possible to deduce the full mechanism for a reaction. For example, consider the reaction of 2-bromo-2-methylpropane with hydroxide ions. The chemical equation is

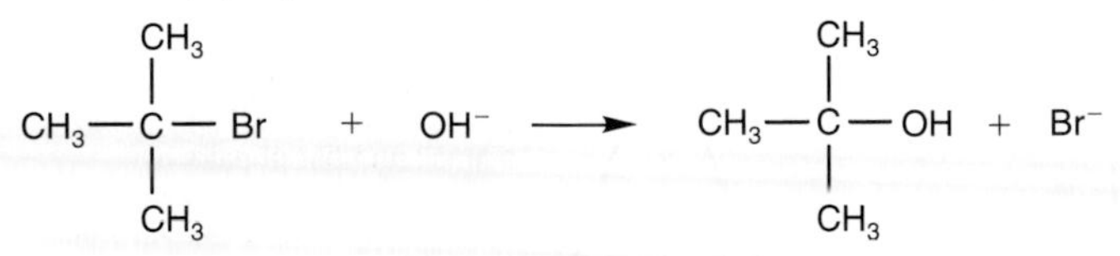

The rate equation is: rate $= k[(CH_3)_3CBr]$

There is no $[OH^-]$ term in the rate equation – this reaction is found to be zero order with respect to OH^-. Therefore, the reaction cannot take place by direct reaction of 2-bromo-2-methylpropane itself with OH^- ions.

The rate equation contains $[(CH_3)_3CBr]$ and this is the only substance in the rate equation, so you can deduce that $(CH_3)_3CBr$ is the only substance in the rate-determining step. Also, the reaction is first order with respect to $(CH_3)_3CBr$, therefore there is only one molecule involved in the rate-determining step. From this information, and your own knowledge of mechanisms, you may now be able to make a prediction of the mechanism for this step – and possibly for the rest of the reaction too.

Chemists have studied this reaction in detail and have found that it takes place in *two* steps. First, the C–Br bond breaks heterolytically:

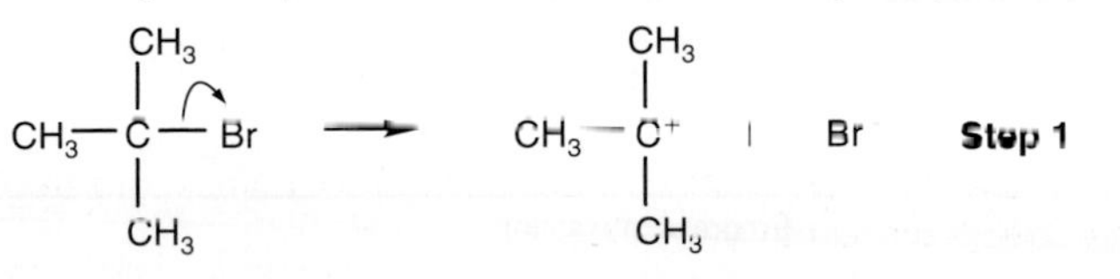

Because this step only involves $(CH_3)_3CBr$, its rate depends only on $[(CH_3)_3CBr]$, not on $[OH^-]$.

The second step involves reaction of the carbocation, $(CH_3)_3C^+$, with OH^-:

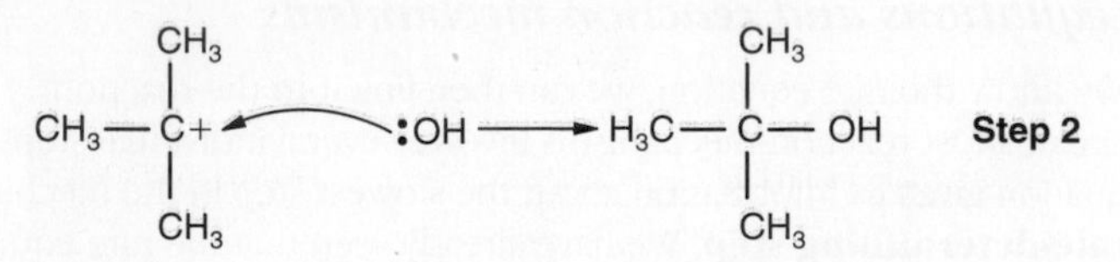

Like most ionic reactions, this process in Step 2 is very fast – certainly faster than Step 1. So the rate of Step 1 controls the rate of the whole reaction. That is why the overall rate of the reaction depends *only* on the concentration of $(CH_3)_3CBr$ – the reaction is first order with respect to $(CH_3)_3CBr$, but zero order with respect to OH^-. Step 1 is said to be the rate-determining step and *its* rate equation becomes the rate equation for the whole reaction.

The mechanism in this example involves two steps. Some simple reactions occur in a single step; more complex ones may involve more than two steps. But in every case, once you have broken the reaction down into steps, you can write the rate equation for each step from its chemical equation – but the *overall* rate equation for the reaction can only be found by experiment.

Rate-determining steps

Why are some steps in a reaction slow and others fast? One reason is that the steps have different energy barriers (activation enthalpies). If the energy barrier is big, few pairs of colliding molecules have enough energy to pass over it and the rate of conversion of reactants into products is slow. Figure 13 compares the enthalpy profiles for reactions with large and small activation enthalpies.

The enthalpy profile will be different for a reaction involving two steps. There will be two activation enthalpies – one for each step.

Note: The rate-determining step may be the only step in the complete reaction mechanism. However, in a multi-step mechanism, the rate-determining step can come at any point in the reaction. It may be the first step, in which case every step afterwards will happen more quickly and the mechanism will not be held up in forming the end product, after the first step. Alternatively, the rate-determining step may be the final step, causing a 'bottleneck' as the previous steps happen quickly and are waiting for this step to form the product. Alternatively, it may come somewhere in the middle.

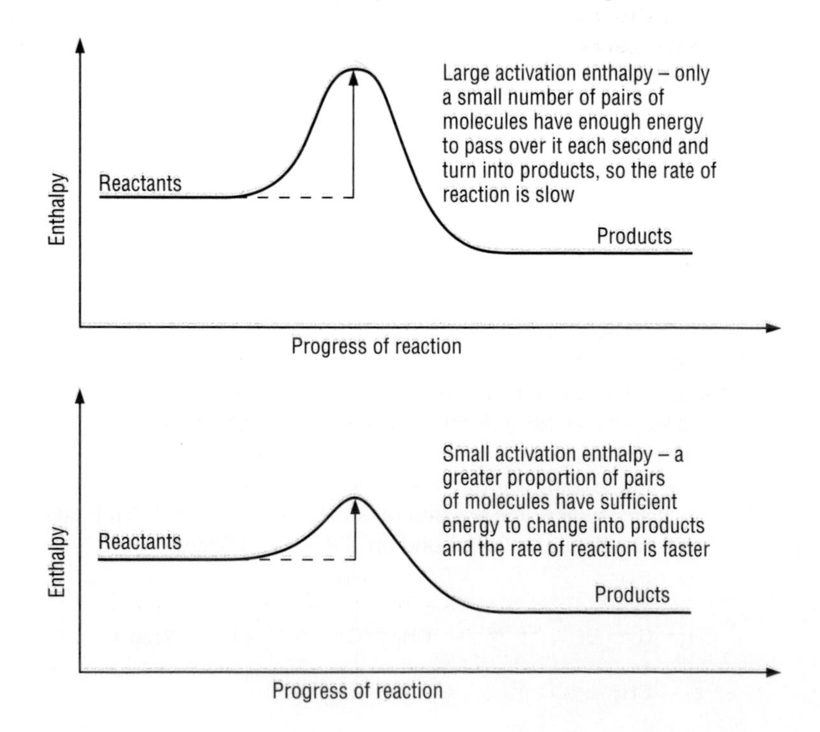

► **Figure 13** The rate of a reaction depends on the size of the activation enthalpy.

In the case of the reaction of 2-bromo-2-methylpropane with OH^- ions, Step 1 (the rate-determining step) has the larger activation enthalpy (Figure 14). Usually, the chemicals in the middle of a reaction mechanism – the **intermediates** – are at a higher energy than the reactants or products, because they have unusual structures or bonding – like the carbocation $(CH_3)_3C^+$.

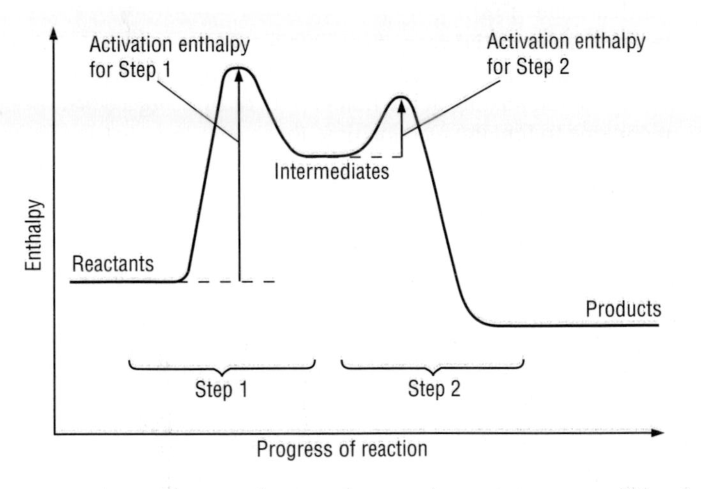

◄ **Figure 14** For a reaction involving two steps, there are two activation enthalpies.

In any reaction with several steps, the rate-determining step will be the one with the largest activation enthalpy.

Summary

- The **rate of a reaction** is the change in concentration of a reactant or product divided by the time taken for the change to occur.
- The rate of a reaction changes during the reaction as the reactants are used up.
- Reactions are followed by measuring the change, with time, of a property that is proportional to the concentration of a reactant or product – e.g. pH of the reaction mixture, colour intensity if the reaction involves a coloured compound, volume of a gas given off.
- A plot of this property against time is the **progress curve** for the reaction.
- The **half-life** of a reaction is the time taken for the concentration of a reactant to decrease to one half of its initial value.
- If the rate at which a reactant is used up in a reaction displays a constant half-life, the reaction is first order with respect to that reactant.
- The **rate equation** for a reaction shows how the rate of reaction varies with the concentration of each of the reactants.
- Many reactions occur in a series of simple individual steps. If one step in a reaction mechanism is much slower than the others it is called the **rate-determining step**.
- Rate equations can only be predicted from the chemical equations for individual steps in a reaction mechanism. They cannot be predicted from the overall equation for the reaction and *must always be found experimentally*.
- For the reaction of A with B, the rate equation is given by **rate = $k[A]^m[B]^n$**, where k is the **rate constant** and m and n are the **orders of the reaction** with respect to A and B. The overall order of the reaction is $m + n$.
- In rate experiments it is important to make measurements at a constant temperature and to investigate only one variable at a time. Other concentrations must be kept constant – one way of doing this is to use an excess of the reactant not under investigation.

- To find the order of reaction with respect to a reactant A, the rate of reaction is found at different concentrations of A. This may be done in various ways:
  - by drawing tangents at various points on the progress curve for the reaction (in which [A] is changing and everything else is constant). The gradient of the tangent is proportional to the rate of reaction for that value of [A].
  - by finding the **initial rate** of reaction for different reaction mixtures – for example, by drawing tangents to the progress curve for each reaction mixture at the point $t = 0$, or by using a clock technique to find the reaction time for a small amount of reaction to take place.

A graph of the rate of reaction (or a quantity that is proportional to the rate of reaction) is then plotted against [A]. If the plot is a straight line, which shows that the rate is proportional to [A], the reaction is **first order** with respect to A. A horizontal straight line shows that the rate is independent of [A] and so the reaction is **zero order** with respect to A. Other orders give a curve. If the reaction is **second order** with respect to A, a plot of the rate against $[A]^2$ is a straight line.

- An alternative method of showing that a reaction is first order is to calculate the values of several half-lives for the reaction from the progress curve. If the values of the half-lives are equal then the reaction is first order.
- The rate constant for a reaction at a particular temperature can be found by substituting values for the rate of the reaction and values of [A] and [B] into the rate equation – make sure that you use the correct units. If you have been following the reaction by measuring a property such as pH, the rate of reaction in terms of this property must be converted to the rate of reaction in terms of a concentration.

Problems for 10.3

1 Use the rate equations for the following reactions to write down the order of the reaction with respect to each of the reactants (and catalyst where present).

a The elimination of hydrogen bromide from bromoethane:

$$CH_3CH_2Br + OH^- \rightarrow CH_2{=}CH_2 + Br^- + H_2O$$

rate = $k[CH_3CH_2Br]$

b The acid-catalysed hydrolysis of methyl methanoate:

$$HCOOCH_3 + H_2O \xrightarrow{H^+(aq)} HCOOH + CH_3OH$$

rate = $k[HCOOCH_3][H^+]$

c The hydrolysis of urea, NH_2CONH_2, in the presence of the enzyme urease:

$$NH_2CONH_2(aq) + H_2O(l) \xrightarrow{\text{urease}} 2NH_3(aq) + CO_2(g)$$

rate = $k[NH_2CONH_2][\text{urease}]$

d One of the propagation steps in the radical substitution of an alkane by chlorine:

$$CH_3{\cdot}(g) + Cl_2(g) \rightarrow CH_3Cl(g) + Cl{\cdot}(g)$$

rate = $k[CH_3{\cdot}][Cl_2]$

e The formation of the World War I poison gas, phosgene, from carbon monoxide and chlorine:

$$CO(g) + Cl_2 \rightarrow COCl_2(g)$$

rate = $k[CO]^{1/2}[Cl_2]$

f The decomposition of nitrogen dioxide to oxygen and nitrogen monoxide:

$$2NO_2(g) \rightarrow 2NO(g) + O_2(g)$$

rate = $k[NO_2]^2$

2 Write down the rate equations for the following reactions.

a Experiments show that the reaction of 1-chlorobutane with aqueous sodium hydroxide is first order with respect to 1-chlorobutane and first order with respect to hydroxide ion:

$$CH_3CH_2CH_2CH_2Cl + OH^- \rightarrow CH_3CH_2CH_2CH_2OH + Cl^-$$

b The hydrolysis of sucrose, $C_{12}H_{22}O_{11}$, is first order with respect to sucrose and first order with respect to acid catalyst, $H^+(aq)$:

$$C_{12}H_{22}O_{11} + H_2O \rightarrow 2C_6H_{12}O_6$$

3 The initial rate method was used to investigate the reaction

$2H_2(g) + 2NO(g) \rightarrow 2H_2O(g) + N_2(g)$

The results of some studies which were made at 973 K are shown below.

$[H_2]$/ 10^{-2} mol dm^{-3}	[NO]/10^{-2} mol dm^{-3}	Rate/ 10^{-6} mol dm^{-3} s^{-1}
2.0	2.50	4.8
2.0	1.25	1.2
2.0	5.00	19.2
1.0	1.25	0.6
4.0	2.50	9.6

a What is the order of reaction with respect to
 i $H_2(g)$?
 ii NO(g)?

b Write down the rate equation for this reaction.

c Calculate a value for the rate constant for this reaction at 973 K.

4 The decomposition of nitrogen(V) oxide, carried out at 318 K, was investigated.

$N_2O_5 \rightarrow 2NO_2 + \frac{1}{2}O_2$

The following results were obtained:

Time/s	$[N_2O_5]$/mol dm^{-3}
0	2.33
184	2.08
319	1.91
526	1.67
867	1.36
1198	1.11
1877	0.72
2315	0.55
3144	0.34
3500	0.21

a Plot a graph of $[N_2O_5]$ against time.

b Find the first three half-lives in these results. What do these values suggest about the order of the reaction with respect to N_2O_5?

c Find the reaction rate for five of the values of $[N_2O_5]$ by drawing tangents at five points on your graph, and then calculating the gradient at each point.

d Plot a graph of reaction rate against $[N_2O_5]$. What order with respect to $[N_2O_5]$ is suggested by this graph? Does this agree with the order deduced in part **b**?

e Write the rate equation for the reaction.

f Calculate the value of the rate constant, including the units, by finding the gradient of the graph plotted in part **d**.

5 When cyclopropane gas is heated it isomerises to propene gas.

$$\text{cyclo-}(H_2C{-}CH_2{-}CH_2) \longrightarrow CH_2{=}CH{-}CH_3$$

The following data were obtained by heating cyclopropane in a sealed container. The temperature was 700 K and the initial pressure was 0.5 atm.

Time/10^3 s	0	13	36	56	83	108
% cyclopropane	100	84	63	49	35	25

a What type of isomerism is shown by this pair of compounds?

b What would be the pressure in the reaction vessel at the end of the reaction?

c Plot a graph of % cyclopropane against time.

d Use your graph to measure some half-life values, and so determine the order of the reaction with respect to cyclopropane.

6 The equation for the reaction for the iodination of propanone in acidic solution is

$$I_2(aq) + CH_3COCH_3(aq) \xrightarrow{H^+(aq)} CH_2ICOCH_3(aq) + H^+(aq) + I^-(aq)$$

The rate equation for the reaction has been determined experimentally – it is first order with respect to propanone, zero order with respect to iodine and first order with respect to hydrogen ions.
The following mechanism for the reaction has been proposed:

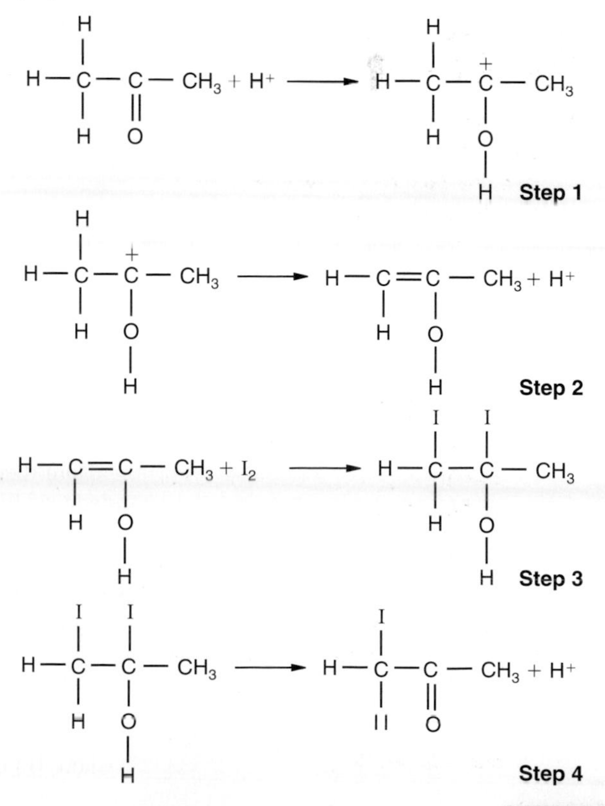

a Write a rate equation for the reaction.
b Is this proposed mechanism consistent with the rate equation? Justify your answer.
c Which step is the rate-determining step?
d If you wished to remove a sample during the reaction for analysis, how would you quench the reaction?

7 Consider the following reaction between substances A and B:

$$A + 2B \rightarrow AB_2$$

The rate equation for the reaction is
rate = $k[B]^2$
Consider the two proposed mechanisms:
Mechanism 1
$B + B \rightarrow B_2$ *Step 1*
$A + B_2 \rightarrow AB_2$ *Step 2*
Mechanism 2
$A + B \rightarrow AB$ *Step 1*
$AB + B \rightarrow AB_2$ *Step 2*
Which mechanism do you think is correct? Explain your answer.

10.4 *Enzymes and rate*

The mechanism of enzyme catalysed reactions

Enzymes are **specific** because they have a precise tertiary structure which exactly matches the structure of the **substrate** – the molecule that is reacting. It is an example of molecular recognition. Enzymes have a cleft on their surface formed by the way the protein chain folds. The shape of the cleft is tailored for the substrate molecules to fit into. Within the cleft are chemical groups – some of the side chains on the amino acid residues – which bind the substrate and possibly react with it. This region of the enzyme is called its **active site**.

The bonds which bind the substrate to the active site have to be weak so that the binding can be readily reversed when the products need to leave the active site after the reaction. The bonds are usually hydrogen bonds or interactions between ionic groups. The binding may cause bonds within the substrate to weaken or it may alter the shape of the substrate so that atoms are brought into contact to help them to react.

After reaction, the product leaves the enzyme which is then free to start again with another molecule of substrate. The whole process is summarised in Figure 1.

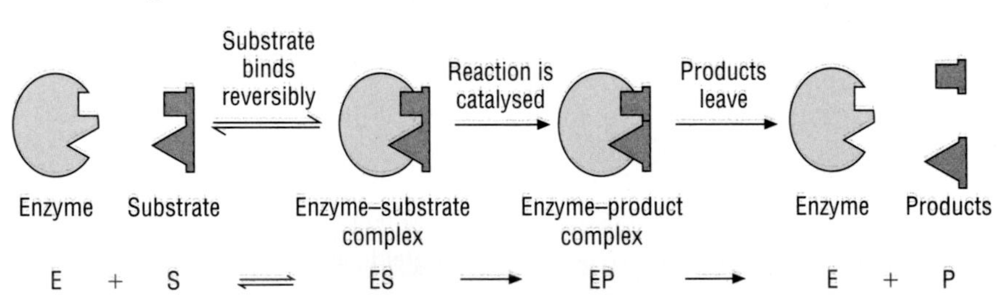

▶ **Figure 1** Illustration of the 'lock and key' model of enzyme catalysis.

Our theories about the mechanism of enzyme-catalysed reactions were deduced from studies of the orders of the reactions.

When the substrate concentration is low in this reaction

$$\underset{\text{substrate}}{S} \xrightarrow{\text{enzyme}} \underset{\text{products}}{P}$$

it is found that the rate equation for the reaction is

rate = $k[E][S]$

where [E] is the concentration of the enzyme.

From this, we deduce that the rate-determining step must involve one enzyme molecule and one substrate molecule:

$$\mathrm{E} + \mathrm{S} \xrightarrow{\text{rate-determining step}} \underset{\text{enzyme–substrate complex}}{\mathrm{ES}}$$

The steps that follow this are faster:

$$\mathrm{ES} \xrightarrow{\text{fast}} \underset{\text{enzyme–product complex}}{\mathrm{EP}} \xrightarrow{\text{fast}} \mathrm{E} + \mathrm{P}$$

This is because when the substrate concentration is low not all enzyme active sites will have a substrate molecule bound to them. The overall reaction rate depends on how frequently enzymes encounter the substrate, which depends on how much substrate there is – twice as much substrate means twice as many encounters. The reaction rate is, therefore, first order with respect to the substrate. You can see this in Figure 2.

However, if the substrate concentration is high, the rate equation becomes

$$\text{rate} = k[\mathrm{E}]$$

because, with more than enough substrate around, the first step is no longer the rate-determining one. All the enzyme active sites are occupied and [ES] is constant. What matters now is *how fast* ES or EP can break down to form the products. The rate of breakdown of ES depends on its concentration, which is effectively [E] since almost all the enzyme is present as ES. In this situation, the reaction rate does not depend on the substrate concentration – the reaction is zero order with respect to substrate.

Enzymes are specific in their action. They are very sensitive to substrate, temperature, pH and inhibition. You can find out about this in the **Thread of Life** module.

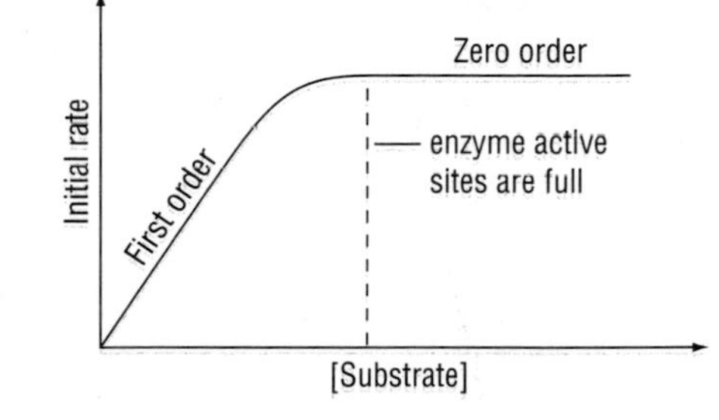

▲ **Figure 2** Graph showing initial rate against concentration of substrate.

Problems for 10.4

1 *Canavalia ensiformis* (jack bean) is a common bean in America. It contains the enzyme urease. Urea, $(NH_2)_2CO$, is decomposed into ammonia and carbon dioxide in the presence of water and urease.

a Write a balanced equation for the reaction.

b Draw a line graph to show the change in rate with urea concentration, using the lock and key model to account for the shape of the graph.

c What would happen to the rate of reaction if a small amount of mercury(II) chloride was added to a mixture of urea and jack bean in aqueous solution? Explain your answer.

d What experimental method would you use to determine the rate of reaction for this reaction?

2 When a slice of apple is exposed to air, it turns brown quickly. This is because the enzyme **o-diphenol oxidase** catalyses the oxidation of phenols in the apple to dark-coloured products. The first stage in the reaction is to convert the catechol to o-quinone:

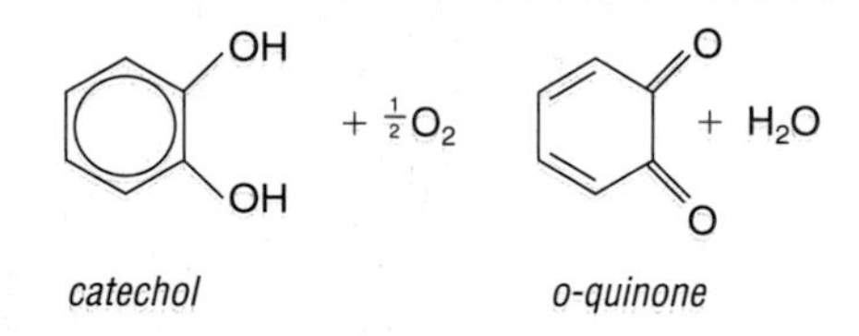

Explain why the conversion of catechol to o-quinone is significantly reduced when 3-hydroxybenzoic acid is added into a mixture of catechol and o-diphenol oxidase.

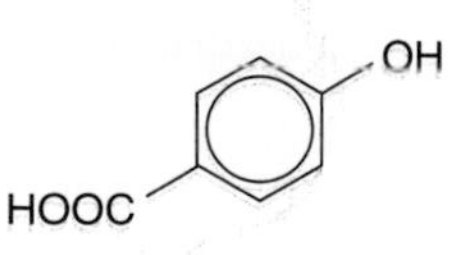

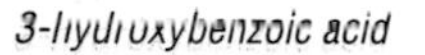
3-hydroxybenzoic acid

AS

10.5 *What is a catalyst?*

A **catalyst** is a substance which speeds up a reaction but can be recovered chemically unchanged at the end. The process of speeding up a chemical reaction using a catalyst is called **catalysis**.

Catalysts do not undergo any permanent chemical change, though sometimes they may be changed *physically*. For example, the surface of a solid catalyst may crumble or become roughened. This suggests that the catalyst is taking some part in the reaction, but is being regenerated.

Only small amounts of a catalyst are needed usually. The catalyst does not affect the *amount* of product formed, only the *rate* at which it is formed.

Catalysts in living systems are called enzymes. You can find out more about enzymes in **Section 10.4**.

A catalyst does not appear as a reactant in the overall equation for a reaction.

Types of catalysts

If the reactants and catalyst are in the same physical state (for example, both are in aqueous solution) then the reaction is said to involve **homogeneous catalysis**. Enzyme-catalysed reactions in cells take place in aqueous solution and are examples of this type of catalysis.

In contrast, many important industrial processes involve **heterogeneous catalysis**, where the reactants and the catalyst are in different physical states. This usually involves a mixture of gases or liquids reacting in the presence of a solid catalyst.

Heterogeneous catalysts

When a solid catalyst is used to increase the rate of a reaction between gases or liquids, the reaction occurs on the surface of the solid. Figure 1 shows an example. The reactants form bonds with atoms on the surface of the catalyst – we say they are **adsorbed** onto the surface. As a result, bonds in the reactant molecules are weakened and break. New bonds form between the reactants, held close together on the surface, to form the products. This in turn weakens the bonds to the catalyst surface and the product molecules are released.

▼ **Table 1** Examples of the use of heterogeneous catalysts in industrial processes.

Industrial process	Catalyst	Detail in
reforming of petroleum fractions to produce high octane petrol components e.g. hexane → cyclohexane	platinum finely dispersed on the surface of aluminium oxide	Developing Fuels
hydrogenation of unsaturated oils to give saturated fats e.g. vegetable oil + H_2 → margarine	nickel powder	Section 12.2
manufacture of ammonia from nitrogen and hydrogen in the Haber process $N_2 + 3H_2 \rightarrow 2NH_3$	finely divided iron	Agriculture and Industry

It is important that the catalyst has a large surface area for contact with reactants. For this reason, solid catalysts are used in a finely divided form or as a fine wire mesh. Sometimes the catalyst is supported on a porous material to increase its surface area and prevent it from crumbling. This happens in the catalytic converters fitted to car exhaust systems (see the **Developing Fuels** module).

Zeolites (crystalline aluminosilicate materials – see the **Developing Fuels** module) are widely used in industry as heterogeneous catalysts – for example, in the cracking of petroleum fractions.

Many of the heterogeneous catalysts used in industrial processes are transition metals (the metals in the central block of the Periodic Table) or transition metal compounds (see **Section 11.5**). You will meet a number of examples at different stages in the course.

Catalyst poisoning

Catalysts can be **poisoned** so that they no longer function properly. Many substances which are poisonous to humans operate by blocking an enzyme-catalysed reaction.

In heterogeneous catalysis, the 'poison' molecules are adsorbed more strongly to the catalyst surface than the reactant molecules. The catalyst cannot catalyse a reaction of the poison and so becomes inactive with poison molecules blocking the active sites on its surface. This is the reason why leaded petrol cannot be used in cars fitted with a catalytic converter; lead is strongly adsorbed to the surface of the catalyst.

Catalyst poisoning is also the reason why it is not possible to replace the very costly metals (platinum and rhodium) in catalytic converters by cheaper metals (such as copper and nickel). These metals are vulnerable to poisoning by the trace amounts of sulfur dioxide always present in car exhaust gases. Once the catalyst in a converter becomes inactive it cannot be regenerated. A new converter has to be fitted.

Catalyst poisoning can be a problem in industrial processes. In the UK, nearly all the hydrogen for the Haber process is prepared by steam reforming of methane. Methane reacts with steam in the presence of a nickel catalyst:

$$CH_4(g) + H_2O(g) \rightleftharpoons CO(g) + 3H_2(g)$$

If the feedstock for the process contains sulfur compounds, these must be removed first to prevent severe catalyst poisoning.

Sometimes it is possible to clean or **regenerate** the surface of a catalyst. In the catalytic cracking of long-chain hydrocarbons, for example, carbon is produced and the surface of the zeolite catalyst becomes coated in a layer of soot. This blocks the adsorption of reactant molecules and the activity of the catalyst is reduced. The catalyst is constantly recycled through a separate container where hot air is blown through the zeolite powder. The oxygen in the air converts the carbon to carbon dioxide and cleans the catalyst surface. You can read more about catalyst regeneration in the **Developing Fuels** module.

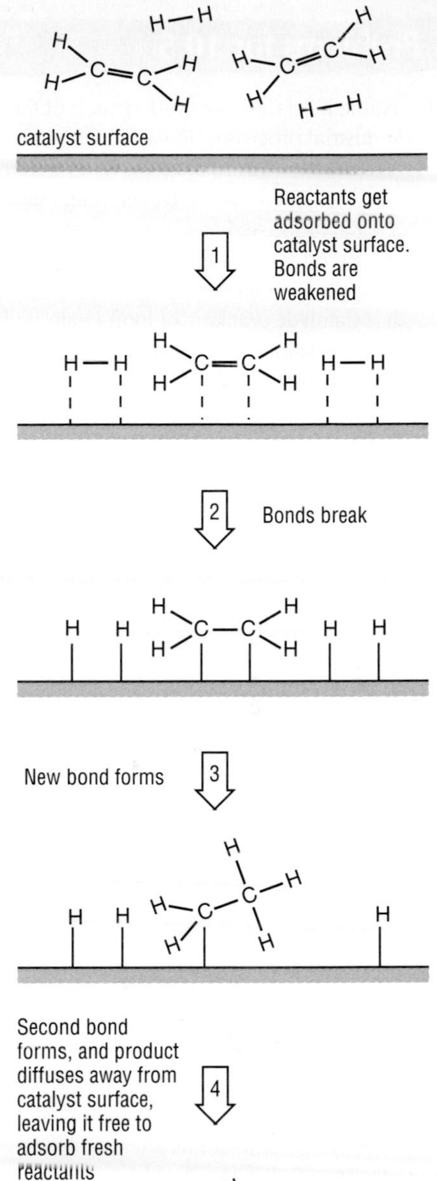

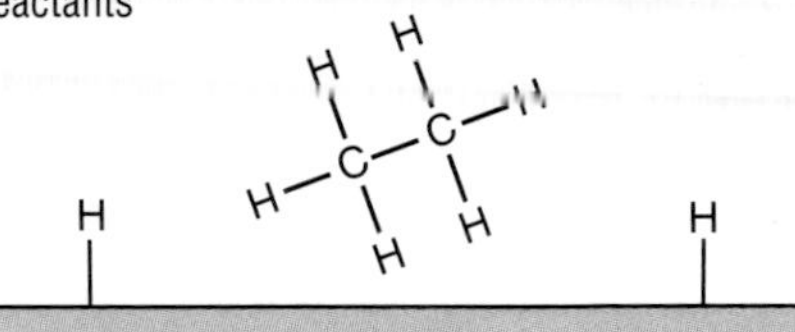

▲ **Figure 1** An example of heterogeneous catalysis. The diagrams show a possible mechanism for nickel catalysing the reaction between ethene and hydrogen to form ethane.

Problems for 10.5

1 Name a catalyst involved in each of the following industrial processes. In each case, state whether the process involves homogeneous or heterogeneous catalysis. (Refer to the **Developing Fuels** module to find the catalysts used.)
  a Reforming gasoline fractions to produce high octane petrol components.
  b Catalytic cracking of long-chain hydrocarbons.
  c Oxidation of CO and unburnt petrol in a car exhaust.

2 When carbon monoxide and nitrogen monoxide in car exhaust gases pass through a catalytic converter, carbon dioxide and nitrogen are formed.
  a Write a balanced chemical equation for this reaction.
  b Explain why it is important to reduce the quantities of carbon monoxide and nitrogen monoxide released into the atmosphere.
  c Figure 1 in this section (page 233) shows a possible mechanism for the nickel-catalysed reaction between ethene and hydrogen. The reactants are adsorbed onto the surface of the nickel catalyst, where the reaction takes place.
    i Explain the meaning of the term *adsorbed*.
    ii Using (N)–(O) to represent nitrogen monoxide and (C)–(O) to represent carbon monoxide, draw out a possible mechanism for the formation of carbon dioxide, (O)–(C)–(O).
  d Suggest why catalytic converters do not work effectively until a car engine has warmed up.

AS

10.6 *How do catalysts work?*

In **Section 10.5** you met examples of heterogeneous catalysts that are used to speed up chemical reactions. In this section, you will see how the collision theory and the use of enthalpy profiles help us to understand how catalysts work.

In a chemical reaction, existing bonds in the reactants must first stretch and break. Then new bonds can form as the reactants are converted to products.

Bond breaking is an endothermic process. A pair of reacting molecules must have enough energy between them to collide with an energy greater than the activation energy before reaction can occur. If the energy barrier is very high then relatively few pairs of molecules will have enough energy to react when they collide – so the reaction is slow.

Catalysts speed up reactions by providing an alternative reaction pathway for the breaking and remaking of bonds that has a lower activation enthalpy. Now that the energy barrier is lower, more pairs of molecules can react when they collide – this means that the reaction proceeds more quickly. Note that the enthalpy change, ΔH, is the same for the catalysed and uncatalysed reactions. Figure 1 shows the enthalpy profiles for an uncatalysed and a catalysed reaction.

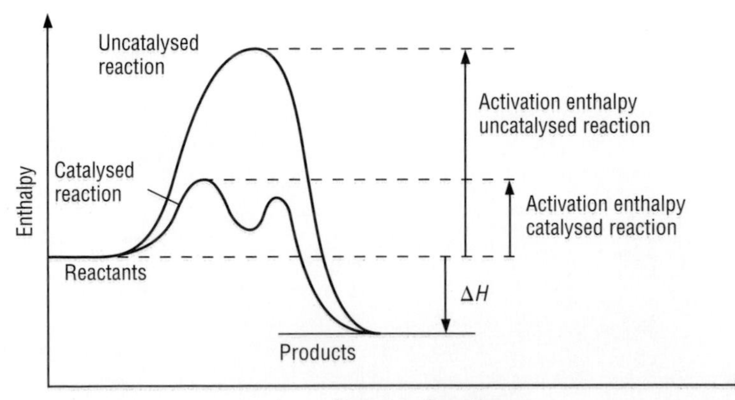

► **Figure 1** The effect of a catalyst on the enthalpy profile for a reaction.

Catalysts and equilibrium

Catalysts do not affect the position of equilibrium in a reversible reaction. They alter the *rate* at which the equilibrium is attained, but not the *composition* of the equilibrium mixture.

Homogeneous catalysts

Homogeneous catalysts normally work by forming an intermediate compound with the reactants. That is why the enthalpy profile for the catalysed reaction in Figure 1 has two humps – one for each step. The intermediate compound then breaks down to give the product and reform the catalyst.

An example of homogeneous catalysis in the gas state is the destruction of ozone in the stratosphere by CFCs. Ozone molecules are broken down to form oxygen molecules. The reaction is catalysed by chlorine atoms, which form an intermediate, ClO. This then reacts with an oxygen atom to produce oxygen molecules and regenerate the Cl atom:

$$Cl + O_3 \rightarrow O_2 + ClO$$

Then $\quad ClO + O \rightarrow Cl + O_2$

Overall change: $O_3 + O \rightarrow 2O_2$

One chlorine atom can take many molecules through this **catalytic cycle**.

In industry, most processes involve heterogeneous rather than homogeneous catalysis. Homogeneous reactions, however, can be more specific and controllable, so that fewer unwanted products are formed. A good example of a homogeneous catalyst helping to produce a single product is the manufacture of ethanoic acid from methanol. A soluble compound of rhodium is used as the catalyst leading to a 99% conversion of methanol to ethanoic acid.

You will meet many examples of homogeneous **acid** and **base catalysis** in the course – for example, in the hydrolysis of esters and amides.

There is more about homogeneous catalysis in **Section 11.5**.

Problems for 10.6

1 The activation enthalpy for the decomposition of hydrogen peroxide to oxygen and water is $+36.4\,kJ\,mol^{-1}$ in the presence of an enzyme catalyst and $+49.0\,kJ\,mol^{-1}$ in the presence of a very fine colloidal suspension of platinum. The overall reaction is exothermic.

- **a** Sketch, on the same enthalpy diagram, the enthalpy profiles for both catalysts.
- **b** How will the rate of the decomposition of hydrogen peroxide differ for the two catalysts at room temperature? Explain your answer.

2 The breakdown of ozone in the stratosphere is catalysed by chlorine atoms. The overall reaction is

$$O_3 + O \rightarrow 2O_2$$

- **a** The Cl atoms act as a homogeneous catalyst in this process. What is meant by the term *homogeneous*?
- **b** The enthalpy profiles for the catalysed and uncatalysed reactions are shown in Figure 1 in this section (page 234).
  - **i** Explain why the enthalpy profile for the catalysed reaction has two humps.
  - **ii** Describe what is happening at the peaks and troughs on the two curves.

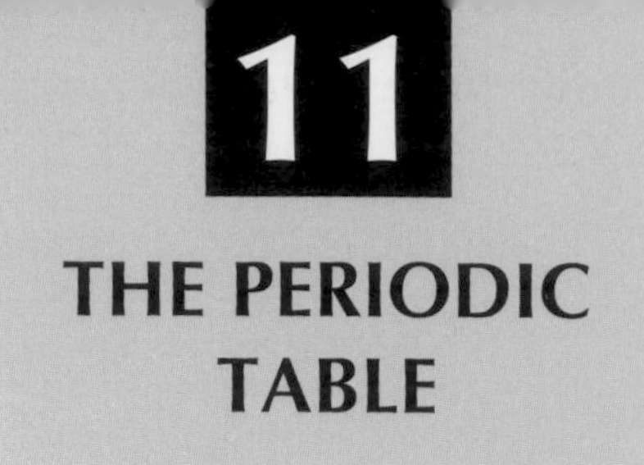

11 THE PERIODIC TABLE

AS

11.1 *Periodicity*

The modern Periodic Table

Figure 1 shows a modern form of the Periodic Table.

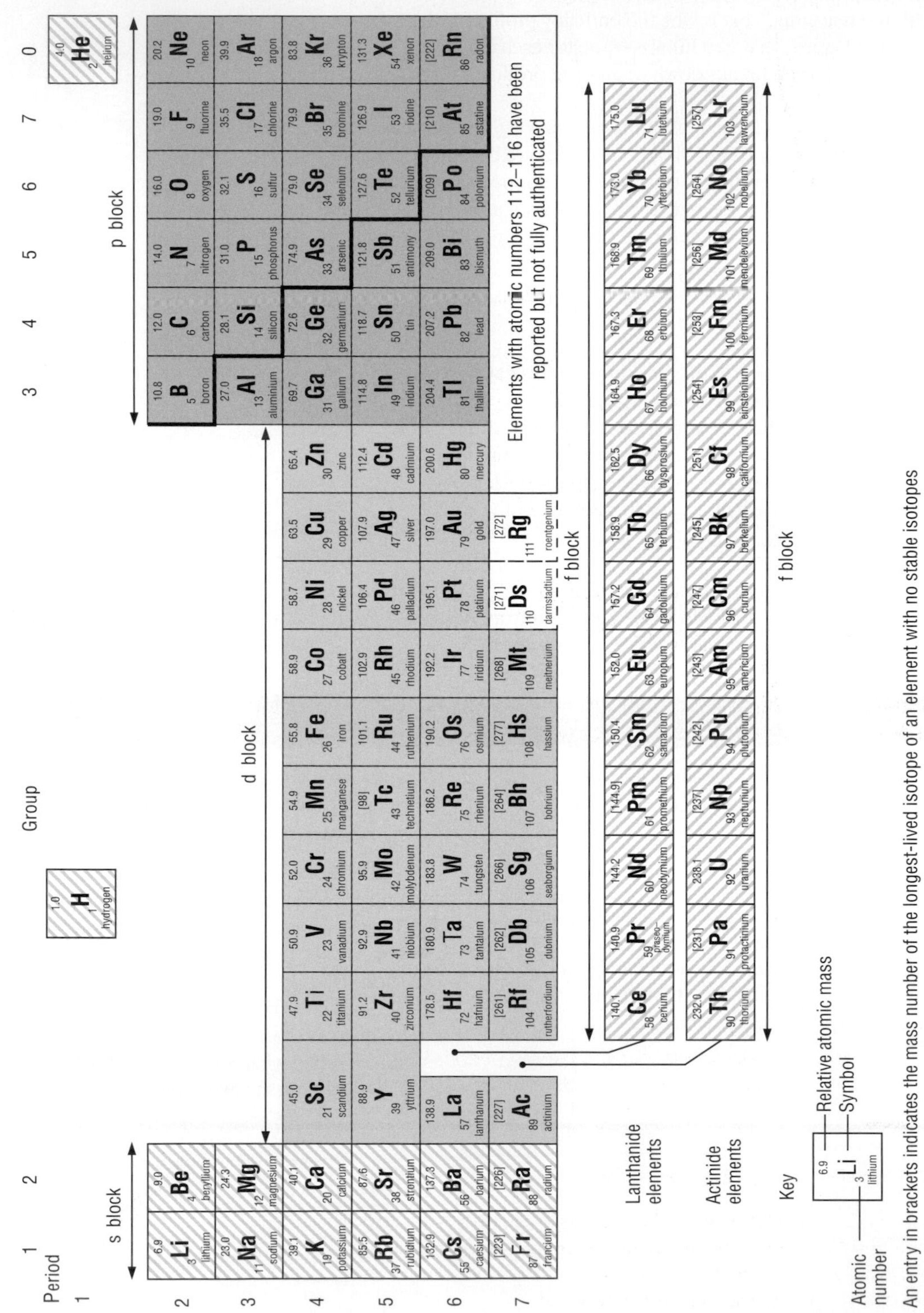

An entry in brackets indicates the mass number of the longest-lived isotope of an element with no stable isotopes

▲ **Figure 1** The modern Periodic Table.

The modern Periodic Table is based on one proposed by the Russian chemist **Mendeleev** in 1869. He organised the 61 elements known at that time into a table made up of horizontal rows and vertical columns. Elements with similar properties were placed underneath each other in the columns. Mendeleev left gaps in his Periodic Table. Not only did he predict that new elements would be discovered to fill the gaps, he also predicted the properties of these yet undiscovered elements.

In Mendeleev's Periodic Table the elements were arranged in order of increasing relative atomic mass. At first glance, the same seems to be true of the modern version – but not quite.

Some pairs of elements are 'out of order' – for example, tellurium (Te; A_r = 128) comes before iodine (I; A_r = 127). It is **atomic number** – the number of protons in the nucleus – which really determines the place of an element in the Periodic Table.

Figure 2 shows how, with the exception of hydrogen, the elements can be organised into four **blocks**, labelled **s**, **p**, **d** and **f**. (You will find out why these letters are used when you study **Section 2.4**.)

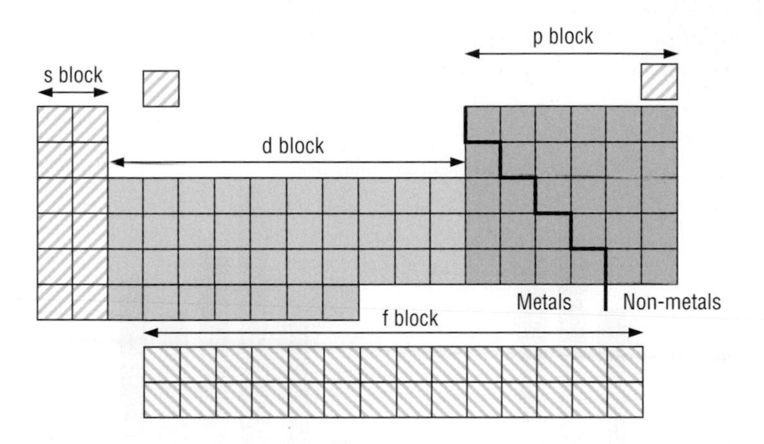

◄ **Figure 2** Blocks in the Periodic Table.

Elements in the same block show general similarities. For example, all the non-metals are in the p block; many of the reactive metals (like sodium, potassium and strontium) are in the s block.

Vertical columns in the Periodic Table are called **groups**. The elements in a group show specific similarities. You may already know some of the common features of the elements of Group 1 (the alkali metals), Group 7 (the halogens) and Group 0 (the noble gases).

Horizontal rows in the Periodic Table are called **periods**. Because they cut across the groups, there are fewer common features among the elements of a period. After hydrogen and helium, there are two *short periods* (Li to Ne and Na to Ar) and four *long periods* (K to Kr etc.). There are 8 elements in a short period, but 18 in the first long period because of the inclusion of 10 d-block elements.

Elements change from metallic to non-metallic across a period. They become increasingly metallic *down* a group – and increasingly non-metallic *up* a group.

Physical properties and the Periodic Table

The arrangement of elements by rows and columns in the Periodic Table is a direct result of the electronic structure of atoms (see **Sections 2.3** and **2.4**). When we arrange the elements in order of atomic number, we see the patterns in physical properties which Mendeleev noted.

There are *trends* in properties as you go down a group, but also many properties vary in a fairly regular way as you move across a period from left to right – the pattern is then repeated as you go across the next period. The occurrence of periodic patterns is called **periodicity**.

Look again at Figure 2. One of the most obvious periodic patterns is the change from metals to non-metals as you go across the periods. The zig-zag line across the p block in Figure 2 marks the change. So if you investigated the **electrical conductivity** of the elements, for example, you would find that the metallic elements on the left of the period are good conductors of electricity – the non-metals on the right of the period are non-conductors.

Let us look at the periodic patterns in some other physical properties.

Density

The **density** of an element is its mass per unit volume. Elements with high densities contain heavy particles which are closely packed together so that they occupy a relatively small volume. The opposite, of course, is true for low density elements. Figure 3 shows the variation in the densities of the elements across Periods 2 and 3. Across Period 2 (Li to Ne) the densities of the elements increase to a maximum at boron (B) in Group 3 and then fall. A similar pattern is observed in Period 3 (Na to Ar), with the maximum at the Group 3 element aluminium (Al).

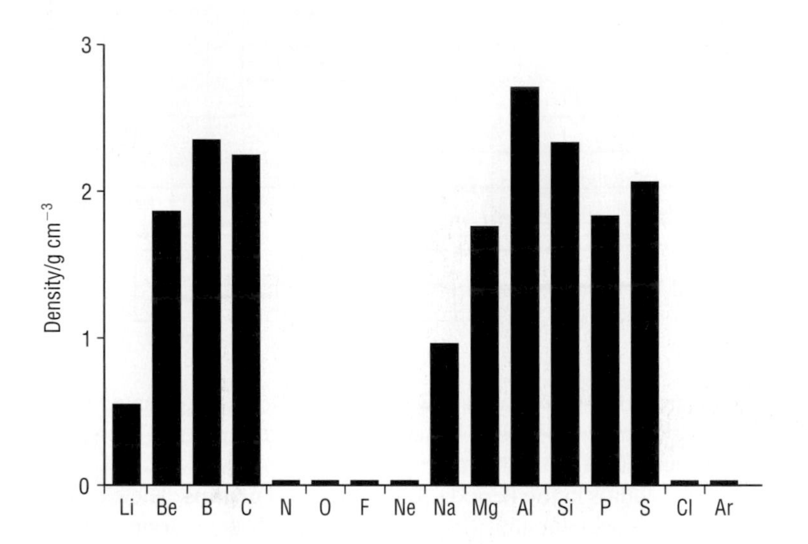

► **Figure 3** Densities of elements in Periods 2 and 3.

Melting points and boiling points

When elements are melted, and then boiled, the bonds between their constituent particles must be overcome. The strength of these bonds influences whether an element has a high or low melting point or boiling point.

Figure 4 shows the variation in melting and boiling points of the elements across Period 3 (Na to Ar). Both the melting and boiling points initially increase as you go across the period, and then fall dramatically from silicon (Si) to phosphorus (P). This means that the bonds between the particles in phosphorus must be very much weaker and easier to overcome than those between the particles in silicon. Similar patterns in melting and boiling points are observed for the elements in other periods.

You will find out more about the types of bonds that are broken on melting and boiling in **Chapter 5**.

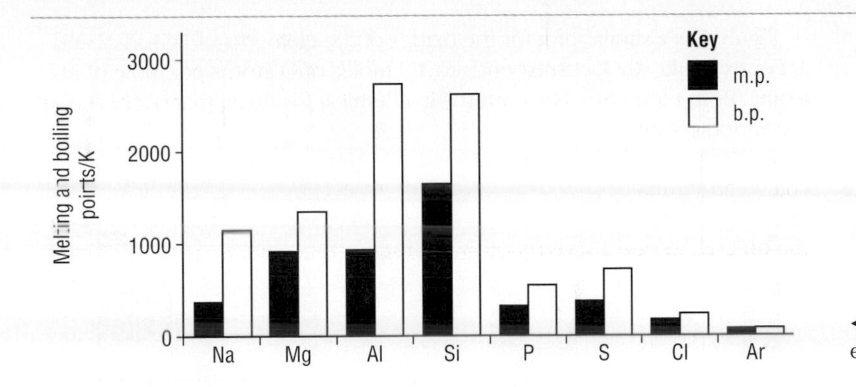

◀ **Figure 4** Melting and boiling points of elements in Period 3.

Atomic size

The size of an atom is determined by the space its electrons occupy. So you might expect that as the number of electrons in an atom increases so will its atomic size. This is certainly true for the atoms in a group, where another electron shell is added as we move from one element to the next below it.

However, you can see from Figure 5 that atoms become smaller on crossing a period, despite the increasing number of electrons. This is because the successive electrons added on crossing a period enter the *same* electron shell. In addition, the number of protons in the nucleus steadily increases across a period and the increasing positive nuclear charge attracts the electrons more and more strongly. As a result, the atoms get smaller even though they have more electrons.

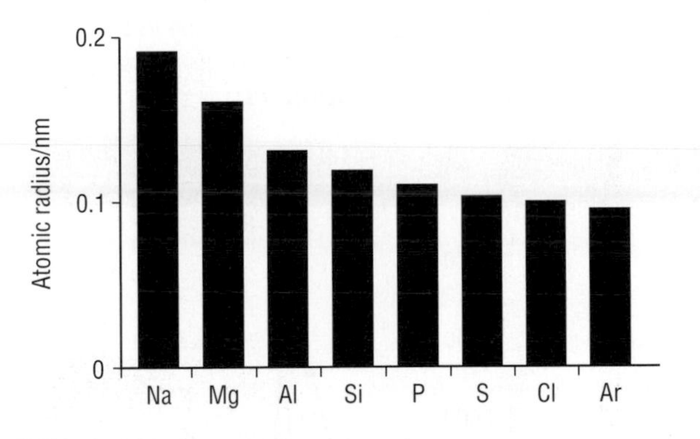

◀ **Figure 5** Variation in atomic radius across Period 3.

Periodicity in chemical formulae

Figure 6 shows the variation in the formulae of the chlorides of the elements of Periods 2 and 3. The vertical axis represents the maximum number of moles of chlorine atoms which combine with 1 mole of atoms of the element: for example, CCl_4 corresponds to 4 moles of Cl atoms per mole of C atoms.

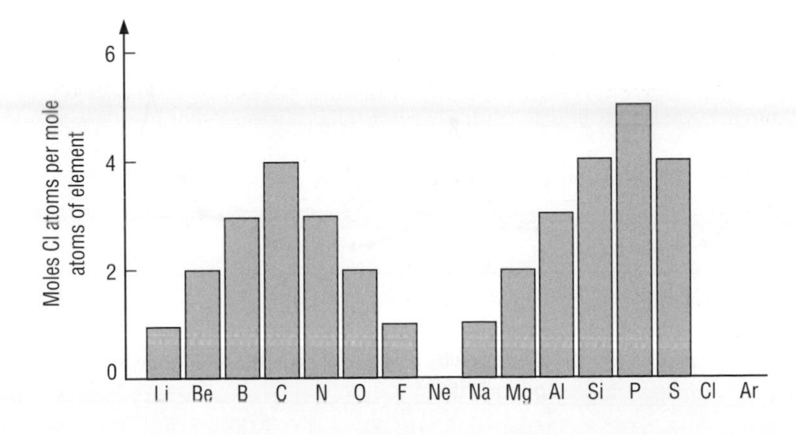

◀ **Figure 6** Periodicity in the formulae of chlorides – where more than one chloride exists, the highest is shown here.

Figure 7 is a similar plot for the oxides of the elements of Periods 2 and 3. For example, Al_2O_3 corresponds to 1.5 moles of O atoms per mole of Al atoms. Both plots show the patterns in chemical formulae that occur across the Periodic Table.

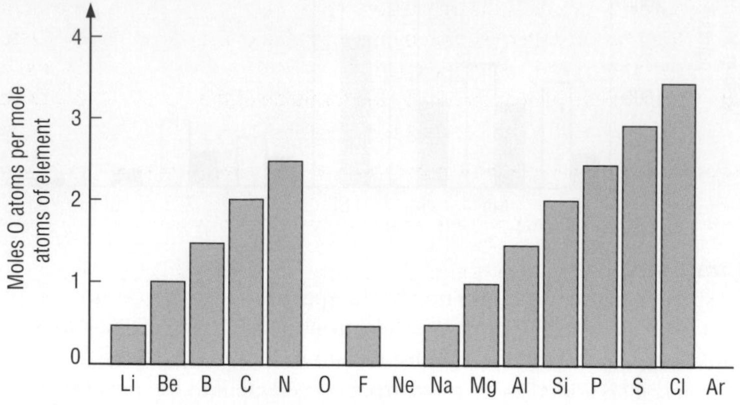

► **Figure 7** Periodicity in the formulae of oxides – where more than one oxide exists, the highest is shown here.

Periodicity in ionisation enthalpies

The amount of energy needed to remove an electron from an atom is known as its **ionisation enthalpy.** Each element has a different *first* ionisation enthalpy that corresponds to the energy needed to remove the *outermost* electron from the atom. These also display periodicity and the property is discussed in detail in **Section 2.5**.

Periodicity in chemical properties

As you would expect, there are also periodic patterns in the chemical properties of the elements and their compounds as you go across a period. You will study these in the rest of **Chapter 11**.

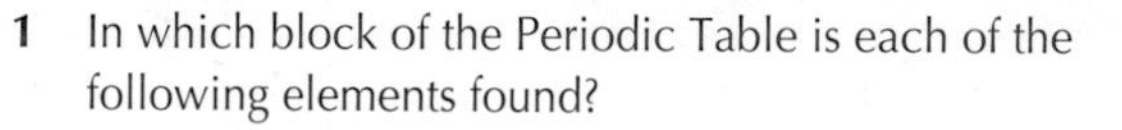

Problems for 11.1

1 In which block of the Periodic Table is each of the following elements found?

a tungsten, W **b** antimony, Sb

c rubidium, Rb **d** holmium, Ho

2 The graph on the right shows the variation in the melting points of the first 20 elements with atomic number.

a Which elements are at the *peaks* of the melting point graph? Which elements are in the *troughs*?

b Explain why this graph provides evidence for periodicity.

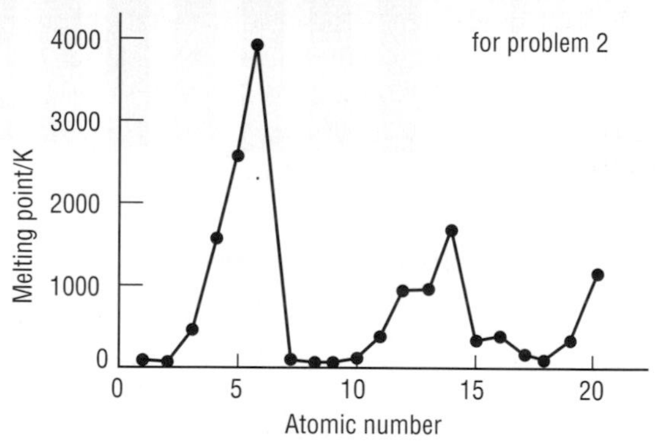

3 Molar atomic volume is the volume occupied by 1 mole of a solid or liquid element. The molar atomic volumes for elements 1–56 are plotted in the graph below.

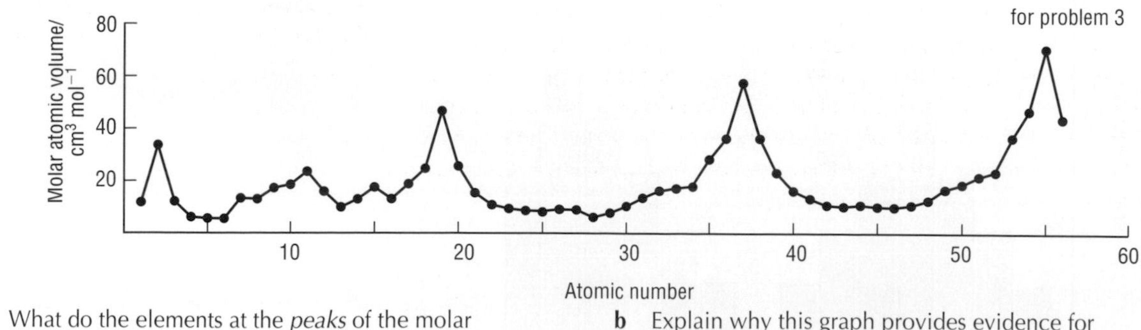

a What do the elements at the *peaks* of the molar atomic volume graph have in common?

b Explain why this graph provides evidence for periodicity.

4 Look at Figure 6 in this section (page 239).
 a Work out the formulae for each of the chlorides of the elements in Figure 6.
 b Make an outline sketch of the Periodic Table for these elements and fill in the formulae from part **a**.
 c Describe the pattern shown by the formulae of the chlorides as you go across a period.
 d Describe the pattern shown by the formulae of the chlorides as you go down a group.

5 Look at Figure 7 in this section (page 240).
 a Work out the formulae for each of the oxides of the elements in Figure 7.
 b Make an outline sketch of the Periodic Table for these elements and fill in the formulae from part **a**.
 c Describe the pattern shown by the formulae of the oxides as you go across a period.
 d Describe the pattern shown by the formulae of the oxides as you go down a group.

11.2 *The s block: Groups 1 and 2*

AS

The s block contains two groups of reactive metals. Group 1 metals (Figure 1) are also called the **alkali metals**. Group 2 metals (Figure 2) are also called the **alkaline earth metals**.

These groups illustrate two trends that also apply to other groups in the Periodic Table:

- *elements become more metallic as you go down a group*. For this reason, the most reactive metals in Groups 1 and 2 are found at the bottom of each group.
- *elements become less metallic as you go across a period from left to right*. For this reason, the Group 1 are more reactive than the Group 2 metals in the same period.

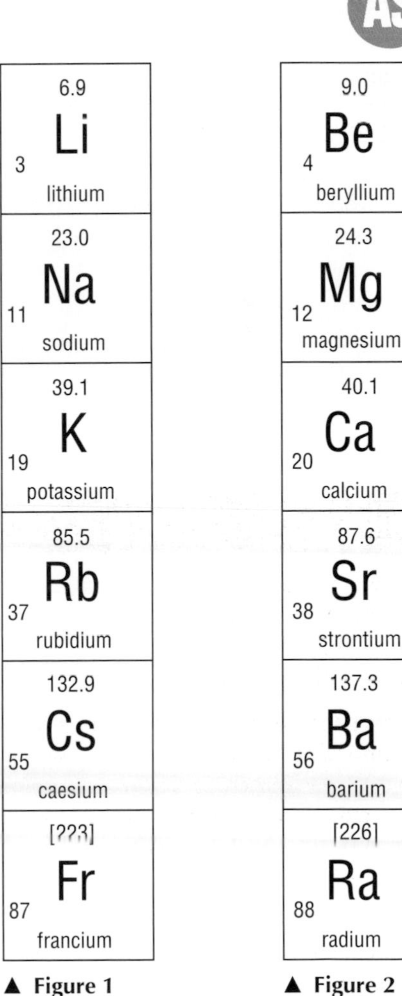

▲ **Figure 1** The elements of Group 1.

▲ **Figure 2** The elements of Group 2.

Physical properties

Although we talk about the s-block elements as being typical metals, this really only applies to their *chemical* properties. These metals are not as widely used (as elements) as the familiar metals of the d block – such as iron, copper and chromium. The s-block metals tend to be soft, weak metals with low melting points, and the metals themselves are too reactive with water and oxygen to have many uses. However, the *compounds* of the s-block elements are very important.

Chemical reactions

As Group 1 and 2 metals are all very reactive, they are never found in nature in their native, uncombined state. However, compounds of s-block metals are very common throughout nature. Indeed, much of the ground beneath your feet is made from compounds of s-block metals such as magnesium and calcium.

Like all groups in the Periodic Table, Groups 1 and 2 show patterns of reactivity as you go down the group. There are *similarities* between the reactions of the elements within a group, but also *differences* that show up as patterns, or trends (see **Section 11.1**). The *similarities* happen because the elements in a particular group all have similar arrangements of electrons in their atoms (see **Section 2.3**). The *differences* happen because as you go down the groups the size of the atoms increases. We can illustrate these similarities and trends by looking at Group 2 in a little detail.

Some chemical properties of Group 2 elements and their compounds

Magnesium, calcium, strontium and barium are the most well known elements in Group 2 and you may already have met them in your experimental work.

The elements of Group 2 are all reactive. They form compounds containing ions with a 2+ charge, such as Mg^{2+} and Ca^{2+}.

Reactions of the elements with water

All the elements react with water to form hydroxides and hydrogen, with a steady increase in the vigour of the reaction as you move down the group.

Magnesium reacts only slowly, even when the water is heated. Barium reacts rapidly, giving a steady stream of hydrogen. (Even so, none of the elements reacts as vigorously with water as the Group 1 elements such as sodium and potassium do.)

The general equation, using M to represent a typical Group 2 metal reacting with water, is

$$M(s) + 2H_2O(l) \rightarrow M(OH)_2(aq) + H_2(g)$$

Oxides and hydroxides

The general formula of the oxides is MO, and that of the hydroxides is $M(OH)_2$.

In water, the oxides and hydroxides form alkaline solutions, although they are not very soluble. Forming alkaline solutions is typical of metal oxides and hydroxides, in contrast to non-metals whose oxides are usually acidic. The most strongly alkaline oxides and hydroxides are those at the bottom of the group.

As you would expect, the oxides and hydroxides react with acids to form salts:

$$MO(s) + 2HCl(aq) \rightarrow MCl_2(aq) + H_2O(l)$$
$$M(OH)_2(s) + H_2SO_4(aq) \rightarrow MSO_4(aq) + 2H_2O(l)$$

This neutralising effect is used by farmers when they put lime (calcium hydroxide) on their fields to neutralise soil acidity.

Effect of heating carbonates

The general formula of Group 2 carbonates is MCO_3. When you heat the carbonates, they decompose forming the oxide and releasing carbon dioxide:

$$MCO_3(s) \rightarrow MO(s) + CO_2(g)$$

The carbonates become more difficult to decompose as you go down the group. For example, magnesium carbonate is quite easily decomposed by heating in a test tube over a Bunsen burner flame, but calcium carbonate needs much stronger heating directly in the flame before it will decompose. We say that the **thermal stability** of calcium carbonate is greater than that of magnesium carbonate.

The decomposition of calcium carbonate (limestone) is an important process, used to manufacture calcium oxide (quicklime).

Solubilities of compounds

The solubilities of compounds of Group 2 elements in water show clear trends as you go down the group. The solubilities of the hydroxides and carbonates are summarised in Table 1.

	Solubility (g per 100 g water)	
Hydroxides		
$Mg(OH)_2$	0.0012	increasing solubility ↓
$Ca(OH)_2$	0.12	
$Sr(OH)_2$	1.0	
$Ba(OH)_2$	3.7	
Carbonates		
$MgCO_3$	0.06	decreasing solubility ↓
$CaCO_3$	0.0013	
$SrCO_3$	0.001	
$BaCO_3$	0.002	

◀ **Table 1** Solubilities of Group 2 hydroxides and carbonates.

Solubility of hydroxides

The pattern is for the hydroxides to become *more* soluble as you go down the group – as Table 1 shows, $Mg(OH)_2$ is much less soluble than $Ba(OH)_2$. This pattern is repeated for most Group 2 compounds where the negative ion has a single charge (1−).

Solubility of carbonates

The pattern is for the carbonates to become *less* soluble as you go down the group – as Table 1 shows, $MgCO_3$ is more soluble than $CaCO_3$. This general pattern is repeated for most Group 2 compounds where the negative ion has a double charge (2−).

Notice that the figures in Table 1 do not show a perfect trend – for example the pattern in solubilities of carbonates is uneven between Ca and Ba. This usually happens with patterns in the Periodic Table – they are rarely perfect, because there are so many factors operating to decide the properties of elements and their compounds.

When we describe a pattern such as 'carbonates become less soluble as you go down a group' we are stating a rule that works generally and is useful for predicting properties. But, as always in science, predictions can only be confirmed by doing experiments, or looking up the results of experiments done by others.

Summary

The trends in properties of some compounds of Group 2 elements are summarised below.

Element	Trend in reactivity with water	Trend in thermal stability of carbonate	Trend in pH of hydroxide in water	Trend in solubility of hydroxide	Trend in solubility of carbonate
Mg	increasing reactivity ↓	decomposes at increasingly higher temperature ↓	increasing pH ↓	increasing solubility ↓	decreasing solubility ↓
Ca					
Sr					
Ba					

Problems for 11.2

1 Predict the trend in solubility of the sulfates as you go down Group 2.

2 Write a word equation and a balanced chemical equation (with state symbols) for each of the following reactions
  a the action of heated magnesium on steam
  b the neutralisation of hydrochloric acid with calcium oxide
  c the thermal decomposition of beryllium carbonate
  d the action of sulfuric acid on barium hydroxide.

3 On the basis of what you know about the elements in Group 2, predict the following concerning the elements in Group 1
  a the element that will react most vigorously with water
  b the element with the most thermally stable carbonate
  c the charge on a Group 1 metal ion
  d the general formulae of the following compounds (use 'M' to represent the element)
    i Group 1 oxides
    ii Group 1 hydroxides
    iii Group 1 carbonates.

4 Write a word equation and a balanced chemical equation (with state symbols) for each of the following reactions (if any takes place)
  a the action of lithium on water
  b the action of sodium hydroxide on hydrochloric acid
  c the thermal decomposition of potassium carbonate
  d the action of sodium oxide on sulfuric acid.

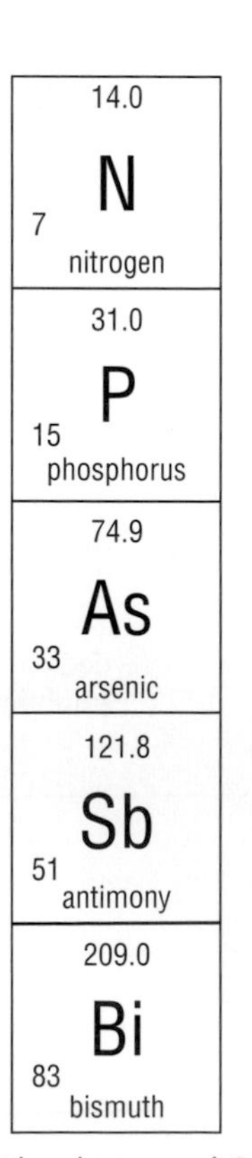

▲ **Figure 1** The elements of Group 5.

11.3 *The p block: nitrogen and Group 5*

Group 5 of the Periodic Table (Figure 1) is a group in the centre of the p block. At the top are non-metals – nitrogen and phosphorus. At the bottom are metalloids (metallic elements with some non-metal character) – antimony (Sb) and bismuth (Bi).

The electronic configurations of the Group 5 elements are shown below:

nitrogen [He] $2s^2 2p^3$
phosphorus [Ne] $3s^2 3p^3$
arsenic [Ar] $3d^{10} 4s^2 4p^3$
antimony [Kr] $4d^{10} 5s^2 5p^3$
bismuth [Xe] $4f^{14} 5d^{10} 6s^2 6p^3$

Atoms of these elements can form three covalent bonds by sharing the three unpaired electrons. This gives compounds in which the oxidation state of the Group 5 element is +3 or −3.

Each atom also has a lone pair of electrons. These enable the atoms to form dative covalent bonds. When they do this, the Group 5 elements can form some compounds in which their oxidation state is +5, for example the nitrate(V) ion, NO_3^-. You will see later how nitrogen does this.

Nitrogen and phosphorus are very important members of Group 5. Both are constituent elements in living things, and both are essential for healthy plant growth. Although nitrogen gas is all around us in the air, we have problems getting it into a reactive form that can be used in agriculture.

Nitrogen

Nitrogen gas may be abundant, but it is remarkably unreactive. Thousands of litres of it pass unreacted through your lungs every day. It even passes through the hot cylinders of a motor car mostly unreacted.

The low reactivity of the N_2 molecule arises from the strong triple bond holding the atoms together (Figure 2).

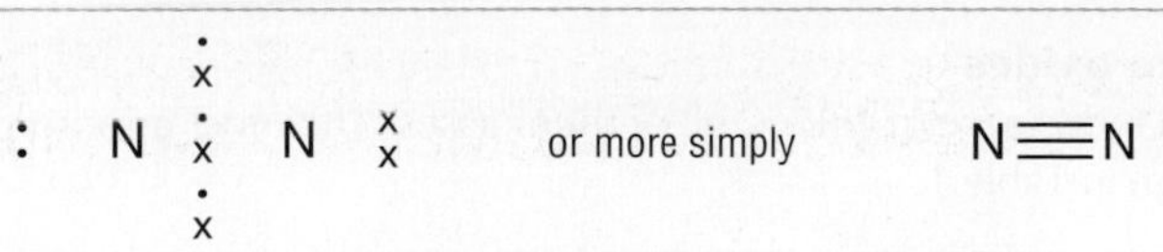

: N $\overset{\times}{\underset{\times}{\cdot}}$ N ${}^{\times}_{\times}$ or more simply N≡N

Bond enthalpy of N≡N bond is $+945\ kJ\ mol^{-1}$
(Bond enthalpy of N—N bond is $+158\ kJ\ mol^{-1}$)

◀ **Figure 2** The bonding in the N_2 molecule.

Before nitrogen can react, the triple bond between the atoms must be broken, or partly broken. The bond enthalpy of N≡N is very large, at $+945\ kJ\ mol^{-1}$, and most reactions of molecular N_2 have high activation enthalpies and require high temperatures and catalysts to make them occur. For example, in the Haber process for making ammonia, a high temperature is needed for N_2 to react with H_2. In nature, it takes a highly energetic lightning flash in a thunderstorm, or temperatures of about 2000 °C, to provide enough energy to make nitrogen react with oxygen to form nitrogen(II) oxide (commonly known as nitrogen monoxide).

Once nitrogen has reacted, though, it can form many compounds. The most important of these are *ammonia, nitrogen oxides* and *nitrates*. All are involved in *the nitrogen cycle*.

Ammonia

Ammonia is nitrogen hydride. The bonding of the ammonia molecule is shown in Figure 3. Notice that the lone pair of electrons on the N atom is not involved in the bonding, so it is available to form dative covalent bonds.

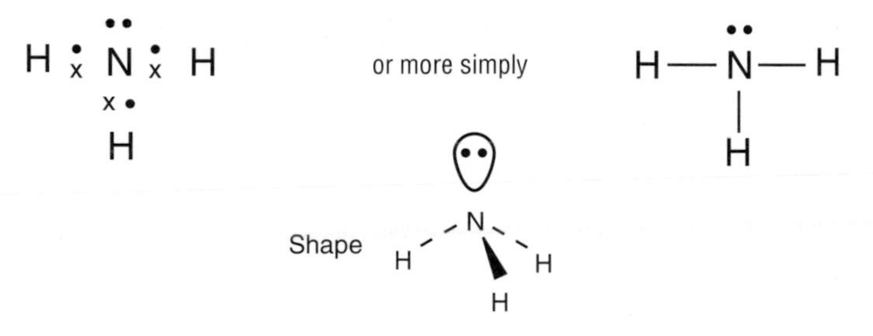

◀ **Figure 3** The bonding in the NH_3 molecule.

Ammonia readily acts as a base and forms dative covalent bonds to H^+ ions to give the ammonium ion:

$$NH_3(g) + H^+(aq) \rightleftharpoons NH_4^+(aq)$$

Figure 4 shows the bonding in the ammonium ion.

Ammonia also forms dative bonds to transition metal ions, which makes it a good ligand in complexes. There is more about ligands in **Section 11.6**.

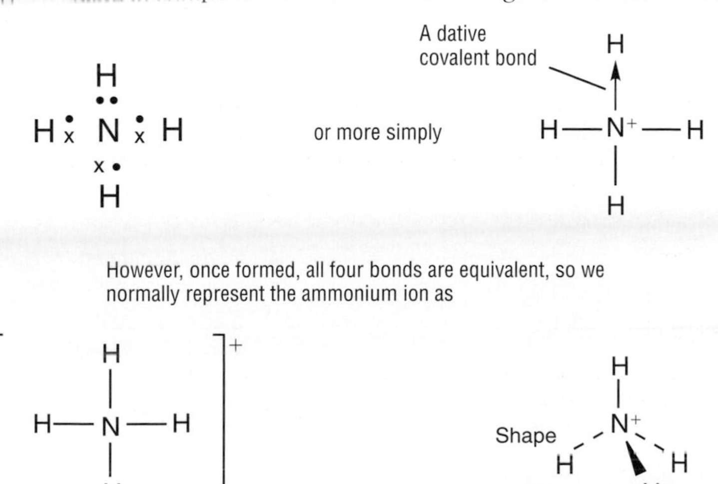

◀ **Figure 4** The bonding in the NH_4^+ ion.

Nitrogen oxides

Nitrogen forms several oxides, all of them gases. The most important ones are shown in Table 1.

► **Table 1** Oxides of nitrogen.

Name and formula	Appearance	Where it comes from
nitrogen(II) oxide, NO	colourless gas, turns to brown NO_2 in air	combustion processes, especially vehicle engines; thunderstorms; formed in the soil by denitrifying bacteria
nitrogen(IV) oxide, NO_2	brown gas (toxic)	from oxidation of NO in atmosphere
dinitrogen(I) oxide, N_2O	colourless gas	formed in the soil by denitrifying bacteria

Nitrates

Two kinds of nitrate ions are involved in the nitrogen cycle – nitrate(III), NO_2^- and nitrate(V), NO_3^-. Notice that they are both named 'nitrates' but they are distinguished from one another by showing the oxidation state of the nitrogen. Figure 5 shows the bonding in nitrate(III) and nitrate(V) ions. (See **Section 9.1** for tips on how to work out oxidation states.)

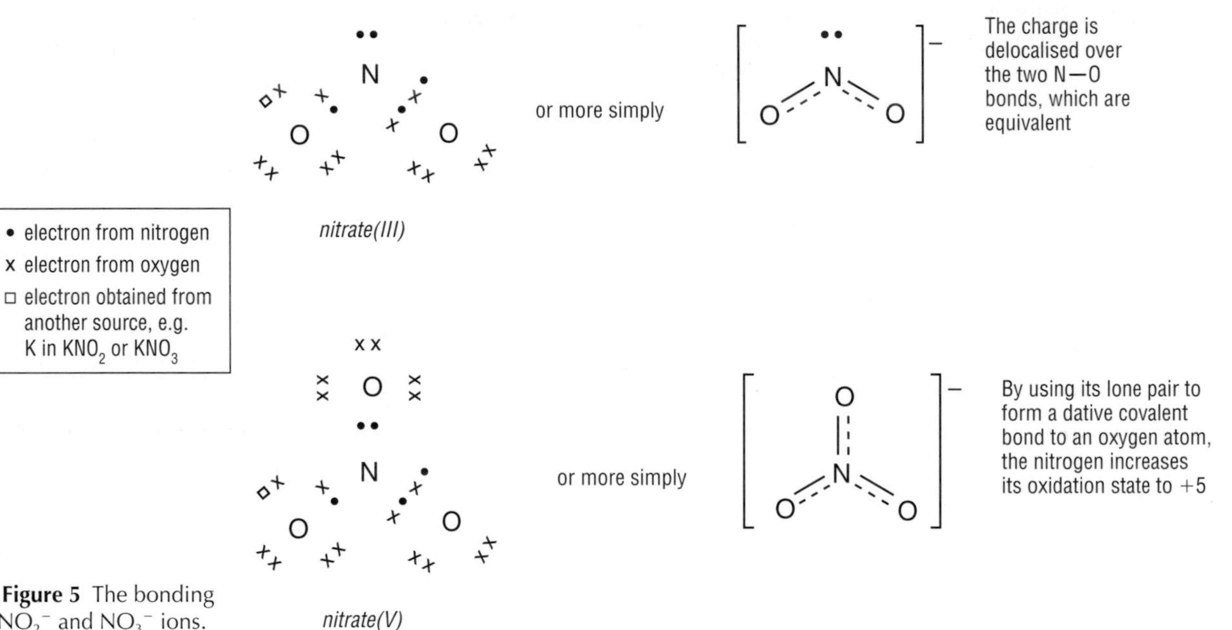

▲ **Figure 5** The bonding in NO_2^- and NO_3^- ions.

► **Table 2** Nitrates.

Name and formula	Commonly called	Properties
nitrate(V) ion, NO_3^-	nitrate	oxidising agent; metal nitrate(V) compounds are very soluble in water
nitrate(III) ion, NO_2^-	nitrite	can be reducing or oxidising agent; metal nitrate(III) compounds are soluble in water

Problems for 11.3

1 Nitrogen, N_2, at the top of Group 5, is very unreactive. The next member of Group 5, phosphorus, is highly reactive – white phosphorus, P_4, catches fire spontaneously in air. The shape of the P_4 molecule is shown below (see also Figure 3a in **Section 5.2**).

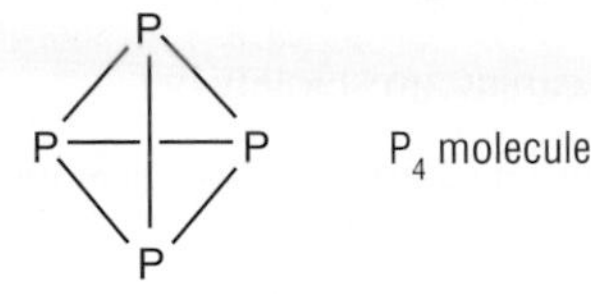

Suggest an explanation for the difference in reactivity between these two close neighbours in Group 5.

2 Look at the following pairs of nitrogen compounds – the conversions from the first compound to the second are part of the nitrogen cycle. (See the **Agriculture and Industry** module for further details.)

a $NO_2^- \rightarrow NO_3^-$ nitrification
b $NO_3^- \rightarrow N_2$ denitrification
c $NH_4^+ \rightarrow NH_3$ putrefaction
d $N_2 \rightarrow NH_3$ ammonia synthesis
e $NO \rightarrow NO_2$ in the atmosphere

For each conversion:

i give the oxidation state of N in each of the two compounds
ii say whether the conversion of the first compound to the second involves:
A oxidation
B reduction
C neither oxidation nor reduction.

3 One of the stages in the denitrification process brought about by bacteria in the soil involves the conversion of NO into N_2O.

a What is the change in the oxidation state of nitrogen in this conversion?
b Write a balanced half-equation for the conversion

$$NO \rightarrow N_2O$$

You will need to add electrons, H_2O molecules and H^+ ions to balance the equation.

4 Write a balanced half-equation for each of the conversions **a–e** in problem 2.

5 Nitric(V) acid is manufactured by the catalytic oxidation of ammonia. This produces NO, which is further oxidised and then dissolved in water to produce HNO_3:

$$4NH_3(g) + 5O_2(g) \rightarrow 4NO(g) + 6H_2O(g) \quad \textit{Reaction 1}$$
$$2NO(g) + O_2(g) \rightarrow 2NO_2(g) \quad \textit{Reaction 2}$$
$$3NO_2(g) + H_2O(l) \rightarrow 2HNO_3(aq) + NO(g) \quad \textit{Reaction 3}$$

a The final stage of this process, reaction 3, produces NO as well as HNO_3. How would the NO be dealt with?
b Name one compound, useful in agriculture, that is manufactured using nitric acid.
c **i** Starting with 1000 kg of ammonia, what is the maximum mass of HNO_3 that could be produced?
ii Give two reasons why the mass of nitric(V) acid actually produced will be less than your answer in part **i**.
d What particular environmental protection measures might be needed in a chemical plant using this manufacturing process?

11.4 *The p block: Group 7*

AS

What are the halogens?

The *halogens* are the elements in Group 7 of the Periodic Table (Figure 1, page 248). All halogen atoms have seven electrons in the outer shell. The halogens are the most reactive group of non-metals, and none of them is found naturally in the element form. They are all found in compounds, often as *halide* ions (a singly negatively charged ion, e.g. Br^-). Calcium fluoride and sodium chloride are naturally occurring halides. Iodine is also found as sodium iodate, where it is in the form of iodate(V) ions, IO_3^-.

Fluorine and chlorine are the most abundant halogens, bromine occurs in smaller quantities, iodine is quite scarce and astatine is an artificially produced, short-lived, radioactive element.

All the halogen elements occur as *diatomic molecules* – for example F_2 and Br_2. The two atoms are linked by a single covalent bond (Figure 2, page 248).

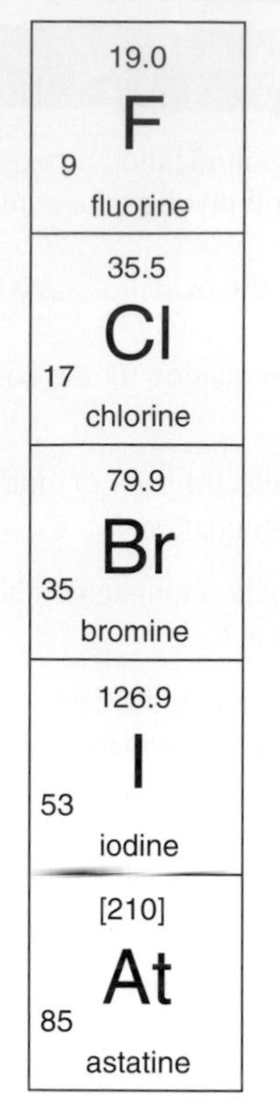

▲ **Figure 1** The elements of Group 7.

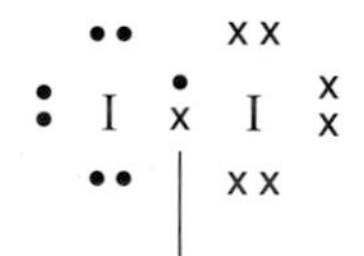

▲ **Figure 2** Covalent bonding in the I_2 molecule.

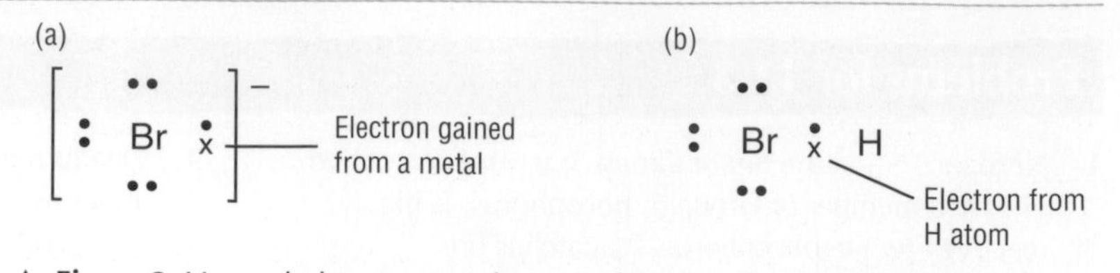

▲ **Figure 3** How a halogen can achieve eight outer electrons during compound formation. **a** Forming an ion (Br^-) via ionic bonding. **b** Forming a molecule (e.g. HBr) via covalent bonding.

In compounds, a halogen atom can attain stability by:

- gaining an electron from a metal to form a halide ion in an ionically bonded compound (Figure 3a)
- sharing an electron from another atom in a covalently bonded compound (Figure 3b).

In both cases, the halogen has an oxidation state of −1 in the compound.

It is also possible for halogens (other than fluorine) to expand their outer shell of electrons so that it holds more than eight electrons. This makes it possible for the halogens to reach higher oxidation states, such as +5 or +7. For example, in the chlorate(V) ion, chlorine has expanded its outer shell to hold 12 electrons (Figure 4).

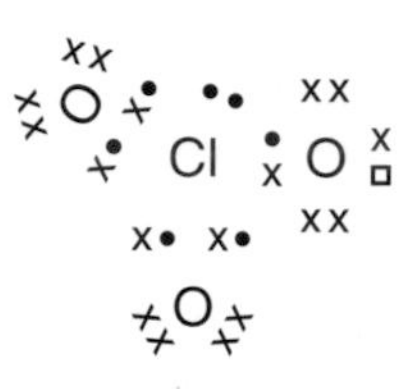

◀ **Figure 4** In the ClO_3^- ion, chlorine has expanded its outer shell to hold 12 electrons.

Physical properties of the halogens

Some properties of the halogens are shown in Table 1.

The halogens are more soluble in organic solvents such as hexane than they are in water. The appearances of their solutions are shown in Table 2.

The physical state of the halogens at room temperature changes from gas, to liquid, to solid as the group is descended (as shown in Table 1). This change is caused by an increase in the strength of the intermolecular bonds between halogen molecules. All the halogens form diatomic, non-polar molecules held together by a single covalent bond. The strongest type of intermolecular bond that can form between two halogen molecules is

▼ **Table 1** Some physical properties of the halogens.

	Fluorine	Chlorine		Bromine		Iodine
Isotopes and abundances	^{19}F 100%	^{35}Cl 75%	^{37}Cl 25%	^{79}Br 50%	^{81}Br 50%	^{127}I 100%
Appearance at room temperature	pale yellow gas	green gas		dark red volatile liquid		shiny black solid; sublimes to give purple vapour on warming
Melting point/K	53	172		266		387
Boiling point/K	85	239		332		457
Solubility at 298 K/g per 100 g of water	reacts with water	0.6		3.5		0.03

► **Table 2** Colours of halogens in solution.

Halogen	In water	In hexane
chlorine	pale green	pale green
bromine	red-brown	red
iodine	brown	violet

instantaneous dipole–induced dipole forces of attraction (see **Section 5.3**). Fluorine is the most volatile halogen – this is because it has the smallest molecules, with the least number of electrons, and so forms the weakest intermolecular bonds. The molecules get bigger as the group is descended. With more electrons in each molecule, the strength of the intermolecular bonds increases, accounting for the change in physical state of the halogens.

Chemical properties of the halogens

The halogens are a group of reactive elements. They tend to remove electrons from other elements – they are oxidising agents. The elements at the top of the group are the most reactive and are the strongest oxidising agents.

To explain this, we can compare fluorine with chlorine. Both of these halogens have atoms that consist of a core with a charge of +7, surrounded by an outer electron shell containing 7 electrons (see **Section 2.3**). Both react by gaining one electron to complete their outer shell (see **Section 3.1**). Fluorine atoms are very small, so the attraction from the core for the extra electron to complete its outer shell is very strong. In chlorine, the outer shell that requires the extra electron is further from the core and the attraction for the extra electron is weaker. This means that fluorine gains an extra electron more readily to become a negative ion than chlorine does. This trend continues down the halogen group, as each successive member of the group has one more complete electron shell than the previous one. Overall, this means that fluorine is the most reactive member of the halogen group, and reactivity decreases as the group is descended.

Reactions with metals

With many metals (such as those in the s block) the halogens react to form compounds containing halide ions – for example potassium bromide, KBr, and calcium chloride, $CaCl_2$. In all these compounds the halogen is in oxidation state −1. See **Section 9.1** for information on oxidation states.

Reactions with non-metals

With non-metals, and some p-block and transition metals, halogens form molecular compounds containing covalent bonds – for example carbon tetrachloride (tetrachloromethane) CCl_4, tin tetrachloride (tin(IV) chloride) $SnCl_4$, phosphorus trichloride (phosphorus(III) chloride) PCl_3. In most of these compounds the halogen is in oxidation state −1.

Halogens also react with one another to form interhalogen covalent compounds – for example ICl and BrF_3. The less reactive of the two halogens is in a positive oxidation state in these compounds.

Reactions with halide ions

If you add a solution containing chlorine to a solution of iodide ions there is a chemical reaction. The solution (which would be almost colourless if nothing happened) turns brown and iodine is produced.

$$\underset{\text{pale green}}{Cl_2(aq)} + 2K^+I^-(aq) \rightarrow 2K^+Cl^-(aq) + \underset{\text{brown}}{I_2(aq)}$$

We can simplify this equation – the K^+ ions can be left out because they are unchanged (spectator ions):

$$Cl_2(aq) + 2I^-(aq) \rightarrow 2Cl^-(aq) + I_2(aq)$$

A similar thing happens if bromine solution is added to iodide ions:

$$\underset{\text{red-brown}}{Br_2(aq)} + 2I^-(aq) \rightarrow 2Br^-(aq) + \underset{\text{brown}}{I_2(aq)}$$

These are examples of displacement reactions (see **Section 9.1**).

Notice that we are using **ionic equations** here. Full chemical formulae have not been used for some of the reagents, e.g. I^-. That's because the equations represent *general* reactions – chlorine will react in the same way with any aqueous solution of iodide ions, whether they come from NaI, KI, CaI_2 and so on. So, although the ionic equation is shorter than the full chemical equation it tells us more – it tells us, for example, how chlorine reacts with a whole range of iodides.

These are **redox reactions**. Bromine, for example, oxidises I^- (oxidation state −1) to iodine (oxidation state 0). In this process, Br_2 (oxidation state 0) is reduced to Br^- (oxidation state −1). In general, a more reactive halogen will oxidise the halide ions of a less reactive one.

The reactions are not reversible – iodine will not liberate bromine from potassium bromide solution because iodine is less reactive than bromine.

In all these displacement reactions, one of the halogens is acting as the oxidising agent and the other as the reducing agent. This is more easily seen if we write a reaction as two half-equations. For example, with bromine and iodide ions the half-equations are

$$Br_2 + 2e^- \rightarrow 2Br^- \quad \text{and} \quad 2I^- \rightarrow I_2 + 2e^-$$

The bromine molecule gains electrons from the iodide ions – in doing this, it is acting as an oxidising agent. The iodide ions lose electrons to the bromine molecule – acting as a reducing agent in the process. Fluorine is the most powerful oxidising agent in the halogen group, with oxidising ability decreasing as the group is descended.

You can read about the redox changes that take place when aqueous halide solutions are electrolysed in the **Elements from the Sea** module.

Reactions of halide ions

Oxidation to form halogens

Halide ions are oxidised to halogens, provided a strong enough oxidising agent is available. Using X^- to represent a general halide:

$$2X^- \rightarrow X_2 + 2e^- \text{ (electrons removed by an oxidising agent)}$$

The oxidising agent might be another halogen (see above) or a strong oxidising agent such as potassium manganate(VII) or concentrated sulfuric acid. For example, potassium manganate(VII) oxidises chloride ions to chlorine. This is a useful way to prepare chlorine in the laboratory.

Reactions with silver ions

Silver halides are precipitated when a solution of silver ions is added to a solution containing Cl^-, Br^- or I^- ions. The general equation is

$$Ag^+(aq) + X^-(aq) \rightarrow AgX(s)$$

These are examples of **precipitation reactions** (see **Section 5.1**). The appearances of the silver halides are

silver chloride	white
silver bromide	cream
silver iodide	yellow

Silver bromide AgBr is decomposed by light to produce silver and bromine. The decomposition is the basis of most traditional photographic processes, although this has been somewhat superseded by the use of digital cameras.

Silver bromide dissolves in sodium thiosulfate solution. The reaction is used to clear unreacted silver bromide crystals from films in the fixing process.

Silver halides still continue to be used in photochromic lenses.

Problems for 11.4

1 a Predict the colour changes (if any) that you would observe when the following reactants are mixed. Where you predict that a displacement reaction will occur, write a balanced ionic equation with state symbols.
  i Chlorine water is mixed with aqueous sodium iodide.
  ii Bromine water is mixed with aqueous potassium fluoride.
  iii Bromine water is mixed with aqueous potassium iodide.
  iv Chlorine water is mixed with aqueous sodium bromide.

 b Write balanced ionic equations, with state symbols, for the following precipitation reactions when aqueous silver nitrate is added to
  i aqueous potassium chloride
  ii aqueous sodium bromide
  iii aqueous sodium iodide.

 c Write balanced equations, with state symbols, for the reaction between the following halogens and s-block metals.
  i Sodium and bromine
  ii Magnesium and chlorine
  iii Potassium and iodine
  iv Calcium and chlorine.

2 Give the oxidation state of the halogen atoms in the following:
 a ICl
 b BrF_3
 c BrO^-
 d IO_3^-
 e IO_4^-

3 Chlorine is used to manufacture a wide range of chemicals. One of the most familiar of these is the bleach used as a powerful domestic disinfectant. Household bleach is an aqueous solution of sodium chloride and sodium chlorate(I), NaClO, in a 1 : 1 mole ratio. It is produced by dissolving chlorine gas in cold dilute aqueous sodium hydroxide.

 a Write a balanced equation, with state symbols, for the reaction of chlorine with sodium hydroxide to form bleach.
 b i in the reaction in part **a**, which element(s) change(s) oxidation states?
  ii Write down these oxidation state changes, and identify which involves an oxidation and which a reduction.
 c The active component of bleach is sodium chlorate(I). This unstable compound cannot be isolated as a solid. Bleach left in sunlight slowly decomposes releasing oxygen, O_2.
  i Write an equation for this decomposition, including state symbols.
  ii Which element(s) change(s) oxidation states?
  iii Write down these oxidation state changes, and identify which involves an oxidation and which a reduction.

4 a Make a copy of Table 1 in this section (page 248) adding a column for astatine, At. Astatine only exists as artificial radioactive isotopes so you will not be able to record isotopic abundances. Astatine-210 is one of 23 known isotopes.
 From the trends in physical properties down Group 7, predict approximate values for the solubility of astatine in water, and its melting and boiling points.
 b Use your knowledge of the colours and the reactions of the halogens to predict the changes (if any) that you might *observe* when the following reagents are mixed. Where a reaction is predicted, write a balanced ionic equation with state symbols.
  i Aqueous bromine water is added to aqueous sodium astatide, NaAt(aq).
  ii Aqueous hydroastatic acid, HAt(aq), is added to aqueous iodine.
  iii Aqueous silver nitrate is added to aqueous sodium astatide.
  iv Hot astatine vapour, $At_2(g)$, is passed over heated metallic sodium.

11.5 *The d block: characteristics of transition metals*

The d block consists of three horizontal series in Periods 4, 5, and 6 – each series contains ten elements (see Figure 2 in **Section 11.1**, page 237). Their chemistry is different from those elements in other parts of the Periodic Table. It results from the special electronic configurations of d-block elements and the energy levels associated with their electrons. Differences between elements within a group in the d block are less sharp than those in the s block and p block, and similarities across the period are greater, so we can discuss the d block as a collection of elements with many features in common.

Electronic configuration

Across the first row of the d block (the ten elements from Sc to Zn) each element has one more proton in the nucleus and one more electron than the previous element. Each 'additional' electron enters the 3d shell. Remember this is *not* the outermost shell, because the outer 4s orbital has already been filled (see **Section 2.4**).

The electronic configurations of the atoms of the first row of the d block are shown in Figure 1. Only the two outer sub-shells are shown because the elements all have an identical core of electrons. The core is the electronic configuration of the noble gas argon, Ar, $1s^2 2s^2 2p^6 3s^2 3p^6$.

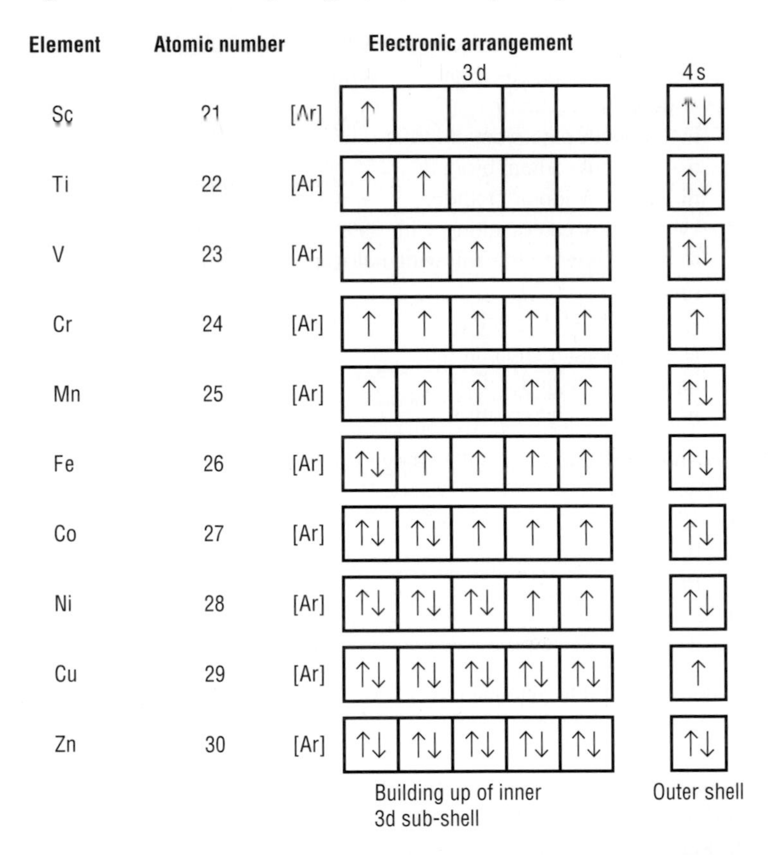

► **Figure 1** Arrangement of electrons in the ground state of elements of the first row of the d block. [Ar] represents the electronic configuration of argon.

These elements have essentially the same *outer* electronic arrangement as each other, in the same way as the elements in a vertical group all have the same outer shell structure. Moreover, unlike the elements in a group, they do not differ by a complete electron shell but by having one more electron in the *inner, incomplete* 3d sub-shell.

Why are chromium and copper different?

Look carefully at Figure 1. The electronic configurations of Cr and Cu don't fit the pattern of building up the 3d sub-shell.

In the ground state of an atom, electrons are always arranged to give the lowest total energy. Because of their negative charge, electrons repel one another so a lower total energy is obtained with electrons singly in orbitals than if they are paired in an orbital.

The energies of the 3d and 4s orbitals are very close together in Period 4. At chromium, the orbital energies are such that putting one electron into each 3d and 4s orbital gives a lower energy than having two in the 4s orbital (Figure 2). At copper, putting two electrons into the 4s orbital would give a higher energy than filling the 3d orbitals.

You might expect Cr to have the electronic arrangement

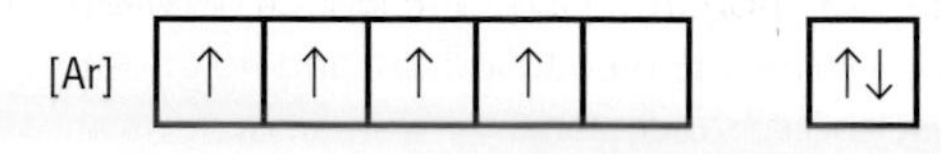

(continuing the pattern of the previous elements). However, its electronic arrangement is in fact

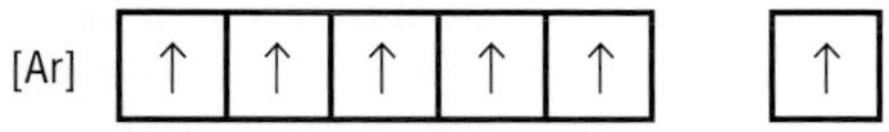

[Ar] ↑ ↑ ↑ ↑ ↑   ↑

because this has a lower energy by avoiding having a pair of electrons in the same orbital.

◀ **Figure 2** The electronic configuration of chromium.

Are all d-block elements transition metals?

The characteristic properties of transition metals, such as coloured compounds and variable oxidation states, are due to the presence of an inner incomplete d sub-shell. Electrons from both the inner d sub-shell and outer s sub-shell can be involved in compound formation.

Not all d-block elements have an incomplete d sub-shell – for example, zinc has the electronic configuration [Ar] $3d^{10} 4s^2$ in which the d sub-shell is full. Zinc does not show typical transition metal properties. Also, its ion Zn^{2+} ([Ar] $3d^{10}$) is not a typical transition metal ion.

Similarly, scandium forms the Sc^{3+} ion which has the stable electronic configuration of argon. This ion, the only one formed by scandium, has no 3d electrons and is unlike transition metal ions in its properties.

For this reason, a transition metal is defined as:

An element which forms at least one ion with a partially filled sub-shell of d electrons.

In the first row of the d block, only the eight elements from titanium to copper are classed as transition metals.

Note that, when d-block elements form ions, it is the s electrons that are lost first. So ions of d-block elements contain only d electrons in the outer shell.

What are transition elements like?

Transition elements are all metals. They are similar *to each other* but show distinct differences from s-block metals such as sodium or magnesium, and p-block metals such as aluminium and tin. Some of the physical properties of chromium, iron and cobalt, together with those of sodium and magnesium, are listed in Table 1 to illustrate this point.

Property	d-block element			s-block element	
	Cr	Fe	Co	Na	Mg
Metallic (atomic) radius/nm	0.13	0.12	0.13	0.19	0.16
Melting point/K	2130	1808	1768	371	922
Boiling point/K	2945	3023	3143	1156	1363
Density/g cm^{-3}	7.2	7.9	8.9	1.0	1.7

◀ **Table 1** Physical properties of some d-block and s-block metals.

Transition elements are dense metals with high melting and boiling points. They tend to be hard and durable, with high tensile strength and good mechanical properties. These properties are a result of strong **metallic bonding** between the atoms in the metal lattice (see **Section 3.1**).

Transition metals can release electrons into the 'pool' of mobile delocalised electrons from both the outer and inner shells. For example, an element such as iron from the first row of the d block can use both 3d and 4s electrons, forming strong metallic bonds between these electrons and the positive ions of the metal. Sodium and magnesium, on the other hand, can only use their 3s electrons.

As a result, we use transition metals to make things such as cars, cooking utensils and tools and to construct buildings and bridges.

The metals are sometimes used pure, but more often they are used in the form of **alloys** – particularly with iron. This is important in engineering because the properties of an alloy can be controlled as desired. Transition metals form a wide range of alloys with each other. Their atoms are often similar in size and behaviour and so the lattice structure may not be altered greatly as a result of substituting one atom for another. Even so, alloying modifies the properties and usually makes the metal harder and less malleable.

The effect of alloying on metal properties

The bonds between the atoms in a metal are strong but are not directed between particular atoms. When a force is applied to a metal crystal, the layers of atoms can 'slide' over one another. This is known as **slip**. After slipping, the atoms settle once again into a close-packed structure. Figure 3 shows the positions of the atoms before and after slip has taken place.

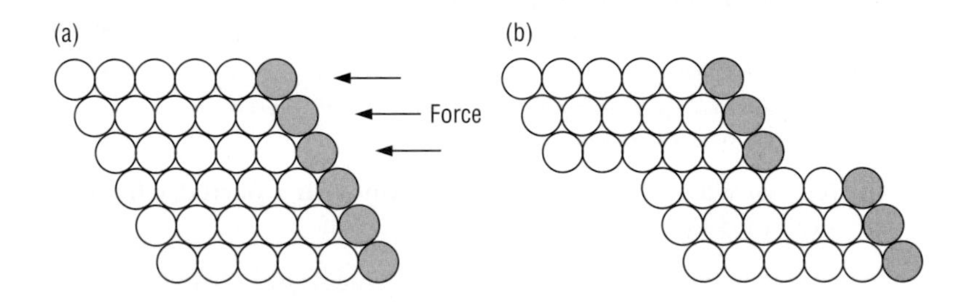

▶ **Figure 3** The arrangement of metal atoms in a crystal: **a** before and **b** after slip. The shaded circles represent the end row of atoms.

This is why metals can be hammered into different shapes or drawn into wires without breaking – they are **malleable** and **ductile**.

In an alloy, differently sized atoms interrupt the orderly arrangement of atoms in the lattice and make it more difficult for the layers to slide over one another (Figure 4) – the metal becomes harder and less malleable.

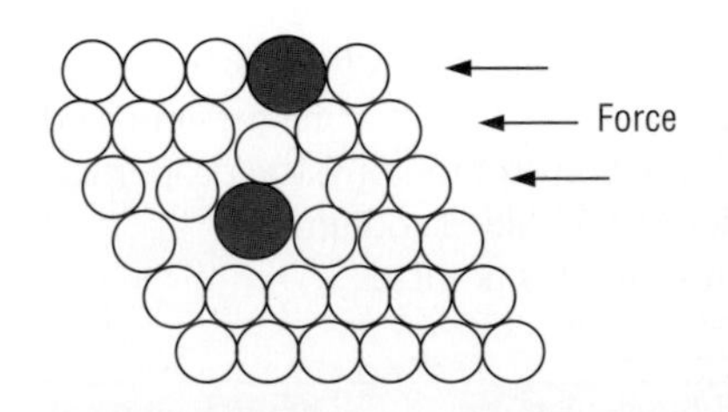

▶ **Figure 4** The arrangement of metal atoms in an alloy.

Smaller atoms, often of non-metals such as carbon or nitrogen, can be taken into the metal lattice and fit into the holes between the metal atoms – this happens in some types of steel. The carbon atoms distort the lattice, making slip between layers more difficult. In some steels, the carbon forms crystals of iron carbide which are very hard. The regions of iron carbide in the softer iron make the steel very strong.

Characteristic chemical properties of transition elements

As with physical properties, the chemical properties of transition metals are very different from those of s-block metals. The typical chemical properties can be summarised as follows:

- the formation of compounds in a variety of oxidation states
- the catalytic activity of the elements and their compounds
- there is a strong tendency to form complexes
- the formation of coloured compounds.

Now let's look at these special properties in more detail. The first two are dealt with in this section; the others are covered in **Section 11.6**, which deals with complex formation.

Variable oxidation states

Transition elements show a great variety of oxidation states in their compounds. By comparison, s-block metals are limited to oxidation states of +1 (for Group 1) and +2 (for Group 2). The reason for this can be seen by comparing successive ionisation enthalpies for an s-block metal (calcium) and a d-block metal (vanadium), as shown in Table 2.

◀ **Table 2** Successive ionisation enthalpies for calcium and vanadium.

	Ionisation enthalpies/kJ mol^{-1}			
	$\Delta H_i(1)$	$\Delta H_i(2)$	$\Delta H_i(3)$	$\Delta H_i(4)$
Ca [Ar] $4s^2$	+596	+1152	+4918	+6480
V [Ar] $3d^3 4s^2$	+656	+1420	+2834	+4513

$\Delta H_i(1)$, $\Delta H_i(2)$, $\Delta H_i(3)$ and $\Delta H_i(4)$ are enthalpy changes for the following reactions:

$$M(g) \rightarrow M^+(g) + e^- \quad \Delta H_i(1)$$
$$M^+(g) \rightarrow M^{2+}(g) + e^- \quad \Delta H_i(2)$$
$$M^{2+}(g) \rightarrow M^{3+}(g) + e^- \quad \Delta H_i(3)$$
$$M^{3+}(g) \rightarrow M^{4+}(g) + e^- \quad \Delta H_i(4)$$

These enthalpy changes are called the **first**, **second**, **third** and **fourth ionisation enthalpies**, respectively, and are a measure of the energy needed to remove a particular electron from the atom or ion (see **Section 2.5**).

Both calcium and vanadium always lose the two 4s electrons. For calcium, the first and second ionisation enthalpies are relatively low since these correspond to removing two s electrons from the outer shell. There is then a sharp increase to the third ionisation enthalpy because it is much more difficult to remove further electrons from the 3p sub-shell.

For vanadium, there is a more gradual increase in successive ionisation energies as first the 4s and then the 3d electrons are removed. You can see from Table 2 that removing a third electron from vanadium does not involve as big a jump in ionisation enthalpy as it does for calcium.

Figure 5 summarises the most common oxidation states of the d-block elements from scandium to zinc. The most important oxidation states are in boxes.

The oxidation state +1 is only important in the case of copper. For all the other metals, the sum of the first two ionisation enthalpies is low enough for two electrons to be removed.

Except for scandium and titanium, all the elements show oxidation state +2, in which the 4s electrons have been lost by ionisation.

All the elements except zinc show an oxidation state of +3.

The number of oxidation states shown by an element increases from Sc to Mn. In each of these elements, the highest oxidation state is equal to the total number of 3d and 4s electrons in atoms of the metal.

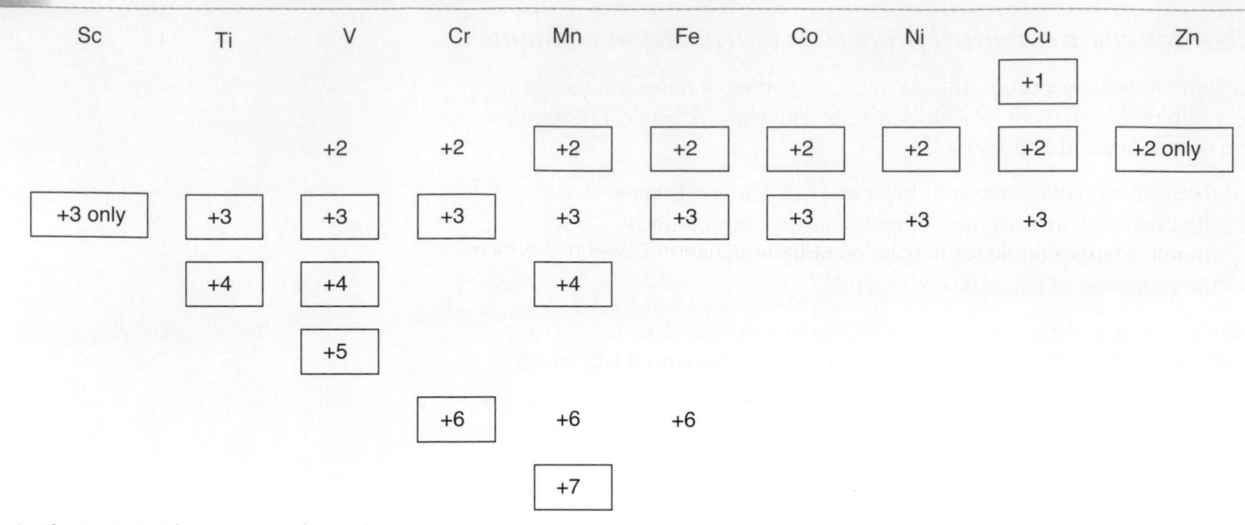

▲ **Figure 5** Oxidation states shown by elements in the first row of the d block – the most important oxidation states are in boxes

After Mn there is a decrease in the number of oxidation states shown by each element. The highest oxidation states become lower and progressively less stable. It seems that the increasing nuclear charge binds the d electrons more strongly in these elements, making them harder to remove.

In general, the lower oxidation states are found in simple ionic compounds – for example, compounds containing Cr^{3+}, Mn^{2+}, Fe^{3+} and Cu^{2+} ions. The metals in their higher oxidation states are usually bound covalently to an electronegative element, such as O or F, often in an anion – for example the vanadate(V) ion, VO_3^-, and the manganate(VII) ion MnO_4^-. Simple ions with high oxidation states, such as V^{5+} and Mn^{7+}, are not formed.

Stability of oxidation states

The change from one oxidation state of a metal to another involves a redox reaction. Thus, the relative stability of different oxidation states can be predicted by looking at the standard electrode potentials ($E^\ominus$ values) for these reactions (see **Sections 9.2** and **9.3**).

The general trends that emerge are:

- higher oxidation states become less stable relative to lower ones on moving from left to right across the series.
- compounds containing metals in high oxidation states tend to be oxidising agents (e.g. manganate(VII) ions, MnO_4^-) whereas compounds with metals in low oxidation states are often reducing agents (e.g. V^{2+} and Fe^{2+} ions).
- the relative stability of the +2 state with respect to the +3 state increases across the series. For elements early in the series, the +2 state is highly reducing – solutions of V^{2+} and Cr^{2+} ions are strong reducing agents. Later in the series, the +2 state is stable and the +3 state is highly oxidising – for example, Co^{3+} is a strong oxidising agent, and Ni^{3+} and Cu^{3+} do not exist in aqueous solution.

Catalytic activity

A catalyst alters the rate of a chemical reaction without being used up in the process. Transition metals and their compounds are effective and important catalysts both in industry and in biological systems.

Chemists believe that catalysts offer a new reaction pathway that has a lower activation enthalpy barrier than that of the uncatalysed reaction (see **Section 10.6**).

Once again, it is the availability of 3d as well as 4s electrons and the ability to change oxidation state that are among the vital factors in making transition metals such good catalysts.

Heterogeneous catalysis
In heterogeneous catalysis, the catalyst is in a different phase from the reactants. In the case of transition metals, this usually means a solid metal catalyst with reactants in the gas phase or liquid phase.

Transition metals can use the 3d and 4s electrons of atoms on the metal surface to form weak bonds to reactants. Once the reaction has occurred on the surface, these bonds can break to release the products.

An important example is the reaction of hydrogenation of alkenes, which is catalysed by nickel or platinum. A suggested mechanism for this reaction is given in Figure 1 in **Section 10.5**.

Homogeneous catalysis
In homogeneous catalysis, the catalyst is in the same phase as the reactants. In the case of transition metals, this often means the reaction takes place in the aqueous phase – the catalyst being an aqueous transition metal ion.

Homogeneous catalysis usually involves the transition metal ion forming an **intermediate compound** with one or more of the reactants, which then breaks down to form the products. An example is the reaction between 2,3-dihydroxybutanoate ions and hydrogen peroxide, which is catalysed by Co^{2+} ions.

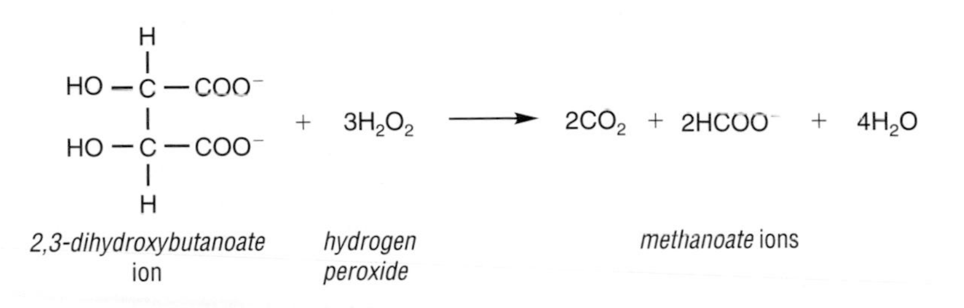

Figure 6 shows a suggested mechanism for this catalysis. Figure 7 shows the corresponding enthalpy profile.

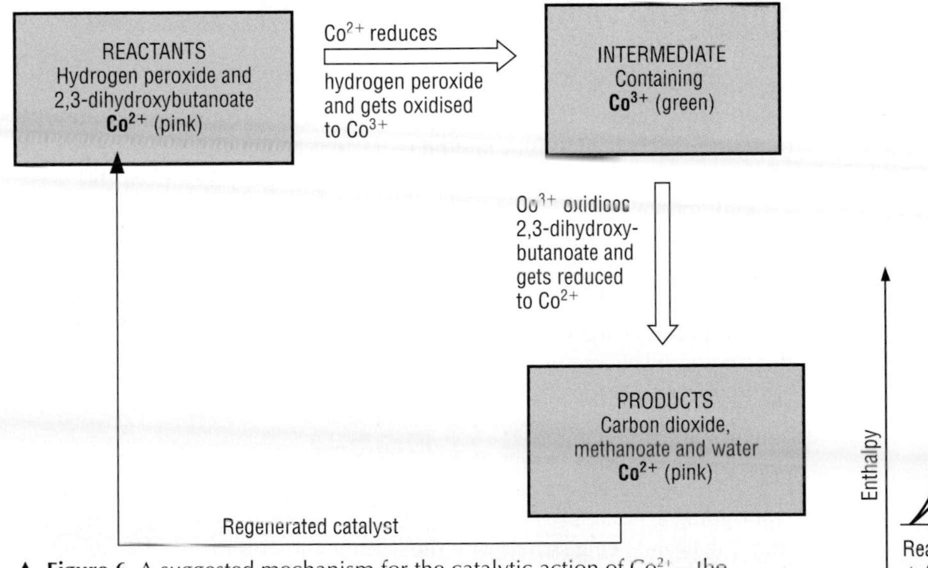

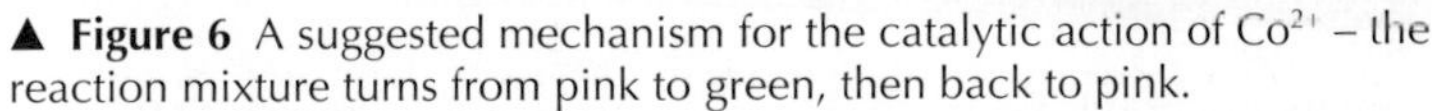
▲ **Figure 6** A suggested mechanism for the catalytic action of Co^{2+} – the reaction mixture turns from pink to green, then back to pink.

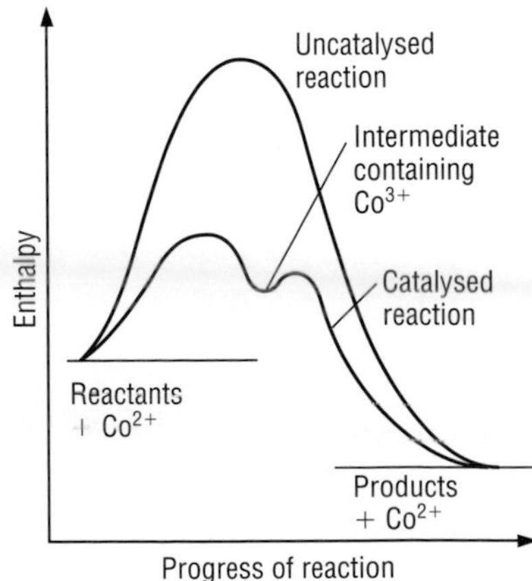

▲ **Figure 7** The enthalpy profiles for the catalysed and uncatalysed reactions.

Transition metal ions are particularly effective catalysts in redox reactions. This is because they can readily move from one oxidation state to another, in the way that Co readily moves between the +2 and +3 states in the reaction above.

Problems for 11.5

1 Use Figure 1 in this section (page 252) to write out the electronic configuration for each of the elements in the series scandium to zinc (e.g. Sc [Ar] $3d^1 4s^2$).

2 a Write the electronic configurations of the following ions (e.g. Fe^{2+} [Ar] $3d^6 4s^0$)
i Cu^{2+} iii Fe^{3+} v Cr^{3+}
ii Cu^+ iv V^{3+} vi Ni^{2+}
b Explain why Cu^{2+} behaves as a typical transition metal ion, but Cu^+ does not.

3 Look at the information in Table 1 in this section (page 253).
a What do you notice about the size of the three transition metal atoms compared with those of sodium and magnesium?
b How do the melting and boiling points of transition metals compare with those of non-transition metals?
c How do the densities of s-block and d-block metals compare?
d Write out the electronic configurations of the elements Cr, Fe, Co, Na and Mg (e.g. Na $1s^2 2s^2 2p^6 3s^1$).
e Suggest why certain properties are common to transition metals, but different from the properties of s-block metals like sodium and magnesium.

4 A $Fe^{3+}(aq) + e^- \rightarrow Fe^{2+}(aq)$ $E^\ominus = +0.77\,V$
B $Mn^{3+}(aq) + e^- \rightarrow Mn^{2+}(aq)$ $E^\ominus = +1.56\,V$
C $O_2(g) + 4H^+(aq) + 4e^- \rightarrow 2H_2O(l)$ $E^\ominus = +1.23\,V$
a Use the given values of standard electrode potentials to predict what will happen when
i half-cells **A** and **C** are connected
ii half-cells **B** and **C** are connected.
In each case, write a balanced overall equation for the reaction.
b Use your answers to part **a** to help you predict whether
i an acidified solution of iron(II) will be oxidised by air
ii an acidified solution of manganese(II) will be oxidised by air.
c Predict what will happen when the following half-cells are connected.

$Cu^{2+}(aq) + e^- \rightarrow Cu^+(aq)$ $E^\ominus = +0.16\,V$
$Cu^+(aq) + e^- \rightarrow Cu(s)$ $E^\ominus = +0.52\,V$

Use your answers to explain the following. When red copper(I) oxide is added to dilute sulfuric acid, a blue solution of copper(II) sulfate is formed, along with a red-brown precipitate of copper metal.
Write a balanced overall equation for the reaction.

11.6 *The d block: complex formation*

What are complexes?

A **complex** consists of a central metal atom or ion surrounded by a number of negatively charged ions or neutral molecules possessing a lone pair of electrons. These surrounding anions or molecules are called **ligands**.

A complex may have an overall positive charge, a negative charge or no charge at all. For example,

$$[Fe(H_2O)_6]^{3+} \quad [NiCl_4]^{2-} \quad Ni(CO)_4$$

If a complex is charged, it is called a **complex ion** – the overall charge is the sum of the charge on the central metal ion and the charges on the ligands.

For $[NiCl_4]^{2-}$, charge $= (2+) + 4(1-) = 2-$

In reality, the charges on a complex ion are delocalised over the whole ion.

Bonding in complexes is complicated – it usually involves electron pairs from the ligand being shared with the central ion. This means that ligands are *electron donors*.

The number of bonds from the central ion to ligands is known as the **coordination number** of the central ion. The most common coordination numbers are six and four but two does occur, for example in complexes of Ag(I) and Cu(I).

Shapes of complexes

The shape of a complex depends on its coordination number. (There is more about the shapes of molecules in **Section 3.2**.)

Complexes with **coordination number 6** usually have an **octahedral** arrangement of ligands around the central metal ion (Figure 1) – this is the most common coordination number.

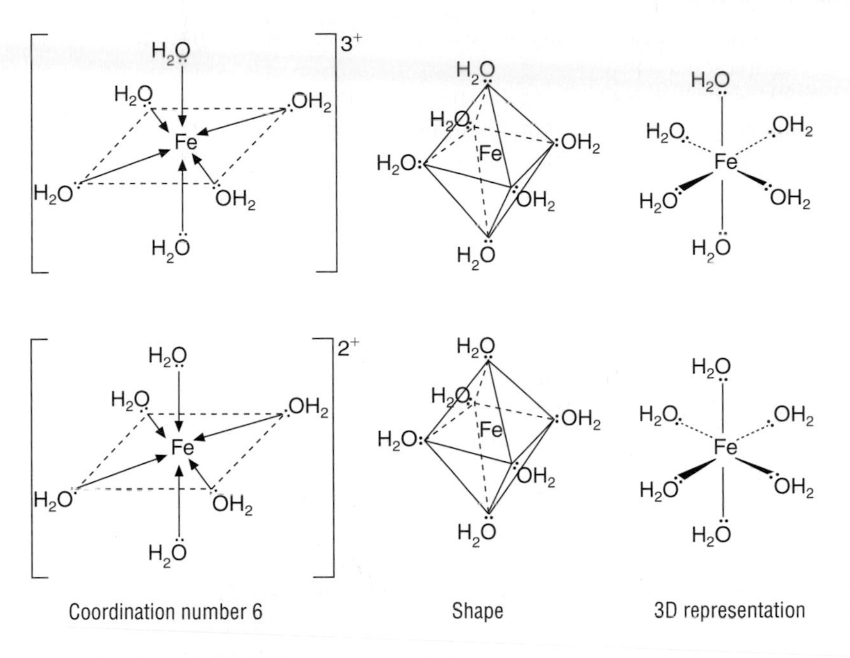

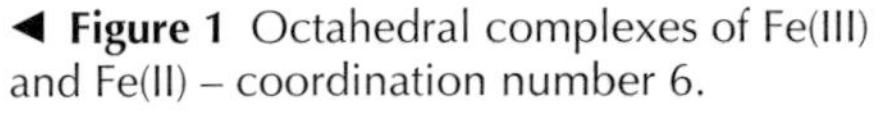

◀ **Figure 1** Octahedral complexes of Fe(III) and Fe(II) – coordination number 6.

Complexes with **coordination number 4** usually have a **tetrahedral** arrangement of ligands around the central metal ion (Figure 2).

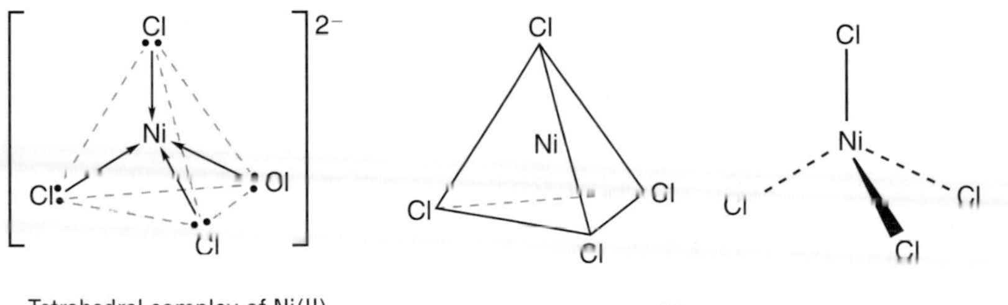

◀ **Figure 2** A tetrahedral complex of Ni(II) – coordination number 4.

Some four-coordinate complexes have a **square planar** structure (Figure 3).

◀ **Figure 3** A square planar complex of Ni(II) – coordination number 4.

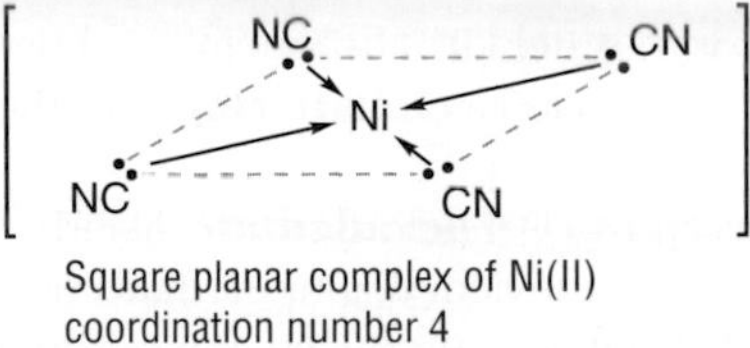

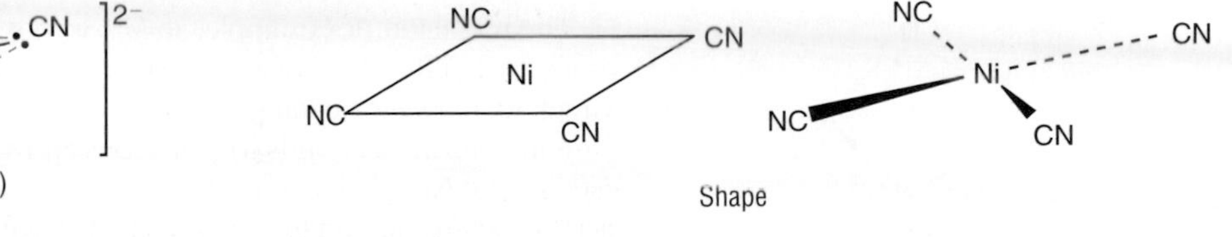

Complexes with **coordination number 2** usually have a **linear** arrangement of ligands (Figure 4).

▲ **Figure 4** A linear complex of Ag(I) – coordination number 2.

▶ **Table 1** The shapes of complexes.

Coordination number	Shape of complex	Examples
6	octahedral	$[Fe(CN)_6]^{3-}$ $[Ni(NH_3)_6]^{2+}$
4	tetrahedral	$[NiCl_4]^{2-}$
4	square planar	$[Ni(CN)_4]^{2-}$
2	linear	$[Ag(NH_3)_2]^+$

Table 1 gives a summary of the shapes of complexes with some examples.

Most complexes have regular shapes like those in Figures 1–4 but some important complexes do not. For example, in the $[Cu(H_2O)_6]^{2+}$ ion four of the water ligands are held more strongly than the remaining two, so that four of the copper–oxygen bonds are shorter. This produces a distorted octahedral arrangement.

Naming complex ions

The systematic naming of complex ions is based on four rules.

1. Give the **number of ligands** of the same type around the central cation using the Greek prefixes mono-, di-, tri-, tetra-, penta- and hexa-.
2. **Identify the ligands** (in alphabetical order if there is more than one type) using the ending '-o' for anions – for example F^- fluoro, Cl^- chloro, CN^- cyano, OH^- hydroxo. Neutral ligands keep their name, except for H_2O aqua and NH_3 ammine.
3. Name the **central metal ion**. Use the English name for a positively charged or neutral complex. If the overall charge of the complex is negative then use the Latinised name for the cation and add the suffix *–ate* – for example cuprate for copper, ferrate for iron, zincate for zinc.
4. Indicate the **oxidation number** of the central metal in brackets. For example:

   $[Cr(H_2O)_6]^{3+}$ hexaaquachromium(III) ion
   $[Fe(CN)_6]^{4-}$ hexacyanoferrate(II) ion
   $[CuCl_4]^{2-}$ tetrachlorocuprate(II) ion
   $[Cu(NH_3)_4]^{2+}$ tetraamminecopper(II) ion.

Polydentate ligands

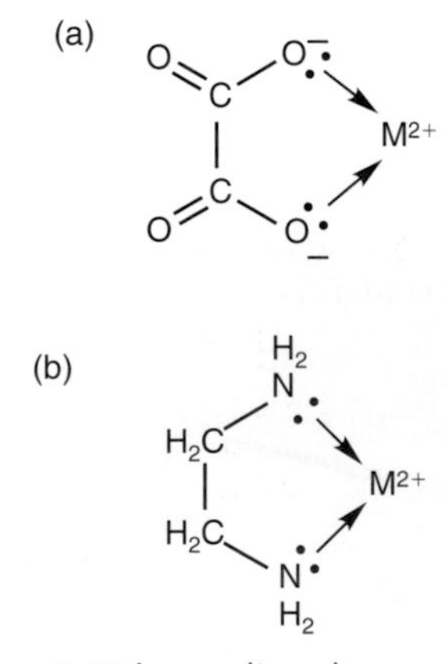

▲ **Figure 5** Bidentate ligands.
a The ethanedioate ion.
b The 1,2-diaminoethane molecule.

Some ligand molecules, such as NH_3 and H_2O, can only bond to a metal ion through a single atom or ion and are called **monodentate** ligands.

Others can bond through more than one atom and are **polydentate** ('many-toothed') ligands. For example, the ethanedioate ion and the 1,2-diaminoethane molecule, shown in Figure 5, are **bidentate** ('two-toothed') ligands because they can form *two* bonds with a metal ion by using pairs of electrons from two oxygen or nitrogen atoms.

The metal ion (for example, an M^{2+} ion) is held in a five-membered ring as if it was in the claws of a crab. The ring is called a **chelate ring** after the Greek word *chele* meaning 'crab'.

$edta^{4-}$ forms six bonds to metal ions and so is a **hexadentate** ligand (see Figure 6a) – 'edta' is an abbreviation for **e**thylene**d**iamine**t**etra**a**cetic acid, which is the acid from which the ion $edta^{4-}$ is produced. $edta^{4-}$ forms complexes using the two nitrogen atoms and the four oxygen atoms of the COO^- groups. $edta^{4-}$ acts as a kind of cage and traps the metal atom inside, as shown for the complex with nickel in Figure 6b.

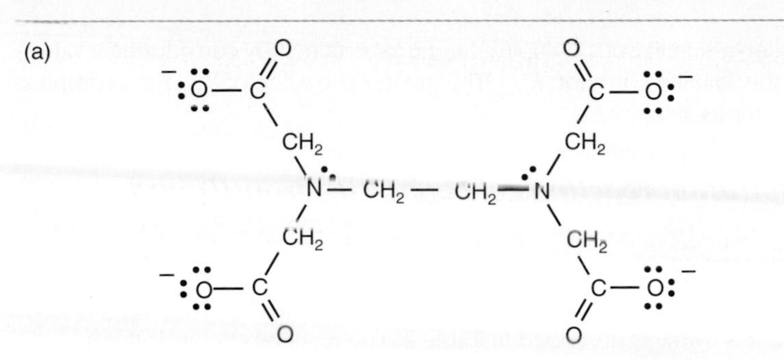

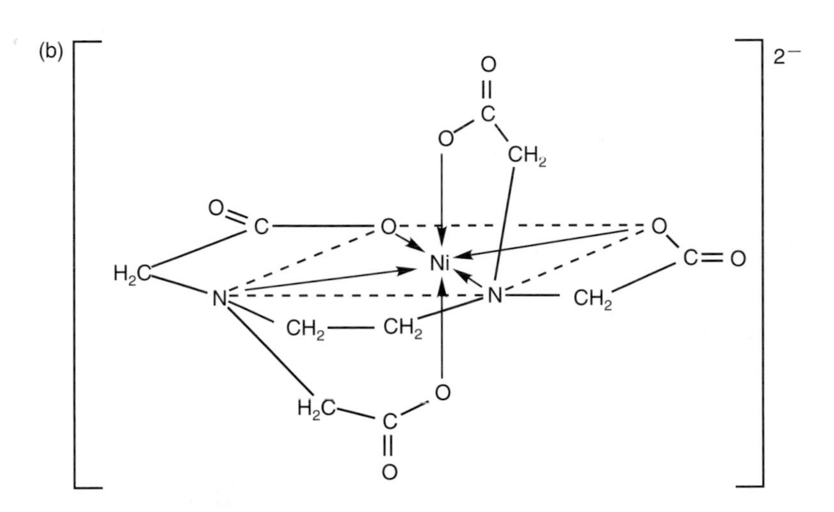

◀ **Figure 6 a** edta^{4-}, a hexadentate ligand. **b** The nickel–edta complex ion $[Ni(edta)]^{2-}$.

Ligand substitution reactions

When copper sulfate is dissolved in water, the copper ions form a complex ion with a coordination number of 6 (Figure 7).

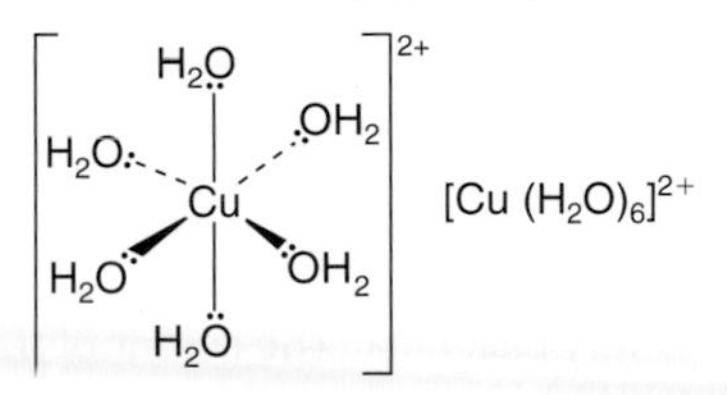

◀ **Figure 7** The Cu^{2+} ion forms a complex ion with a coordination number of 6.

However, reactions can occur where one ligand displaces another. For example, if you add concentrated hydrochloric acid to $[Cu(H_2O)_6]^{2+}$ then chloride ions will displace the water ligands and the $[CuCl_4]^{2-}$ complex forms. This is an example of a **ligand substitution** reaction:

$$[Cu(H_2O)_6]^{2+}(aq) + 4Cl^- \rightarrow [CuCl_4]^{2-} + 6H_2O$$

blue octahedral → yellow tetrahedral

But if concentrated aqueous ammonia is added to the same complex ion, ammonia molecules will displace the water ligands.

$$[Cu(H_2O)_6]^{2+}(aq) + 4NH_3 \rightarrow [Cu(NH_3)_4(H_2O)_2]^{2+} + 4H_2O$$

blue octahedral → deep blue/violet octahedral

These two reactions are summarised in Figure 8.

Ligand substitution occurs if the new complex formed is more stable than the previous complex. Therefore the stability of a complex depends on its ligands. For example, the complex ion of copper(II) with ammonia ligands is more stable than with water ligands.

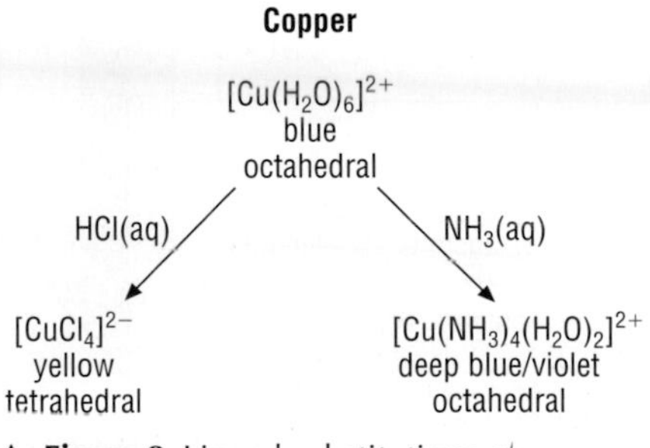

▲ **Figure 8** Ligand substitutions of copper(II) complexes.

The relative stability of a complex can be ascertained by considering a value called the stability constant, K_{stab}. The greater the value of K_{stab} for a complex, the greater its stability.

Precipitation reactions of copper(II), iron(II) and iron(III) ions

If any copper(II), iron(II) and iron(III) ions are in solution and sodium hydroxide solution is added, characteristically coloured hydroxide precipitates form, as described in Table 2. The reactions are as follows.

$$Cu^{2+}(aq) + 2OH^-(aq) \rightarrow Cu(OH)_2(s)$$
$$Fe^{2+}(aq) + 2OH^-(aq) \rightarrow Fe(OH)_2(s)$$
$$Fe^{3+}(aq) + 3OH^-(aq) \rightarrow Fe(OH)_3(s)$$

▶ **Table 2** Colours of copper and iron hydroxides.

Metal hydroxide	Description of precipitate
$Cu(OH)_2$	pale blue solid
$Fe(OH)_2$	green gelatinous precipitate
$Fe(OH)_3$	orange gelatinous precipitate

Coloured compounds

Compounds of d-block transition metals are frequently coloured, both in the solid state and in solution. You can get an idea of the wide range of colours from Table 3.

▶ **Table 3** Colours of some transition metal ions in aqueous solution.

Ion	Outer electrons	Colour	Ion	Outer electrons	Colour
Ti^{3+}	$3d^1$	purple	Fe^{3+}	$3d^5$	yellow
V^{3+}	$3d^2$	green	Fe^{2+}	$3d^6$	green
Cr^{3+}	$3d^3$	violet	Co^{2+}	$3d^7$	pink
Mn^{3+}	$3d^4$	violet	Ni^{2+}	$3d^8$	green
Mn^{2+}	$3d^5$	pale pink	Cu^{2+}	$3d^8$	blue

The intensity of colour varies greatly – for example, MnO_4^- is an intensely deep purple but Mn^{2+} is very pale pink. The colour of transition metal compounds can often be related to the presence of unfilled or partly filled d orbitals (and so to unpaired electrons) in the metal ion.

When white light falls on a substance, some may be absorbed, some transmitted and some reflected (see **Section 6.7**). If light in the visible region of the spectrum is absorbed then the compound will appear coloured.

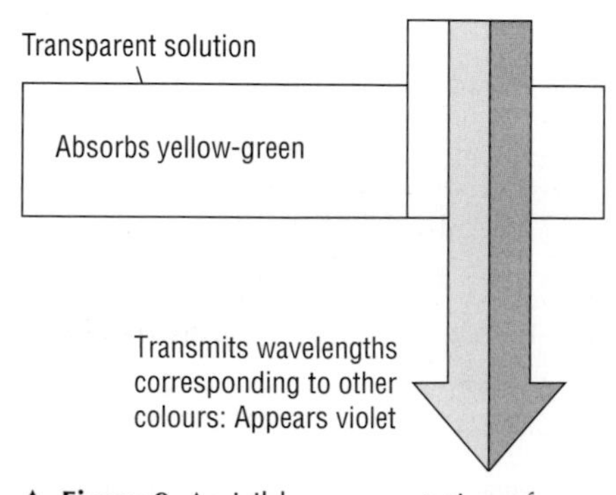

▲ **Figure 9** A visible representation of absorption and transmission of a coloured solution.

What happens when a d-block compound absorbs light?

Light is absorbed by an atom only if the energy of the light matches the energy gap between two energy states in the atom. If it does then an electron is promoted from an orbital of lower energy to one of higher energy. So the atom or ion absorbing the radiation changes from its ground state to an excited state.

The frequency (ν) of the light absorbed depends on the energy difference between these two levels, ΔE. ($\Delta E = h\nu$, where h is Planck's constant.) There are five d orbitals in d-block metals; however, the ligands around the metal atom or ion cause these orbitals to split in such a way that some are slightly higher than others (this is covered in more detail in **Section 6.9**). The difference between the two levels, ΔE, is now such that the light absorbed falls in the visible part of the spectrum. The transmitted light is the light not absorbed, and is the colour we see (Figure 9).

The absorption spectrum for the $[Ti(H_2O)_6]^{3+}$ ion is shown in Figure 10. The complex absorbs green–yellow light – therefore the solution looks violet.

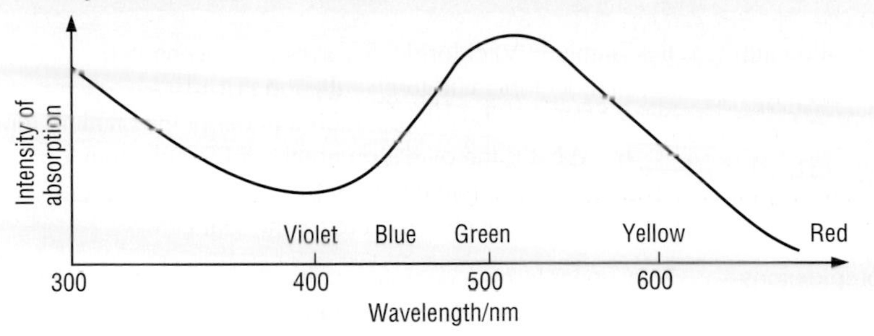

◀ **Figure 10** The absorption spectrum of the hydrated Ti^{3+} ion, $[Ti(H_2O)_6]^{3+}$.

$[Ti(H_2O)_6]^{3+}$ is a simple example because the ion contains only one d electron. The situation is more complicated when there are several d electrons and many more transitions are possible between d levels. However, the basic principles are the same.

The colour of a transition metal complex depends on:

- the number of d electrons present in the transition metal ion
- the arrangement of ligands around the ion, because this affects the splitting of the d sub-shell
- the nature of the ligand, because different ligands have a different effect on the relative energies of the d orbitals in a particular ion.

For example, NH_3 ligands cause a larger difference than H_2O ligands in splitting d-orbital energies. The blue colour of hydrated Cu^{2+} ions, for example, changes to deep blue/violet when NH_3 is added (see Figure 11).

$$4NH_3(aq) + [Cu(H_2O)_6]^{2+} \rightarrow [Cu(NH_3)_4(H_2O)_2]^{2+} + 4H_2O(l)$$

blue → deep blue/violet

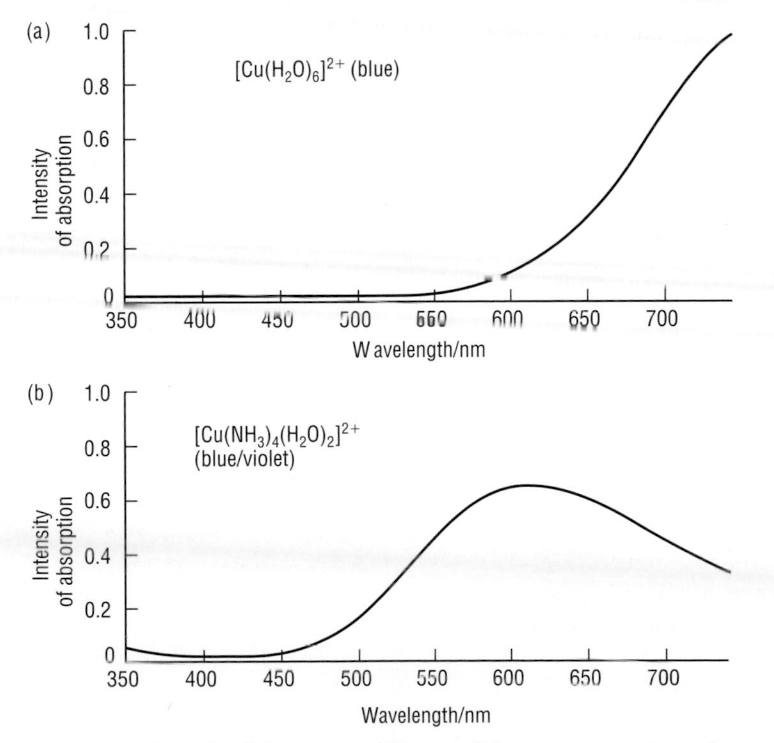

◀ **Figure 11 a** Absorption spectrum for $[Cu(H_2O)_6]^{2+}$.
b Absorption spectrum for $[Cu(NH_3)_4(H_2O)_2]^{2+}$. In the ammonia complex, the ammonia ligands cause a bigger splitting of d-orbital energies. This means that light of higher energy, and lower wavelength, is absorbed.

You can see from the following complexes of chromium that the colour also depends on the number of each kind of ligand present:

$[Cr(H_2O)_6]^{3+}$	$[Cr(H_2O)_5Cl]^{2+}$	$[Cr(H_2O)_4Cl_2]^{+}$
violet	green	dark green

Problems for 11.6

1 Give the coordination number for each of the following complex ions:
- **a** $[Ag(H_2O)_2]^+$
- **b** $[CoCl_4]^{2-}$
- **c** $[Co(CN)_6]^{3-}$
- **d** $[Cr(H_2O)_5OH]^{2+}$

2 Write down the formulae of the following complex ions – make sure you include the charge on the ion:
- **a** hexaaquamanganese(II) ion (ligand = H_2O; metal ion = Mn^{2+})
- **b** tetraamminezinc(II) ion (ligand = NH_3; metal ion = Zn^{2+})
- **c** hexafluoroferrate(III) ion (ligand = F^-; metal ion = Fe^{3+})
- **d** pentaaquamonohydroxochromium(III) ion (ligands = H_2O and OH^-; metal ion = Cr^{3+})

3 What is the oxidation state of the central metal ion of each of the complexes **a**–**d** in problem 1?

4 Give systematic names for the following complex ions:
- **a** $[V(H_2O)_6]^{3+}$
- **b** $[Fe(CN)_6]^{4-}$
- **c** $[CoCl_4]^{2-}$
- **d** $[Ag(NH_3)_2]^+$
- **e** $[Cr(H_2O)_4Cl_2]^+$

5
- **a** Explain why titanium(IV) oxide (the white pigment in white paint) is not coloured.
- **b** Explain why compounds of Sc^{3+}, Zn^{2+} and Cu^+ are not coloured.

6 Titanium(IV) chloride dissolves in concentrated hydrochloric acid to give the ion $[TiCl_6]^{2-}$.
- **a** What is the coordination number of the titanium ion?
- **b** What is the oxidation number of the titanium ion?
- **c** Suggest a name for $[TiCl_6]^{2-}$.
- **d** Draw a likely structure for the ion.

7 Copper forms a complex with 1,2-diaminoethane ('en') and with a similar bidentate ligand, 1,3-diaminopropane ('pn'), $NH_2CH_2CH_2CH_2NH_2$.
The lg K_{stab} values for these two complexes are shown below

	$[Cu(en)_2]^{2+}$	$[Cu(pn)_2]^{2+}$
lg K_{stab}	20.03	17.17

Look at models of the two complexes of copper. What size of chelate ring seems to lead to the more stable structure?

8 Classify the following ligands as mono-, bi- or polydentate ligands:

a H_2O **b**

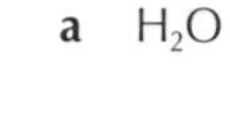

c

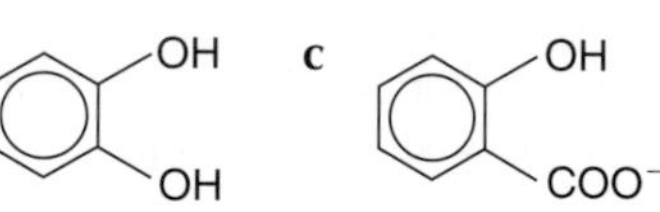

9 Complexes formed by $edta^{4-}$ involve pairs of electrons on nitrogen and oxygen atoms, in the same way as complexes formed by NH_3 and H_2O. Explain why stability constants of edta complexes in aqueous solution are generally so much larger than those of corresponding complexes with NH_3 and H_2O.

12 ORGANIC CHEMISTRY: FRAMEWORKS

12.1 *Alkanes*

AS

Many carbon compounds are found in living organisms, which is why their study got the name **organic chemistry**. Today, organic chemistry includes all carbon compounds whatever their origin – except CO, CO_2 and the carbonates, which are traditionally included in inorganic chemistry studies.

Only carbon can form the diverse range of compounds necessary to produce the individuality of living things.

Carbon is unique

About seven million compounds containing carbon and hydrogen are known to chemists. This is far more than the number of compounds from all the other elements put together.

Why carbon?

Carbon's electron structure is shown in Figure 1. This electron structure makes it the first member of Group 4 in the centre of the Periodic Table, and is responsible for its special properties.

A carbon atom has four electrons in its outer shell. It could achieve stability by losing or gaining four electrons; but this is too many electrons to lose or gain. The resulting carbon ions would have charges of 4+ or 4−, respectively, and would be too highly charged. So when carbon forms compounds the bonds are *covalent* rather than ionic.

In methane (CH_4), for example, the carbon atom achieves stability by *sharing* its outer electrons with four hydrogen atoms (Figure 2) forming four C–H covalent bonds.

Carbon forms strong covalent bonds with itself to give *chains* and *rings* of its atoms, joined by C–C covalent bonds. This property is called *catenation* and leads to the limitless variety of organic compounds possible.

Each carbon atom can form four covalent bonds, so the chains may be straight or branched and can have other atoms or groups substituted on them.

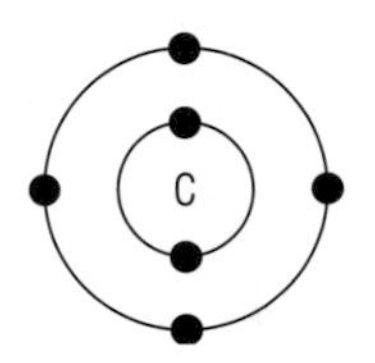

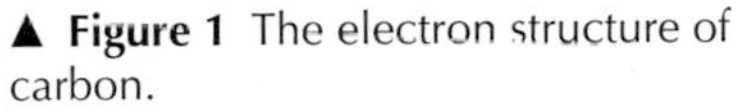

▲ **Figure 1** The electron structure of carbon.

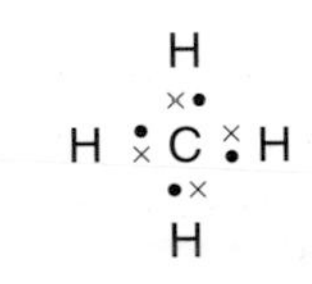

▲ **Figure 2** Covalent bonds in methane.

Hydrocarbons

Chemists cope with the vast number of organic compounds by dividing them into groups of related compounds.

Hydrocarbons are compounds containing *only* carbon atoms and hydrogen atoms; they are represented by the general molecular formula C_xH_y. There are different types of hydrocarbons. For example

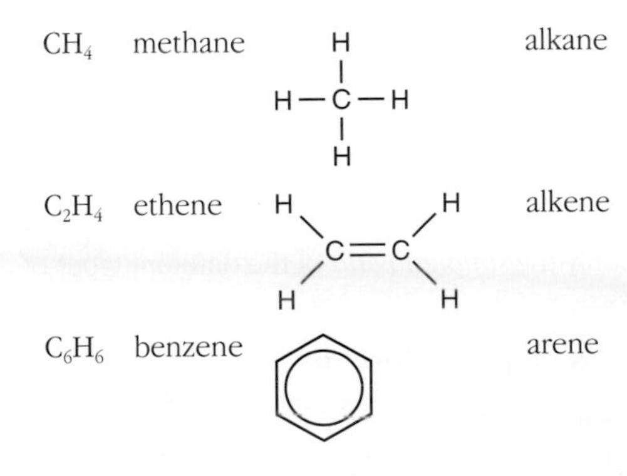

The ring of six carbon atoms in benzene has special properties. (You will find out about these later in the course in **Sections 12.3** and **12.4**.) Compounds that contain a benzene ring are called **aromatic** compounds. Compounds that do not contain a benzene ring are called **aliphatic** compounds.

Hydrocarbons are relatively unreactive – this is particularly true of alkanes and arenes. They form the unreactive framework of organic compounds. But when you attach other groups, such as OH, to the hydrocarbon framework its properties are modified. So we can think of organic compounds as having hydrocarbon *frameworks*, with *modifiers* attached.

A modifier such as the hydroxyl group, OH, is also called a **functional group**.

The C=C double bond in an alkene is much more reactive than a C–C single bond. It is often regarded as a functional group, even though it is part of the hydrocarbon framework.

In this section we are concerned with frameworks.

Alkanes

Alkanes are **saturated** hydrocarbons. 'Saturated' means that they contain the maximum number of hydrogen atoms possible, with no double or triple bonds between carbon atoms.

The general molecular formula of the alkanes is C_nH_{2n+2} where $n = 1, 2, 3 \dots$. Table 1 shows the names and formulae of some simple alkanes.

▶ **Table 1** Some simple alkanes.

n	Molecular formula	Name
1	CH_4	methane
2	C_2H_6	ethane
3	C_3H_8	propane
4	C_4H_{10}	butane
5	C_5H_{12}	pentane

Look at the names in Table 1. The first part indicates the number of carbon atoms in each molecule; *alk-* is the general term for *meth-, eth-, prop-* and so on. The second part, *-ane*, indicates that the compounds are alkanes.

The names of all alkanes end in **-ane**.

A series of compounds related to each other in this way is called an **homologous series**. All the members of the series have the same general molecular formula, and each member differs from the next by a $-CH_2-$ unit. All the compounds in a series have similar chemical properties, so chemists can study the *properties of the group* rather than those of individual compounds. However, physical properties, such as melting point, boiling point and density, do change but gradually in the series as the number of carbon atoms in the molecules increases.

Finding the formulae of alkanes

Section 1.1 explains the difference between empirical and molecular formulae. The molecular formula is more useful, because it tells you how many atoms of each type are present in a molecule of the compound. The molecular formula may or may not be the same as the empirical formula for the compound, which gives only the *simplest ratio* of the different types of atoms present.

molecular formula = *m* × empirical formula
where $m = 1, 2, 3, \dots$

For ethane
molecular formula
C_2H_6
empirical formula
CH_3

The empirical formula of a compound is worked out from its composition by mass.

For a hydrocarbon, the composition by mass is easily found by burning a known mass in oxygen and measuring the amounts of carbon dioxide and water formed. This is called *combustion analysis* and can be performed automatically by a machine.

Example

0.100 g of a hydrocarbon **X** on complete combustion gave 0.309 g CO_2 and 0.142 g H_2O. Calculate the empirical formula of the compound **X**.

A_r: H, 1.0; C, 12.0; O, 16.0
$M_r(CO_2)$ = 44.0
$M_r(H_2O)$ = 18.0

Answer

First, calculate the masses of C and H in 0.100 g of the compound.

44.0 g CO_2 contains 12.0 g C
$\therefore$ mass of C in 0.100 g **X** = (12.0/44.0) × 0.309 g
= 0.0843 g
18.0 g H_2O contains 2.0 g H
$\therefore$ mass of H in 0.100 g **X** = (2.0/18.0) × 0.142 g
= 0.0158 g

	C	:	H
ratio by mass	0.0843	:	0.0158
ratio by moles	0.00703	:	0.0158
simplest ratio (divide by smaller)	1	:	2.25
whole number ratio	4	:	9

$\therefore$ empirical formula of X is C_4H_9

Once we have the empirical formula, we can work out the molecular formula, provided we know the relative molecular mass of the compound.

Example

The relative molecular mass of **X** was found to be 114.0 by using a mass spectrometer. Find the molecular formula of **X**.

Answer

Empirical formula of **X** is C_4H_9.

But M_r (C_4H_9) = 57.0. This is half of 114.0. So the molecular formula of **X** must be $(C_4H_9)_2 = C_8H_{18}$.

Structure of alkanes

Figure 2 in this section (page 265) shows a dot–cross formula for methane. It shows all the outer electrons in each atom and how electrons are shared to form the covalent bonds. For larger molecules, dot–cross formulae are rather cumbersome and the shared electron pairs can be replaced by lines representing the covalent bonds (Figure 3).

Represents a covalent bond, formed by sharing a pair of electrons, one from each atom

H
|
H—C—H
|
H

▲ **Figure 3** Using lines to represent covalent bonds.

Figure 3 is called a **full structural formula** – it shows all the atoms and all the bonds in the molecule.

We can also write an abbreviated version known as a *shortened structural formula*. Take heptane, for example. Its full structural formula is

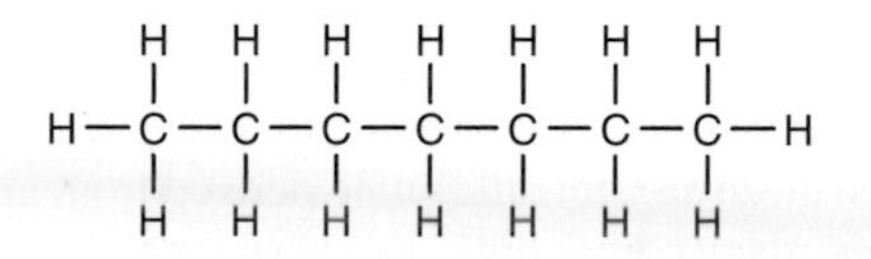

while its shortened structural formula is

$CH_3–CH_2–CH_2–CH_2–CH_2–CH_2–CH_3$

which is sometimes further shortened to

$CH_3CH_2CH_2CH_2CH_2CH_2CH_3$

Look at Table 2 (page 268). It gives the full structural formulae and shortened formulae for some alkanes. Note that each carbon atom is bonded to four other atoms. Each hydrogen atom is bonded to only one other atom.

▶ **Table 2** Structural formulae of alkanes.

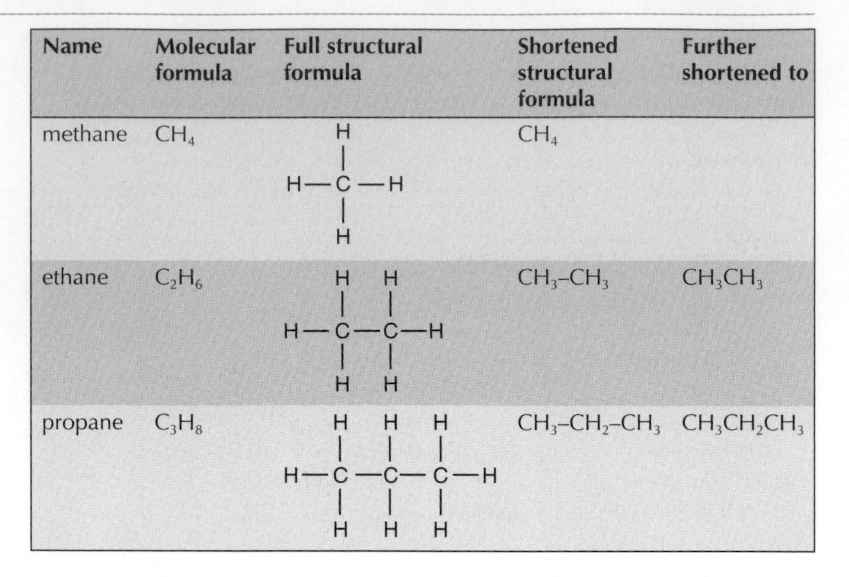

Name	Molecular formula	Full structural formula	Shortened structural formula	Further shortened to
methane	CH_4	H H—C—H H	CH_4	
ethane	C_2H_6	H H H—C—C—H H H	$CH_3–CH_3$	CH_3CH_3
propane	C_3H_8	H H H H—C—C—C—H H H H	$CH_3–CH_2–CH_3$	$CH_3CH_2CH_3$

Branched alkanes

Alkanes may have straight or branched chains. So it is often possible to draw more than one structural formula for a given molecular formula. There is often a straight-chain compound and one or more branched-chain compounds with the same molecular formula.

For C_4H_{10}, there are two possible structural formulae:

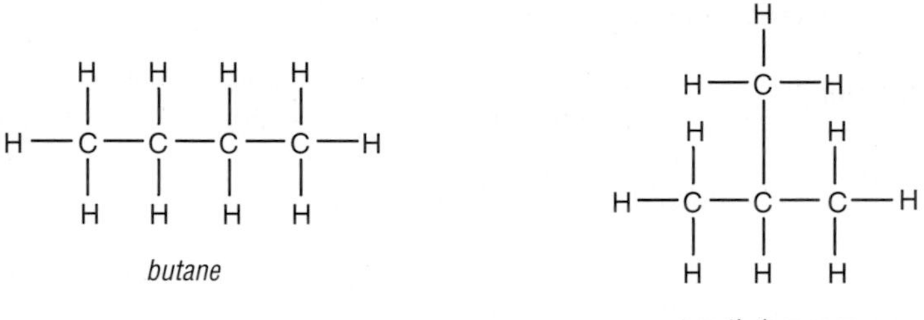

butane

methylpropane

These two compounds are **isomers** because they have the same molecular formulae but different structural formulae. There is more about this type of isomerism in **Section 3.3**.

Note how the branched-chain isomer is named. It is regarded as being formed from a straight-chain alkane, propane, by attaching a $–CH_3$ side group to the second carbon atom. It is therefore called methylpropane.

The $–CH_3$ group is just methane with a hydrogen atom removed so that it can join to another atom. It is called a **methyl group**.

Side groups of this kind are called **alkyl groups**. They have the general formula C_nH_{2n+1} and are often represented by the symbol **R** (see Table 3).

▼ **Table 3** Some common alkyl groups.

Alkyl group	Formula
methyl	$CH_3–$
ethyl	$CH_3CH_2–$
propyl	$CH_3CH_2CH_2–$
butyl	$CH_3CH_2CH_2CH_2–$
pentyl	$CH_3CH_2CH_2CH_2CH_2–$

An alkyl group R has the general formula C_nH_{2n+1}

Butane and *methylpropane* are **systematic names**. Every organic compound can be given a systematic name derived from an internationally agreed set of rules. Many compounds also have common names. The systematic name is important because it allow us to draw the full structural formula, and vice versa.

Cycloalkanes

As well as open-chain alkanes, it is also possible for alkane molecules with cyclic structures to exist. These molecules are called **cycloalkanes** and have the general formula C_nH_{2n}. They have two fewer hydrogen atoms than the corresponding alkane, because there are no $–CH_3$ groups at the ends of the chain.

Table 4 shows some different ways of representing cycloalkanes. The **skeletal formula** shows only the shape of the carbon framework. Each line represents a C–C bond. The carbon atoms are at the corners. The C–H bonds are not shown but it is easy to work out how many there are – in saturated compounds, carbon always forms four covalent bonds.

◀ **Table 4** Cycloalkanes.

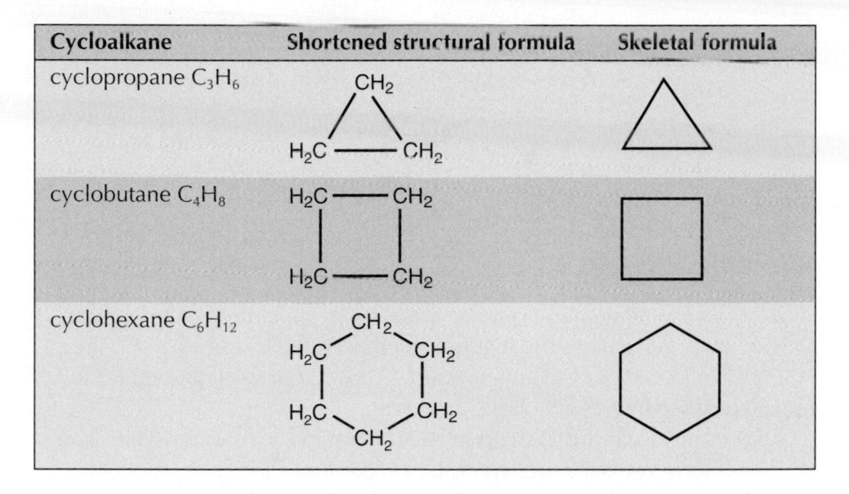

Cycloalkane	Shortened structural formula	Skeletal formula
cyclopropane C_3H_6	CH_2, H_2C—CH_2 (ring)	
cyclobutane C_4H_8	H_2C—CH_2, H_2C—CH_2 (ring)	
cyclohexane C_6H_{12}	CH_2, H_2C, CH_2, H_2C, CH_2, CH_2 (ring)	

Shapes of alkanes

Representing structures in a two-dimensional way on paper can give a misleading picture of what the molecule looks like.

The pairs of electrons in the covalent bonds repel one another, and so arrange themselves round the carbon atom as far apart as possible (see **Section 3.2**). Thus, the C–H bonds in methane are directed so they point towards the corners of a regular tetrahedron (Figure 4). The carbon atom is at the centre of the tetrahedron, and the H–C–H bond angles are 109° (109° 28′ to be precise).

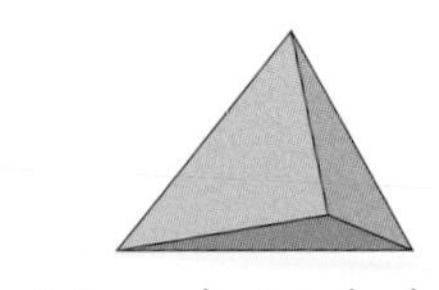

▲ **Figure 4** A regular tetrahedron.

The best way to show this is to use a molecular model. Figure 5 shows how we represent the three-dimensional shape of the methane molecule on paper.

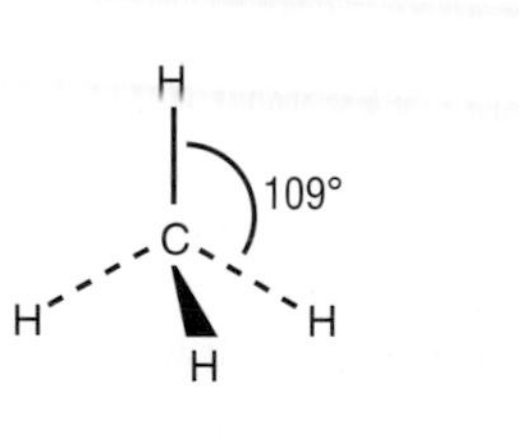

Represents a bond in the plane of the paper

Represents a bond in a direction behind the plane of the paper

Represents a bond in a direction in front of the plane of the paper

◀ **Figure 5** The three-dimensional shape of methane.

The structure of ethane in three dimensions is shown in Figure 6. Each carbon atom is at the centre of a tetrahedral arrangement.

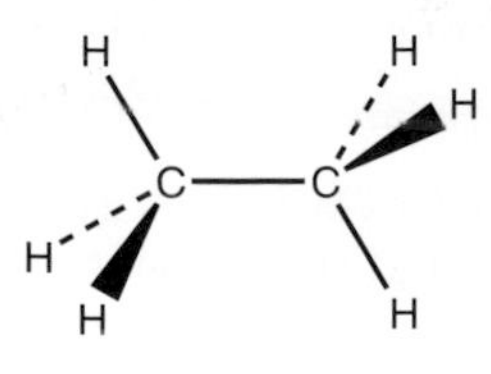

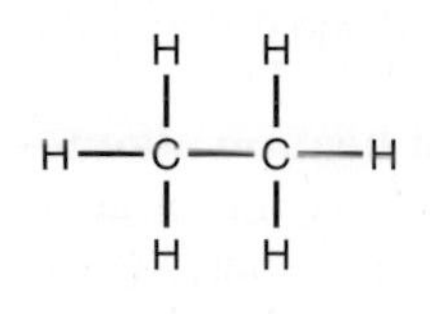

A simpler way of drawing ethane which shows the shape less accurately

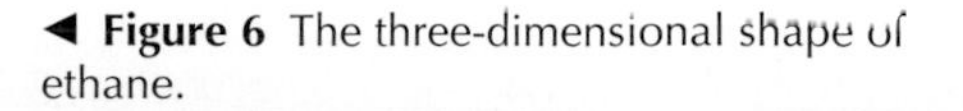

◀ **Figure 6** The three-dimensional shape of ethane.

Figure 7 shows the three-dimensional structure of butane. You can see that hydrocarbon chains are not really 'straight', but a zig-zag of carbon atoms. All the bond angles are 109°.

The shape of a hydrocarbon chain is often represented by a skeletal formula, also shown in Figure 7.

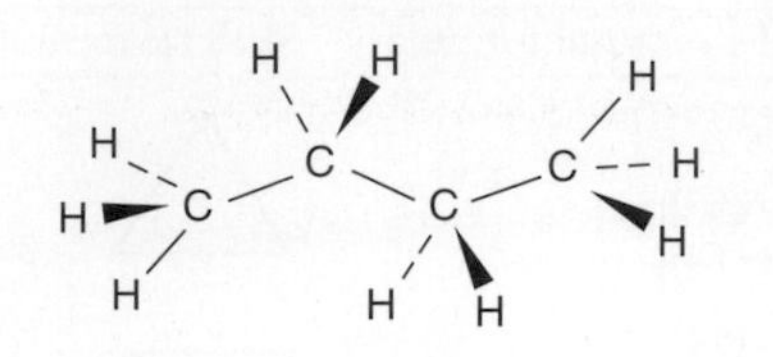

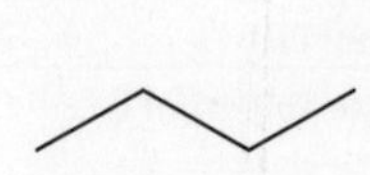

► **Figure 7** The three-dimensional shape of butane.

What are alkanes like?

Whether an alkane is solid, liquid or gas at room temperature depends on the size of its molecules. The first four members of the series (C_1–C_4) are colourless gases. Higher members (C_5–C_{16}) are colourless liquids and the remainder are white waxy solids.

Alkanes mix well with each other, but do not mix with water. The alkanes and water form two separate layers. This is because alkanes contain non-polar molecules but liquids such as water and methanol contain polar molecules which attract each other and prevent the alkane molecules mixing with them. (There is more about the polarity of molecules in **Section 5.3**.)

Chemical reactions of alkanes

Alkanes are very unreactive towards many laboratory reagents. They are unaffected by polar reagents such as acids and alkalis, or oxidising agents such as potassium manganate(VII).

When they do react, it is usually in the gas phase and energy must be supplied to get the reaction started.

Oxidation of alkanes

Alkanes do not react with air at room temperature but, if heated, they burn readily to give carbon dioxide and water. Energy must be supplied to stretch and break the bonds in the alkane and in oxygen before the new bonds in carbon dioxide and water can start to form (see **Section 4.2**).

Alkanes must be vaporised before they will burn, and so the less volatile alkanes ignite less readily. This is why petrol ignites so much more easily than oil.

Once started, the combustion is very exothermic, which is why alkanes are such good fuels.

$$CH_4(g) + 2O_2(g) \rightarrow CO_2(g) + 2H_2O(l); \quad \Delta H_c^{\ominus} = -890\,kJ\,mol^{-1}$$
$$C_7H_{16}(g) + 11O_2(g) \rightarrow 7CO_2(g) + 8H_2O(l); \quad \Delta H_c^{\ominus} = -4817\,kJ\,mol^{-1}$$

If the supply of air is limited, the combustion may be **incomplete** and produce CO and C (soot) along with partially oxidised hydrocarbons.

Action of heat on alkanes

The alkanes in crude oil are separated into fractions by fractional distillation – for example, naphtha (C_6–C_{10}), kerosene (C_{10}–C_{16}) and gas oil (C_{14}–C_{20}) (see **DF3** in the **Developing Fuels** module).

These fractions are then heated under different conditions, when three types of reaction can occur: **isomerisation**, **reforming** and **cracking**. These processes are summarised in Table 5.

▼ **Table 5** The action of heat on alkanes. Skeletal formulae are used in the examples. [benzene ring symbol] is the skeletal formula for a benzene ring.

	Alkanes			
	Isomerisation	**Reforming**	**Cracking**	
			Steam cracking	**Catalytic cracking**
Feedstock	C_4–C_6 alkanes	Naphtha	Naphtha/kerosene	Gas oil
Product molecules	Same molecular formula as reactants; branched	Same number of C atoms as reactants; cyclic	More molecules; fewer C atoms than reactants; small molecules such as C_2H_4, C_2H_6, C_3H_6, C_3H_8, C_4 and C_5 alkenes and alkanes	More molecules; fewer C atoms than reactants; some unsaturated; some branched; some cyclic
Conditions	Pt/Al_2O_3; 150 °C	Pt/Al_2O_3; 500 °C; hydrogen is recycled through mixture to reduce 'coking'	No catalyst; 900 °C; short residence time; steam is a diluent to prevent 'coking'	Zeolite; 500 °C
Example		+ H_2 + $3H_2$	+	+
Uses of products	To improve octane number of petrol	To improve octane number of petrol	To manufacture polymers	To improve octane number of petrol

Problems for 12.1

1 Copy and complete the following table.

Empirical formula	Molecular formula	M_r
?	C_3H_8	44.0
CH_2	?	168.0
?	C_6H_6	?
$C_{10}H_{21}$	?	282.0
?	C_5H_{10}	?
CH	?	26.0
?	$C_{10}H_8$	?

2 A hydrocarbon contains 85.7% C and 14.3% H by mass. Its relative molecular mass is 28.0.

a Find its empirical formula.

b Suggest a molecular formula for this compound.

3 A hydrocarbon contains 82.8% by mass of carbon.

a Work out its empirical formula.

b Suggest its molecular formula, giving your reasons.

4 A hydrocarbon was subjected to combustion analysis. A sample of the compound gave 0.314 g CO_2 and 0.129 g H_2O.

a Calculate the masses of carbon and hydrogen in the sample.

b Find the number of moles of carbon and hydrogen in the sample of the compound.

c What is the empirical formula of the compound?

d If its relative molecular mass is 84.0, what is its molecular formula?

e Draw a possible skeletal formula.

5 Which of the following formulae represent **a** alkanes, **b** cycloalkanes?

A C_6H_{12}
B C_2H_6
C C_5H_{10}
D C_7H_{16}
E C_9H_{20}
F $C_{13}H_{26}$

6 What is the molecular formula of each of the following alkanes?

a heptane (seven carbon atoms)

b cyclooctane (eight carbon atoms)

c hexadecane (sixteen carbon atoms).

7 Draw 'dot–cross' diagrams and the corresponding full structural formulae for:

a ethane

b propane

c cyclopropane.

8 Name the following hydrocarbons:

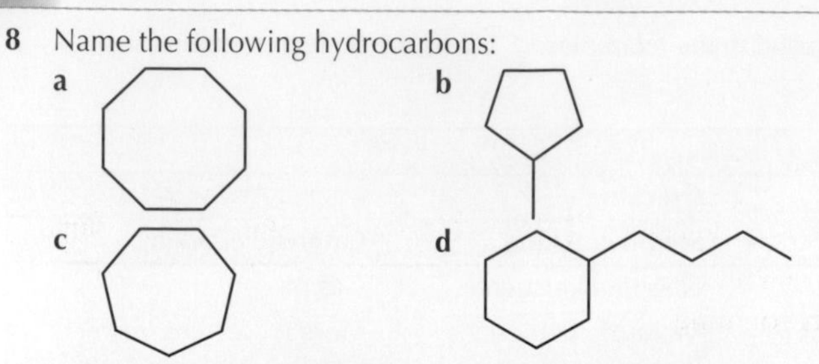

9 Draw skeletal formulae for
 a cyclopentane
 b methylcyclohexane
 c propylcyclopropane
 d decane, $C_{10}H_{22}$
 e heptane, C_7H_{16}
 f methylpropane.

10 There are two different compounds with the molecular formula C_4H_{10}. Draw skeletal formulae for these and give their systematic names.

11 Write balanced equations for each of the following reactions.
 a Pentane, C_5H_{12}, burns in a plentiful supply of air to form carbon dioxide and water.
 b Pentane burns in a limited supply of air to form carbon monoxide and water.

12 When an alkane is cracked, each molecule forms at least two new molecules.
 a What reaction conditions are needed to cause cracking reactions in alkanes?
 b Which of the following rules are true in writing an equation for cracking?
  i There are more total molecules on the reactant side.
  ii There are more total molecules on the product side.
  iii All the crackate molecules are unsaturated.
  iv Some of the crackate molecules are unsaturated.
  v Crackate molecules are always smaller.
 c Write *three* different equations for the cracking of heptane. You will probably find it easier to use skeletal formulae.

13 Look at the following hydrocarbons:
 A $CH_3–CH_2–CH_3$
 B

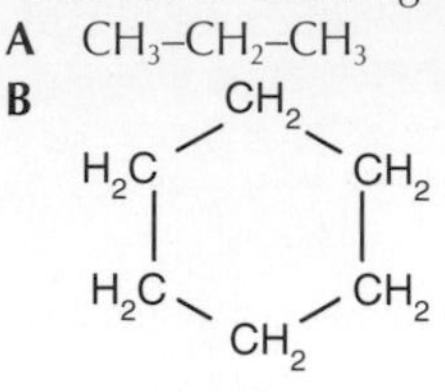

 C

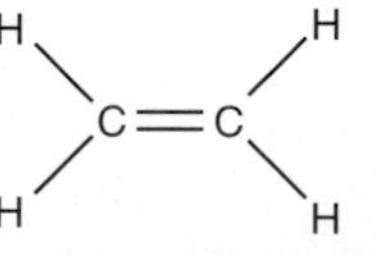

 D $CH_3–(CH_2)_{20}–CH_3$
 E $CH_3–(CH_2)_4–CH_3$
 a Draw a skeletal formula for each of the hydrocarbons **A** to **E**.
 b Which hydrocarbon(s) would you expect to be solid at 298 K?
 c Which hydrocarbon would you expect to be the most volatile?
 d Hydrocarbon **B** can be made from **E** by a reforming reaction. Write a balanced equation for this reaction.
 e Which two hydrocarbons are isomers?
 f Give an example of a hydrocarbon that might have been formed by cracking another of the hydrocarbons shown here. (Identify both alkanes.) Write a balanced equation for this process.
 g For which hydrocarbon(s) are there no isomeric compounds?

12.2 *Alkenes*

What are alkenes?

Ethene is the simplest example of a class of hydrocarbons called **alkenes**. It has the structure

$$\mathrm{H_2C{=}CH_2}$$

Alkenes are distinguished from other hydrocarbons by the presence of the C=C double bond. The double bond implies that they are **unsaturated hydrocarbons**.

Examples of other members of the alkene family are:

propene	$CH_3–CH=CH_2$	
but-1-ene	$CH_3–CH_2–CH=CH_2$	
but-2-ene	$CH_3–CH=CH–CH_3$	
pent-1-ene	$CH_3–CH_2–CH_2–CH=CH_2$	

As with the alkanes, the boiling points of alkenes increase as the number of carbon atoms increases. Ethene, propene and the butenes are gases. After that they are liquids, and eventually solids.

Notice how the alkenes are named. Take but-1-ene as an example – the *but-* part tells you it has the same carbon chain length as butane. The *-ene* suffix (the end part of the name) tells you it is an alkene and has a C=C double bond. The number tells you which carbon atom in the chain is the first to be involved in the double bond, assuming that you start counting from the end of the chain which is closest to the double bond.

The names of all alkenes end in **-ene**.

Note that all non-cyclic alkenes have the general formula, C_nH_{2n}, where $n = 2, 3, 4 \ldots$ This is the same as the general formula of the cycloalkanes – e.g. cyclohexane – but the double bond makes alkenes react very differently from cycloalkanes.

There are **cycloalkenes** – such as cyclohexene, an important intermediate in the production of some types of nylon – and there are **dienes**, such as penta-1,3-diene.

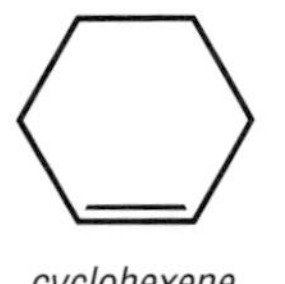

cyclohexene

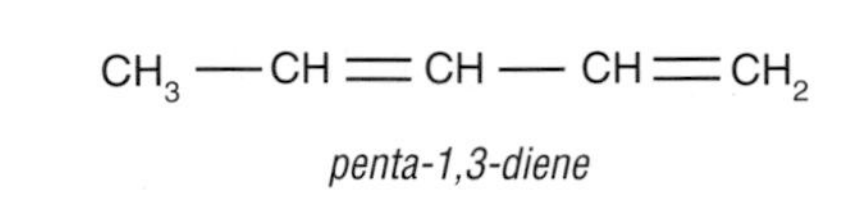

penta-1,3-diene

Shapes of alkenes

Ethene is a flat molecule with the shape and bond angles shown in Figure 1.

The bonds around the C=C group are arranged the same way in all alkenes. Take propene for example – its shape is shown in Figure 2. Although there is a tetrahedral arrangement of bonds around carbon atom 3, the bonds around carbon atoms 1 and 2 are planar.

One way of explaining this is to say that the planar arrangement places the groups of electrons around carbon atoms 1 and 2 as far apart as possible. The groups of electrons constitute regions of negative charge, and repulsions between them are minimised when the angles are 120°. (The four electrons in the double bond count as one region of negative charge.)

There is more about the shapes of molecules in general in **Section 3.2**.

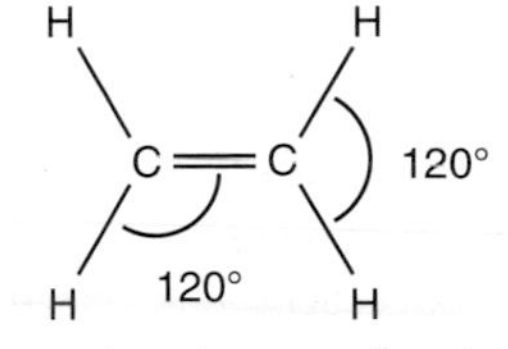

▲ **Figure 1** The ethene molecule is flat.

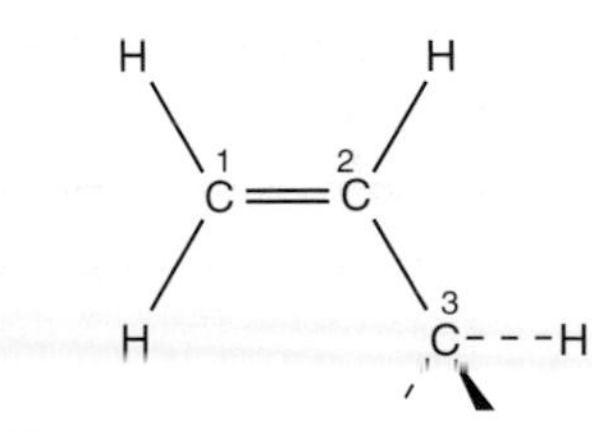

▲ **Figure 2** The three-dimensional shape of propene.

Chemical reactions of ethene

The four electrons in the double bond of ethene give the region between the two carbon atoms a higher than normal density of negative charge. Positive ions, or molecules with a partial positive charge on one of the atoms, will be attracted to this negatively charged region. They may then go on to react by accepting a pair of electrons from the C=C double bond. When they do this we describe them as **electrophiles**. Compare them with *nucleophiles*, which are attracted to regions of positive charge on a carbon atom (**Section 13.1**). Most of the reactions of alkenes involve electrophiles.

An electrophile is a positive ion or a molecule with a partial positive charge that will be attracted to a negatively charged region and react by accepting a lone pair of electrons to form a covalent bond.

Electrophilic addition reactions

Reaction with bromine

When we bubble ethene gas through bromine, the red-brown bromine becomes decolorised – this is a good general **test for unsaturation** in an organic compound.

The ethene reacts with bromine to form 1,2-dibromoethane.

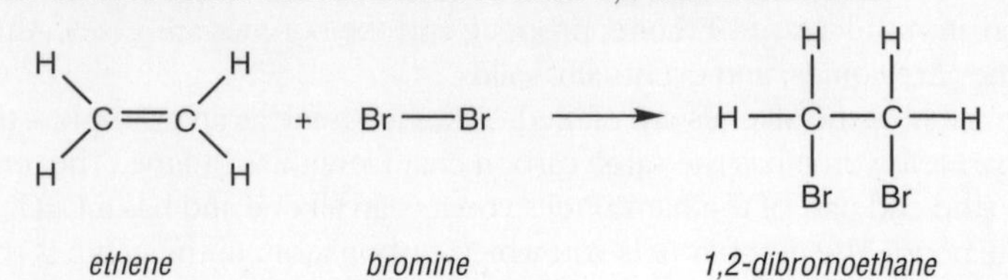

Let's look at the mechanism of this reaction.

Chemists believe that the bromine molecule becomes **polarised** as it approaches the alkene. This means that the electrons in the bromine are repelled by the alkene electrons and are pushed back along the molecule. The bromine atom nearest the alkene becomes slightly positively charged and the bromine atom furthest from the alkene becomes slightly negatively charged.

The positively charged bromine atom now behaves as an electrophile and reacts with the alkene double bond

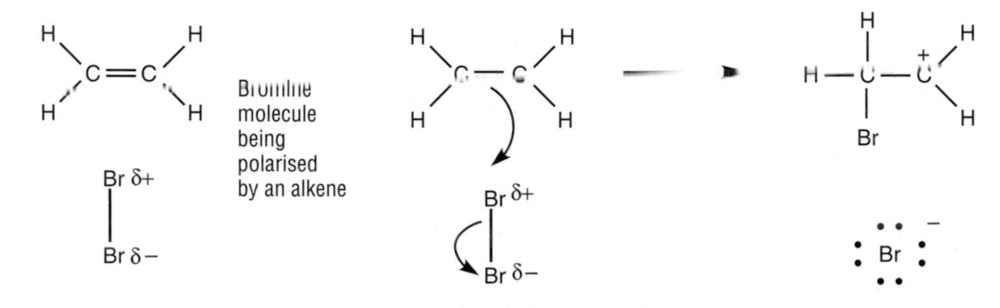

Remember, we use curly arrows like these to represent the movement of pairs of electrons in chemical reactions.

Notice that one of the carbon atoms now has a share in only six outer electrons. It has become positively charged: it is a **carbocation**. Carbocations react very rapidly with anything that has electrons to share – such as the bromide ion. A pair of electrons moves from the Br^- ion to the positively charged carbon to form a new C–Br covalent bond.

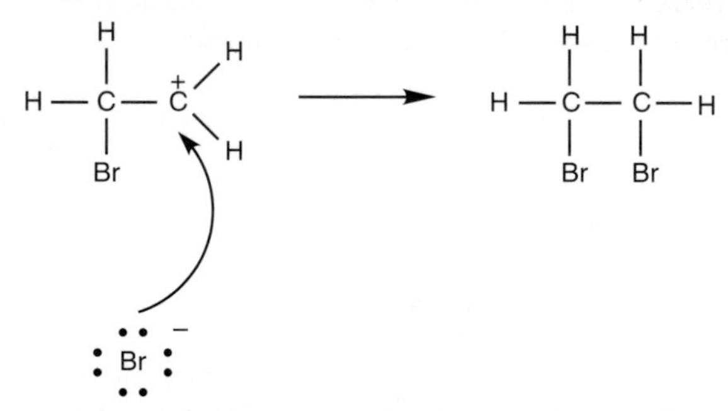

(Note that the Br^- could attack from *either side* of the positively charged carbon atom – here we have shown it attacking from below.)

This is a simplified version of the mechanism. An alternative interpretation involves an intermediate bromonium ion. You may come across this mechanism in post-A level studies.

The overall equation for the two steps in the mechanism is

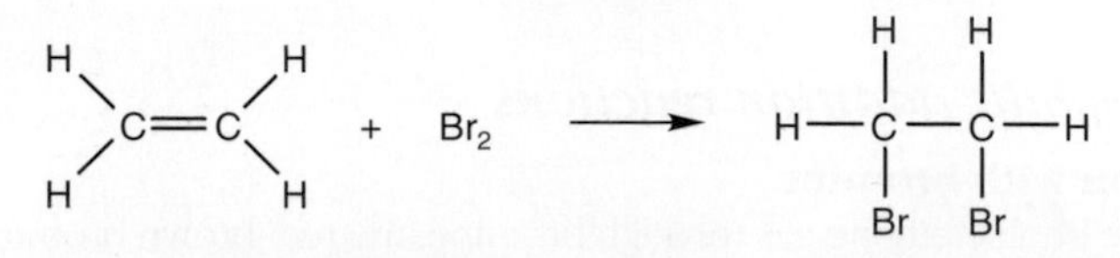

This equation represents an **addition reaction** and, since the initial attack is by an electrophile, the process is called an **electrophilic addition**.

An addition reaction is one where two or more molecules react to form a single larger molecule.

The mechanism for electrophilic addition via a carbocation is supported by experimental evidence. For example, if hex-1-ene reacts with bromine in the presence of chloride ions, two products are formed: 1,2-dibromohexane and 1-bromo-2-chlorohexane.

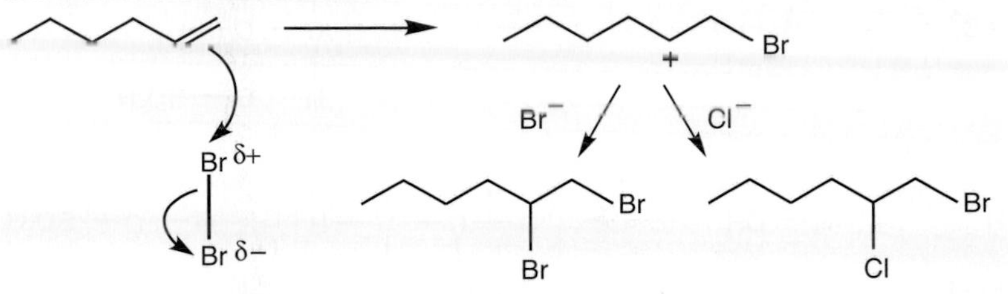

Often the test for an alkene involves shaking the alkene with *bromine water* rather than pure bromine. In this case, there is an alternative to the second stage in the reaction. Water molecules have lone pairs of electrons and can act as nucleophiles in competition with Br^- ions.

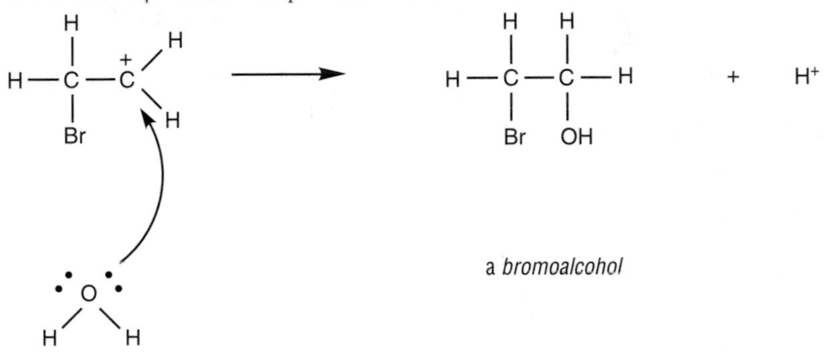

If the bromine water is dilute, there will be many more H_2O molecules than Br^- ions present and the bromoalcohol will be the main product of the reaction. This does not affect what you *see* – the bromine water is still decolorised.

Reaction with hydrogen bromide

Ethene reacts readily at room temperature with a solution of HBr in a polar solvent. It is another example of electrophilic addition.

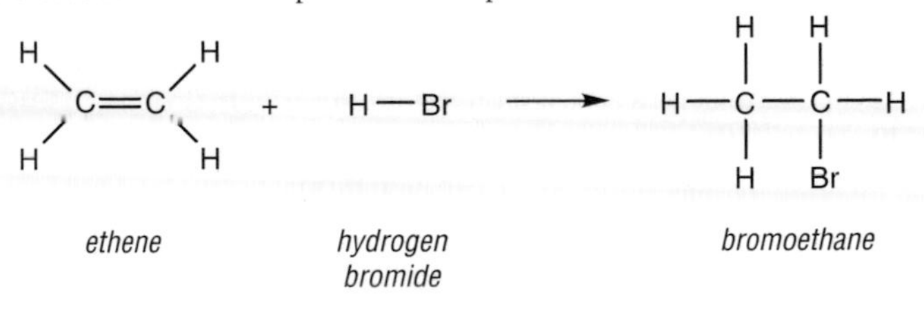

Alkenes also react with *gaseous* HBr but ions are not involved and the mechanism involves a *radical* addition. (The conditions under which a reaction is carried out can be very important in determining the mechanism.)

Reaction with water

In the presence of a catalyst (phosphoric acid adsorbed onto solid silica) and high temperature and pressure, ethene and water (as steam) undergo an addition reaction. The process is used for the industrial manufacture of ethanol.

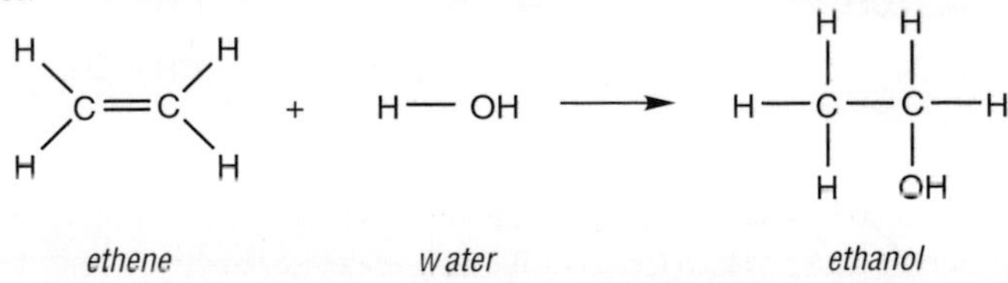

In the laboratory, ethene can be converted to ethanol by first adding concentrated sulfuric acid, and then diluting with water.

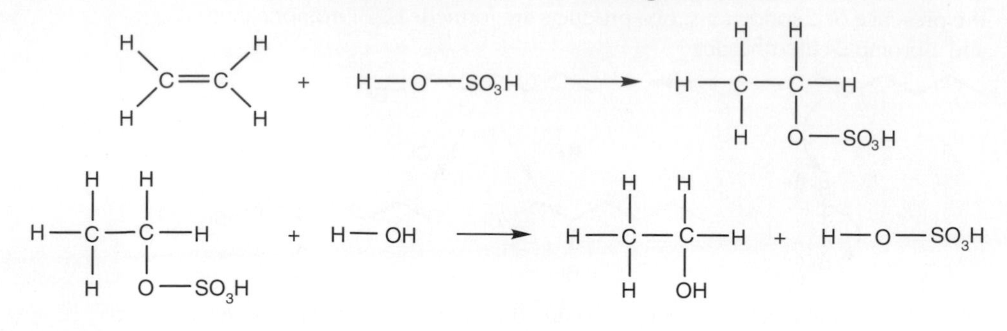

The overall reaction is addition of water across the double bond.

The addition of water to an alkene is an example of a **hydration** reaction.

Industrially, the hydration can be carried out using phosphoric acid (H_3PO_4) adsorbed onto silica and steam, with heating under pressure (approx. 300 °C and 60 atm).

Reaction with hydrogen

This is another example of an addition reaction but here the mechanism involves hydrogen atoms, and takes place on the surface of a catalyst. A catalyst is needed to help break the strong H–H bond and form H atoms before alkenes react with hydrogen. If a platinum catalyst is used the process takes place under normal laboratory conditions. Nickel is a cheaper but less efficient catalyst. It needs to be very finely powdered and the gases need heating to approximately 150 °C under a pressure of 5 atm for **hydrogenation** to occur.

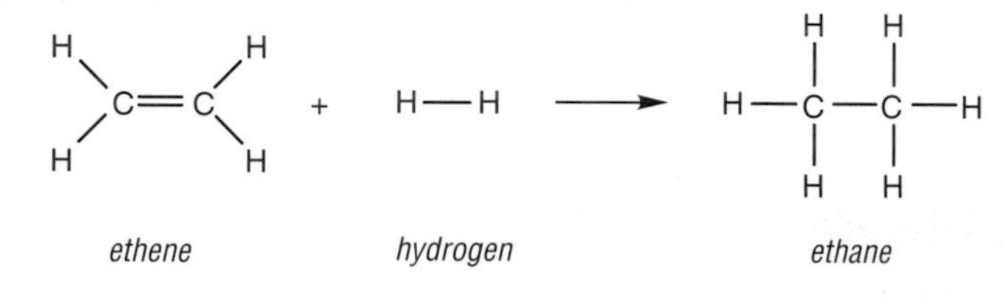

Figure 1 in **Section 10.5** suggests a possible mechanism for this reaction.

Margarine manufacture is based on the addition of hydrogen to C=C bonds. Plant and animal oils, which are unsaturated, are *hardened* (or made more solid and fatty like butter) by hydrogenation.

You have probably come across the term 'unsaturated' in relation to margarines 'high in polyunsaturates'. The oils from which the margarine is made contain molecules with several C=C double bonds. After hydrogenation, the resulting margarines still have some C=C double bonds.

Ethene can also undergo addition reactions to form polymers – see **Section 5.6**.

Problems for 12.2

1 Draw skeletal formulae for the following alkenes:

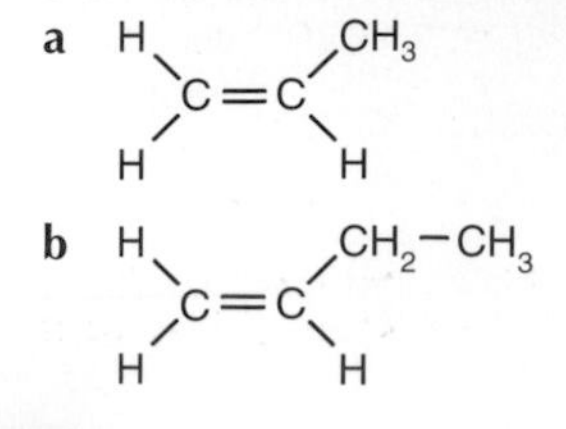

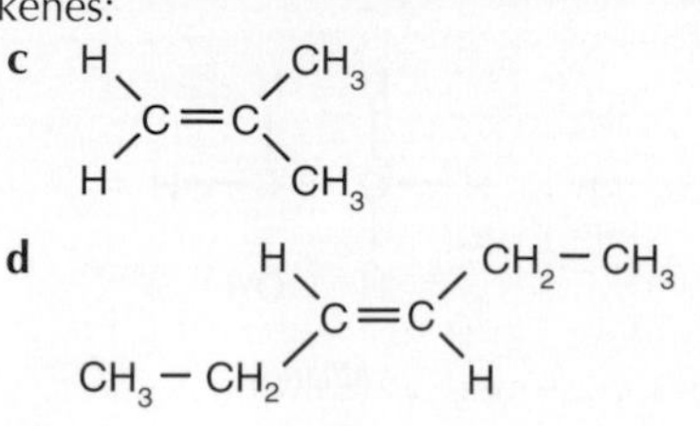

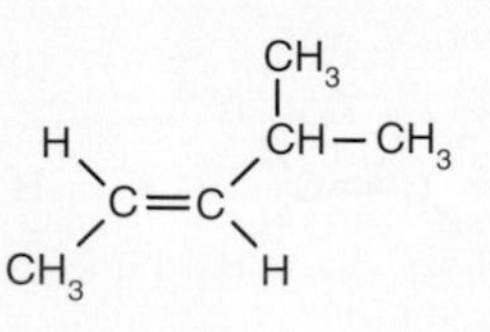

2 Name the following alkenes from their skeletal formulae:

a
b
c
d
e
f

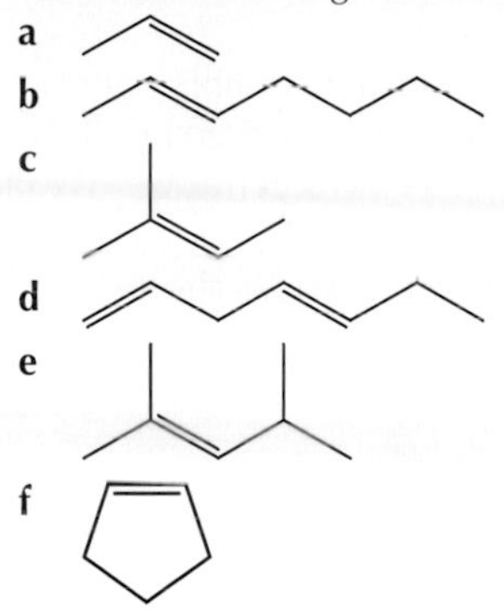

3 Draw skeletal formulae for the following alkenes:
   a pent-2-ene
   b hex-3-ene
   c 2,3-dimethylpent-2-ene
   d cyclopenta-1,3-diene
   e 3-ethylhept-1-ene.

4 Draw the full structural formula of 2,3-dimethylpent-2-ene.

5 Use full structural formulae to write an overall equation for
   a the reaction of propene with bromine
   b the reaction of propene with hydrogen.

6 The reaction of propene with hydrogen bromide can give two different products. Draw the full structural formulae of these products.

7 Use skeletal formulae to write an overall equation for the reaction of
   a pent-2-ene with bromine
   b but-2-ene with bromine water
   c but-2-ene with hydrogen bromide
   d hexa-1,4-diene with excess hydrogen
   e cyclopentene with hydrogen
   f cyclohexa-1,3-diene with bromine.

8 Give the reagents and conditions needed to carry out each of the following reactions.
   a Bromination of alkenes.
   b Industrial production of ethanol from ethene.
   c Laboratory hydrogenation of alkenes to alkanes.
   d Industrial production of margarine from sunflower oil.

9 a Using, as a guide, the steps given on page 274 for the electrophilic addition of bromine to an alkene, use curly arrows to draw out a mechanism for the electrophilic addition of HBr to ethene.
   b Give a description in words for each step.

10 Describe a test you could carry out in the laboratory to show the presence of unsaturation in an organic compound. State what you would *do* and what you would *see*.

12.3 *Arenes*

What's special about benzene?

Benzene, the simplest arene, is a colourless liquid – nothing special there. It has a molecular formula C_6H_6 and so it must be very unsaturated. The puzzle is that benzene resists addition reactions and does not behave like a normal unsaturated compound. It doesn't, for example, decolorise bromine solution. It is much less reactive than you might think and has its own characteristic properties. The reason for this is to do with the rather special cyclic structure of benzene, which is shown in Figure 1.

The benzene ring is a flat hexagon (Figure 2).

All the bond angles are 120° and all the carbon–carbon bonds are the same length – less than for a carbon–carbon single bond but greater than for a carbon–carbon double bond (see Table 1).

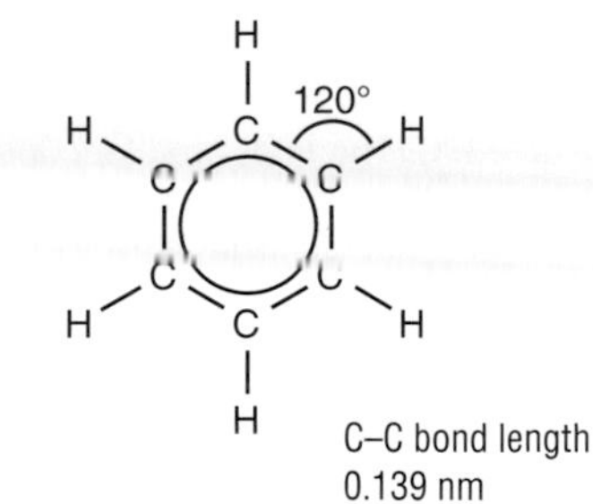

▲ **Figure 1** The structure of the benzene ring.

▼ **Table 1** Some carbon–carbon bond lengths.

	Bond	Bond length/nm
alkane	C—C	0.154
alkene	C═C	0.134
benzene ring	C┅C	0.139

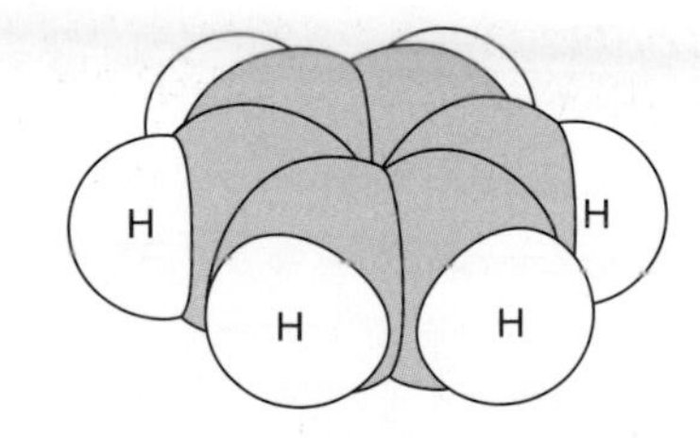

▲ **Figure 2** A diagram of a space-filling model of a benzene molecule, showing its planar shape.

Try drawing a dot–cross formula to represent the bonding in a benzene ring. Each carbon atom has four outer electrons which can be used to form bonds. Three of these electrons are used to form single bonds with the two carbon atoms next to them in the ring and with a hydrogen atom. This leaves one electron on each carbon atom.

Instead of overlapping in pairs to form three separate double bonds, these remaining electrons are spread out evenly and are shared by all six carbon atoms in the ring. The spreading out of electrons in this way is called **electron delocalisation**. You can see one way of drawing a dot–cross diagram for benzene in Figure 3. The shaded ring represents the six delocalised electrons, one from each carbon atom.

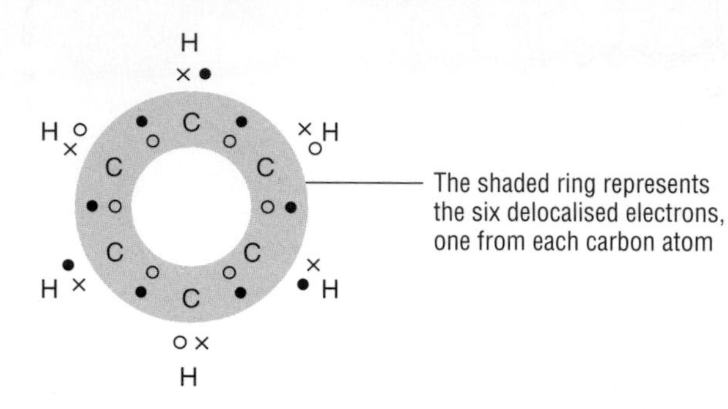

▶ **Figure 3** Dot–cross diagram for benzene – the open and filled circles represent electrons from C atoms; the crosses represent electrons from H atoms.

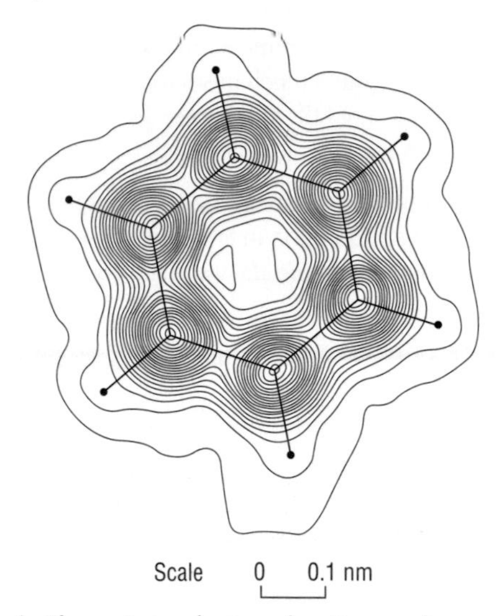

▲ **Figure 4** An electron density map for benzene at −3 °C. The lines are like contour lines on a map – they show parts of the molecule with equal electron density.

Additional evidence for the delocalised structure comes from **electron density maps**, which can be drawn up from X-ray diffraction studies. Figure 4 shows a uniform electron density in all the carbon–carbon bonds.

You can think of the delocalised structure of benzene as halfway between the two extreme structures:

These are sometimes called **Kekulé structures** because they were first proposed by August Kekulé in 1865. Neither of these forms actually exists, and the delocalised arrangement is often represented by drawing a circle inside the ring.

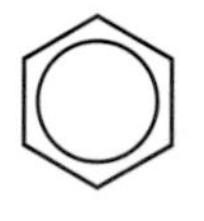

Stability of benzene

The cyclic delocalisation of electrons in benzene makes the molecule more stable than you would expect if it had a Kekulé-type structure, with three double bonds and three single bonds. Electrons repel one another, so a system in which they are delocalised and as far apart from one another as possible will involve minimum repulsion and will be stabilised.

The stability of the benzene ring has an important effect on the way benzene reacts – it tends to undergo reactions in which the stable ring is preserved. Reactions which disrupt the delocalised electron system are less favourable and need higher temperatures and more vigorous conditions. You can find out more about the way benzene reacts in **Section 12.4**.

The results of thermochemical experiments give us a way of estimating just how much more stable a benzene ring is than a hypothetical Kekulé-type structure. One way of doing this is to measure the enthalpy change when benzene reacts with hydrogen to form cyclohexane:

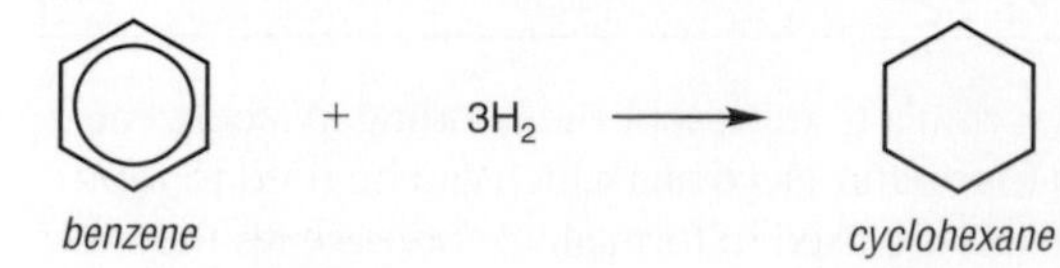

You can compare this with the value you would expect if benzene had the hypothetical Kekulé structure. This value cannot be measured of course, but you can make a reasonable estimate. The enthalpy change when *cyclohexene* reacts with hydrogen can be measured:

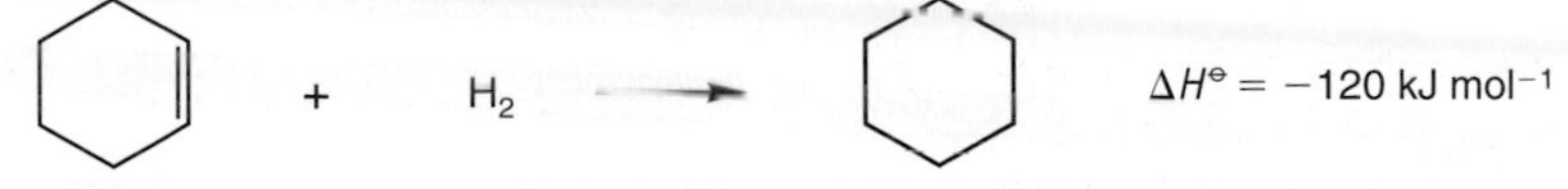

If you assume that the three double bonds in the Kekulé structure behave independently of each other, the enthalpy change of hydrogenation would be three times the value for cyclohexene:

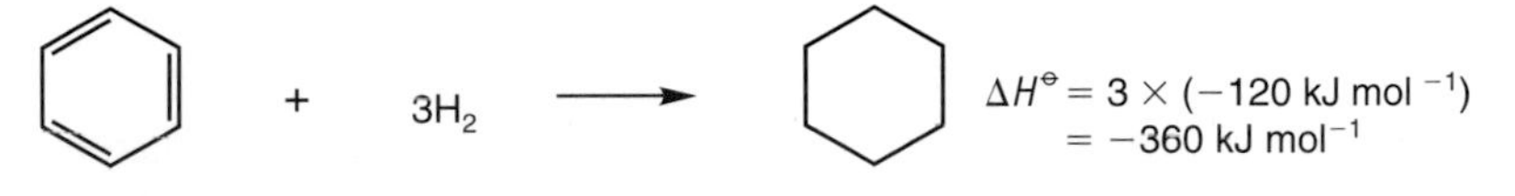

The enthalpy changes for these reactions are summarised in Figure 5.

So when 1 mole of benzene molecules are hydrogenated (360 – 208) kJ mol^{-1} = 152 kJ mol^{-1} less energy is given out than would be expected if they had the Kekulé structure. So, benzene must be more stable than the Kekulé structure by 152 kJ mol^{-1}.

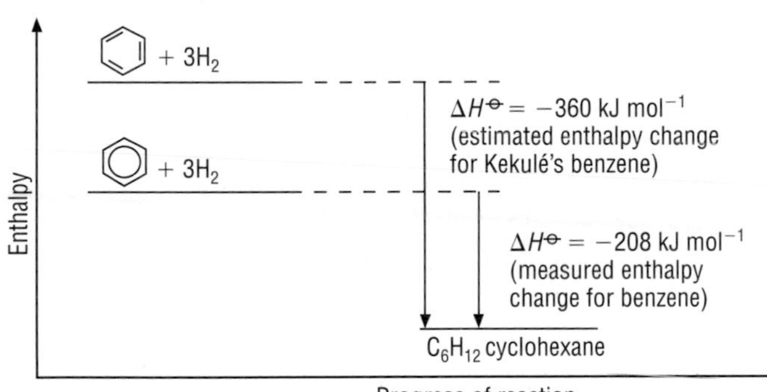

◀ **Figure 5** Enthalpy changes for the hydrogenation of benzene and the hypothetical Kekulé structure.

What are arenes?

Hydrocarbons like benzene, which contain rings stabilised by electron delocalisation, are called **arenes**. The ending '-ene' tells you they are unsaturated, like alkenes. The 'ar-' prefix comes from *aromatic*, which means 'sweet-smelling'. Arenes are sometimes called **aromatic hydrocarbons**.

The term *aromatic* was originally used to describe the characteristic fragrance (aroma) of some naturally occurring oils which contained benzene rings. It is now used in a much wider sense because not all arenes are sweet-smelling. Benzene itself has a rather strong, unpleasant smell.

There are many arenes. Below are some examples in which hydrogen atoms on the benzene ring have been replaced by alkyl groups – note how they are named.

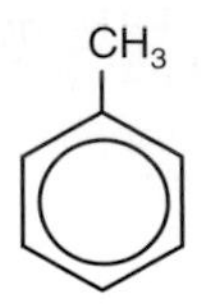

methylbenzene

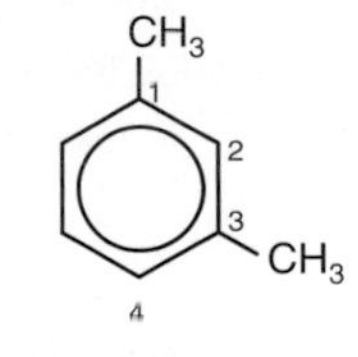

1,3-dimethylbenzene
(numbers as low as possible)

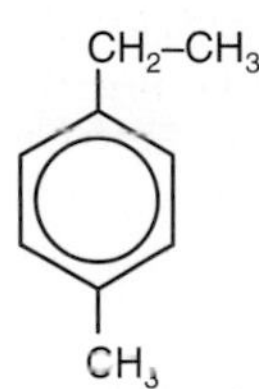

1-ethyl-4-methylbenzene
(groups in alphabetical order)

Phenylethene is used to make the polymer poly(phenylethene), more commonly known as polystyrene.

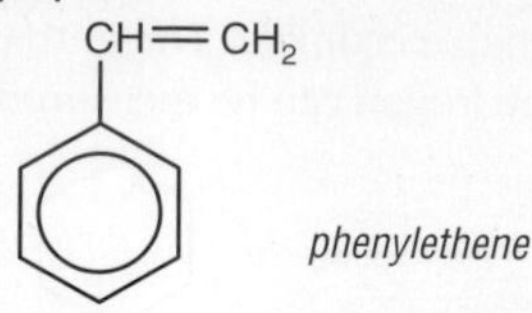

phenylethene

Notice that phenylethene is named from the parent compound ethene, rather than from benzene. A benzene ring in which one hydrogen atom has been substituted by another group is known as a **phenyl group**. The phenyl group has the formula C_6H_5– and the structure

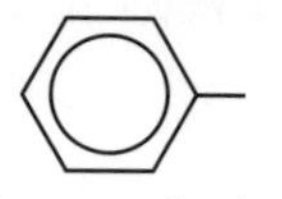

The phenyl group is related to benzene in the same way that a methyl group is related to methane.

Benzene rings can be 'joined together' to give **fused ring systems**, such as **naphthalene** and **anthracene**. Notice that where the rings join, they share a pair of carbon atoms. So naphthalene, with two rings, has 10 carbon atoms rather than 12.

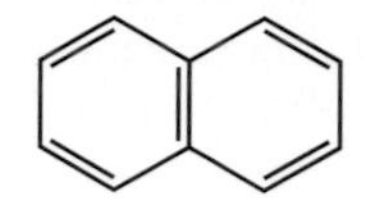

naphthalene, $C_{10}H_8$

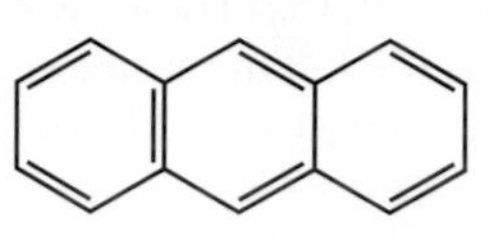

anthracene, $C_{14}H_{10}$

Like benzene, these molecules contain delocalised electrons but the delocalisation extends over all the rings. It is harder to represent the delocalisation in fused rings using circles. For example, the structure

incorrect representation

for naphthalene would suggest that the molecule contains two separate delocalised systems with six delocalised electrons in each. This is not the case, so we draw the rings with double bonds – but remember this is just a convenient representation of a complex delocalised structure.

Compounds derived from arenes

Hydrogen atoms on the benzene ring can be replaced by different functional groups. For example:

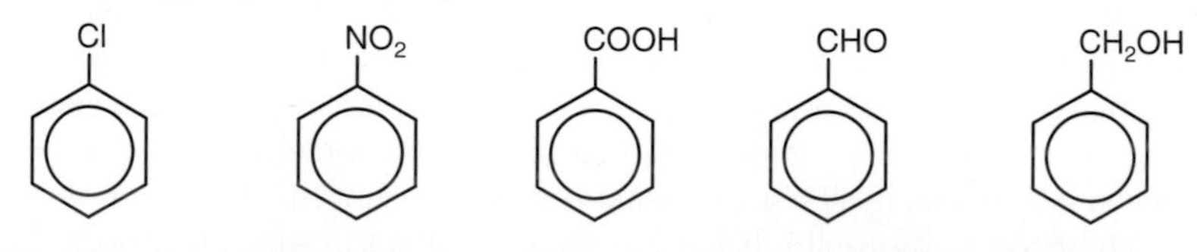

chlorobenzene *nitrobenzene* *benzoic acid* *benzaldehyde* *benzyl alcohol*

Below are two important aromatic compounds whose names are based on the phenyl group rather than on benzene – they are **phenol** (C_6H_5OH) and **phenylamine** ($C_6H_5NH_2$):

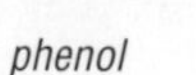

phenol

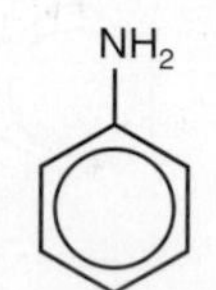

phenylamine

Problems for 12.3

Note that the structure of benzene is represented by

when writing both full structural formulae and skeletal formulae.

1 a Name the following arenes.

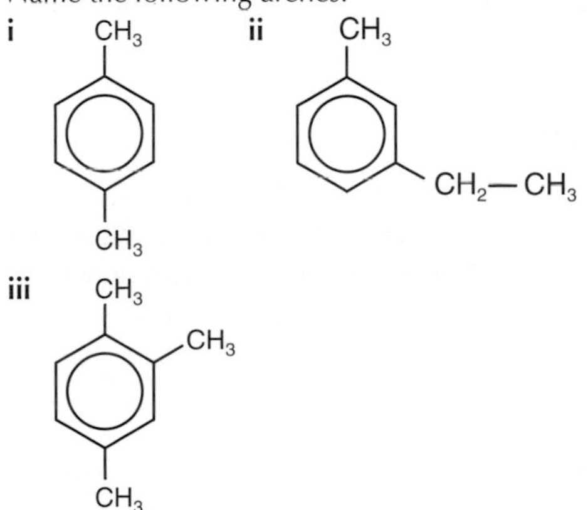

b Draw a skeletal formula for each of the compounds in part **a**.

2 Name the following compounds from their skeletal formulae.

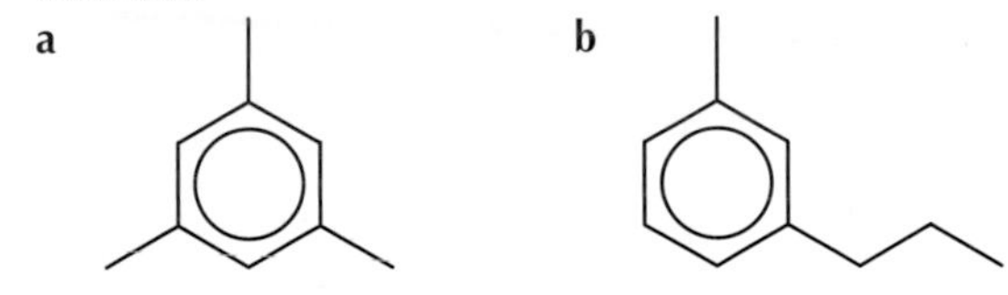

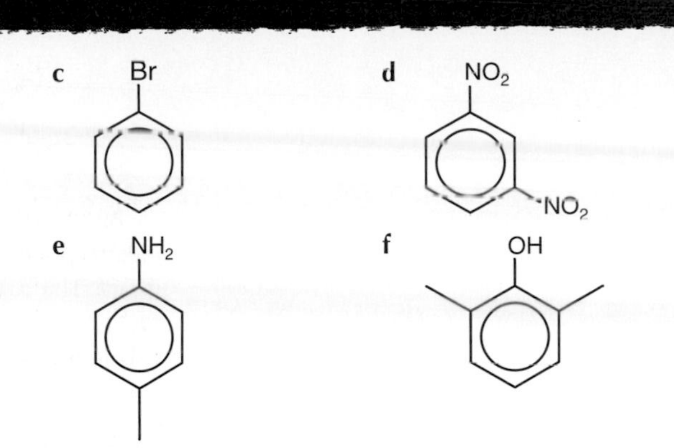

3 Draw skeletal formulae for the following compounds:
a 1,2-dimethylbenzene
b phenylethene
c 4-methylphenol
d 1,3-dichlorobenzene
e 1-ethyl-2,3-dimethylbenzene.

4 a Kekulé first proposed the cyclic structure for benzene in 1865. His ring contained three double bonds and three single bonds. Draw a skeletal formula with the bonds to scale (see Table 1 in this section, page 277). What effect would these bond lengths have on the overall shape of the ring?
b If Kekulé's structure was correct, there should be two isomers of 1,2-dimethylbenzene. Draw skeletal formulae for these isomers.
c In fact, only one form of 1,2-dimethylbenzene is known. Explain why this is so.

12.4 *Reactions of arenes*

The six electrons in the delocalised system in benzene do not belong to any particular carbon atom and are free to move around the ring. They are more loosely held than electrons in normal bonds. They spread out in a cloud which extends above and below the plane of the benzene ring (Figure 1).

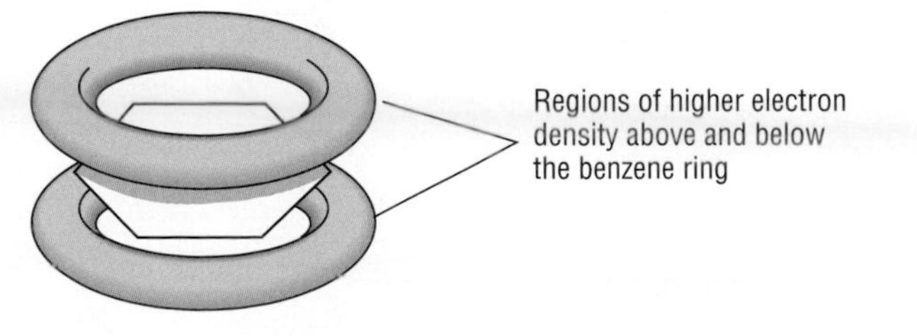

◀ **Figure 1** The regions of higher electron density above and below the benzene ring.

These regions of higher electron density tend to attract positive ions, or atoms with a partial positive charge within molecules – so benzene, like alkenes, reacts with **electrophiles**.

However, there is a significant difference between the behaviour of an alkene and an arene towards electrophiles.

Alkenes react with electrophiles in **addition reactions** (see **Section 12.2**). The product is a saturated molecule. For example,

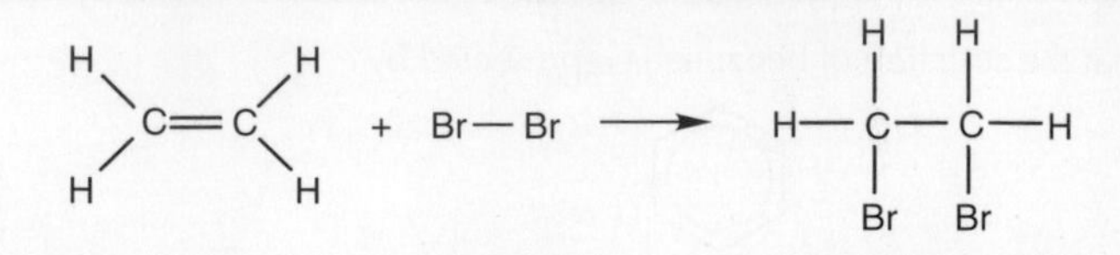

Benzene undergoes **substitution** rather than addition reactions with electrophiles. By reacting in this way, the stable benzene ring system is kept intact. The reactions are relatively slow because the first step in the reaction mechanism disrupts the delocalised electron system, and this requires a substantial input of energy.

Bromination of benzene

Benzene reacts with bromine in the presence of a catalyst, such as iron filings or iron(III) bromide. The bromine is decolorised and fumes of HBr are given off. The reaction which takes place is a substitution reaction:

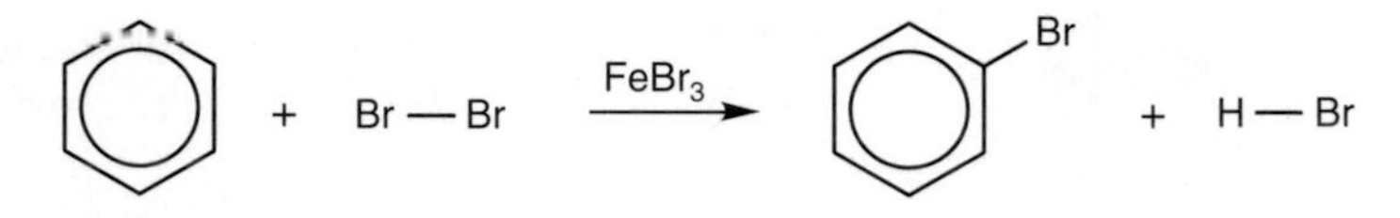

The first step involves reaction of the benzene ring with the electrophile Br^+ – so the process is called **electrophilic substitution**.

The bromine molecule becomes polarised as it approaches the benzene ring:

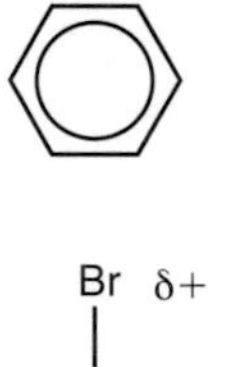

The positively charged end of the bromine molecule is now electrophilic. Even so, these polarised bromine molecules react only very slowly with the stable benzene ring. Iron filings help to speed up this reaction by making the bromine molecules more polarised. First, the iron reacts with bromine to form iron(III) bromide ($FeBr_3$):

$$2Fe + 3Br_2 \rightarrow 2FeBr_3$$

It is thought that the $FeBr_3$ helps to polarise the bromine molecule by accepting a lone pair from one of the bromine atoms. The bromine molecule becomes so polarised that it splits into Br^+ and $FeBr_4^-$:

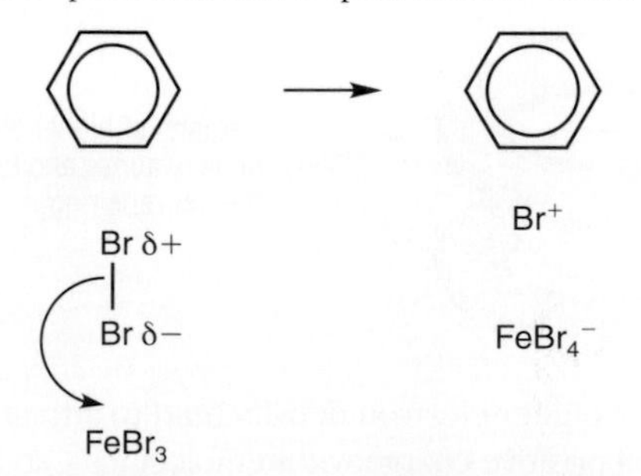

Br^+ then becomes bound to a ring carbon atom and H^+ is lost from the ring. This H^+ reacts with $FeBr_4^-$ to produce HBr and regenerate the $FeBr_3$ catalyst:

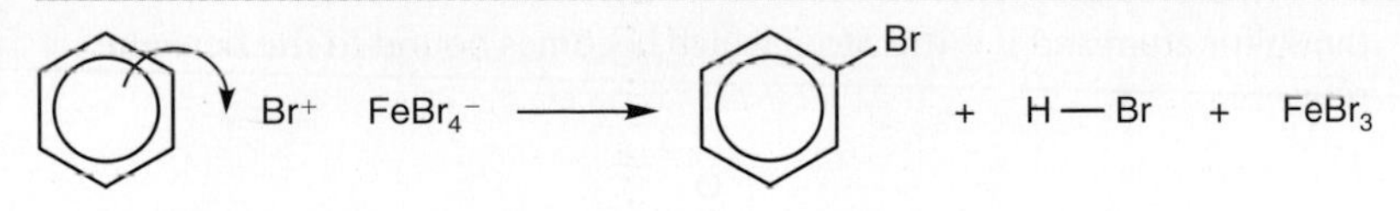

It is easy to see why addition reactions are much more difficult. If a molecule of benzene underwent addition with a molecule of bromine, the stable ring of delocalised electrons would be broken:

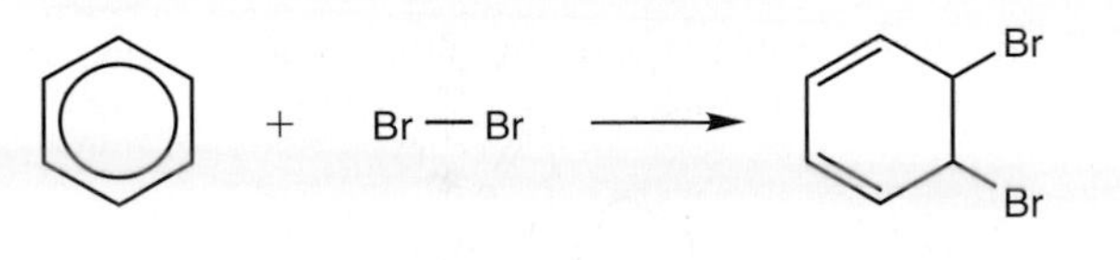

Other electrophilic substitution reactions of benzene

Benzene undergoes many substitution reactions. In every case, there is an electrophile which is attracted to the delocalised electron system in the benzene ring.

Nitration

Benzene reacts with a mixture of concentrated nitric acid and concentrated sulfuric acid – this is called a **nitrating mixture**. If the temperature is kept below 55 °C then the product is nitrobenzene:

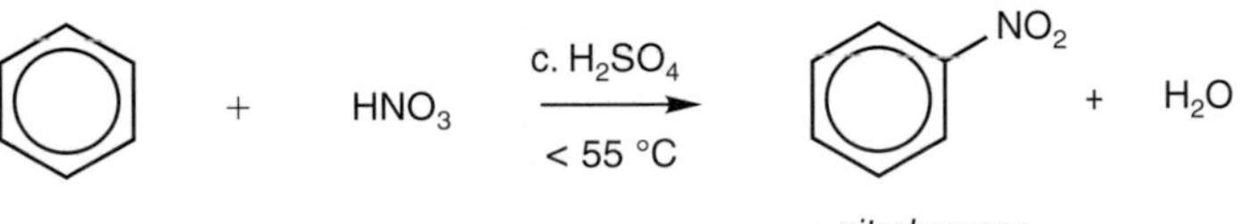

nitrobenzene

At higher temperatures, further substitution of the ring takes place to give the di- and tri-substituted compounds:

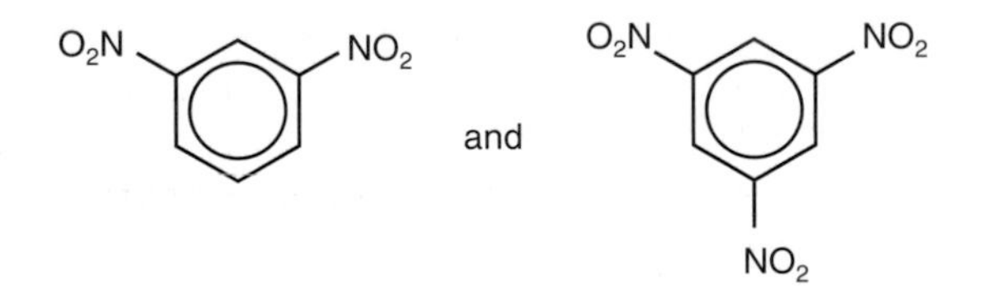

The electrophile which reacts with the benzene ring is the NO_2^+ cation. There is good evidence that this is formed in the nitrating mixture by the reaction of sulfuric acid with nitric acid:

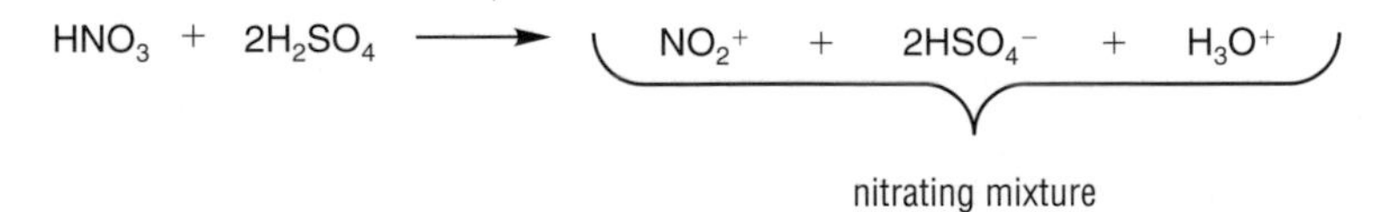

nitrating mixture

Sulfonation

When benzene and concentrated sulfuric acid are heated together under reflux for several hours, benzenesulfonic acid is formed:

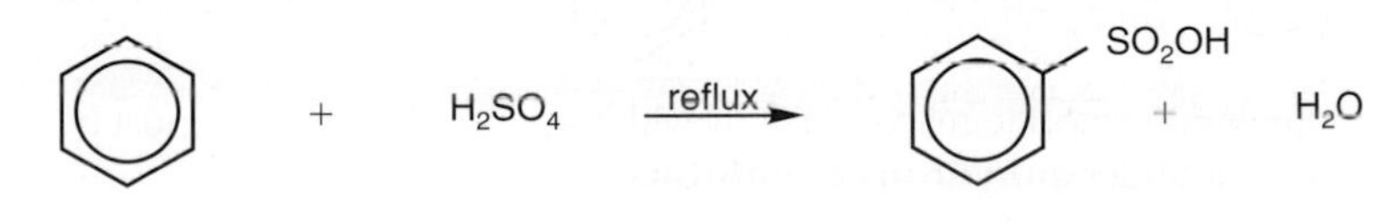

benzenesulfonic acid

The electrophile in this case is thought to be SO_3, which is present in the concentrated sulfuric acid. SO_3 carries a large partial positive charge on the sulfur atom and it is this atom which becomes bound to the benzene ring:

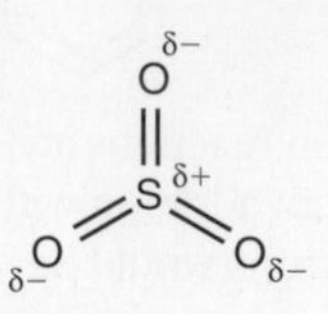

The full structural formula of benzenesulfonic acid is

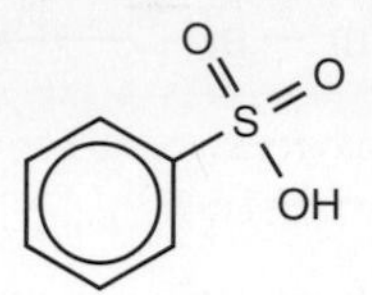

Benzenesulfonic acid is a strong acid and forms salts in alkaline solution:

sodium benzenesulfonate

Most solid detergents contain salts of this kind, with a long alkyl group attached to the benzene ring. The hydrocarbon part of the molecule mixes with fats, and the ionic part mixes with water.

Chlorination

A chlorine atom may be substituted into a benzene ring, in much the same way as a bromine atom. An aluminium chloride catalyst is often used:

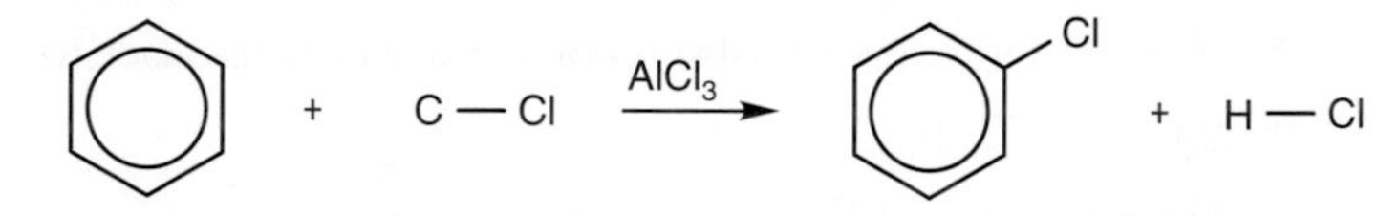

The mechanism is similar to the bromination reaction. The aluminium chloride helps to polarise the chlorine molecule – this produces the electrophile Cl^+, which reacts with the benzene ring to form chlorobenzene:

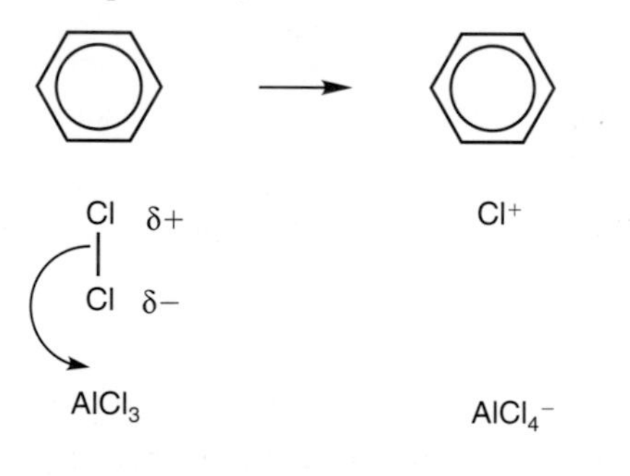

Then:

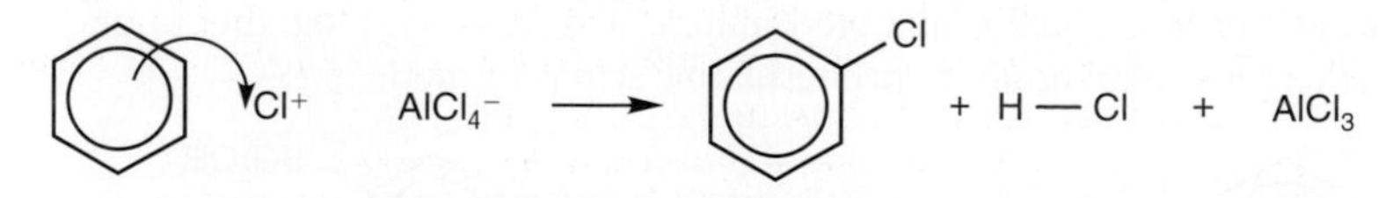

Aluminium chloride reacts violently with water, so the reaction must be carried out under **anhydrous conditions**.

Friedel–Crafts reactions

Aluminium chloride can also be used as a catalyst to help polarise *halogen-containing organic molecules* and cause them to substitute in a benzene ring. This type of reaction is called a **Friedel–Crafts reaction**, after its discoverers.

If benzene is warmed with chloromethane and anhydrous aluminium chloride, a substitution reaction occurs and methylbenzene is formed:

$$C_6H_6 + CH_3{-}Cl \xrightarrow[\text{reflux}]{AlCl_3} C_6H_5CH_3 + H{-}Cl$$

As in the reaction with halogens, the aluminium chloride helps to polarise the chloromethane molecule:

$$\overset{\delta+}{CH_3}{-}\overset{\delta-}{Cl} \quad AlCl_3$$

The positively charged carbon atom then reacts with the benzene ring to form methylbenzene. The reaction is called **alkylation** because an alkyl group is introduced into the ring.

A similar reaction takes place when benzene is treated with an acyl chloride (or an acid anhydride) and aluminium chloride. Here the reaction is an **acylation**:

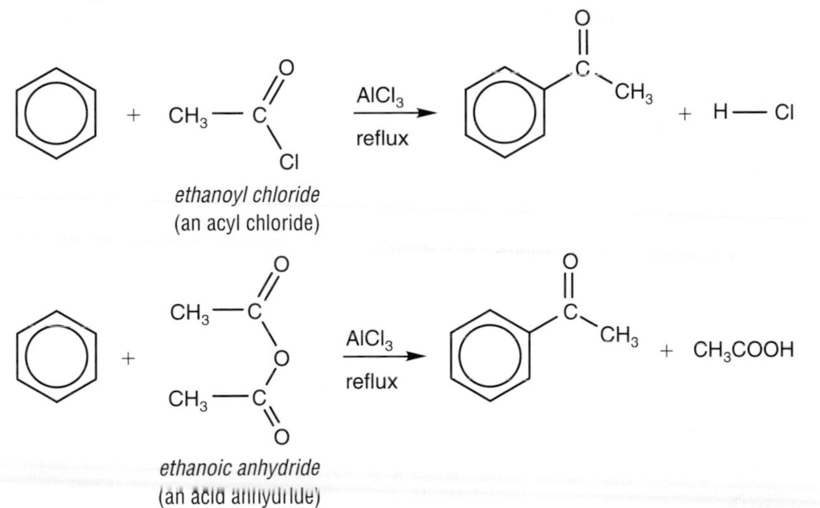

Friedel–Crafts reactions are particularly useful to synthetic chemists because they provide a way of adding carbon atoms to the benzene ring and building up side-chains.

More recently, Friedel–Crafts acylation reactions have been carried out using 'ionic liquids' as combined solvent/catalyst systems.

Ionic liquids are liquids at room temperature that contain only ions. They are usually made up of organic cations with either organic or inorganic anions and are highly conductive.

The formula of an ionic liquid can be written as $[Q]Cl{\cdot}2AlCl_3$, where Q is any of a range of organic cations, for example the 1-ethyl-3-methylimidazolium ion. (Don't worry, you don't have to remember this name!)

These liquids have been promoted as new 'green' alternatives to conventional solvents used in the laboratory because:

1. their low volatility reduces emissions;
2. their low flammability and low toxicity increases safety;
3. the temperatures at which Friedel–Crafts reactions can be carried out are often lower than by conventional methods;
4. the ionic liquid can be easily recycled, saving resources and money.

Summary

Benzene is an important starting material for the synthesis of many useful compounds – such as dyes, pharmaceuticals and perfumes. Electrophilic substitution reactions provide ways of introducing different functional groups into the ring. These groups may then be modified further to build up more complicated molecules.

The important electrophilic substitution reactions are summarised in Figure 2. They are *general* reactions and work on other arenes as well as benzene itself.

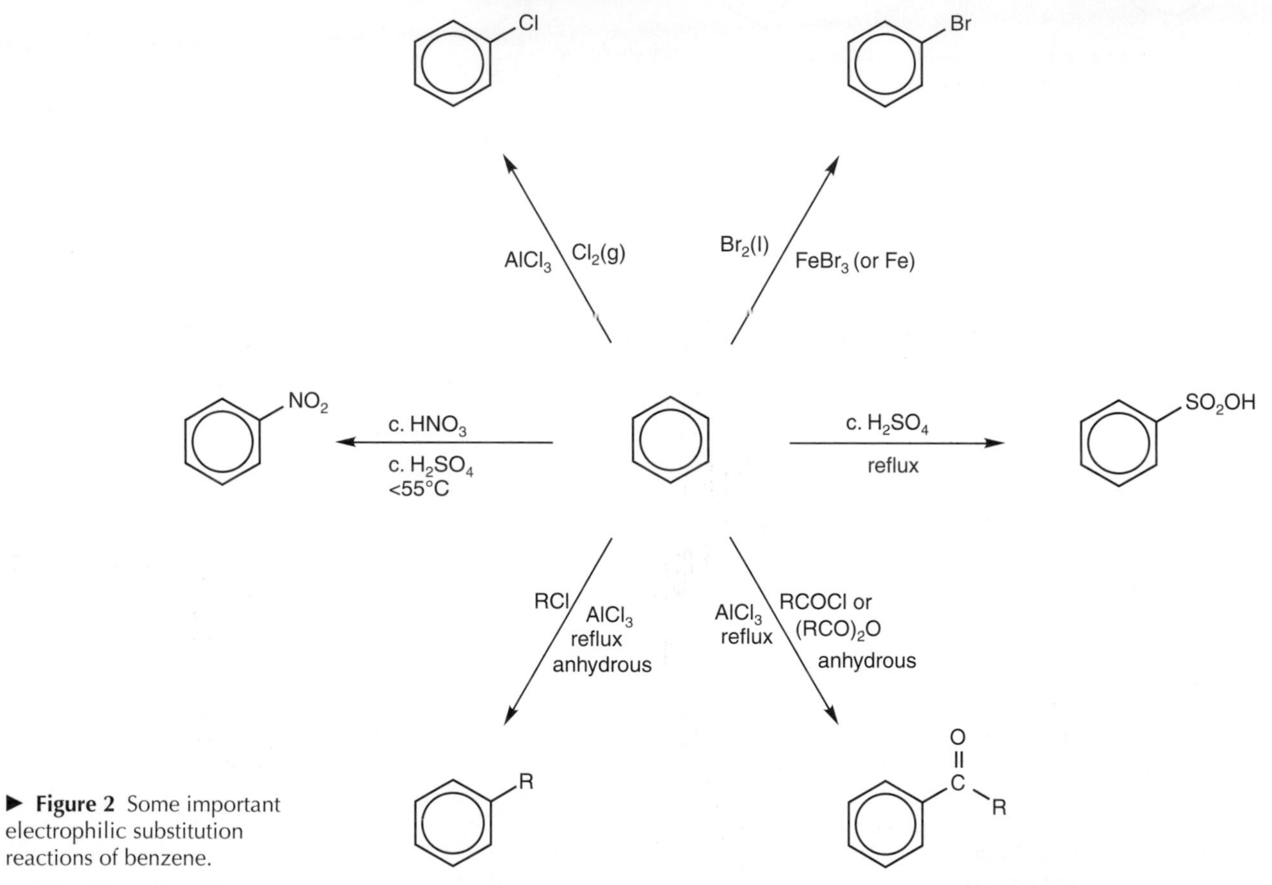

► **Figure 2** Some important electrophilic substitution reactions of benzene.

Problems for 12.4

1 Complete the following reaction schemes by drawing the structure of the missing reactant or product, or by writing the reagent and conditions on the arrow.

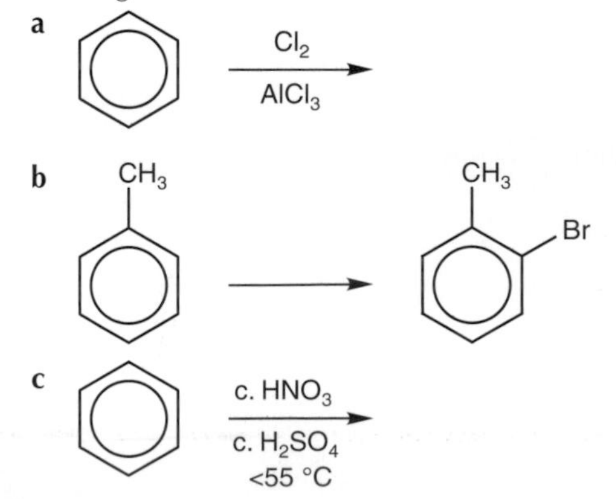

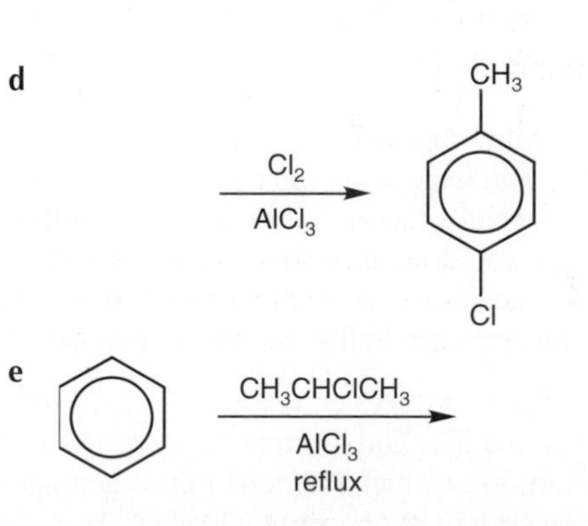

f, g, h

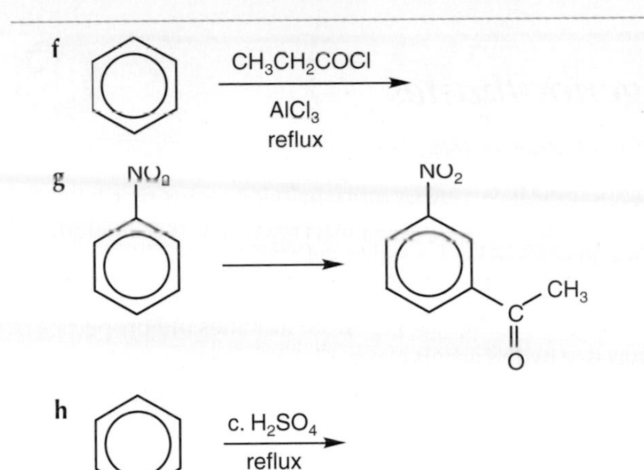

2 Predict the major product of each of the following reactions. In each case, 1 mole of the compound reacts with 1 mole of Cl_2 molecules.

a, **b**

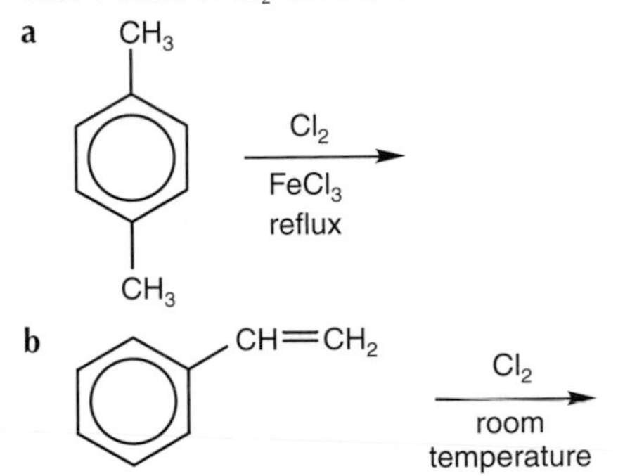

3 Iodine is too unreactive to substitute in benzene, even in the presence of a catalyst. However, iodobenzene can be made by treating benzene with iodine(I) chloride (ICl).

a Why is iodine(I) chloride a polar molecule?

b Write a balanced equation for the reaction of iodine(I) chloride with benzene.

c Explain why a catalyst is not needed.

d Why is chlorobenzene not formed in this reaction?

4 Napthalene undergoes electrophilic substitution reactions in the same way as benzene. Two monosubstituted products can be obtained – in one, substitution is at the 1-position; in the other, it is at the 2-position.

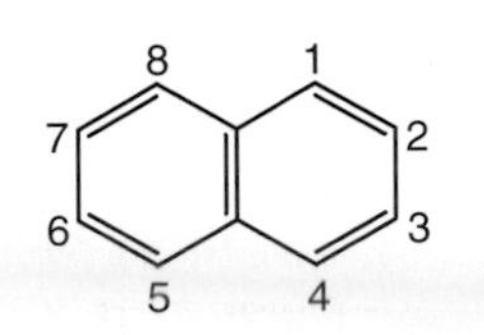

Write a *balanced equation* to show the reaction which takes place when naphthalene is heated with:

a a nitrating mixture at 50°C to give a 1-substituted product

b concentrated sulfuric acid at 160°C to give a 2-substituted product.

5 Benzene undergoes an addition reaction with hydrogen in the same way as alkenes do, but considerably higher temperatures are needed. A special nickel catalyst is used which is very active.

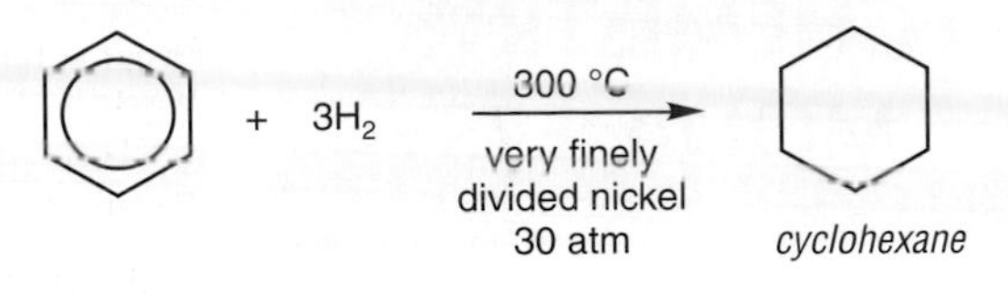

Explain why special conditions are needed for the hydrogenation of benzene.

6 For each of the following reactions, choose the correct *two* terms to describe the mechanism from the list below:

nucleophilic	substitution
electrophilic	addition
radical	elimination

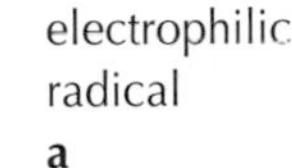

a

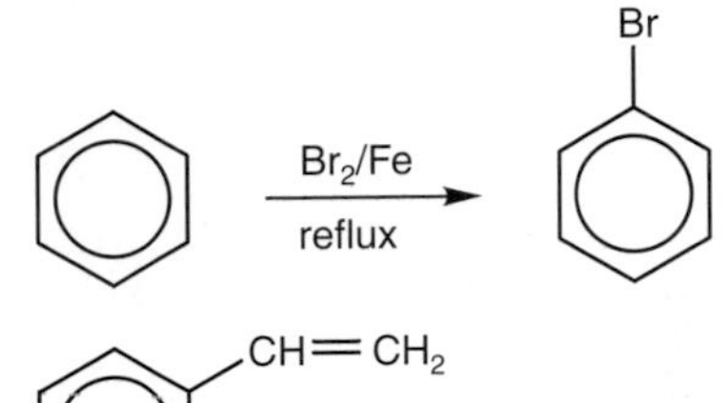

b

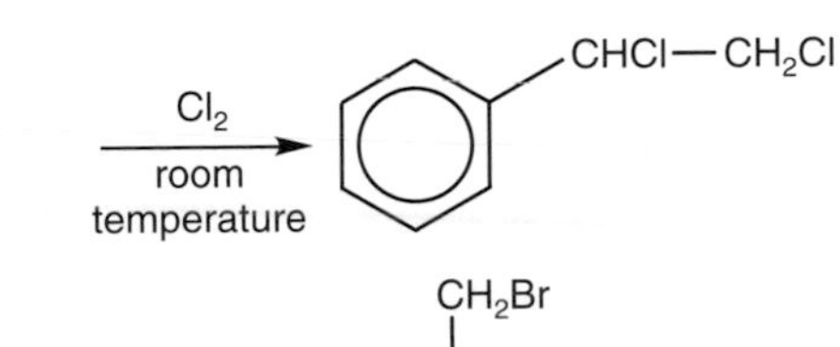

c, **d**

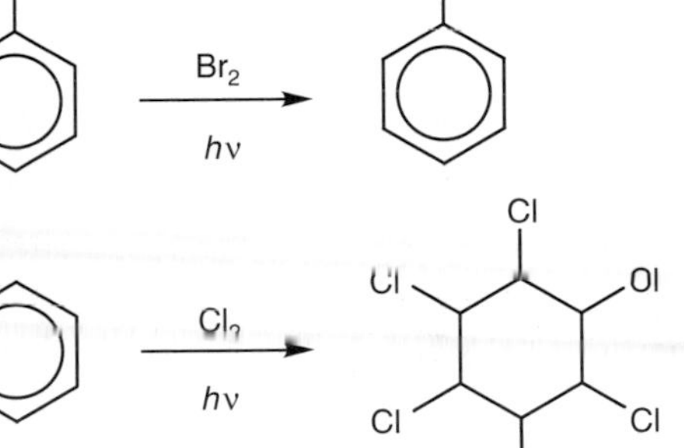

13 ORGANIC CHEMISTRY: MODIFIERS

13.1 *Halogenoalkanes*

AS

Organic halogen compounds

Organic halogen compounds have one or more halogen atoms (F, Cl, Br or I) attached to a hydrocarbon chain. Their occurrence in nature is limited, but they are useful for all sorts of human purposes so chemists make and use them a lot.

As with all functional groups, the halogen atom modifies the properties of the relatively unreactive hydrocarbon chain. The simplest examples are the **halogenoalkanes** (sometimes called **haloalkanes**), with the halogen atom attached to an alkane chain (Figure 1).

$Cl - CH_2 - CH_3$

chloroethane
halogenoalkane

▲ **Figure 1** Chloroethane is a typical halogenoalkane.

Naming halogenoalkanes

The halogenoalkanes are examples of several homologous series – the chloroalkanes, the bromoalkanes and so on. They are named after the parent alkanes, using the same basic rules as for naming alcohols – except in this case the halogen atom is added as a *prefix* to the name of the parent alkane.

Thus, the molecule $CH_3CH_2CH_2Cl$ is called *1-chloropropane* while $CH_3CHClCH_2Cl$ is called *1,2-dichloropropane*. The more complicated molecule, $CH_3CHBrCH_2CH_2Cl$, is called *3-bromo-1-chlorobutane*.

These compounds are shown in Table 1.

► **Table 1** Naming halogenoalkanes.

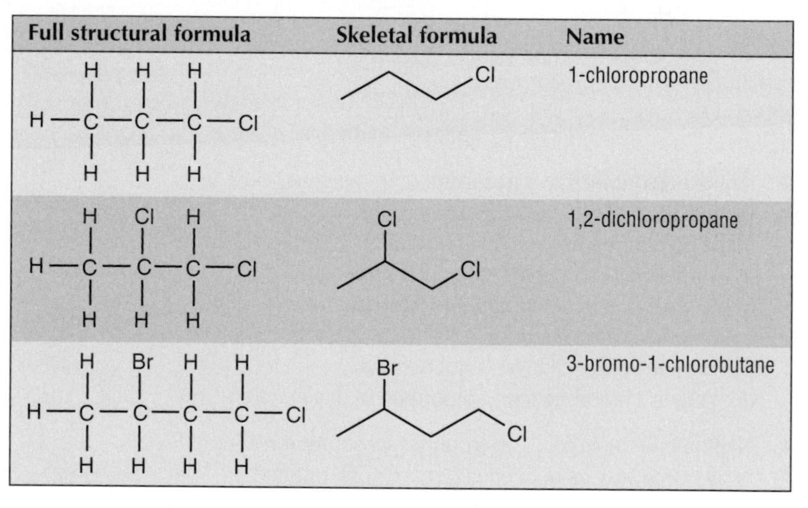

Full structural formula	Skeletal formula	Name
		1-chloropropane
		1,2-dichloropropane
		3-bromo-1-chlorobutane

Notice that the prefixes 'bromo-' and 'chloro-' are listed in alphabetical order. Notice too that the numbers used to show the positions of the bromine and chlorine atoms are the lowest ones possible, e.g. 3 and 1 rather than 2 and 4 in the case of 3-bromo-1-chlorobutane.

As you would expect, the properties of halogenoalkanes depend on which halogen atoms they contain.

Physical properties of halogenoalkanes

Look at the electronegativity values shown in Figure 13 in **Section 3.1**. The carbon–halogen bond is polar, but not polar enough to make a big difference to the physical properties of the compounds. For example, all halogenoalkanes are immiscible with water. Their boiling points depend on the size and number of halogen atoms present – the bigger the halogen atom and the more halogen atoms there are, the higher the boiling point – as you can see from Table 2.

Compound	State at 298 K	Boiling point/K
CH_3F	g	195
CH_3Cl	g	249
CH_3Br	g	277
CH_3I	l	316
CH_2Cl_2	l	313
$CHCl_3$	l	335
CCl_4	l	350
C_6H_5Cl	l	405

◀ **Table 2** Boiling points of some organic halogen compounds.

The influence of halogen atoms on the boiling point is important when it comes to designing halogen compounds for particular purposes. If you want a compound with a high boiling point, you have to include a larger halogen atom such as Cl or Br rather than a smaller one such as F – but it is these larger atoms that can cause the greatest environmental damage. These considerations are very important in the design of replacements for CFCs.

Chemical reactions of halogenoalkanes

Reactions of halogenoalkanes involve breaking the C–Hal bond ('Hal' stands for any halogen atom). The bond can break homolytically or heterolytically (see **Section 6.3**).

Homolytic fission

Homolytic fission forms **radicals**. One way this can occur is when radiation of the right frequency (visible or ultraviolet) is absorbed by the halogenoalkane. For example, with chloromethane:

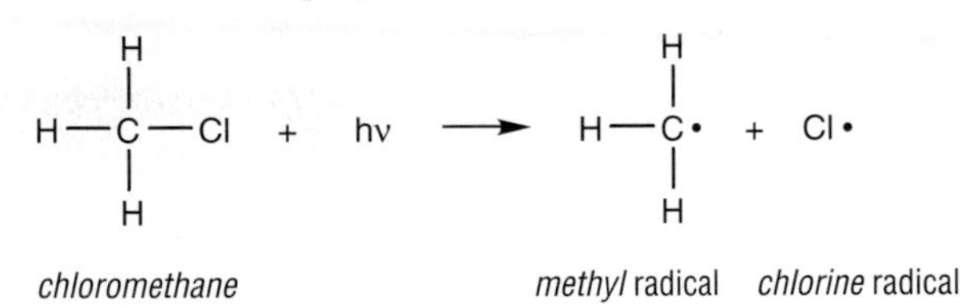

We write this in shortened form as

$$CH_3–Cl + h\nu \rightarrow CH_3\cdot + Cl\cdot$$

This kind of reaction occurs when halogenoalkanes reach the stratosphere, where they are exposed to intense ultraviolet radiation. This is how the chlorine radicals that cause so much trouble for ozone in the stratosphere have formed.

Heterolytic fission

Heterolytic fission is more common under normal laboratory conditions, where reactions of halogenoalkanes tend to be carried out in a polar solvent such as ethanol, or ethanol and water. The C–Hal bond is already polar, and in the right situation it can break forming a negative halide ion and a positive **carbocation**. For example, with 2-chloro-2-methylpropane:

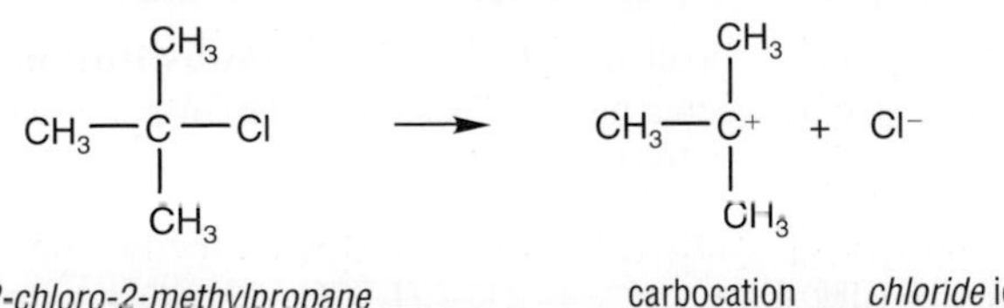

Sometimes, ions are not formed by simple bond fission in this way. Instead, heterolytic bond fission is brought about by another, negatively charged, substance reacting with the positively polarised carbon atom, causing a substitution reaction – there is more about this later in this section.

Importance of reaction conditions

For many molecules, the conditions under which a reaction is carried out can determine how a bond breaks. For example, when bromomethane is dissolved in a polar solvent, such as a mixture of ethanol and water, the C–Br bond breaks heterolytically to form ions. However, when it reacts in a non-polar solvent, such as hexane, or in the gas phase at high temperature or in the presence of light, the C–Br bond breaks homolytically.

Different halogens, different reactivity

Whether homolytic or heterolytic fission occurs, all reactions of halogenoalkanes involve breaking the C–Hal bond. The stronger the bond is, the more difficult it is to break. Figure 2 gives the bond enthalpies of the four different types of C–Hal bond. Remember, the higher the bond enthalpy, the stronger the bond.

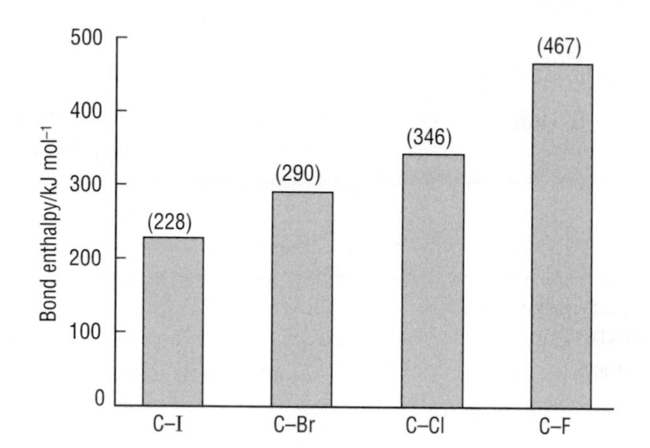

► **Figure 2** Some bond enthalpies.

> You might imagine that a large bond polarity in a C–X bond would result in it breaking easily: that C–F bonds would break more easily than C–Cl bonds. However, it has been shown experimentally that bond enthalpy is the overriding factor in determining reactivity. For example, a chloroalkane hydrolyses *more* readily than the equivalent fluoroalkane. For more details see **Activity ES6.2**.

The great strength of the C–F bond makes it very difficult to break, so fluoro compounds are very unreactive. As you go down the halogen group the C–Hal bond gets weaker and weaker, so the compounds get more and more reactive. Chloro compounds are fairly unreactive, and once they have been released into the troposphere they stay there quite a long time. This means that compounds such as CFCs can stay around long enough to get into the stratosphere and wreak havoc on the ozone layer.

Bromo and iodo compounds are fairly reactive, which makes them useful as intermediates in synthesising other organic compounds.

Substitution reactions of halogenoalkanes

> Halogenoalkanes are usually hydrolysed by heating under reflux with NaOH(aq). Often ethanol is added to act as a solvent.

These are typical of halogenoalkanes. For example, a **substitution reaction** takes place between a halogenoalkane and hydroxide ions, in which the halogenoalkane is **hydrolysed** to form an alcohol. With bromobutane, for example:

$$CH_3–CH_2–CH_2–CH_2–Br + OH^- \rightarrow CH_3–CH_2–CH_2–CH_2–OH + Br^-$$

The C–Br bond is polar:

$$\overset{\delta+}{C}—\overset{\delta-}{Br}$$

The oxygen atom of the hydroxide ion is negatively charged:

$$H—\ddot{\underset{\cdot\cdot}{O}}:^-$$

The partial positive charge on the carbon atom attracts the negatively charged oxygen of the hydroxide ion. A lone pair of electrons on the O atom forms a bond with the C atom as the C–Br bond breaks:

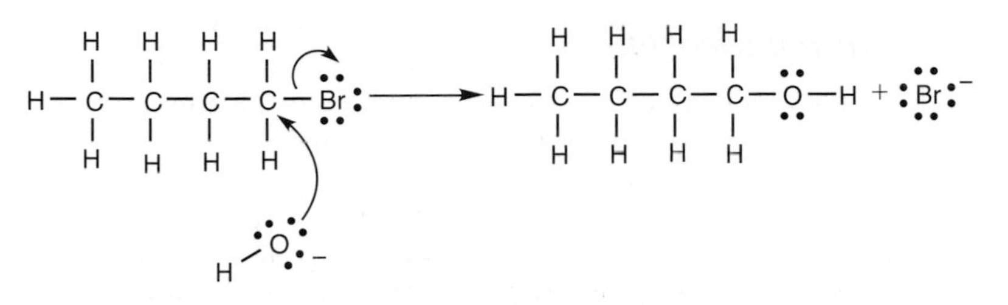

Notice that this reaction involves heterolytic fission – ions, rather than radicals, are formed. A free carbocation is not formed in this case, because the OH^- ion attacks at the same time as the C–Br bond breaks – it's a smooth, continuous process.

Note too that we have used **curly arrows** again to show the movement of electrons during the reaction. In this case a full-headed arrow is used to show the movement of *a pair* of electrons – compare this with the half-arrow used to show the movement of single electrons in radical reactions.

Halogenoalkanes perform substitution reactions with many different reagents, as well as hydroxide ions. What is needed is a group carrying a pair of electrons to start forming a bond to the carbon atom.

Attacking groups like these, which can donate a pair of electrons to a positively charged carbon atom to form a new covalent bond, are called **nucleophiles**. They can be either negative ions, like OH^-, or molecules with a lone pair of electrons available for bonding, for example H_2O and NH_3. Table 3 shows some common nucleophiles.

A nucleophile is a molecule or negatively charged ion with a lone pair of electrons that it can donate to a positively charged atom to form a covalent bond.

Name and formula	Structure, showing lone pairs
hydroxide ion, OH^-	$H\ \ddot{\underset{\cdot\cdot}{O}}:^-$
cyanide ion, CN^-	$:N\equiv C:^-$
ethanoate ion, CH_3COO^-	$CH_3—C(=O)—\ddot{\underset{\cdot\cdot}{O}}:^-$
ethoxide ion, $C_2H_5O^-$	$CH_3CH_2—\ddot{\underset{\cdot\cdot}{O}}:^-$
water molecule, H_2O	$H—\ddot{\underset{\cdot\cdot}{O}}—H$
ammonia molecule, NH_3	$H—\ddot{N}(—H)—H$

◄ **Table 3** Some common nucleophiles (for shape and bond angles, refer to **Section 3.2**).

The carbon atom attacked by the nucleophile may be part of a carbocation and carry a *full* positive charge, or it may be part of a neutral molecule (as with the bromobutane reaction above) and carry a *partial* positive charge as a result of bond polarisation.

If we write Nu^- as a general symbol for any nucleophile, the **nucleophilic substitution** process can be described by:

In general, when any nucleophile Nu^- reacts with a general halogenoalkane R–Hal, the reaction that occurs is

$$R\text{–}Hal + Nu^- \rightarrow R\text{–}Nu + Hal^-$$

Water as a nucleophile

Nucleophiles don't need to have a full negative charge – it is possible for a neutral molecule to act as a nucleophile provided it has a lone pair of electrons which can be used to form a bond to a carbon atom.

For example, the water molecule has a lone pair on the oxygen atom, so water can act as a nucleophile and attack a halogenoalkane molecule such as 1-bromobutane – though this reaction is slower than the reaction with OH^- ions. The substitution reaction goes in two stages – first H_2O attacks the halogenoalkane:

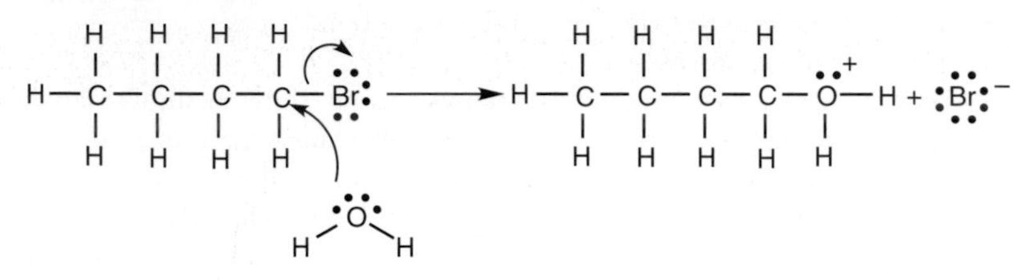

Then the resulting ion loses H^+ to form an alcohol:

The overall equation for the reaction of water with a general halogenoalkane R–Hal is

$$R\text{–}Hal + H_2O \rightarrow R\text{–}OH + H^+ + Hal^-$$

This type of reaction with water is also known as a **hydrolysis** reaction.

Ammonia as a nucleophile

The halogenoalkane is heated with concentrated ammonia solution in a sealed tube.

Ammonia, NH_3, can act as a nucleophile in a similar way to water with the lone pair of electrons on the N atom attacking the halogenoalkane. The product is an **amine** with an NH_2 group. The overall equation is

$$R\text{–}Hal + NH_3 \rightarrow R\text{–}NH_3^+Hal^- \rightleftharpoons R\text{–}NH_2 + H^+ + Hal^-$$

Using nucleophilic substitution to make halogenoalkanes

When a halogenoalkane reacts with OH^- ions, a nucleophilic reaction occurs and an alcohol is formed (see above). We can use the *reverse* of this reaction to produce a halogenoalkane from an alcohol if we choose the reaction conditions carefully. It is another nucleophilic substitution reaction, but this time the nucleophile is Hal^-.

For example, we can make 1-bromobutane using a nucleophilic substitution reaction between butan-1-ol and Br^- ions. The reaction is done in the presence of a strong acid, and the first step involves bonding between H^+ ions and the O atom on the alcohol:

```
    H  H  H  H                          H  H  H  H
    |  |  |  |  ..                      |  |  |  |  .+
H — C— C— C— C— O —H   ———→   H — C — C — C — C — O — H
    |  |  |  |  ..                      |  |  |  |  |
    H  H  H  H   )                      H  H  H  H  H
                H+
```

This gives the C atom to which the O is attached a higher partial positive charge. It is now more readily attacked by Br^- ions, forming bromobutane:

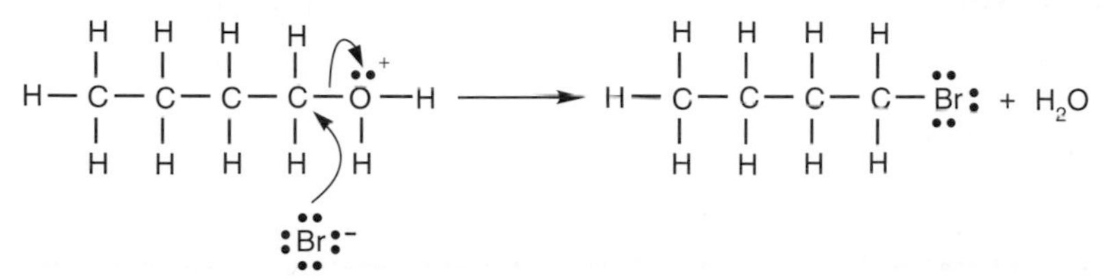

The overall equation for the reaction is:

$$CH_3CH_2CH_2CH_2OH + H^+ + Br^- \rightarrow CH_3CH_2CH_2CH_2Br + H_2O$$

Problems for 13.1

1 Use systematic nomenclature to name the following halogenoalkanes:

a $CHCl_3$

b $CH_3CHClCH_3$

c CF_3CCl_3

d

```
    H   Cl  F
    |   |   |
H — C — C — C — F
    |   |   |
    H   H   F
```

e

```
   H  Cl  Br  H
   |   |   |  |
H— C — C — C — C —H
   |   |   |  |
   H   H  Br  H
```

2 Draw skeletal formulae for compounds **d** and **e** in problem 1 – use Table 1 in this section (page 288) as a guide.

3 When 1-chloropropane is heated under reflux with aqueous sodium hydroxide solution, a nucleophilic substitution reaction occurs, forming propan-1-ol.

a Write a balanced equation to show the overall reaction.

b Explain why this is classed as a substitution reaction.

c Write down the structure of the attacking nucleophile, showing the charge and any lone pairs of electrons.

4 Draw a full structural formula for each of the following:

a 1,1,1-trichloroethane

b 1,1,2,2-tetrafluoropropane

c 4-bromo-1,1-dichloropentane.

5 Look at the following compounds.

A CH_3CH_2Br

B $CH_3CH_2CH_2Cl$

C $CH_3CHClCH_3$

D CH_3I

E (cyclohexane ring)—Cl

F $CHCl_3$

a Which compound reacts most rapidly with alkali?

b When hydrolysis is carried out by water in the presence of silver nitrate, which compound(s) produce a cream precipitate?

c Which compound has the lowest boiling point? (It will help to look back to Table 2 in this section – page 289.)

d Which compound is the least volatile?

e Draw the structure of the product when compound **E** is hydrolysed. Name the product

6 Table 3 in this section (page 291) shows the structure of common nucleophiles. Write a balanced equation to show the nucleophilic substitution reaction that takes place between each of the following pairs of compounds. Your equations should show the structures of the reactants and products clearly. Example: bromomethane and CN^- (cyanide) ions

```
    H                           H
    |                           |
H — C — Br  +  CN⁻   ———→   H — C — CN  +  Br⁻
    |                           |
    H                           H
```

a iodoethane and OH^- ions

b bromoethane and CN^- ions

c chlorocyclopentane and OH^- ions

d 2-chloro-2-methylpropane and H_2O

e 1,2-dibromoethane and OH^- ions

f bromomethane and $C_2H_5O^-$ (ethoxide) ions

g 2 chloropropane and CH_3COO^- (ethanoate) ions.

7 Concentrated ammonia solution reacts with 1-bromoethane when heated in a sealed tube.

- **a** Write an overall equation for the nucleophilic substitution reaction that takes place.
- **b** Using the reaction of halogenoalkanes with water as a guide (page 292), draw out the mechanism for this reaction.
- **c** Write a few sentences to explain this mechanism to a fellow student.
- **d** Give definitions of all the terms in your mechanism.

8 Suggest the starting halogenoalkane, reagents and the conditions you would use to prepare a sample of each of the following compounds:

- **a** $CH_3—CH_2—CH_2—CH_2—CH_2—OH$
- **b** $CH_3—CH(NH_2)—CH_3$
- **c** $CH_3—CH_2—CH_2—OH$

9 Chloromethane has a lifetime in the troposphere of about 1 year, which allows enough time for some of it to be transported into the stratosphere where it helps to destroy ozone.

In contrast, iodomethane has a tropospheric lifetime of about 8 days and much less reaches the stratosphere. This is because iodomethane is rapidly broken down by light in the troposphere – this is called *photolysis*.

- **a** Write equations for the photolysis of iodomethane and chloromethane.
- **b** Explain why iodomethane is photolysed in the troposphere, whereas chloromethane is only photolysed when it reaches the stratosphere.

AS

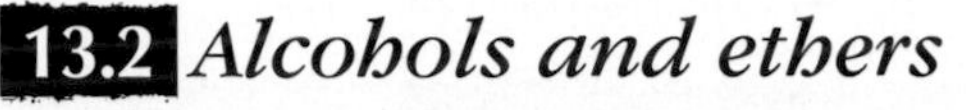

13.2 Alcohols and ethers

There are two isomers with the molecular formula C_2H_6O:

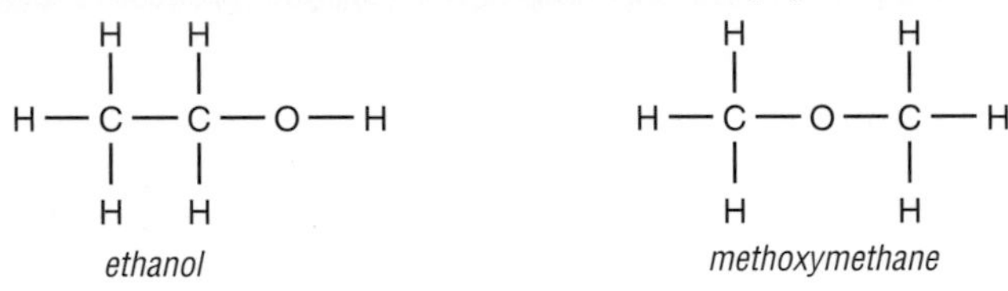

They are isomers with different **functional groups**.

Ethanol belongs to a homologous series called **alcohols**. All members of the series contain the **hydroxyl** functional group, **–OH**, and show similar chemical properties. They differ only in the length and structure of the hydrocarbon chain.

Methoxymethane is a member of a different homologous series, called the ethers. All ethers contain the alkoxy functional group, –OR, and show similar chemical properties.

Hydroxyl group
–OH
Alkoxy group
–OR

Alcohols

Alcohols are derived from alkanes by substituting an –OH group for an –H atom. They are named from the parent alkane by omitting the final *-e* and adding the ending *-ol*. For example:

methanol	$CH_3–OH$
ethanol	$CH_3–CH_2–OH$
propan-1-ol	$CH_3–CH_2–CH_2–OH$
butan-1-ol	$CH_3–CH_2–CH_2–CH_2–OH$

General formula of alcohols is R–OH
The names of alcohols end in **-ol**.

You need to be familiar with the different ways of drawing structures of organic compounds, depending on how much detail is needed. For example, the structure of ethanol may be represented:

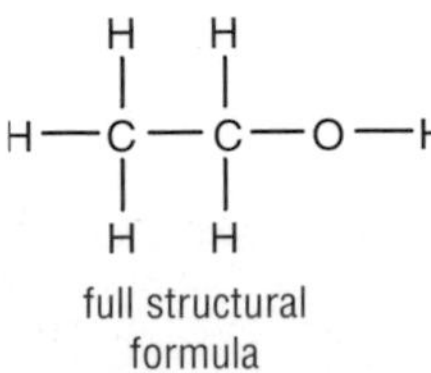

full structural formula

$CH_3—CH_2—OH$ CH_3CH_2OH

shortened structural formulae

OH

skeletal formula

ethanol

Isomeric alcohols are possible for alcohols containing more than two carbon atoms. To distinguish between these, it is necessary to label the position of the –OH group. The hydrocarbon chain is always numbered from the end which gives the lowest number for the position of the functional group. So, CH_3–CH_2–CH_2–CH_2–OH is named butan-1-ol, not butan-4-ol. Table 1 gives some more examples.

◀ **Table 1** Naming alcohols.

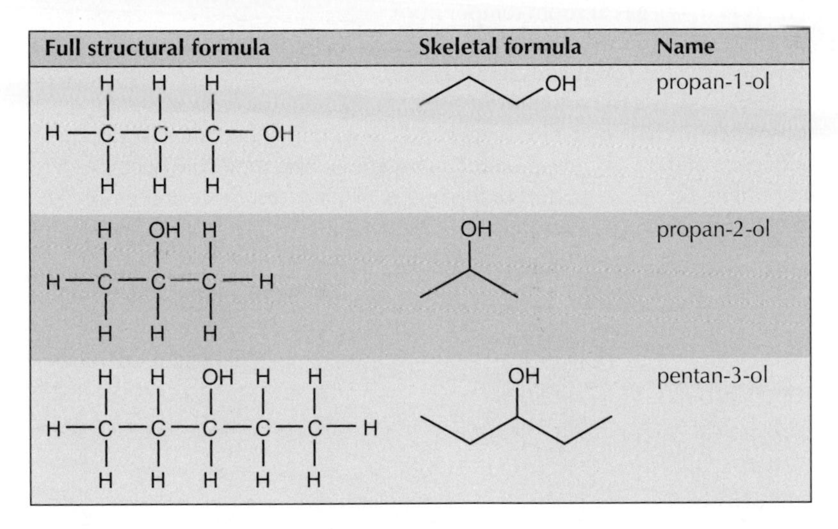

Full structural formula	Skeletal formula	Name
		propan-1-ol
		propan-2-ol
		pentan-3-ol

Some alcohols, particularly biologically occurring ones, contain more than one –OH group in their molecules. They are known as **polyhydric** alcohols. Look how they are named.

Physical properties of alcohols

You can think of alcohols as being derived from water by replacing one of the H atoms with an alkyl group:

Like water molecules, alcohol molecules are polar because of the polarised O–H bond. In both water and alcohols, there is a special sort of strong attractive force *between the molecules* due to **hydrogen bonds** (Figure 1).

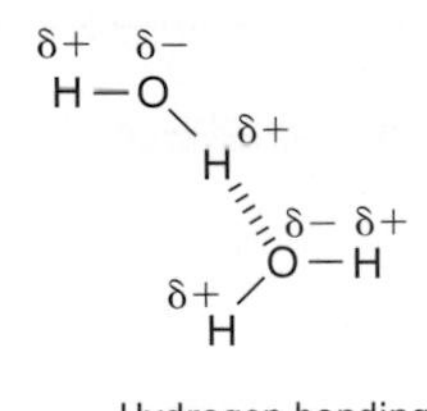

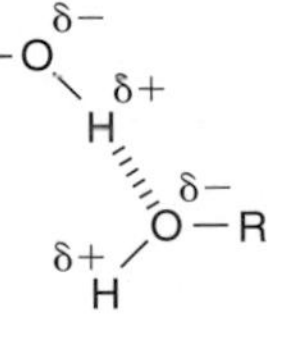

Represents a hydrogen bond

◀ **Figure 1** Hydrogen bonding in water and alcohols. Hydrogen bonds are discussed in more detail in **Section 5.4**.

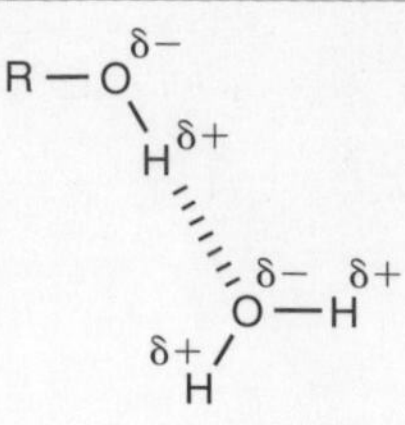

▲ **Figure 2** Hydrogen bonding between water molecules and alcohol molecules.

Hydrogen bonds are not as strong as covalent bonds, but are stronger than other attractive forces *between* covalent molecules. When a liquid boils, these forces must be broken so the molecules escape from the liquid to form a gas. This explains why the boiling points of alcohols are higher than those of corresponding alkanes with similar relative molecular mass (M_r). For example, ethanol ($M_r = 46.0$) is a liquid, while propane ($M_r = 44.0$) is a gas at room temperature.

Hydrogen bonding *between alcohol and water molecules* (see Figure 2) explains why the two liquids mix together.

Table 2 shows the solubility of some alcohols in water. As the hydrocarbon chain becomes longer and the molecule becomes larger, the influence of the –OH group on the properties of the molecule becomes less important. So the properties of the higher alcohols get more and more like those of the corresponding alkane.

► **Table 2** Solubility of alcohols in water.

Name	Formula	Solubility/g per 100 g water
methanol	CH_3OH	miscible in all proportions
ethanol	CH_3CH_2OH	miscible in all proportions
propan-1-ol	$CH_3CH_2CH_2OH$	miscible in all proportions
butan-1-ol	$CH_3CH_2CH_2CH_2OH$	8.0
pentan-1-ol	$CH_3CH_2CH_2CH_2CH_2OH$	2.7
hexan-1-ol	$CH_3CH_2CH_2CH_2CH_2CH_2OH$	0.6

Reactions of alcohols

There are of three types of alcohols – primary, secondary and tertiary – according to the position of the OH group (see Table 3). The reactions of alcohols depend on the type of alcohol involved.

► **Table 3** Primary, secondary and tertiary alcohols.

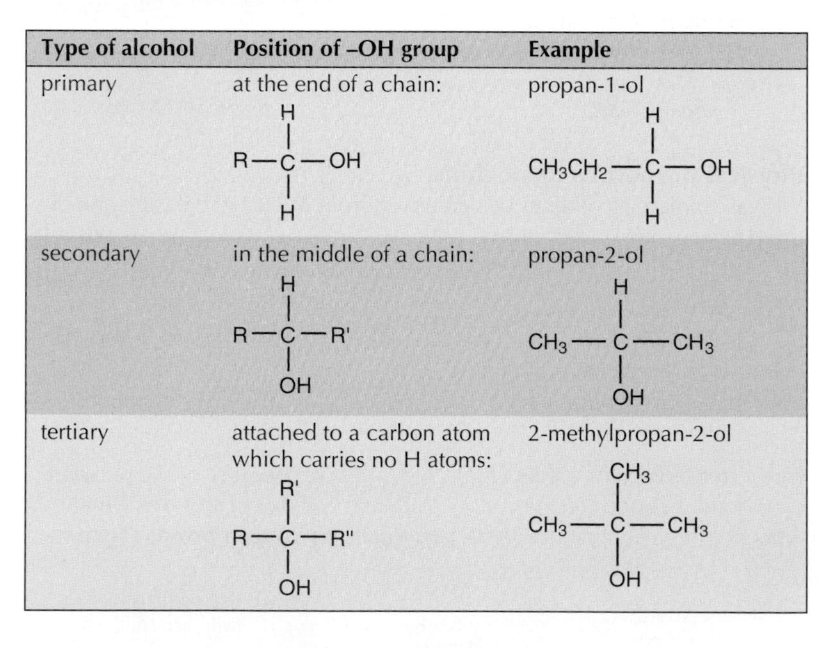

Type of alcohol	Position of –OH group	Example
primary	at the end of a chain: R–CH(H)–OH (R—C(H)(H)—OH)	propan-1-ol: CH_3CH_2—C(H)(H)—OH
secondary	in the middle of a chain: R—C(H)(OH)—R'	propan-2-ol: CH_3—C(H)(OH)—CH_3
tertiary	attached to a carbon atom which carries no H atoms: R—C(R')(OH)—R"	2-methylpropan-2-ol: CH_3—C(CH_3)(OH)—CH_3

Oxidation

The –OH group can be oxidised by strong oxidising agents, such as acidified potassium dichromate(VI). This process is called **oxidation**. The basic reaction is

Alcohols are usually oxidised by heating with an acidified solution of potassium dichromate(VI), $K_2Cr_2O_7$.

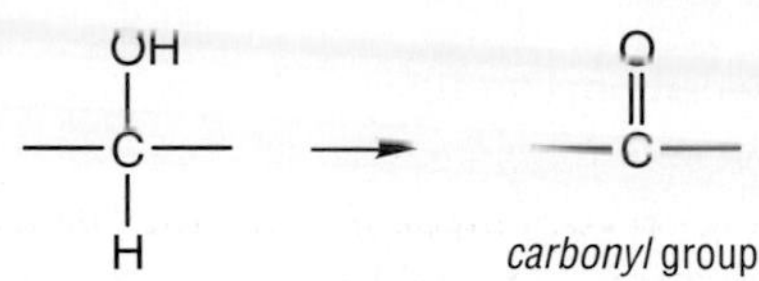

The orange dichromate(VI) ion, $Cr_2O_7^{2-}(aq)$, is reduced to green $Cr^{3+}(aq)$.

Notice that in this reaction two atoms of hydrogen are being removed – one from the oxygen atom, and one from the carbon atom. Oxidation of the –OH group will not take place unless there is a hydrogen atom on the carbon atom to which the –OH is attached.

The product is a **carbonyl** compound – an aldehyde or a ketone. The type of product you get depends on the type of alcohol you start with.

Primary alcohols, such as ethanol, are oxidised to **aldehydes** – but the aldehyde is then itself oxidised to a **carboxylic acid**. With ethanol:

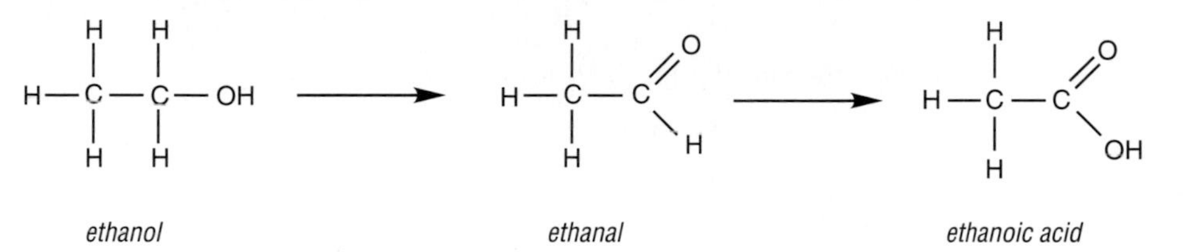

The aldehyde can be extracted as an intermediate product if the reaction is done carefully (see **Section 13.7**).

If the aldehyde is required, the aldehyde is distilled out from the reaction mixture before it is oxidised further.

Secondary alcohols, such as propan-2-ol, are oxidised to **ketones**. For example:

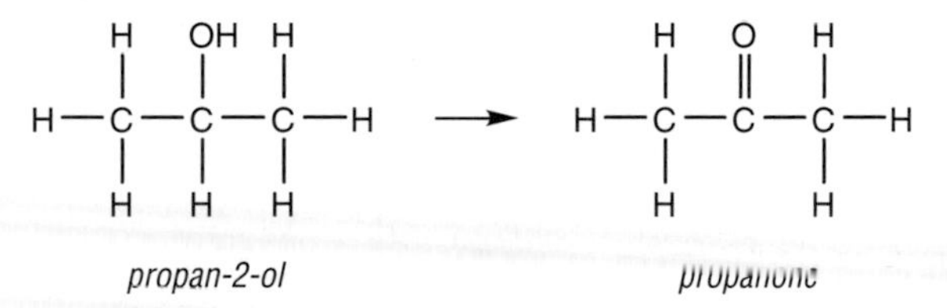

It is difficult to oxidise the ketone further than this, because to do so would involve breaking a C–C bond.

Tertiary alcohols, such as 2-methylpropan-2-ol, are difficult to oxidise because they do not have a hydrogen atom on the carbon atom to which the –OH group is attached.

Table 4 summarises these reactions.

◄ **Table 4** Oxidation of alcohols.

Type of alcohol	Product(s) of oxidation
primary	aldehyde, carboxylic acid
secondary	ketone
tertiary	does not oxidise

Aldehydes and ketones

Aldehydes and ketones contain the **carbonyl group,** $-\overset{\displaystyle O}{\overset{\|}{C}}-$

In an **aldehyde**, the carbonyl group is at the *end* of an alkane chain, so the functional group is

$$R-C(=O)-H$$

Aldehydes are named using the suffix *-al*. For example,

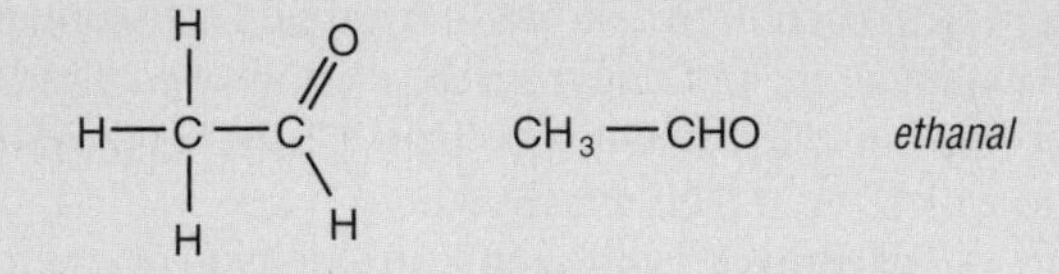

In a **ketone**, the carbonyl group is *inside* an alkane chain, so the functional group is

$$R'(R)C=O$$

Ketones are named using the suffix *-one*. For example,

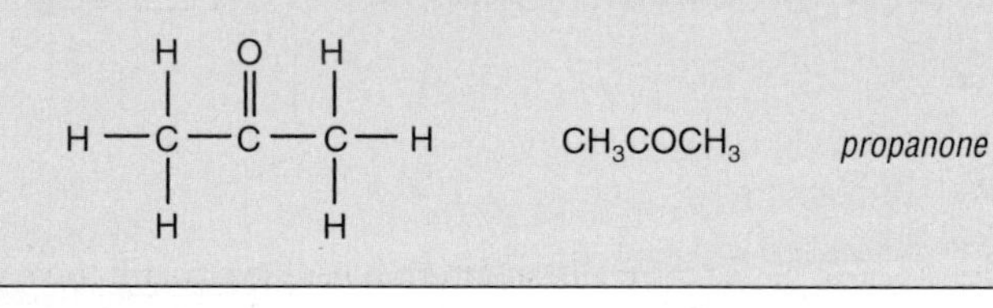

Dehydration of alcohols

Many alcohols can lose a molecule of water to form an alkene.

For example, propene is formed when vapour of propan-1-ol is passed over a hot catalyst of alumina (Al_2O_3) at 300 °C.

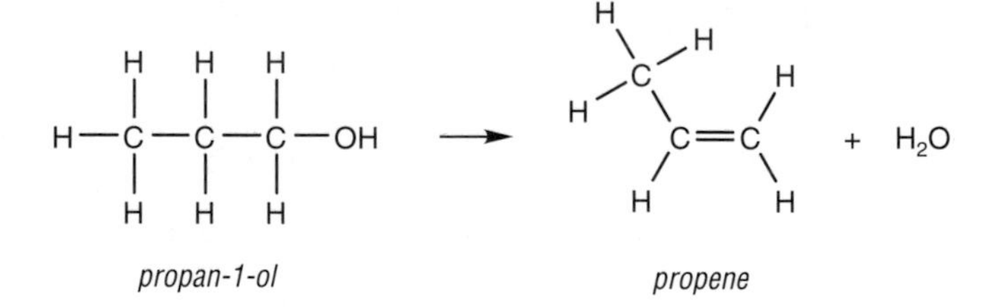

You may find it easier to follow the reaction if you write skeletal formulae:

The reaction is described as **dehydration** since it involves the removal of a *water* molecule from a molecule of the reactant. Alcohols can also be dehydrated by heating with concentrated sulfuric acid. Dehydration is an example of an **elimination reaction**.

You can also think of an elimination reaction as being the reverse of an addition reaction (see **Section 12.2**).

Elimination reactions

An elimination reaction is one in which a small molecule is removed from a larger molecule leaving an unsaturated molecule.

In the case of alcohols the small molecule is water.

Ethers

Ethers are derived from alkanes by substituting an alkoxy group (–OR) for an H atom. For example,

ethoxyethane $CH_3–CH_2–O–CH_2–CH_3$
methoxypropane $CH_3–CH_2–CH_2–O–CH_3$

Note that the longer hydrocarbon chain is chosen as the parent alkane.

General formula of ethers R–O–R′

Physical properties of ethers

You can think of ethers as being derived from water by replacing *both* the H atoms by alkyl groups.

Ether molecules are only slightly polar and the attractive forces *between* the molecules are relatively weak. There are no H atoms attached to the oxygen to form hydrogen bonds between ether molecules.

The boiling point of an ether is similar to that of the alkane with corresponding relative molecular mass. Like alkanes, the lower ethers are very volatile and dangerously flammable.

Ethers are only slightly soluble in water, but mix well with other non-polar molecules such as alkanes.

Problems for 13.2

1 Use systematic nomenclature to name the following compounds from their skeletal formulae.

a
b
c
d
e
f
g

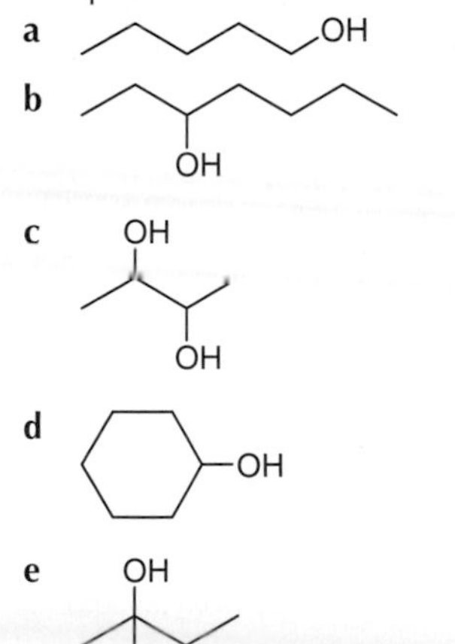

2 Draw skeletal formulae for two structural isomers of pentan-1-ol that are not alcohols and name them.

3 a Why does ethanol mix with water but hexanols do not mix with water?
b Why is ethoxyethane immiscible with water?

4 Look at the following compounds:

A B C D E F

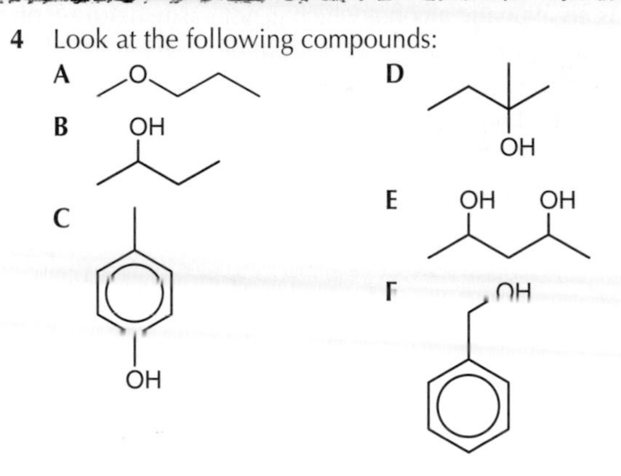

a Which compound(s) is(are) alcohols?
b Which compound(s) is(are) ethers?
c Which compound(s) is(are) phenols?
d Which compound(s) is(are) diols?
e Which compounds are isomers?
f Which compound do you think will be the most volatile?
g Which compound would you expect to be the most soluble in water?

5 Here are the boiling points and relative molecular masses (M_r) of a number of substances:

Substance	Boiling point/°C	M_r
water, H_2O	100	18.0
ethane, CH_3CH_3	−88.5	30.0
ethanol, CH_3CH_2OH	78	46.0
butan-1-ol, $CH_3CH_2CH_2CH_2OH$	117	74.0
ethoxyethane, $CH_3CH_2OCH_2CH_3$	35	74.0

Use ideas about bonds between molecules to explain why

a ethanol has a higher boiling point than ethane
b water has a higher boiling point than ethanol
c butan-1-ol has a higher boiling point than ethanol
d butan-1-ol has a higher boiling point than ethoxyethane.

6 **a** Use systematic nomenclature to name the following compounds:

A $CH_3—CH_2—CH_2—CH_2—OH$

B

```
          OH
          |
CH3 — CH — CH2 — CH3
```

b Draw structures for the following compounds:

C 3-methylpentan-3-ol
D 2-methylpentan-3-ol

7 Which of the compounds **A** to **D** in problem 6

a are secondary alcohols
b is an aliphatic alcohol which is not easily oxidised on heating with acidified potassium dichromate(VI)
c produces a carboxylic acid on refluxing with excess acidified potassium dichromate(VI)?

8 Draw the structure of the oxidation product in each of the following reactions. In each case, name the type of compound formed.

a

```
                  CH3
                  |
CH3 — CH2 — CH — CH2 — OH
```

+ acidified dichromate(VI) (mild conditions and product distilled off as formed)

b

```
         OH
         |
CH3 — CH — CH3
```

+ acidified dichromate(VI) (heat under reflux)

9 Look at the skeletal formulae of the following compounds.

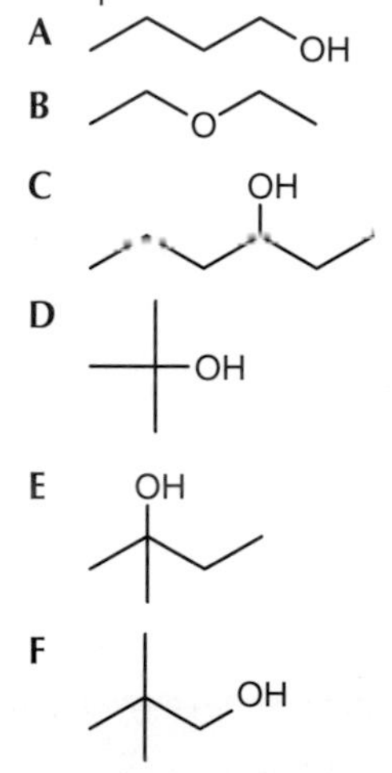

a Which compound(s) is(are) primary alcohols?
b Which compound(s) is(are) secondary alcohols?
c Which compound(s) is(are) tertiary alcohols?
d Which compound(s) cannot be oxidised by acidified dichromate(VI)?
e Which compound(s) give(s) but-1-ene on dehydration?

13.3 *Carboxylic acids and their derivatives*

This section provides an introduction to carboxylic acids and their derivatives. You will find more details in later sections. Carboxylic acids contain the **carboxyl** group

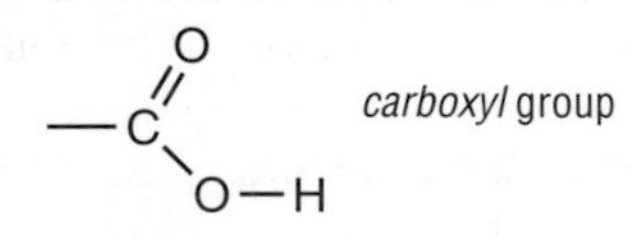

This formula is often abbreviated to –COOH, although the two oxygen atoms are *not* joined together.

The structure of the rest of the molecule can vary widely and this gives rise to a large number of different carboxylic acids. When the remainder of the molecule is an alkyl group, the acids can be represented by the general formula **R–COOH**.

Carboxylic acids are named from the parent alkane by omitting the final *-e* and adding the ending *-oic acid*. For example:

The names of carboxylic acids end in **-oic acid**.

One C atom:	CH_4 methane	H–COOH methanoic acid
Two C atoms:	CH_3–CH_3 ethane	CH_3–COOH ethanoic acid
Three C atoms:	CH_3–CH_2–CH_3 propane	CH_3–CH_2–COOH propanoic acid

Note that the carbon atom in the –COOH group counts as one of the carbon atoms for the name.

The first two members of the series, methanoic acid and ethanoic acid, are often still called by their older names, formic acid and acetic acid, so it is as well to know these too.

Branched-chain carboxylic acids are named in a similar way to alkanes:

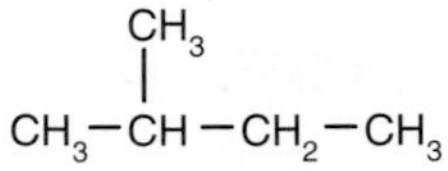

CH_3–CH(CH_3)–CH_2–CH_3

2-methylbutane

CH_3–CH(CH_3)–CH_2–COOH

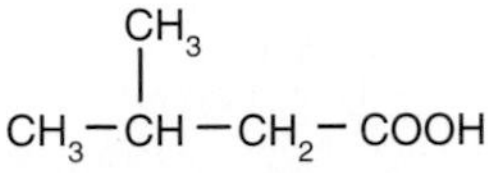

3-methylbutanoic acid

The skeletal formula of 3-methylbutanoic acid is

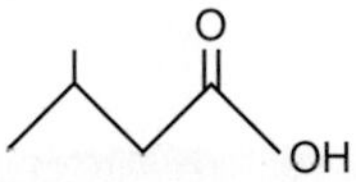

O

OH

When two carboxyl groups are present, the ending **-dioic acid** is used. For example,

COOH–COOH

ethanedioic acid

COOH–CH_2–COOH

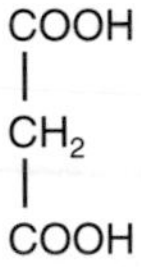

propanedioic acid

A carboxyl group can be attached to a benzene ring. For example,

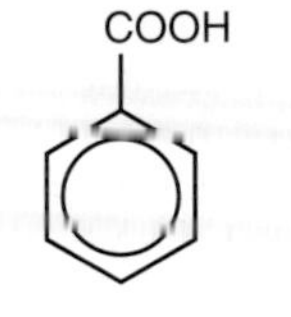

benzenecarboxylic acid
also known as *benzoic acid*

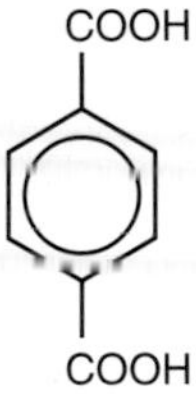

COOH

COOH

benzene-1,4-dicarboxylic acid

You will learn about the chemical reactions of carboxylic acids in **Section 13.4**.

The –OH group in the carboxyl group can be replaced by other groups to give a whole range of **carboxylic acid derivatives**. Some examples of acid derivatives are shown in Table 1 (page 302). You will meet these again in several parts of the course, so there will be plenty of opportunities to get used to recognising them. The table shows the sections which deal with each derivative in more detail.

▶ **Table 1** Some examples of acid derivatives.

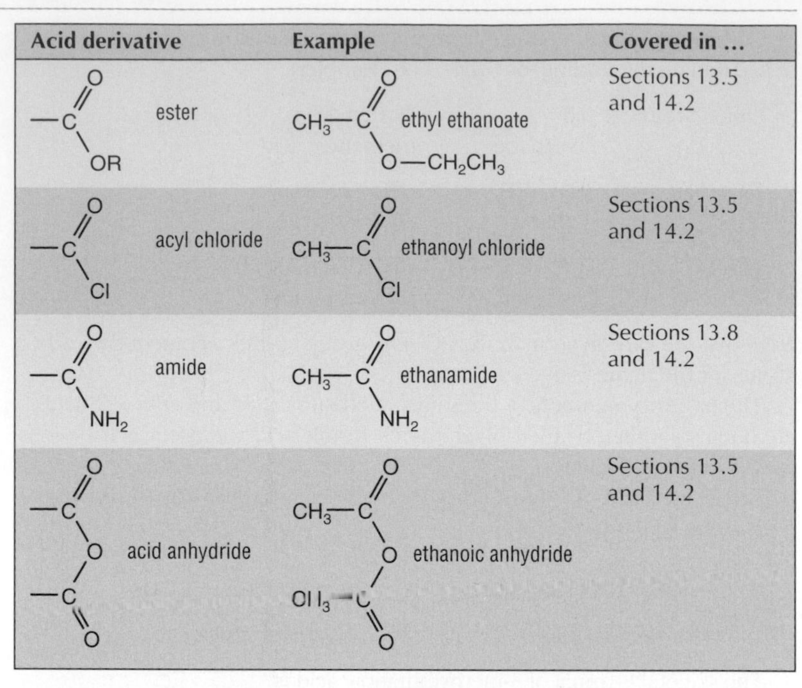

Acid derivative	Example	Covered in ...
ester: $-C(=O)-OR$	ethyl ethanoate: $CH_3-C(=O)-O-CH_2CH_3$	Sections 13.5 and 14.2
acyl chloride: $-C(=O)-Cl$	ethanoyl chloride: $CH_3-C(=O)-Cl$	Sections 13.5 and 14.2
amide: $-C(=O)-NH_2$	ethanamide: $CH_3-C(=O)-NH_2$	Sections 13.8 and 14.2
acid anhydride: $-C(=O)-O-C(=O)-$	ethanoic anhydride: $CH_3-C(=O)-O-C(=O)-CH_3$	Sections 13.5 and 14.2

Problems for 13.3

1 Use systematic nomenclature to name the following carboxylic acids:
  a H–COOH
  b $CH_3–CH_2–CH_2–CH_2–COOH$
  c $CH_3–CH_2–CH(CH_3)–COOH$

2 Name the following carboxylic acids from their skeletal formulae:

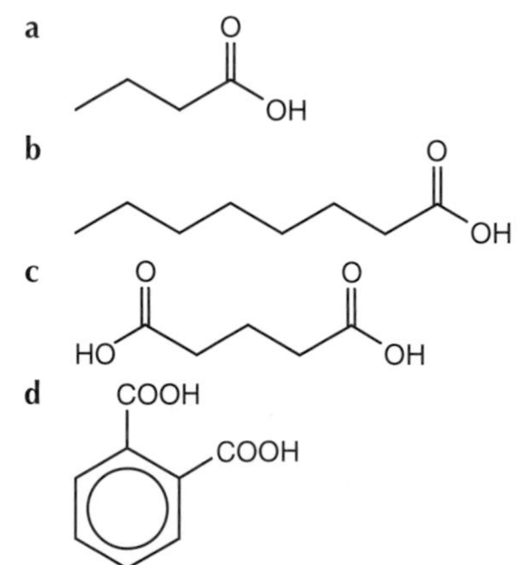

3 Draw full structural formulae for:
  a hexanoic acid
  b hexanedioic acid
  c benzene-1,3-dicarboxylic acid.

4 Write out the structural formula of:
  a the methyl ester of propanoic acid
  b the acyl chloride derivative of benzoic acid
  c the amide of butanoic acid
  d the acid anhydride formed from propanoic acid
  e the acyl chloride of hexanedioic acid

5 Draw out the structural formulae of four isomeric esters that have the molecular formula $C_5H_{10}O_2$. Name each isomer you draw.

13.4 *The –OH group in alcohols, phenols and acids*

The hydroxyl group, –OH, can occur in three different environments in organic molecules:

- attached to an alkane chain in **alcohols**. There are three types of alcohols – primary, secondary and tertiary – according to the position of the OH group (see **Section 13.2**).
- attached to a benzene ring in **phenols**, for example

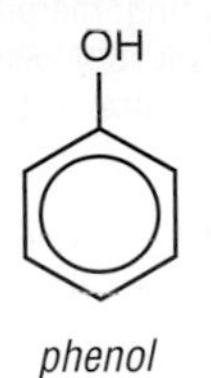

phenol

Although phenols look similar to alcohols, they behave very differently. It is generally true that functional groups behave differently when attached to an aromatic ring from when they are attached to an alkyl group.

- as part of a carboxyl group in **carboxylic acids**, for example

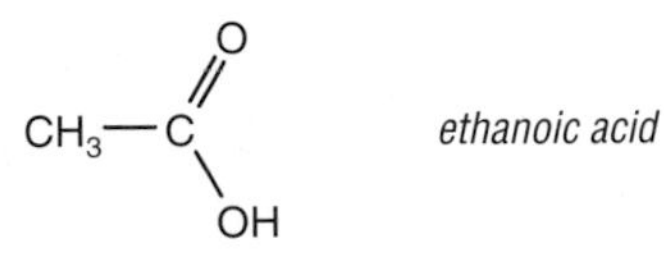

In this section we compare the way the –OH group behaves in these three different kinds of compounds.

Acidic properties

The –OH group can react with water like this:

$$R\text{–}OH + H_2O \rightleftharpoons R\text{–}O^- + H_3O^+$$

where R stands for the group of atoms which makes up the rest of the molecule (see **Section 12.1**).

Water itself does this to a very small extent in the reaction

$$H\text{–}OH + H_2O \rightleftharpoons H\text{–}O^- + H_3O^+$$

So, at any one time a small number of water molecules donate H^+ ions to other water molecules – water behaves as a **weak acid** (see **Section 8.1**). A similar reaction occurs with ethanol, but to a lesser extent. The equilibrium lies further to the left, and ethanol is a weaker acid than water.

With phenol, the equilibrium lies further to the right than in water – phenol is slightly more acidic than water.

Carboxylic acids are even more acidic, though still weak. The order of acid strength is

ethanol < water < phenol < carboxylic acids.

It is the stability of the $R\text{–}O^-$ ion formed from the acid which decides how strong the acid is – i.e. where the position of equilibrium lies.

If the negative charge on the oxygen can be shared with other atoms then the $R\text{–}O^-$ ion will be more stable, and more of it will be made. When an $R\text{–}O^-$ ion is derived from an alcohol, no such sharing is possible. However, if an $R\text{–}O^-$ ion is derived from a phenol or a carboxylic acid then the electric charge gets spread out by a process called **delocalisation**. This involves a spreading out of the electrons over the ion (Figure 1) – there is more about delocalisation in **Section 12.3**.

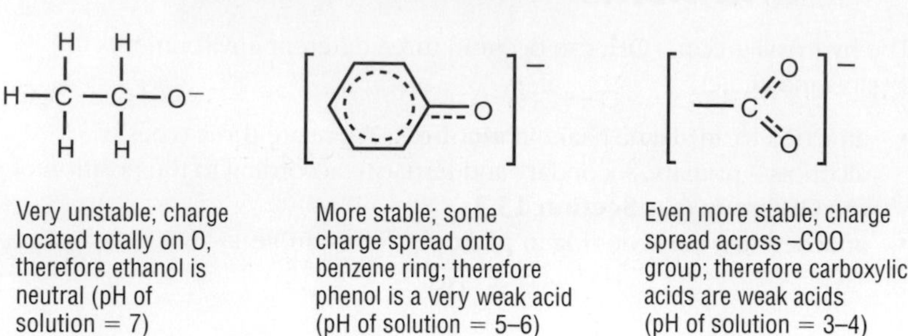

▶ **Figure 1** The strength of an acid depends on the stability of its negative ion.

Phenols and carboxylic acids are strong enough acids to react with strong bases – remember soluble bases are called alkalis, such as sodium hydroxide – to form salts. For example,

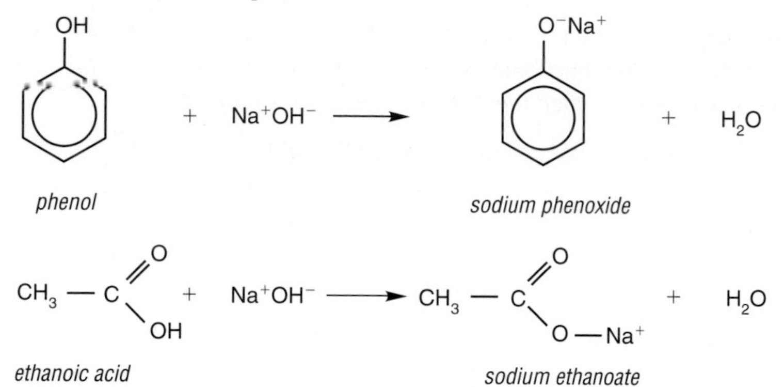

Only with solutions of carboxylic acids is the concentration of H^+(aq) ions high enough to give carbon dioxide when reacted with carbonates. (Note that the H_3O^+ ion is often written as H^+(aq).)

$$CO_3^{2-}(aq) + 2H^+(aq) \rightarrow CO_2(g) + H_2O(l)$$

So carboxylic acids make carbonates fizz, but alcohols and phenols do not.

Summary

- Carboxylic acids and phenols react with NaOH(aq) to form their sodium salts.
- Carboxylic acids, but not phenols, react with Na_2CO_3(aq) – with fizzing and release of carbon dioxide – to form the sodium salt of the acid.
- Alcohols do not react with either NaOH(aq) or Na_2CO_3(aq).

The iron(III) chloride test for phenol and its derivatives

Some groupings of atoms can become closely associated with metal ions and form complexes (see **Section 11.6**).

The –C=C–OH group (called the 'enol' group – can you see why?) can form a purple complex with Fe^{3+} ions in neutral solution. Only phenol and its derivatives have such an arrangement of atoms and they are the only ones to give a colour with iron(III) chloride – this is used as a test for phenol and its derivatives. Similar complexes are used to make the colours of some inks.

Ester formation

Alcohols react with **carboxylic acids** to form **esters**. For example, ethanol reacts with ethanoic acid to form the ester, ethyl ethanoate:

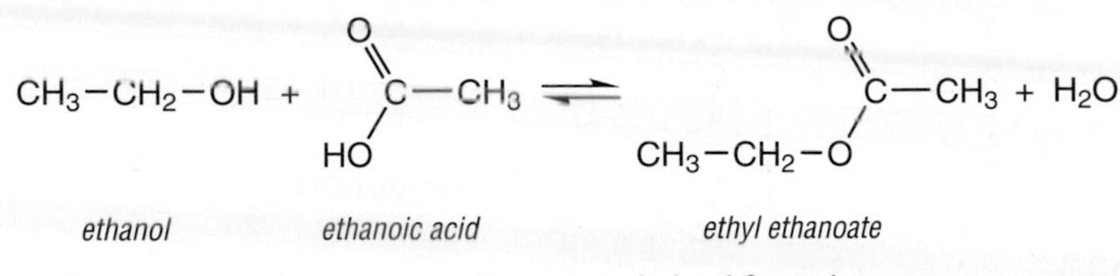

You can write the same reaction using skeletal formulae:

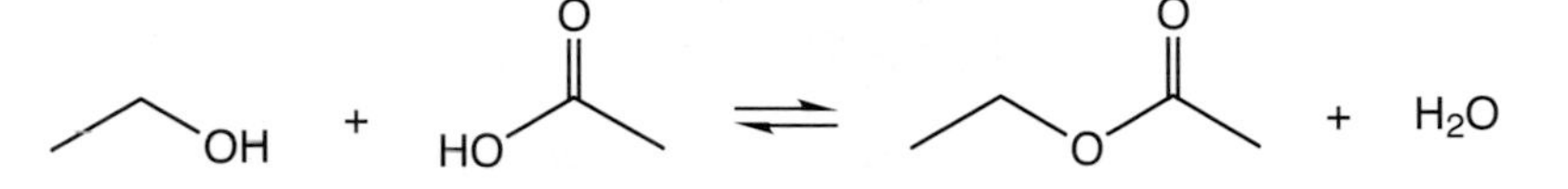

The reaction occurs extremely slowly unless an acid catalyst is present. A small amount of either concentrated sulfuric acid or concentrated hydrochloric acid is generally used, and the reaction mixture is heated. Notice the ⇌ symbol in the equation – the reaction is reversible and comes to an equilibrium, where both reactants and products are present.

Look at the structure of ethyl ethanoate. Esters are named from the alcohol and acid from which they are formed – the section on 'Naming esters' in **Section 13.5** will help you to understand how this is done.

Esters can also be made using **phenols** instead of alcohols – but here an acid derivative, an **acyl chloride** or **acid anhydride** (see Table 1 in **Section 13.3**), is used rather than the carboxylic acid itself.

Esters from salicylic acid

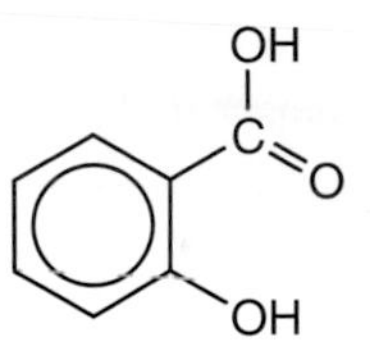

2-hydroxybenzoic acid (salicylic acid)

A look at the structure of 2-hydroxybenzoic acid (salicylic acid) shows that there are two ways of esterifying it. Either the phenol –OH group could be reacted with a carboxylic acid, or the –COOH group could be reacted with an alcohol.

Aspirin is the product of esterifying the phenol –OH group to form 2-ethanoyloxybenzoic acid. It is quite soluble in water, so it can be absorbed into the bloodstream through the stomach wall.

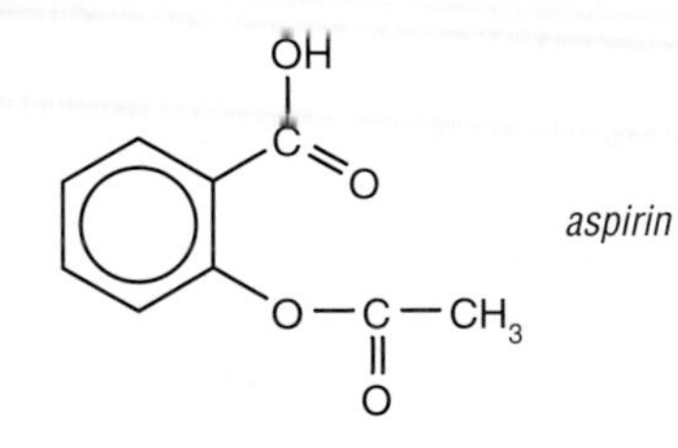

aspirin

The product of reacting the –COOH group with methanol is called methyl 2-hydroxybenzoate. This is better known as *oil of wintergreen* which is used as a liniment. It is soluble in fats rather than water so it is absorbed through the skin – like aspirin, it reduces pain and swelling.

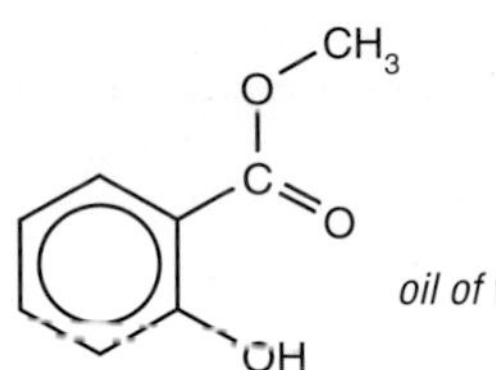

oil of wintergreen

Problems for 13.4

1 a Use systematic nomenclature to name the following compounds:

A $CH_3—CH_2—CH_2—CH_2—OH$

B

```
        OH
        |
CH3—CH—CH2—Cl
```

C

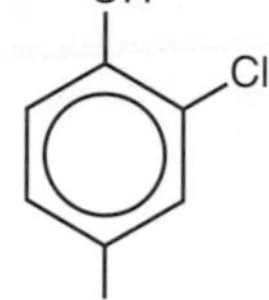

b Draw structures for the following compounds:

D butanal
E butanoic acid
F butanone
G 3-methylpentan-3-ol

2 Which one of compounds **A** to **G** in problem 1:

a is a secondary alcohol
b is an aldehyde
c is a phenol
d is a ketone
e is an aliphatic alcohol which is not easily oxidised on heating with acidified potassium dichromate(VI)
f produces a purple colour with neutral aqueous iron(III) chloride
g gives carbon dioxide with sodium carbonate
h produces a carboxylic acid on refluxing with excess acidified potassium dichromate(VI)?

3 Phenols and carboxylic acids are weak acids and show typical acid properties. Write a balanced equation for each of the following reactions.

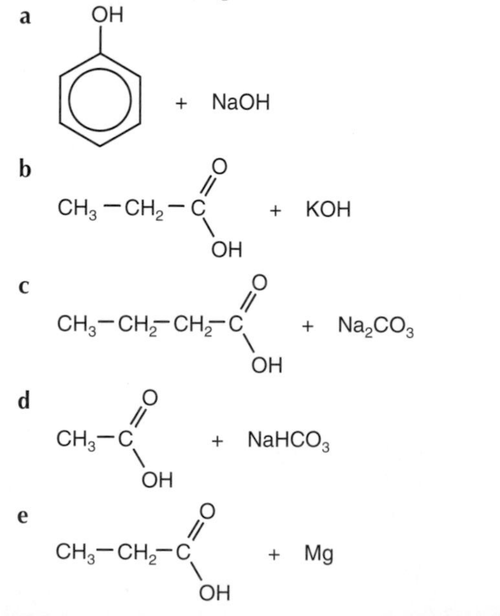

4 Draw the structure of the organic product for each of the following reactions.

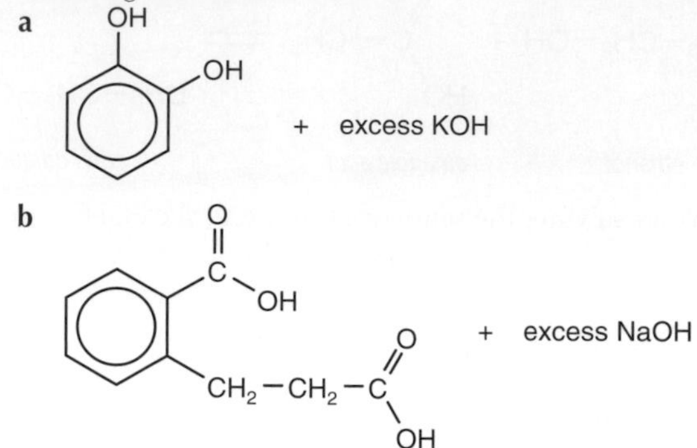

5 Draw the structure of the oxidation product in each of the following reactions – in each case, name the compound formed.

a

```
        CH3
        |
CH3—CH—CH2—OH
```

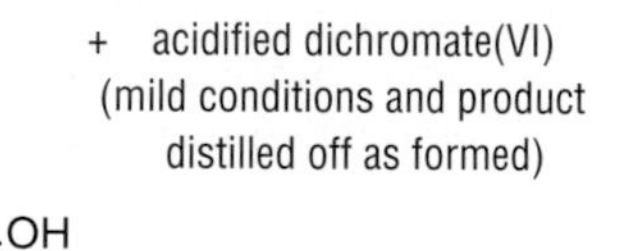

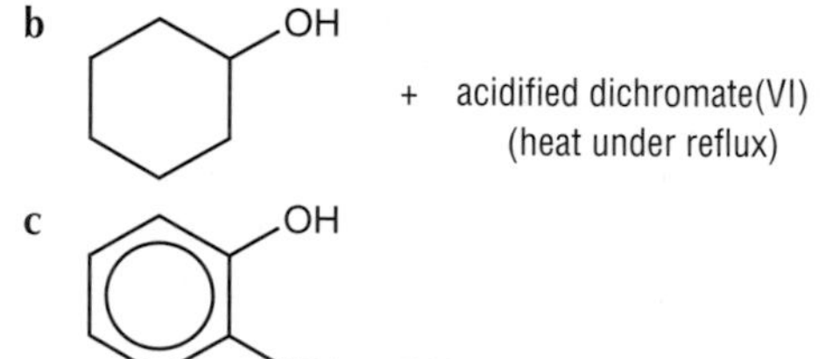

+ excess acidified dichromate(VI) (heat under reflux)

6 a Draw the full structural formula of the ester produced by heating a mixture of methanol and ethanoic acid in the presence of a few drops of concentrated sulfuric acid.

b Name the ester produced.

c Draw the full structural formula of an ester that is an isomer of the ester you have drawn in part **a**.

d Name the isomeric ester.

e Name the alcohol and carboxylic acid needed to make the isomeric ester.

13.5 *Esters*

What are esters and how are they made?

Esters are formed when an alcohol reacts with a carboxylic acid (see **Section 13.4**). The reaction occurs extremely slowly unless an acid catalyst is present – a small amount of concentrated sulfuric acid or concentrated hydrochloric acid is often used.

> **Making an ester from an alcohol**
>
> The alcohol and carboxylic acid are heated together under reflux in the presence of a few drops of concentrated sulfuric acid or concentrated hydrochloric acid.

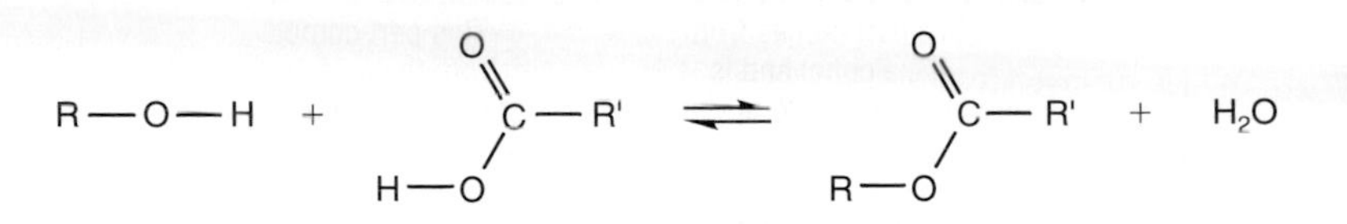

R is an alkyl group such as CH_3 or C_2H_5

Because water is produced, just like when we breathe on a cold surface, the process was christened 'condensation' and the name has stuck. A **condensation reaction** can be thought of as two molecules reacting together to form a larger molecule with the elimination of a small molecule such as water.

The ester link is formed by the condensation reaction of the hydroxyl group in the alcohol and the carboxyl group in the acid – the process is known as **esterification**.

The reaction is reversible and eventually comes to equilibrium. To improve the yield of ester from a given amount of acid, an excess of alcohol is added – or the water can be distilled off as it is formed. These help to drive the equilibrium to the right.

The reaction is easily reversed – the reverse reaction is called **ester hydrolysis** (see later in this section).

Esters have strong, sweet smells which are often floral or fruity. Esters can be used in food flavourings and perfumes. Many naturally occurring esters are responsible for well-known fragrances. Some examples are given in Table 1.

◀ **Table 1** Some ester fragrances.

Ester	Fragrance
ethyl methanoate	raspberries
3-methylbutyl ethanoate	pears
ethyl 2-methylbutanoate	apples
phenylmethyl ethanoate	jasmine

Many organic compounds dissolve readily in esters, so esters are widely used as solvents – for example, in some glues.

Naming esters

Esters are named after the alcohol and acid from which they are derived (Figure 1).

The ending *-oate* instead of the *-oic* from the parent acid shows clearly that the compound is an ester. Thus, ethanol and ethanoic acid give ethyl ethanoate, phenol and benzoic acid give phenyl benzoate.

> The names of esters end in **-oate**.

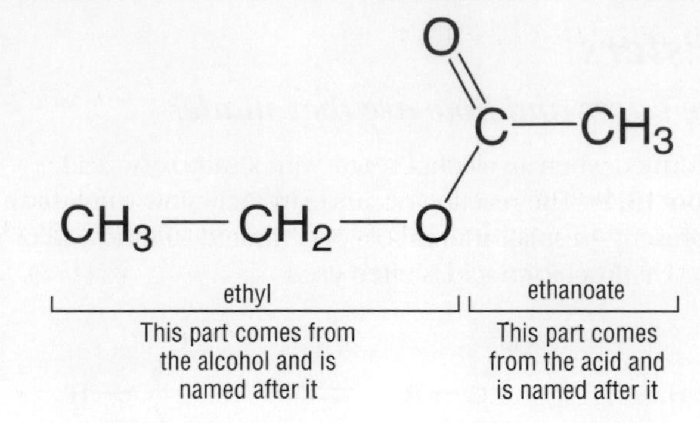

▶ **Figure 1** How to name an ester.

You have to be careful to get the group the right way round. For example, ethyl ethanoate is very different from its isomer, methyl propanoate.

ethyl ethanoate *methyl propanoate*

Making models of these two esters will help you understand how they are named.

Structural formulae

A structural formula, such as those drawn above, is the clearest way of showing the structure of esters, but you will also see the structures written in shortened form.

For example, ethyl ethanoate may be represented by

$$C_2H_5-O-\overset{\overset{\displaystyle O}{\|}}{C}-CH_3$$

Sometimes the acid part is written first and the structure becomes

$$CH_3-\overset{\overset{\displaystyle O}{\|}}{C}-O-C_2H_5$$

You will need to recognise the structure written both ways. The important thing is to identify the group attached to the C=O as this is from the acid while the group attached to –O– is from the alcohol.

Polyesters

Polyesters are condensation polymers (see **Section 5.7**). They are made by reacting a *diol* (containing two –OH groups) with a *dicarboxylic acid* (containing two –COOH groups). For example, a common polyester is made from ethane-1,2-diol and benzene-1,4-dicarboxylic acid.

$HO-CH_2-CH_2-OH$

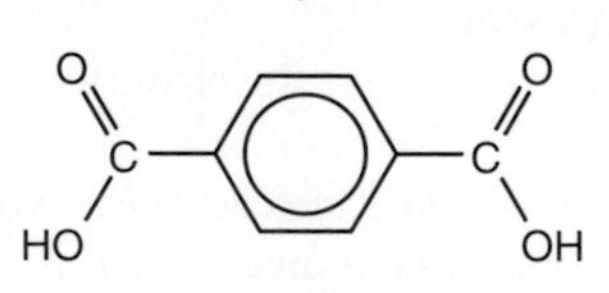

ethane-1,2-diol *benzene-1,4-dicarboxylic acid*

Making esters from phenols

The –OH group in phenol is less reactive to esterification than the –OH of ethanol, so it needs a more vigorous reagent to esterify it.

When ethanoic acid is involved in esterification, the process is sometimes known as **ethanoylation**. A more vigorous ethanoylating agent than ethanoic acid is **ethanoic anhydride**, made by eliminating a molecule of water between two ethanoic acid molecules (see Table 1 in **Section 13.3**):

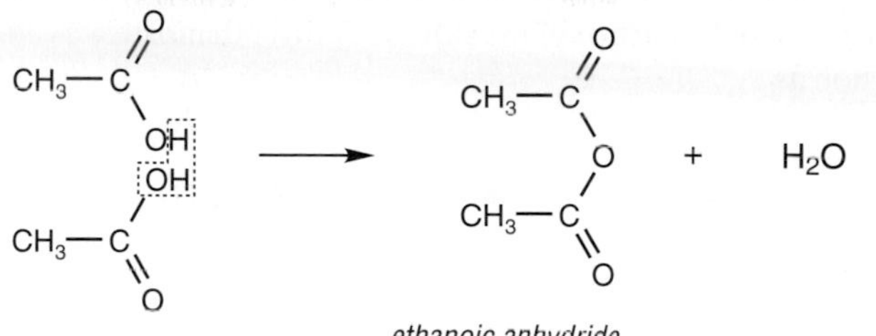

ethanoic anhydride

The equation for the reaction of ethanoic anhydride with 2-hydroxybenzoic acid (salicylic acid) is

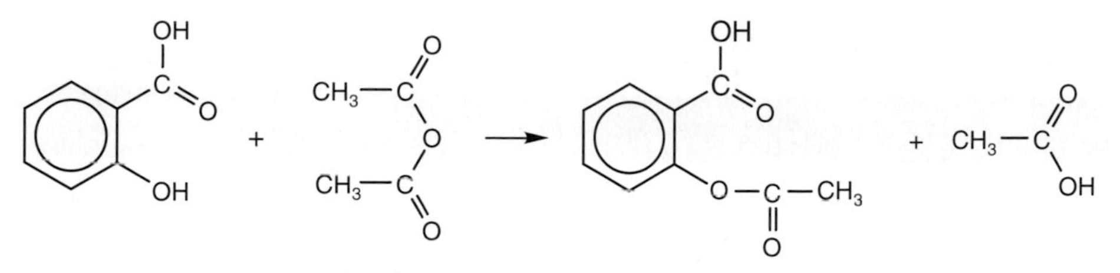

Ethanoic anhydride is often used as an ethanoylating agent because it is reactive but not too unpleasant or dangerous. A much more reactive ethanoylating agent is **ethanoyl chloride** but this is toxic and hazardous to use because it is so reactive.

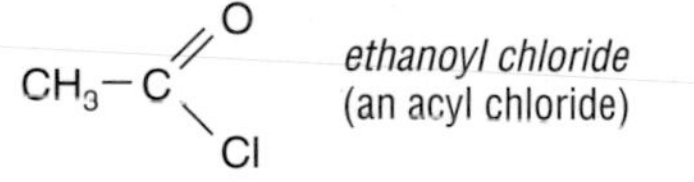

ethanoyl chloride (an acyl chloride)

In this equation ethanoyl chloride is reacting with phenol to produce phenyl ethanoate:

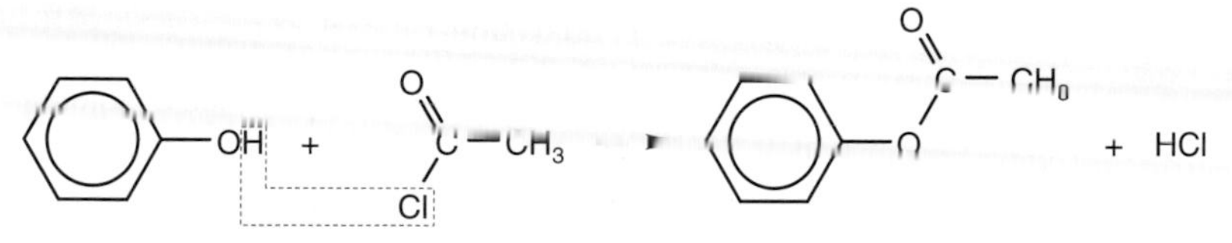

Ethanoic anhydride and ethanoyl chloride are members of a general group of reagents called **acylating agents**. Acylating agents substitute an acyl group, R–CO, for the H on an –OH group.

The *only* way to esterify a phenol is by using an acylating agent. Alcohols can be esterified either by using an acylating agent or by reacting with a carboxylic acid in the presence of an acid catalyst.

Making an ester from a phenol

The phenol is treated with an acid anhydride or an acyl chloride.

With an acid anhydride, the mixture is heated under reflux.

With an acyl chloride, the reaction usually takes place at room temperature.

Both reactions must be carried out in the absence of water.

Ester hydrolysis

The reverse of esterification corresponds to the breakdown of an ester by water – in other words, it is a **hydrolysis**.

$$CH_3-CH_2-O-C(=O)-CH_3 + H_2O \rightleftharpoons CH_3-CH_2-OH + HO-C(=O)-CH_3$$

ethyl ethanoate *ethanol* *ethanoic acid*

On their own, water and an ester react very slowly, but the process can be speeded up by catalysis.

A catalyst is effective for both directions of a reversible reaction, so sulfuric acid (or any other acid) will do – dilute sulfuric acid is used. Excess water displaces the equilibrium to the right and the yield of products is improved.

Another way of hydrolysing an ester is to add an alkali, such as sodium hydroxide solution. When alkali is used, the hydrolysis does not produce a carboxylic acid, but a **carboxylate salt**. For ethyl ethanoate we can write the reaction as

$$CH_3-CH_2-O-C(=O)-CH_3 + OH^- \longrightarrow CH_3-CH_2-OH + {}^-O-C(=O)-CH_3$$

ethyl ethanoate — *ethanol* — *ethanoate* ion

Alkaline hydrolysis (unlike acid hydrolysis) goes to completion and so is usually preferred.

Problems for 13.5

1 Use systematic nomenclature to name the following esters:

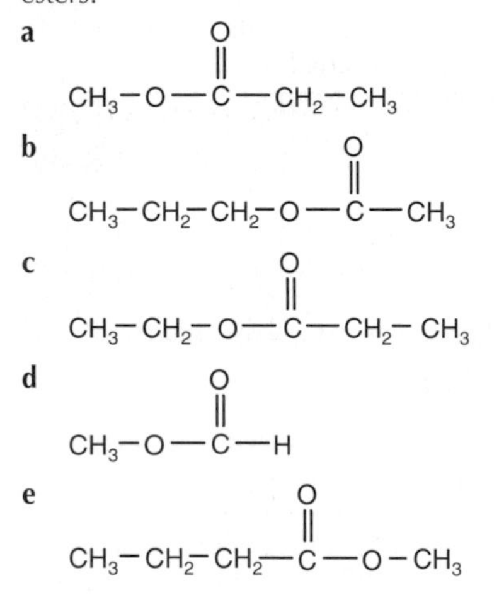

2 Look at the esters in Table 1 in this section (page 307).

a Write down the names of the alcohols and acids from which they are derived.

b Write down the structures of these alcohols, acids and esters.

c How are 3-methylbutyl ethanoate and ethyl 2-methylbutanoate related?

3 Write balanced equations for the reactions that occur when the following pairs of compounds are heated under reflux with a few drops of concentrated sulfuric acid as catalyst.

a propan-2-ol and propanoic acid

b ethanoic acid and ethane-1,2-diol.

4 Draw the structure of a 'polymer chain' formed by joining two molecules of ethane-1,2-diol and two molecules of benzene-1,4-dicarboxylic acid.

5 When ethyl ethanoate is hydrolysed with water enriched with oxygen-18, $H_2{}^{18}O$, the oxygen-18 appears in the ethanoic acid and not in the ethanol.

$$CH_3-CH_2-O-C(=O)-CH_3 + H_2{}^{18}O \xrightarrow{\text{acid catalyst}} CH_3-CH_2-O-H + H-{}^{18}O-C(=O)-CH_3$$

a Use this information to identify which ester bond is broken during hydrolysis of the ester.

b Suggest why it is this bond that is broken in preference to any other bond.

c What role do you think the catalyst might have in this reaction?

6 Write balanced equations for reactions between:
  a ethanoyl chloride and ethanol
  b ethanoyl chloride and 2-methylpropan-2-ol
  c ethanoic anhydride and methanol
  d methyl ethanoate and dilute sulfuric acid
  e phenyl ethanoate and sodium hydroxide solution.

7 Give the reagents and conditions needed to prepare the ester

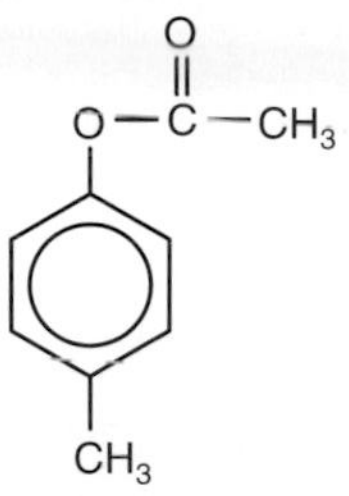

8 A chemist wants to prepare an ester, phenylmethyl ethanoate, for use in a perfume formulation.
  a Name the alcohol needed for this preparation and draw its structure.
  b Suggest *three* compounds that could be used to react with the alcohol to produce the required ester. (Give either the name or formula.)
  c For each of the three possible preparations, give the name and formula of the co-product formed along with the ester.
  d Write a balanced equation for the slow hydrolysis of this ester in moist air.
  e Suggest how the smell of the perfume might change if the ester was partially hydrolysed on exposure to moist air.

13.6 *Oils and fats*

Oils and fats: examples of esters

Oils and fats provide an important way of storing chemical energy in living systems. Most of them have the same basic structure. The only difference is that oils are liquid at room temperature whereas fats are solid. A selection of natural oils and fats is listed in Table 1.

Most oils and fats are **esters** of propane-1,2,3-triol (commonly called *glycerol*) with long-chain carboxylic acids, R–COOH. In *palmitic acid*, for example, R is a $CH_3(CH_2)_{14}$ group:

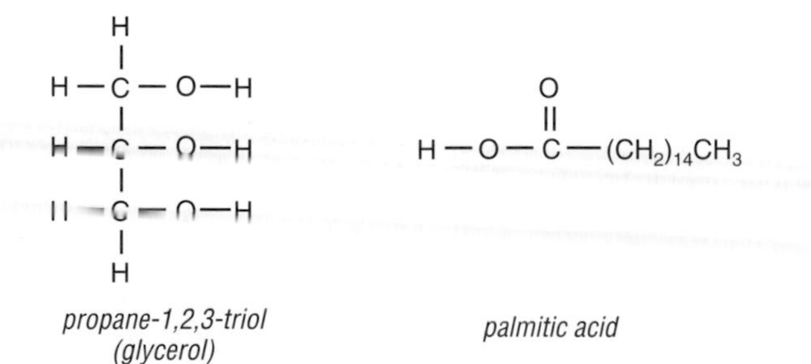

propane-1,2,3-triol (glycerol) *palmitic acid*

▼ **Table 1** Classification of fats and oils, with some examples.

Animal fats	Marine oils	Vegetable oils
beef (dripping)	whale	sunflower
pig (lard)	herring	olive
mutton	cod liver	rapeseed

You can remind yourself of the structure and chemistry of esters by reading **Section 13.5**.

Because glycerol has three alcohol groups in each molecule, *three* carboxylic acid molecules can form ester linkages with each glycerol molecule to form a **triester** (sometimes called a **triglyceride**). The triester formed from glycerol and palmitic acid has the structure

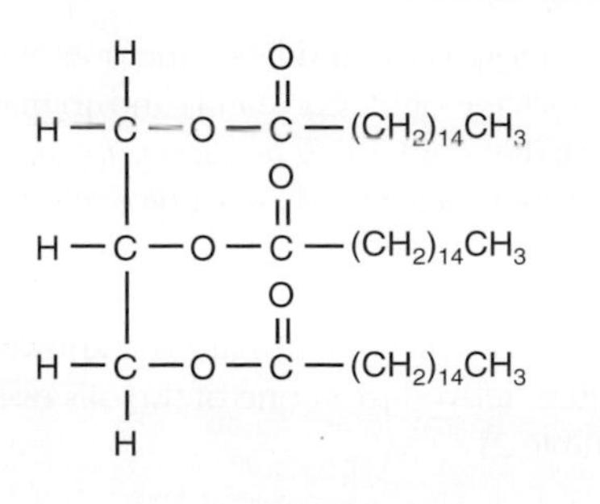

triester formed from *glycerol* and *palmitic acid*

The triesters found in natural oils and fats are often **mixed triesters** in which the three acid groups are not all the same.

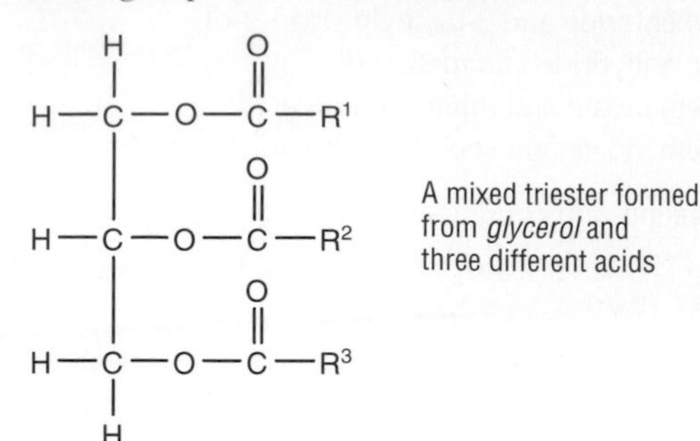

Figure 1 summarises the structure of triesters.

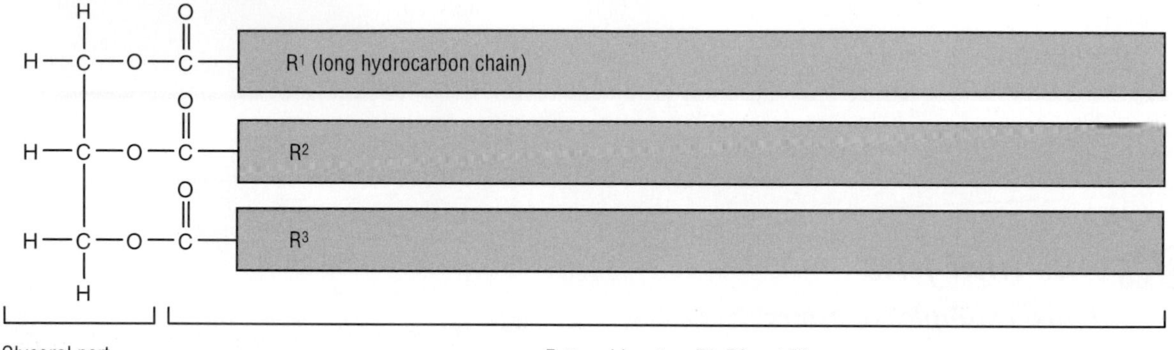

▲ **Figure 1** The general structure of the triesters found in fats and oils.

Fats and fatty acids

The carboxylic acids in fats and oils usually have unbranched hydrocarbon chains. They contain an even number of carbon atoms, ranging from C_4 to C_{24}, but often contain 16 or 18 carbon atoms. They are sometimes called **fatty acids** because of their origin. The alkyl groups, R^1, R^2 and R^3, are either fully saturated or contain one or more double bonds. Table 2 shows some common fatty acids. There are many others, some of which contain three or four double bonds.

► **Table 2** Common fatty acids.

Structure	Traditional name	Origin of name
$CH_3(CH_2)_{14}COOH$	palmitic acid	palm oil
$CH_3(CH_2)_{16}COOH$	stearic acid	suet (Greek: *stear*)
$CH_3(CH_2)_7CH{=}CH(CH_2)_7COOH$	oleic acid	olive oil
$CH_3(CH_2)_4CH{=}CHCH_2CH{=}CH(CH_2)_7COOH$	linoleic acid	oil of flax (Latin: *linum*)

Many doctors believe that the more saturated fats – the ones with few double bonds – tend to cause blockage of blood vessels and so may lead to heart disease. They believe that a diet containing *polyunsaturated* fats with many double bonds may be healthier. Figure 2 shows a typical saturated and unsaturated fatty acid.

The modern systematic names for fatty acids can get quite long (for example, oleic acid is octadec-9-enoic acid), so they are still known by their old names. These names are often derived from one of the oils or fats from which they are obtained (see Table 2).

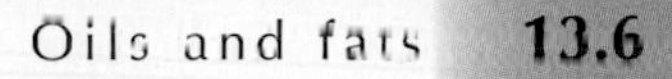

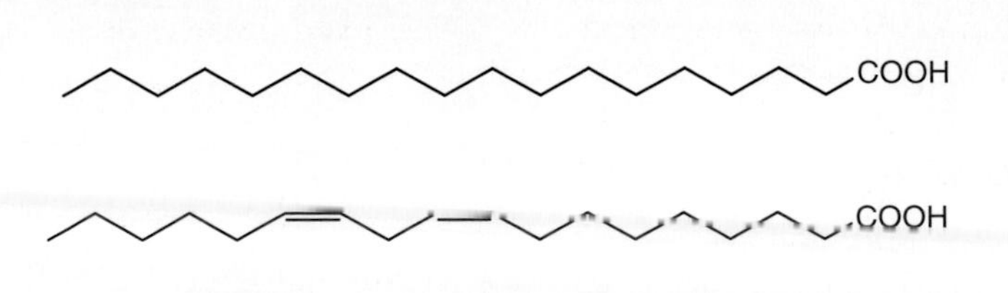

◀ **Figure 2** A typical saturated and unsaturated fatty acid.

Any natural oil or fat contains a mixture of triesters. The nature of the acids present affects the properties and determines whether the substance is a liquid oil or a solid fat.

The *proportions* in which the acid groups occur are more or less constant for a particular oil or fat. Human body fat, for example, contains mainly four acids:

oleic acid	47%	linoleic acid	10%
palmitic acid	24%	stearic acid	8%

An oil or fat can be identified by breaking it down into glycerol and its fatty acids, and then measuring the amount of each acid present.

Like all esters, oils and fats can be split up by **hydrolysis**. This is usually done by heating the oil or fat with concentrated sodium hydroxide solution to give glycerol and the sodium salts of the acids. For example

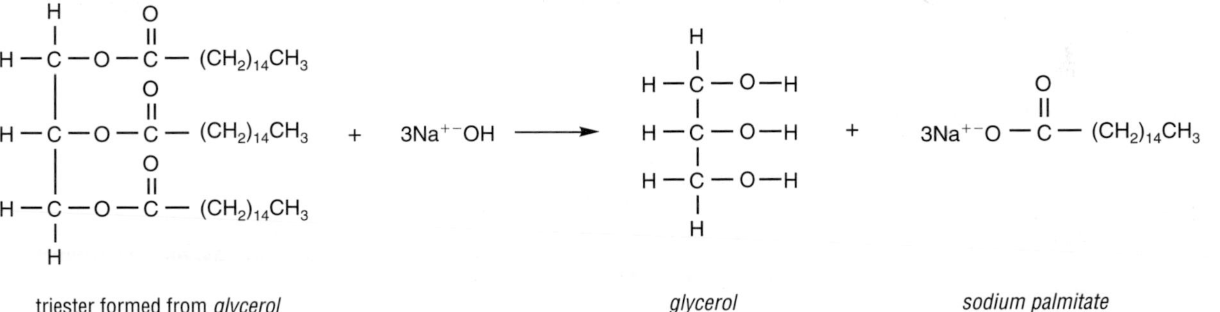

This is how soap is made – soaps are sodium or potassium salts of fatty acids, and they are made by heating oils and fats with sodium or potassium hydroxide. What oils do you think *Palmolive* soap is made from?

The free fatty acids can be released from the sodium salts by adding a dilute mineral acid, such as hydrochloric acid.

Solid or liquid?

Any natural oil or fat contains a mixture of triesters. The nature of the fatty acids present affects the properties and determines whether the substance is a liquid oil or a solid fat.

The shape of saturated and unsaturated triglyceride molecules can be represented as shown in Figure 3 (page 314).

The saturated triglyceride molecules can pack together closely, with intermolecular bonds holding the molecules together. The unsaturated triglyceride molecules cannot pack so closely because the presence of *Z* double bonds causes the molecules to 'kink'. This means that the attractive bonds between molecules will be weaker than between the saturated molecules.

The unsaturated molecules require less energy to separate them and so will have a lower melting point than the corresponding saturated molecular fats. As a result, the more unsaturated fatty acid molecules there are in a triglyceride mixture the more likely it is to be a liquid at room temperature – i.e. an oil. The oils shown in Table 1 have a higher proportion of unsaturated triglycerides than the fats.

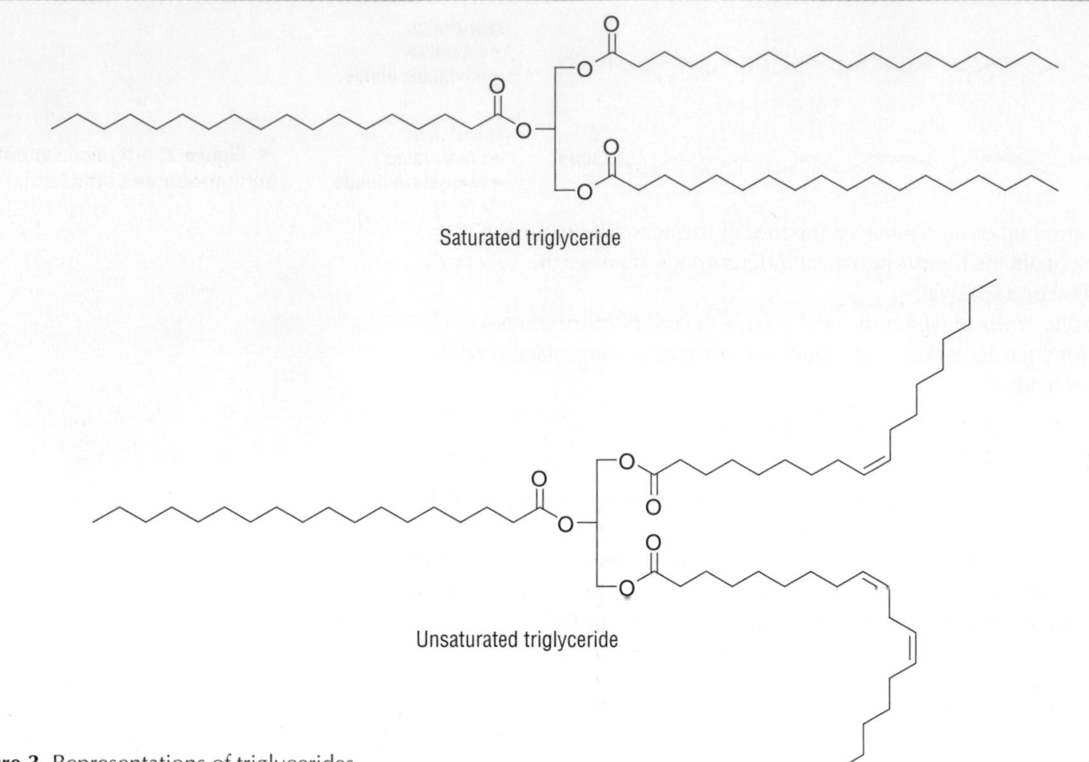

▲ **Figure 3** Representations of triglycerides.

Converting oils to fats

Although many natural oils and fats can be eaten directly, most of them require processing in order to make them fit a number of requirements we have regarding taste, texture and so on. One of the commonest processes is the **hydrogenation** of unsaturated oils to make margarine.

The addition of hydrogen to unsaturated molecules reduces the number of double bonds – i.e. the molecules become less unsaturated and more saturated.

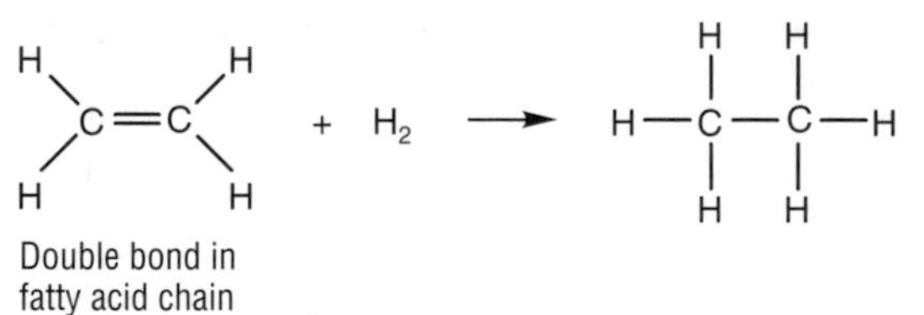

This in turn means that the substance containing the molecules becomes solid. Too much hydrogenation causes the resulting fats to be too hard and brittle, so the hydrogenation has to be carefully controlled.

Modern margarines are made by passing hydrogen through heated oil containing a nickel catalyst. Fat-free milk is added, along with flavourings, salt and colour, and sometimes vitamins. Margarine manufacture is summarised in Figure 4.

Hydrogenation is an example of an addition reaction. You can read about the addition reactions of alkenes in **Section 12.2**.

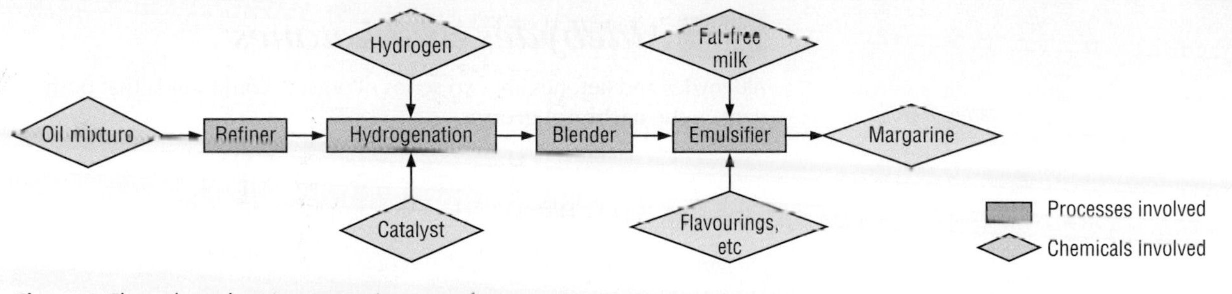

▲ **Figure 4** Flow chart showing margarine manufacture.

Problems for 13.6

1 a What structural feature of a molecule identifies a compound as an ester?

b Write down the full structural formula of the triester made from propane-1,2,3-triol (glycerol) and ethanoic acid.

c Write an equation to show what happens when this ester is heated with concentrated sodium hydroxide solution.

2 Below is the structure of a triester present in a naturally occurring oil:

$$\mathrm{H{-}\underset{|}{\overset{H}{\overset{|}{C}}}{-}O{-}\overset{O}{\overset{\|}{C}}{-}(CH_2)_7CH{=}CH(CH_2)_7CH_3}$$
$$\mathrm{H{-}\underset{|}{\overset{|}{C}}{-}O{-}\overset{O}{\overset{\|}{C}}{-}(CH_2)_7CH{=}CHCH_2CH{=}CH(CH_2)_4CH_3}$$
$$\mathrm{H{-}\underset{\underset{H}{|}}{\overset{|}{C}}{-}O{-}\overset{O}{\overset{\|}{C}}{-}(CH_2)_7CH{=}CH(CH_2)_7CH_3}$$

The triester was hydrolysed by heating with concentrated sodium hydroxide solution, and the resulting solution was neutralised by addition of hydrochloric acid.

a Name the hydrolysis products obtained.

b How many moles of each product would be obtained from 1 mole of the triester?

3 Explain the difference between the following terms, which often appear on food labels:

a saturated fat

b monounsaturated fat

c polyunsaturated fat.

4 Animal fat is made up almost entirely of saturated fats. One compound present in sheep fat produces only glycerol and stearic acid on hydrolysis.

a Draw the skeletal formula of this animal fat.

b Stearic acid reacts with sodium hydroxide to make a crude soap – draw a skeletal formula for this soap.

c Indicate on your structure in part **b** which part of the molecule would be attracted to oil and grease. What *type(s)* of intermolecular bonds are involved?

d Why is this soap soluble in water? Explain your answer in terms of bonding.

5 A food company buys a vegetable oil and converts it to margarine by treatment with hydrogen in the presence of a metal catalyst. During the process some of the double bonds in the oil are hydrogenated.

a Name a suitable catalyst and give conditions under which the hydrogenation could be carried out.

b Assume that the oil contains only the triester in problem 2.

i What is the M_r of the triester in the oil?

ii 1 tonne (1000 kg) of the oil requires 4.90 kg of hydrogen to make margarine. Calculate how many moles of oil are reacting with how many moles of hydrogen.

iii From its structure, how many double bonds are there per molecule, and hence per mole?

iv How many moles of hydrogen would be required to *fully* saturate 1 mole of this oil?

v Refer to your answer in part **iii**. What percentage of the double bonds in the oil has been hydrogenated?

vi Suggest an advantage of not fully saturating the oil.

13.7 *Aldehydes and ketones*

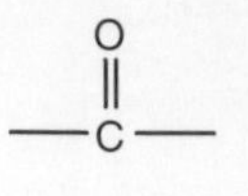

Aldehydes and ketones are two series of organic compounds that both contain the **carbonyl group**.

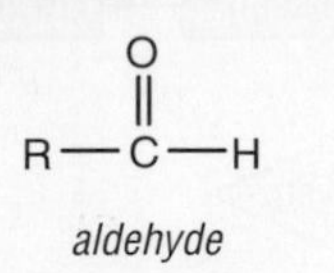

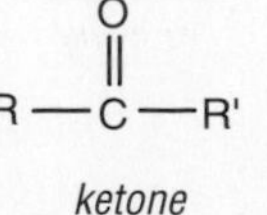

The carbonyl functional group has characteristic properties that are shown by both classes of compound, so it is convenient for the two homologous series to be considered together. However, the presence of a hydrogen atom attached to the carbonyl group gives aldehydes certain properties which ketones do not possess, and enables the two classes of compound to be distinguished from one another.

Aldehydes

The names of aldehydes end in **-al**.

Aldehydes are named using the suffix *-al*. For example:

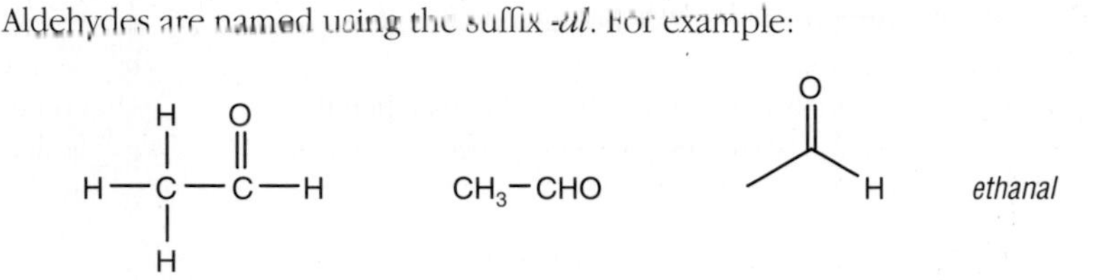

Figure 1 shows the relationship between the name of an aldehyde and the names of the related alcohol and alkane.

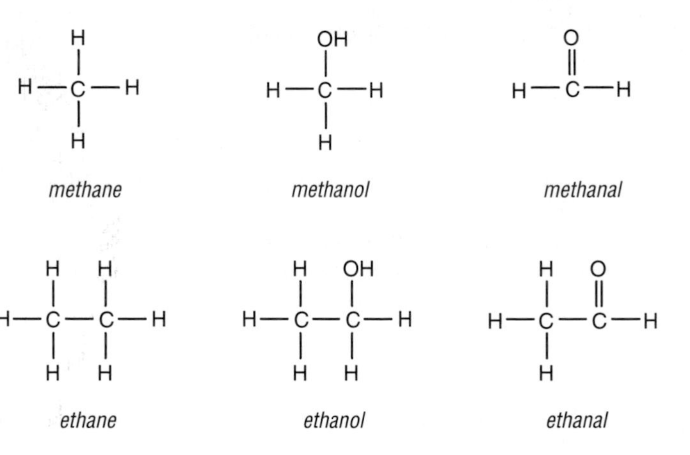

► **Figure 1** Naming aldehydes.

Ketones

The names of ketones end in **-one**.

Ketones are named using the suffix *-one*. For example:

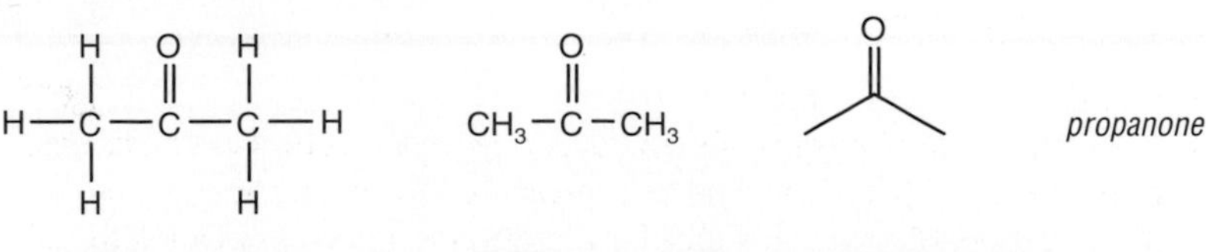

Figure 2 shows the relationship between the name of a ketone and the names of the related alcohol and alkane.

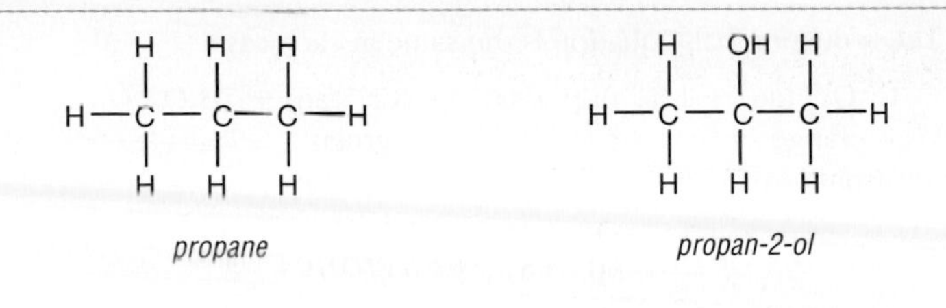

propanone

$CH_3-CH_2-CH_2-CH_2-CH_3$

pentane

$CH_3-CH(OH)-CH_2-CH_2-CH_3$

pentan-2-ol

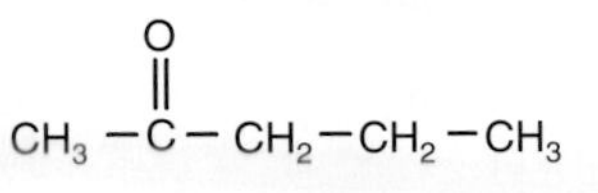

▲ **Figure 2** Naming ketones.

Ketones with five or more carbon atoms have structural isomers. The position of the carbonyl group is specified by inserting the number of its carbon atom – counting from the nearer end of the chain. For example:

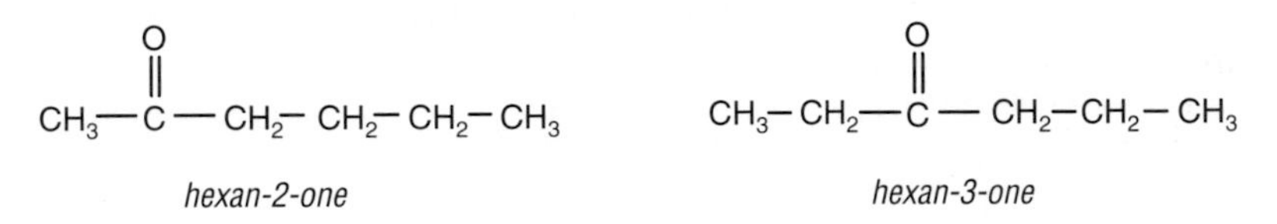

When naming aldehydes and ketones, note that the carbon atom in the carbonyl group counts as one of the carbon atoms for the name.

Preparation of aldehydes and ketones

Aldehydes and ketones can be produced in the laboratory by oxidation of alcohols (see **Section 13.2**). An acidified solution of potassium dichromate(VI) is often used as the oxidising agent and the mixture is heated. The orange dichromate(VI) ion, $Cr_2O_7^{2-}$, is reduced to green Cr^{3+} in solution. The basic reaction is

Aldehydes are produced from *primary* alcohols, and ketones are formed from *secondary* alcohols. For example:

ethanol $CH_3-CH_2-OH \longrightarrow CH_3-CHO$ ethanal

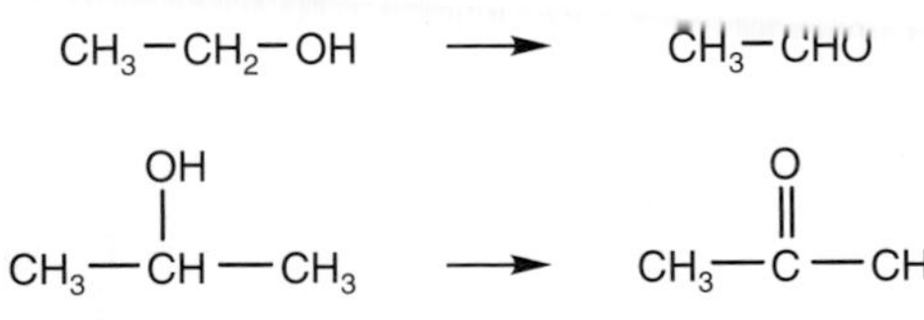

Notice how the nomenclature changes from *-ol* to either *-al* or *-one*.

The **oxidation** of primary alcohols to aldehydes must be done carefully, otherwise the aldehyde will be further oxidised to the carboxylic acid. This is not a problem when oxidising secondary alcohols because ketones do not readily undergo further oxidation.

Preparing aldehydes

The oxidising solution is dripped slowly into the hot alcohol, and the aldehyde is distilled off as it is formed, before it has time to be oxidised further to an acid.

Writing balanced equations

The oxidation reactions of alcohols and aldehydes are **redox** reactions (see **Sections 9.1** and **9.2**). They can be represented by two half-equations.

For example, the half-equations for the oxidation of ethanol and ethanal are:

$CH_3–CH_2–OH \rightarrow CH_3–CHO + 2H^+ + 2e^-$
$CH_3–CHO + H_2O \rightarrow CH_3–COOH + 2H^+ + 2e^-$

The reduction half-equation is the same in each case:

$$Cr_2O_7^{2-}(aq) + 14H^+(aq) + 6e^- \rightarrow 2Cr^{3+}(aq) + 7H_2O$$

orange solution → green solution

Reactions of aldehydes and ketones

Oxidation

Many of the reactions of aldehydes and ketones are similar – because they both contain a carbonyl functional group. However, there are key differences because aldehydes contain a hydrogen atom attached to the carbonyl group.

The hydrogen atom attached to the carbonyl group means that aldehydes are easily oxidised, even by weak oxidising agents such as the copper(II) ion, $Cu^{2+}(aq)$, in Fehling's solution (a test solution containing $Cu^{2+}(aq)$ ions and alkali). The aldehyde is oxidised to a carboxylic acid and the blue copper(II) ions, in solution, are reduced to copper(I) ions. Copper(I) oxide (Cu_2O) is precipitated as an orange brown solid when the mixture is heated.

The reaction with **Fehling's solution** is used as a test to distinguish between aldehydes and ketones.

For example, ethanal is oxidised to ethanoic acid by Fehling's solution:

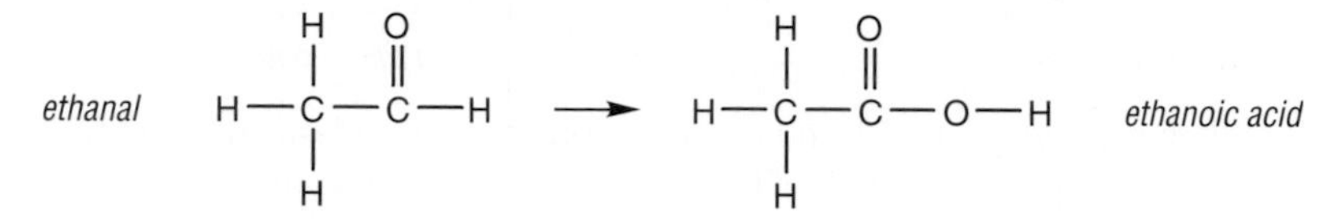

The same reaction can be brought about by heating the alcohol under reflux with an excess of acidified potassium dichromate(VI) solution. Ketones, on the other hand, are not readily oxidised by either Fehling's solution or an acidified solution of potassium dichromate(VI).

Reduction

Sometimes, when devising organic synthetic routes, it is necessary to reduce aldehydes and ketones back to alcohols. This does not take place readily and requires a powerful reducing agent. A complex metal hydride called sodium tetrahydridoborate(III), $NaBH_4$, is often used.

Aldehydes are reduced to primary alcohols, and ketones are reduced to secondary alcohols. For example

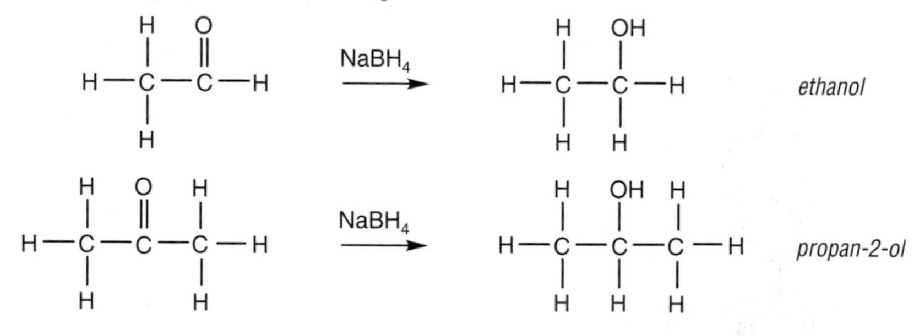

Addition reactions

The carbonyl groups in both aldehydes and ketones undergo addition reactions. For example, hydrogen cyanide, HCN, in the presence of alkali, adds across the C=O bond to form 2-hydroxynitriles (sometimes called cyanohydrins).

For ethanal, the overall reaction is

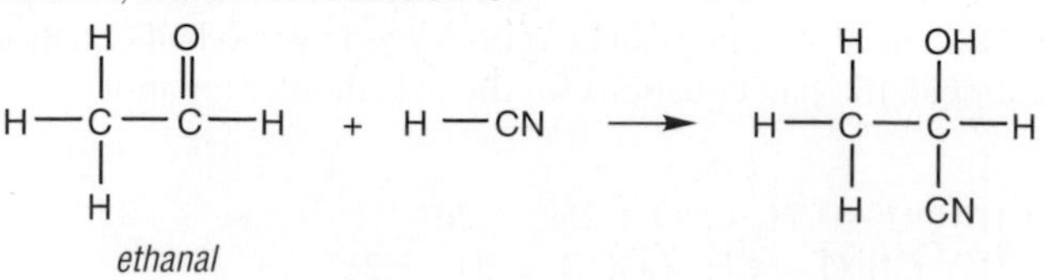

For propanone, the overall reaction is

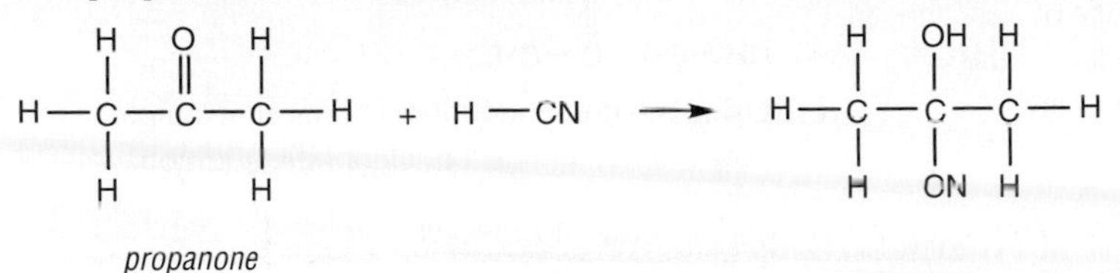

The reaction occurs in two stages. The cyanide ion is a *nucleophile* (see **Section 13.1**) and it is attracted to the carbon atom in the carbonyl group which carries a partial positive charge. A new C–C bond forms, and a pair of electrons in the C=O bond moves onto the oxygen atom, which then carries a negative charge.

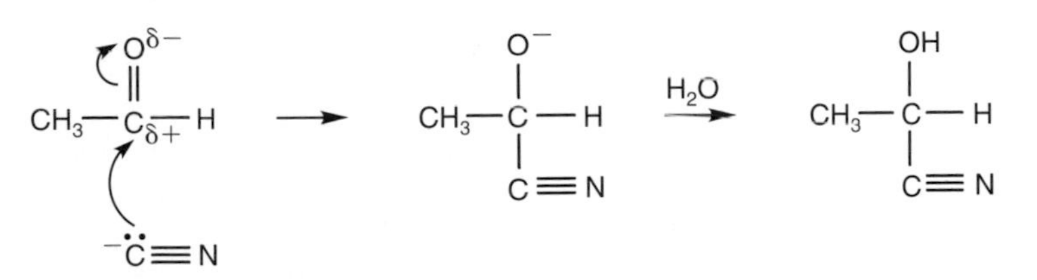

The new negatively charged ion then takes up a proton (H^+) from the solvent, water.

The overall reaction is addition of H–CN across the C=O bond.

The mechanism is described as **nucleophilic addition** because the first step involves attack by a nucleophile. Compare this with the electrophilic addition reactions to a C=C bond where the first step involves attack by an electrophile (see **Section 12.2**).

The reaction of aldehydes and ketones with HCN is useful in organic synthesis because a new C–C bond is formed – it is one way of increasing the length of a carbon chain.

Problems for 13.7

1 Use systematic nomenclature to name the following aldehydes:

a H–CHO

b CH_3-CH_2-CHO

c $CH_3-CH(CH_3)-CH_2-CH_2-CHO$

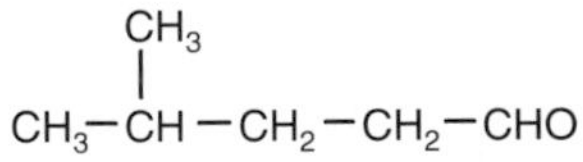

2 Use systematic nomenclature to name the following ketones:

a $CH_3-CO-CH_2-CH_3$

b $CH_3-CH_2-CO-CH_2-CH_3$

3 Draw the full structural formulae and the skeletal formulae of the following compounds:

a pentanal b pentan-2-one.

4 Which of the following alcohols can be oxidised to an aldehyde or a ketone? In each case, state which type of carbonyl compound would be formed and draw its structural formula.

a $CH_3-CH_2-CH(OH)-CH(CH_3)-CH_3$

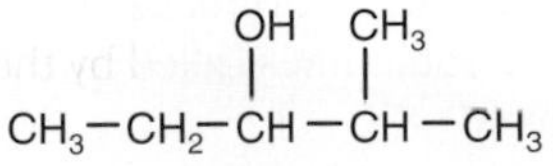

b $CH_3-C(CH_3)_2-CH_2-OH$

c $CH_3-CH_2-CH(CH_3)-CH(OH)-CH_3$

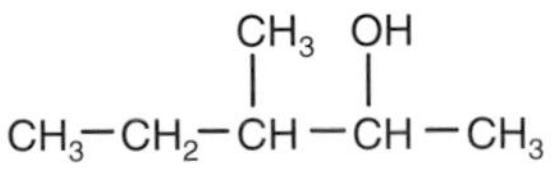

5 The following carbonyl compounds can be reduced to alcohols. In each case draw the structural formula of the alcohol formed and state whether it is a primary or secondary alcohol.

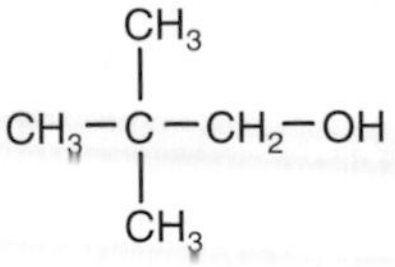

a $CH_3-CH_2-CH_2-CHO$

b $CH_3-CO-CH_2-CH_3$

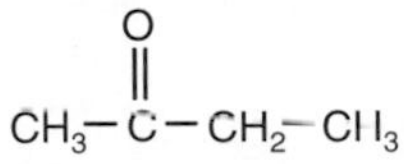

c $CH_3-C(CH_3)_2-CO-CH_3$

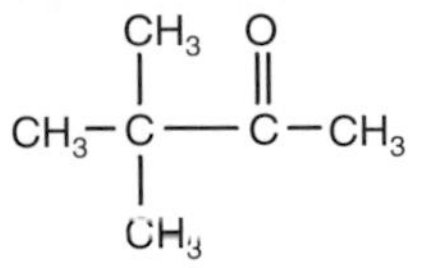

6 Aldehydes and ketones undergo nucleophilic addition reactions with hydrogen cyanide in the presence of alkali. Draw the structural formula for the product of the reaction between hydrogen cyanide and each of the following compounds.

a CH_3-CH_2-CHO

b $CH_3-CH_2-C(=O)-CH_3$

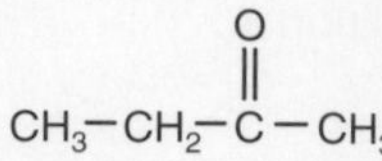

c Using the compound in part **a** as an example, explain how the reaction is a nucleophilic addition.

13.8 *Amines and amides*

What are amines and how are they named?

Amines are the organic chemistry relatives of ammonia. Their structures resemble ammonia molecules in which alkyl groups take the place of one, two or all three hydrogen atoms.

$H-N(H)-H$ *ammonia*

$R-N(H)-H$ *primary amine*

$R-N(R')-H$ *secondary amine*

$R-N(R')-R''$ *tertiary amine*

Notice the following points about naming amines.

- Amines with *one* alkyl group are called **primary amines**; **secondary amines** have *two* alkyl groups, and **tertiary amines** have *three* alkyl groups.
- Lower primary amines are named as follows:

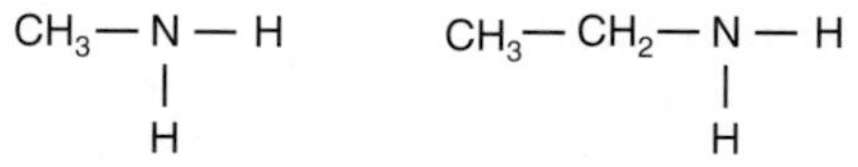

  $CH_3-N(H)-H$ *methylamine*

  $CH_3-CH_2-N(H)-H$ *ethylamine*

- Simple secondary and tertiary amines are named as follows:

  $CH_3-N(H)-CH_3$ *dimethylamine*

  $CH_2-N(CH_3)-CH_3$ *trimethylamine*

- The alkyl groups in secondary and tertiary amines need not all be identical.
- Higher homologues are often named using the prefix 'amino' and the name of the alkane from which they appear to be derived. For example,

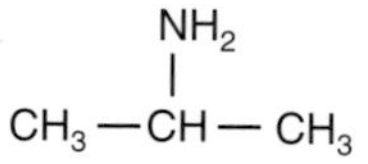

  $CH_3-CH(NH_2)-CH_3$ *2-aminopropane*

You may have noticed the amines that you used in your work on the **Materials Revolution** module to make nylon have amino groups at both ends of the molecule. Amines with two amino groups attached are called **diamines** and are named as follows:

$H_2N-CH_2-CH_2-CH_2-NH_2$ *1,3-diaminopropane*

$H_2N-CH_2-CH_2-CH_2-CH_2-CH_2-CH_2-NH_2$ *1,6-diaminohexane*

Amines with low relative molecular masses are gases or volatile liquids. The volatile amines also resemble ammonia in having strong smells. The characteristic smell of decaying fish comes from amines such as ethylamine and trimethylamine. Rotting animal flesh gives off the diamines $H_2N(CH_2)_4NH_2$ and $H_2N(CH_2)_5NH_2$, which are sometimes called by the names putrescine and cadaverine, respectively.

Properties of amines

The properties of amines are similar to those of ammonia, but modified by the presence of alkyl groups. Most of the properties are due to the lone pair of electrons on the nitrogen atom.

The bonding around the nitrogen atom of an amine is similar to that in ammonia – three pairs of electrons form localised covalent bonds, while the other two electrons form a lone pair (Figure 1).

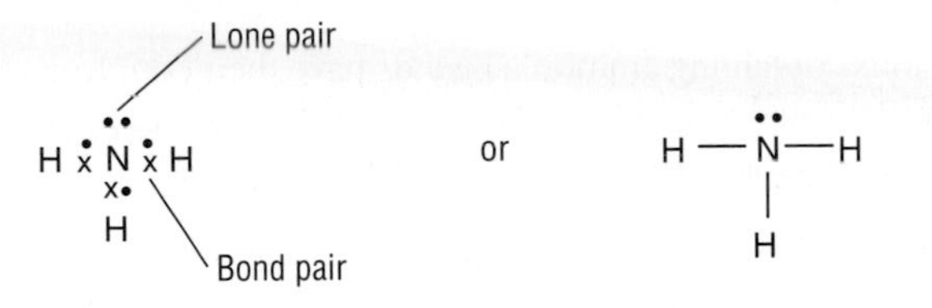

◀ **Figure 1** The arrangement of electrons in the ammonia molecule.

The lone pair electrons is responsible for ammonia being:

- very soluble in water
- a base
- a ligand
- a nucleophile.

We find these properties in amines too.

Solubility of amines

Like ammonia, amines can form hydrogen bonds to water (Figure 2).

Because of this strong attraction between amine molecules and water molecules, amines with small alkyl groups are soluble. Amines with larger alkyl groups are less soluble because the alkyl groups disrupt the hydrogen bonding in water.

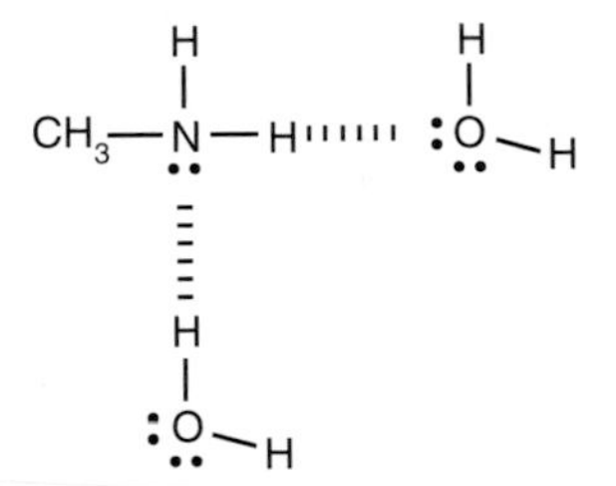

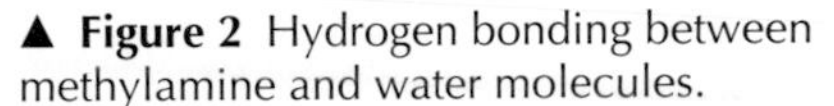

▲ **Figure 2** Hydrogen bonding between methylamine and water molecules.

Amines as bases

The lone pair on the nitrogen atom can take part in dative covalent bonding. When the electron pair is donated to an H^+, ammonia acts as an H^+ acceptor and is a *base*:

$$NH_3(aq) + H_2O(l) \rightleftharpoons NH_4^+(aq) + OH^-(aq)$$

ammonium ion

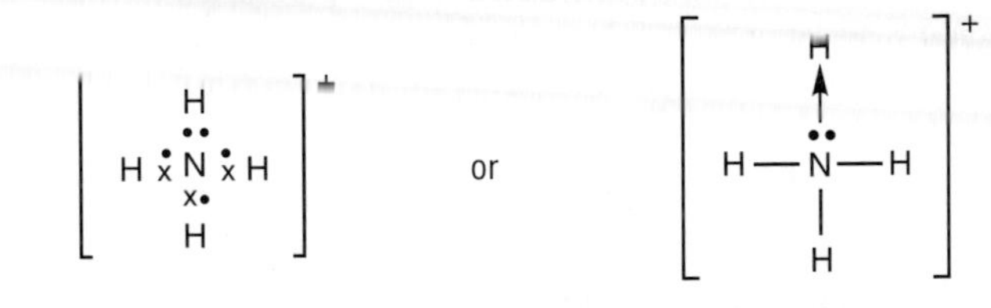

◀ **Figure 3** Dative covalent bond formation between N and H^+ in the ammonium ion.

A similar thing happens with amines. Like ammonia, they can accept H^+ from water. Using propylamine as an example, the reaction with water is

$$CH_3{-}CH_2{-}CH_2{-}NH_2(aq) + H_2O(l) \rightleftharpoons CH_3{-}CH_2{-}CH_2{-}NH_3^+(aq) + OH^-(aq)$$

propylammonium ion

The presence of hydroxide ions makes the solution alkaline – so solutions of amines, like ammonia solution, are alkaline.

Like ammonia, amines also react with acids. The H_3O^+ ions in acidic solutions are more powerful H^+ donors than water. Their reaction with amines goes to completion, and the solution therefore loses its strong amine smell. For example,

$$CH_3{-}CH_2{-}NH_2(aq) + H_3O^+(aq) \rightarrow CH_3{-}CH_2{-}NH_3^+(aq) + H_2O(l)$$

ethylamine *ethylammonium* ion

The above are both examples of neutralisation reactions. You can read more about these acid–base reactions in **Section 8.1**.

Amines as ligands

Ammonia is an effective ligand (see **Section 11.6**) because the lone pair of electrons on the N atom can bond to metal ions. For example, ammonia forms a deep-blue complex ion,

$$[Cu(NH_3)_4(H_2O)_2]^{2+}$$

$CH_3–CH_2–CH_2–CH_2–NH_2$

$C_4H_9NH_2$

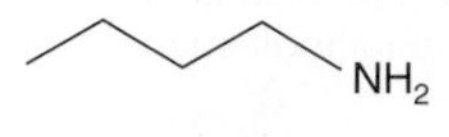

butylamine

Complex ions containing ammonia ligands have their counterparts in amine chemistry. For example, adding butylamine to aqueous copper(II) sulfate produces a dark blue complex ion

$$[Cu(C_4H_9NH_2)_4(H_2O)_2]^{2+}$$

with a structure is similar to $[Cu(NH_3)_4(H_2O)_2]^{2+}$.

Amines as nucleophiles

Ammonia can act as a nucleophile (see **Section 13.1**) with the lone pair of electrons on the N atom attacking electrophiles, such as the positively polarised carbon atom in halogenoalkanes:

$$R–Cl + NH_3 \rightarrow R–NH_3^+\ Cl^- \rightleftharpoons R–NH_2 + H^+ + Cl^-$$

The product of this reaction is an amine – but amines can themselves behave as nucleophiles, because they have a lone pair of electrons on the N atom, just the same as ammonia. So amines undergo substitution reactions with halogenoalkanes, to form secondary and tertiary amines.

$$R'—Cl + R—NH_2 \longrightarrow [R'—NH(H)(R)]^+\ Cl^- \rightleftharpoons R'—NH(R) + H^+ + Cl^-$$

Amide formation

A similar type of reaction occurs when we use an acyl chloride instead of a halogenoalkane. (Acyl chlorides are covered in **Section 13.5**.) This type of reaction is known as acylation, because an acyl group is added to the amine. When an amine is acylated an amide is formed. For example,

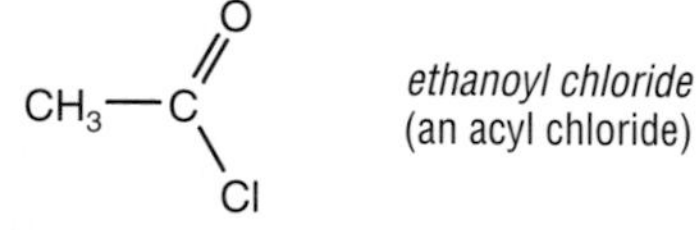

reacts with ammonia to form an **amide**

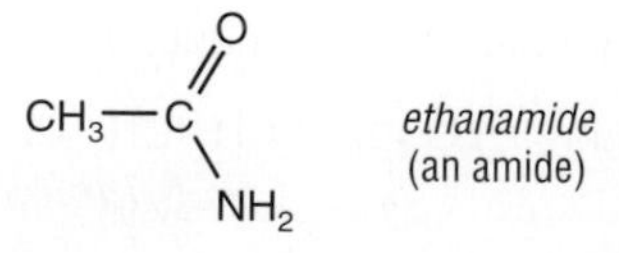

The reaction is very vigorous, even at room temperature, because acyl chlorides are very reactive. They also react violently with water to form the carboxylic acid, so anhydrous conditions must be used.

Figure 4 shows the reactions of ammonia and a primary amine with ethanoyl chloride.

(a) Reaction of ammonia

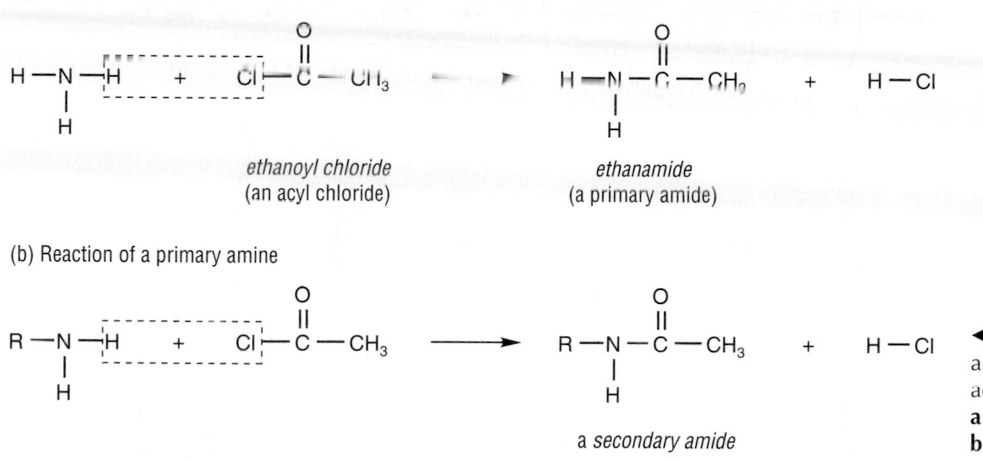

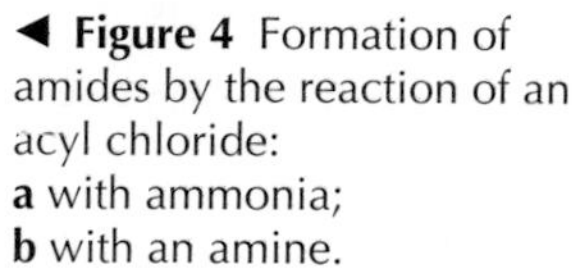
◀ **Figure 4** Formation of amides by the reaction of an acyl chloride:
a with ammonia;
b with an amine.

What are amides?

Amides contain the functional group

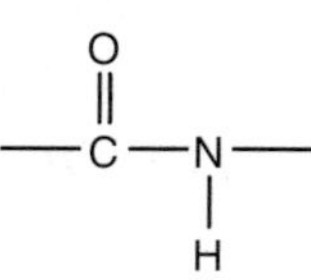

Primary amides have the formula

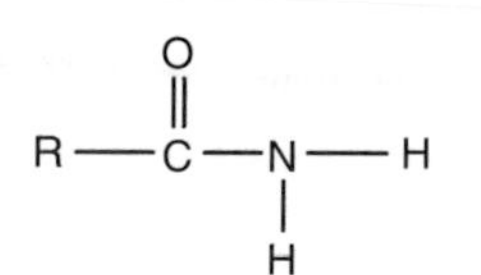

They are formed by the reaction of ammonia with an acyl chloride, as in Figure 4.

Primary amides are named from the corresponding carboxylic acid. For example,

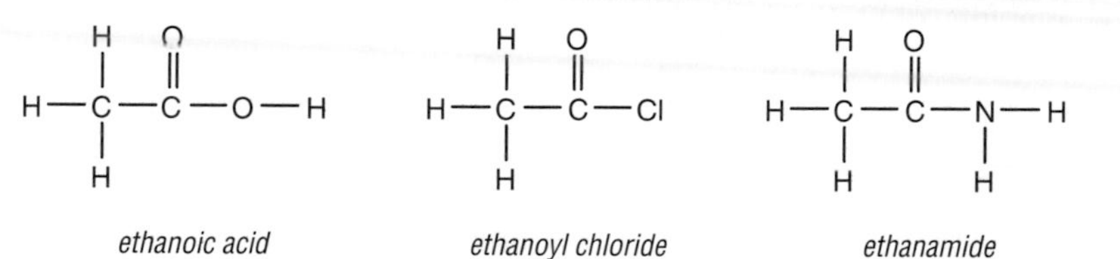

Secondary amides have the formula

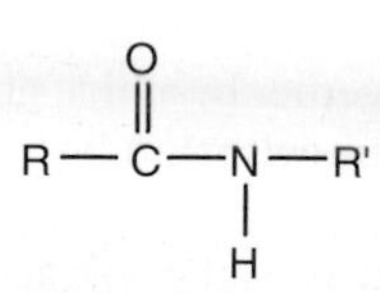

They can be formed by the reaction of an amine with an acyl chloride, as in Figure 4.

Hydrolysis of amides

The C–N bond in amides can be broken by **hydrolysis** – reaction with water. The reaction is catalysed by either acid or alkali, as shown in Figure 5 (page 324).

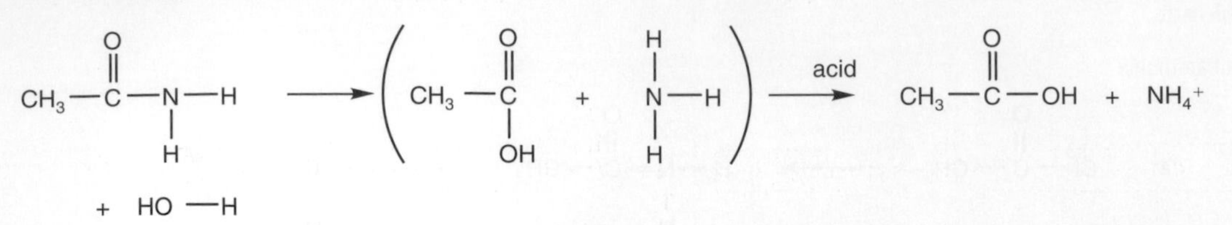

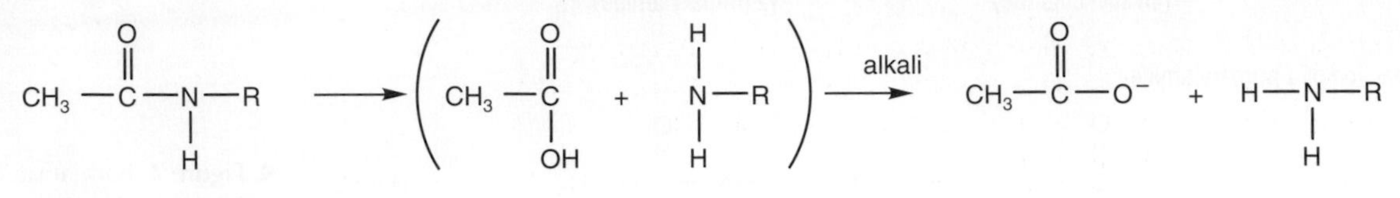

▲ **Figure 5** Hydrolysis of amides:
a with an acid;
b with an alkali.

The amide is heated under reflux with the aqueous acid or alkali – moderately concentrated solutions of the acid or alkali are usually needed.

The products depend upon whether an acid or alkali catalyst is used. Ammonia reacts with acid, so if an acid catalyst is used then the product contains ammonium ions. If an alkali catalyst is used then the carboxylic acid loses an H^+ and the product contains carboxylate ions.

Condensation polymers involving the NH_2 group

When an $-NH_2$ group reacts with the –COOH group in a carboxylic acid (or the –CO–Cl group in an acyl chloride), a **secondary amide group** is formed, with the structure

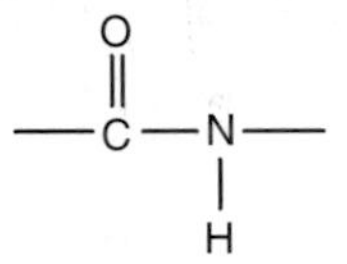

This type of reaction, in which two molecules react together to form a larger molecule with the elimination of a small molecule such as water or HCl, is called a **condensation reaction** (see **Section 5.7**). There is another example of a condensation reaction in **Section 13.5** concerning ester formation.

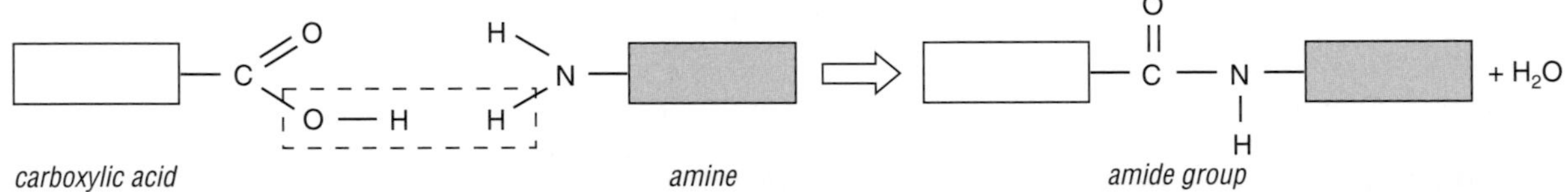

▲ **Figure 6** Making a secondary amide.

If we use *di*amines and *di*carboxylic acids which have reactive groups in *two* places in their molecules, we can make a polymer chain, in which monomer units are linked together by amide groups (Figure 7). The process is called **condensation polymerisation** because the individual steps are condensation reactions.

Examples of a diamine and a dicarboxylic acid that can be made to polymerise in this way are

$H_2N–CH_2–CH_2–CH_2–CH_2–CH_2–CH_2–NH_2$ 1,6-diaminohexane
$HOOC–CH_2–CH_2–CH_2–CH_2–COOH$ hexanedioic acid

Because the group linking the monomer groups together is an amide group, these polymers are called **polyamides**. More usually, though, they are known as **nylons**.

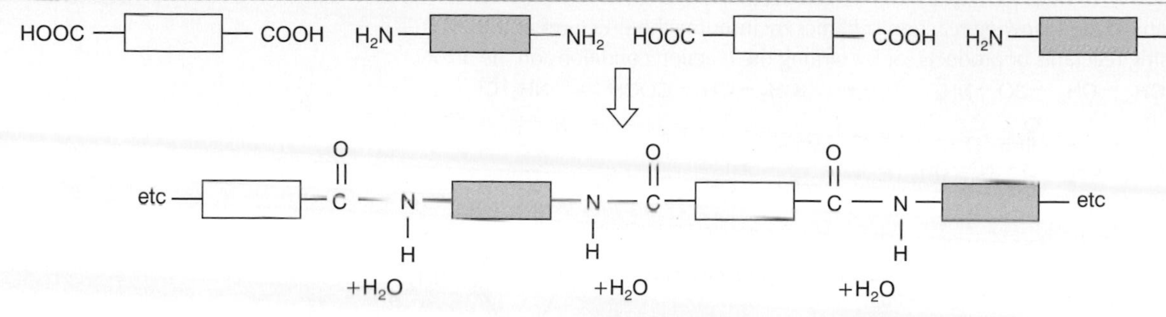

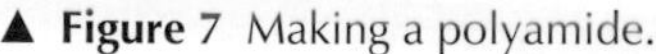

▲ **Figure 7** Making a polyamide.

The industrial preparation of nylon from a diamine and a dicarboxylic acid is quite slow. It is easier to demonstrate the process in the laboratory if an *acyl chloride* derivative of the acid is used. Thus 1,6-diaminohexane and decanedioyl dichloride react readily. The equation is

$$nH_2N\text{–}(CH_2)_6\text{–}NH_2 \;+\; nClCO\text{–}(CH_2)_8\text{–}COCl$$

1,6-diaminohexane   *decanedioyl dichloride*

$$\downarrow$$

$$\text{–}[NH(CH_2)_6\text{–}NH\text{–}CO\text{–}(CH_2)_8\text{–}CO]_n\text{–} \;+\; 2nHCl$$

Notice that in this condensation reaction, HCl is eliminated instead of H_2O.

Proteins are also condensation polymers and also contain the –CO–NH– group. In proteins this group is called the **peptide** link, and the monomers are amino acids.

Problems for 13.8

1 Name the amines with the following structures:

a $CH_3-CH_2-NH_2$

b $CH_3-N(CH_3)-H$

c $CH_3-CH(CH_3)-NH_2$

d $CH_3-N(CH_3)-CH_2-CH_3$

e (cyclohexane ring)$-NH_2$

2 Draw structures for the following amines:

a propylamine

b phenylamine

c diethylmethylamine

d butylethylmethylamine

e 3-aminopentane

f 2,4-diaminopentane.

3 Draw structures for the products formed when 2-aminopropane reacts with:

a hydrochloric acid

b ethanoyl chloride

c chloroethane.

4 Write equations for the reactions of the following pairs of substances:

a $CH_3-CH_2-CH_2-CH_2-NH_2 \;+\; HCl$

b (cyclohexane ring)$-NH_2 \;+\; CH_3-C(=O)Cl$

c (piperidine ring) N–H $+\; CH_3-CH_2-C(=O)Cl$

d (benzene ring)$-NH_2 \;+\; H_2SO_4$

e $CH_3-CH_2-NH_2 \;+\; CH_3-Br$

f $CH_3-CH_2-C(=O)-N(H)-CH_3 \;+\; NaOH(aq)$

5 a Which of the reactions of amines described in this section could not be undergone by a tertiary amine such as triethylamine? Briefly explain your answer.

b Explain, in terms of intermolecular forces and with the aid of a diagram, why butylamine is soluble in water.

c Explain, using an equation, why a deep-blue colour is formed when butylamine is added to copper(II) sulfate solution.

6 Complete the following reaction schemes by inserting the structures of the missing reactants or products, or by writing the reaction conditions on the arrow:

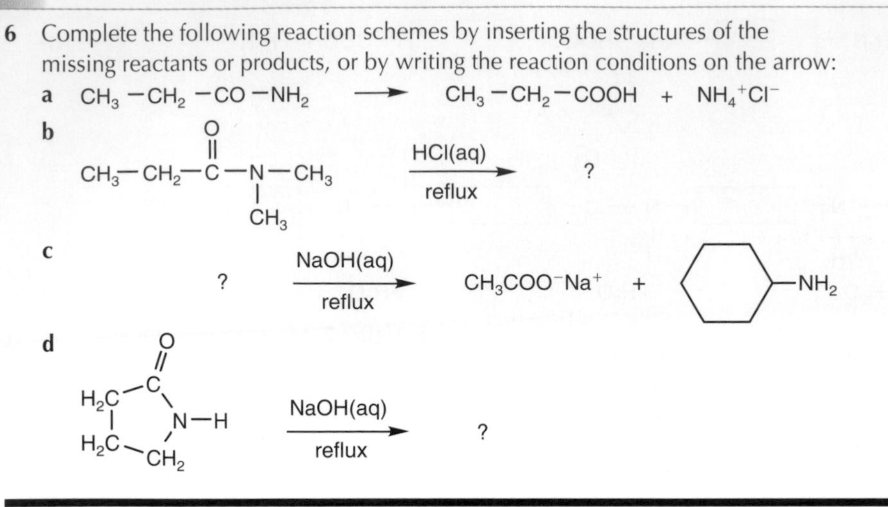

Naming amino acids

The systematic name for α-amino acids, like the ones shown in Figure 1, is *2-aminocarboxylic acid*. However, the name 'α-amino acid' is more commonly used.

Each amino acid has a systematic name, but normally its shorter common name is used instead. Thus the amino acid whose formula is

$$H_2N-\underset{H}{\overset{CH_3}{C}}-COOH$$

has the systematic name *2-aminopropanoic acid*, but its common name *alanine* is normally used.

13.9 *Amino acids and proteins*

Amino acids contain at least one amino group and one carboxylic acid group. The **α-amino acids** are particularly important in living systems. Figure 1 shows the general structure of an α-amino acid.

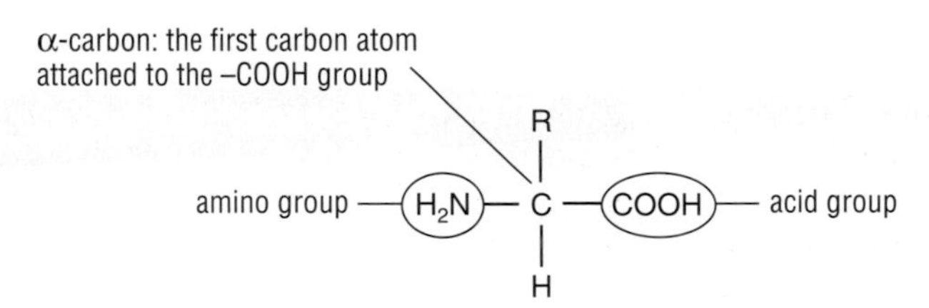

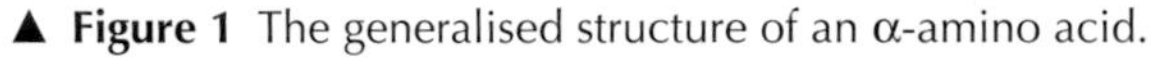

▲ **Figure 1** The generalised structure of an α-amino acid.

Amino acids are examples of **bifunctional compounds** – compounds with *two* functional groups. The properties of bifunctional compounds are sometimes simply the same as the properties of the two separate functional groups. This is not the case with amino acids because the functional groups interact. The proton-donating –COOH and proton-accepting $–NH_2$ groups can react with one another, forming **zwitterions** – particles containing both negatively charged and positively charged groups (Figure 2).

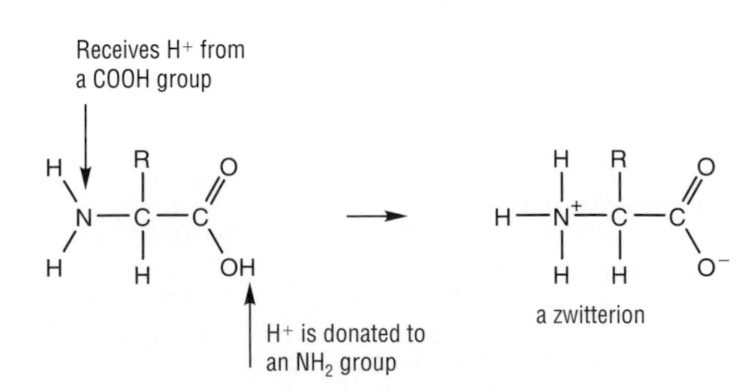

▲ **Figure 2** How an amino acid forms a zwitterion.

An aqueous solution of an amino acid consists mainly of zwitterions, with very few molecules containing the un-ionised groups. Amino acids are very soluble in water because they are effectively ionic.

Unless there is an extra –COOH or $-NH_2$ group in the molecule (as there is in some naturally occurring amino acids) they are neutral in aqueous solution.

Adding small quantities of acid or alkali to an amino acid solution causes little change to the pH because the zwitterions neutralise the effect of the addition.

$$HO^- + H_3N^+\text{–CHR–COO}^- \rightarrow H_2O + H_2N\text{–CHR–COO}^-$$
$$H_3O^+ + H_3N^+\text{–CHR–COO}^- \rightarrow H_2O + H_3N^+\text{–CHR–COOH}$$

So, amino acids exist in three different ionic forms, depending on the pH of the solution they are in:

H_3N^+–CHR–COOH	H_3N^+–CHR–COO^-	H_2N–CHR–COO^-
in acid solution	in neutral solution	in alkaline solution

Solutions which can withstand the addition of small amounts of acid or alkali are called **buffer solutions** (see **Section 8.3**).

Making peptides and proteins

When an $-NH_2$ group reacts with the –COOH group in a carboxylic acid, a **secondary amide group** is formed with the structure

secondary *amide* group

In this process, a molecule of water is eliminated, so it is a condensation reaction (see **Sections 13.5** and **13.8**).

When two amino acids join together in this way, the secondary amide group formed is called a **peptide link**.

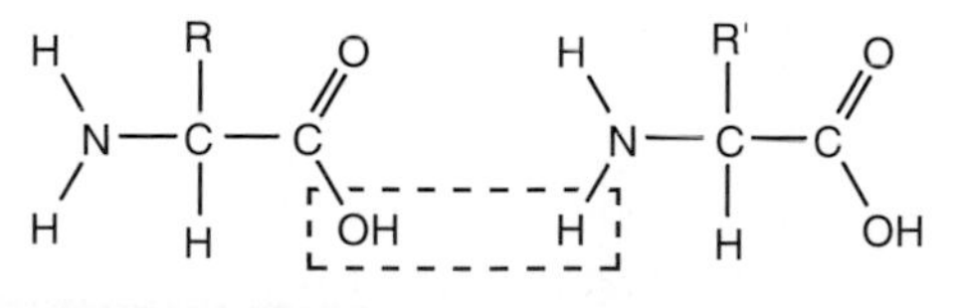

two amino acids produce a dipeptide
(R and R' represent different side chains)

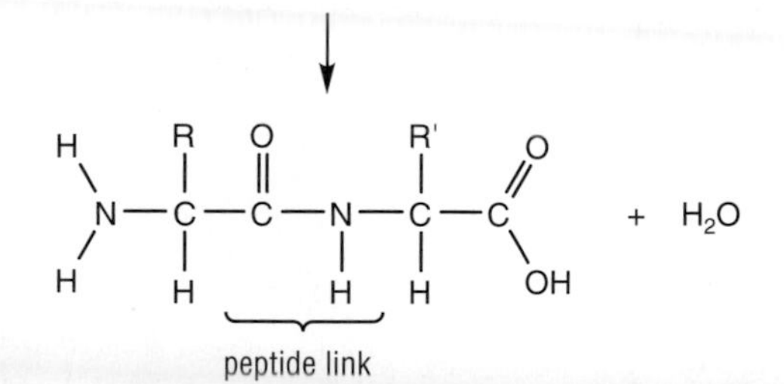

peptide link

The dipeptide formed from glycine (Gly, R = H) and alanine (Ala, R = CH_3) is

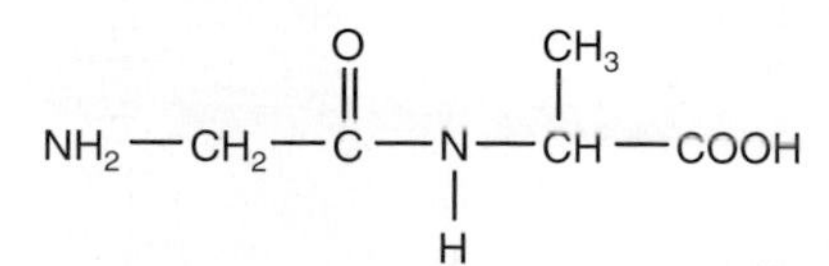

This dipeptide is abbreviated to GlyAla.

From a dipeptide, scientists can make a tripeptide by adding another α-amino acid molecule. A **polypeptide** can contain up to 40 amino acid residues.

Proteins are naturally occurring condensation polymers made from amino acid monomers joined by peptide links. A protein contains more than about 40 amino acid residues, although the distinction between a polypeptide and a protein is an arbitrary one (the term 'residue' is used for an α-amino acid which has lost the elements of water in forming a peptide or protein).

Insulin, for example, is made from 51 amino acid monomers, 14 of them being different. In fact, all proteins are constructed from just 20 amino acids. What makes each protein different is the *order* in which the amino acids are joined to one another – this is called the **primary structure**. Each protein has its own unique primary structure.

Secondary and tertiary structure of proteins

As well as their primary structure, most proteins have a precise shape which arises from the folding together of the chains. Four types of interaction are important in chain folding:

- instantaneous dipole–induced dipole bonds between non-polar side chains
- hydrogen bonds between polar side chains
- ionic bonds between ionisable side chains
- covalent bonding.

The chains in a protein are often folded or twisted in a regular manner as a result of hydrogen bonding. Two arrangements of the protein chain are common:

- tightly *coiled* into a **helix** where the C=O group of one peptide link forms a hydrogen bond to an N–H group *four* peptide links along the chain
- stretched out into regions of extended chain, which lie alongside one another and hydrogen bonded to form a **sheet**.

The two components of a protein's shape – helix and sheet – are sometimes referred to as its **secondary structure**. This is shown in Figure 3.

The chains may then fold up further. The overall shape is stabilised by instantaneous dipole–induced dipole attractive forces, by hydrogen bonding, by ionic attractions and by covalent bonding. The overall shape of a protein is sometimes called its **tertiary structure**.

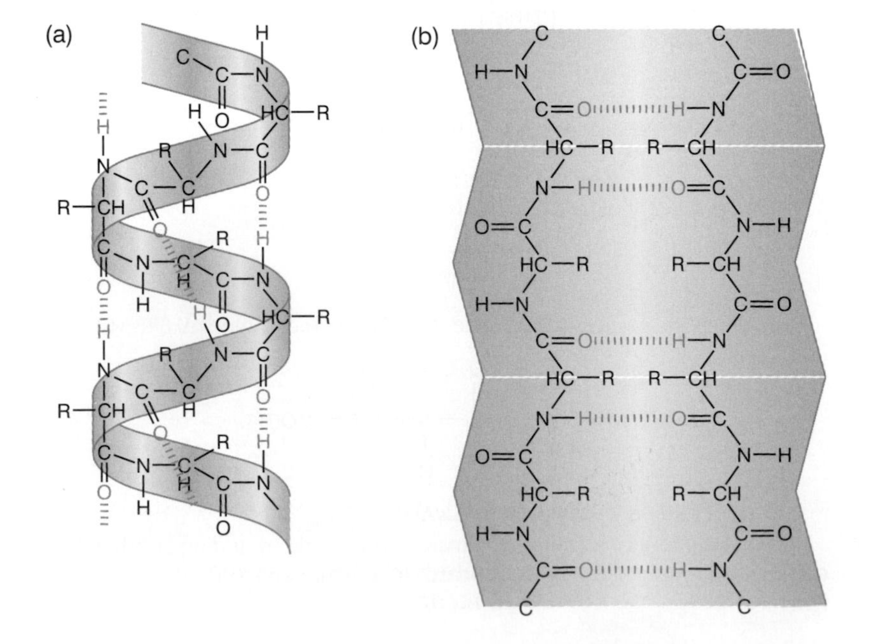

► **Figure 3** The secondary structure of a protein involves folding as a result of hydrogen bonding.
a A helix.
b A sheet.

The primary, secondary and tertiary structures of proteins account for the diversity of properties of proteins and the diversity of protein activity in living things.

Hydrolysis of peptides and proteins

The peptide link in peptides and proteins can be hydrolysed to release the individual amino acids. Peptides are secondary amides, and hydrolysis can be carried out by heating with moderately concentrated acid or alkali (see **Section 13.8**).

The breakdown of proteins in the laboratory is routinely carried out by boiling with moderately concentrated hydrochloric acid to hydrolyse the amide C–N bonds. In living organisms, the hydrolysis of proteins is catalysed by enzymes rather than acid or alkali (see **Section 10.4**).

Paper chromatography can be used to identify the individual amino acids present in a peptide. The peptide is hydrolysed under reflux and the product compared to known samples of pure amino acids using chromatography (see **Chemical Ideas Appendix 1: Experimental Techniques** (Techniques 1 and 4).

Problems for 13.9

1 The general structure of an α-amino acid is

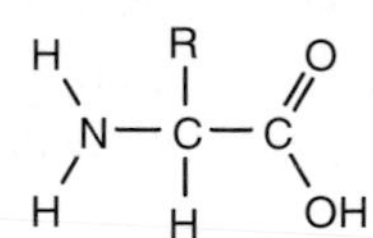

a Draw the structure of the zwitterion present in solutions of the amino acid.

b Write equations for the following reactions:

i alanine (R is CH_3) with hydrochloric acid

ii serine (R is CH_2OH) with sodium hydroxide solution

iii lysine (R is $(CH_2)_4NH_2$) with excess hydrochloric acid

iv aspartic acid (R is CH_2COOH) with excess sodium hydroxide solution.

2 A protein chain contains the following section

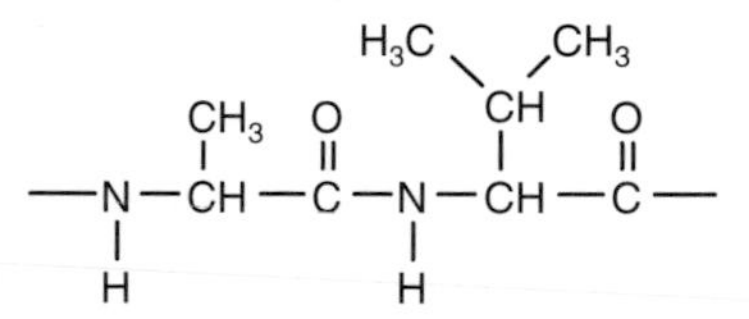

Draw the structures of the amino acids obtained from the hydrolysis of this section of the protein chain.

3 Write equations for the following reactions (glycine R is H):

a Hydrolysis of the dipeptide AlaGly using $4\,mol\,dm^{-3}$ (moderately concentrated) hydrochloric acid.

b Hydrolysis of the dipeptide SerGly using $2\,mol\,dm^{-3}$ (dilute) sodium hydroxide solution.

c Describe how you would use paper chromatography to investigate the amino acids formed when a dipeptide is hydrolysed.

13.10 *Azo compounds*

Azo compounds contain the –N=N– group.

R—N=N—R'

azo group

Compounds in which the groups R and R[1] are arene groups are more stable than those in which R and R[1] are alkyl groups. This is because the –N=N– group is stabilised by becoming part of an extended delocalised system involving the arene groups. These aromatic azo compounds are highly coloured and many are used as dyes.

They are formed as a result of a **coupling reaction** between a **diazonium salt** and a **coupling agent**.

Diazonium salts

The only stable diazonium salts are aromatic ones, and even these are not particularly stable. Benzenediazonium chloride has the following structure:

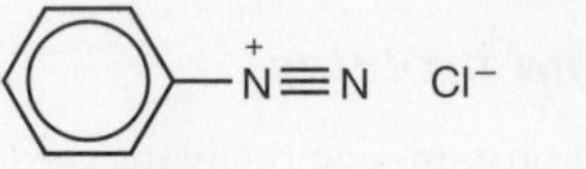

Diazonium salts are unstable because they tend to lose the $-\overset{+}{N}\equiv N$ group as $N_2(g)$. The presence of the electron-rich benzene ring stabilises the $-\overset{+}{N}\equiv N$ group, but even so benzenediazonium chloride decomposes above about 5 °C in aqueous solution – and the solid compound is explosive. For this reason, diazonium salts are prepared in ice-cold solution and are used immediately.

Diazonium salts are prepared by adding a cold solution of sodium nitrite [sodium nitrate(III)] ($NaNO_2$), to a solution of an arylamine in dilute acid kept below 5 °C. This type of reaction is known as **diazotisation**.

The acid (usually hydrochloric acid or sulfuric acid) reacts with sodium nitrite to form unstable nitrous acid (nitric(III) acid):

$$NaNO_2(aq) + HCl(aq) \rightarrow HNO_2(aq) + NaCl(aq)$$

sodium nitrite *nitrous acid*

The nitrous acid then reacts with the arylamine. For example,

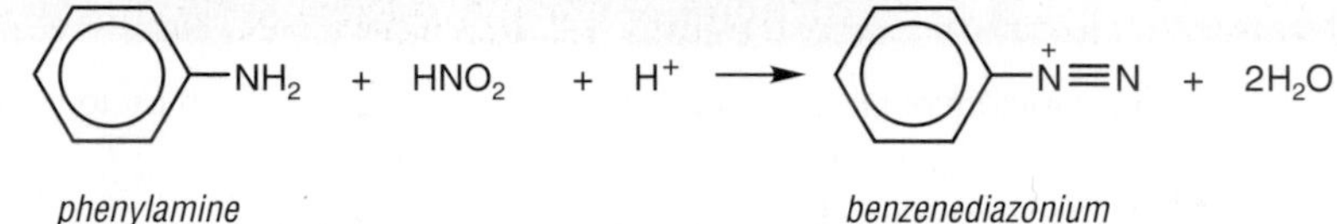

Diazo coupling reactions

In a diazo coupling reaction, a diazonium salt reacts with another compound containing a benzene ring, called a **coupling agent**. The diazonium ion acts as an electrophile and reacts with the benzene ring of the coupling agent. Figure 1 summarises a general coupling reaction.

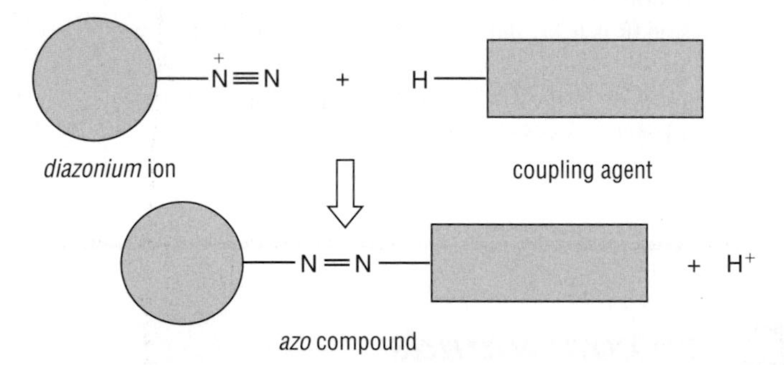

▶ **Figure 1** A generalised coupling reaction.

When the ice-cold solution of the diazonium salt is added to a solution of a coupling agent, a coloured precipitate of an azo compound immediately forms. Many of these coloured compounds are important dyes.

Coupling with phenols

When a solution of the benzenediazonium salt is added to an alkaline solution of phenol, a yellow–orange azo compound is formed:

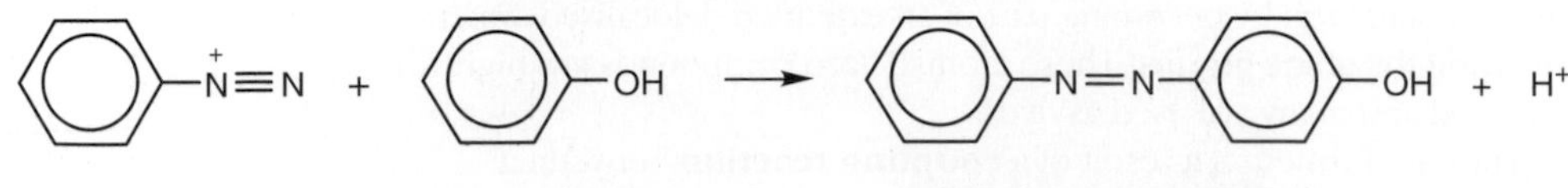

With an alkaline solution of naphthalen-2-ol, a red azo compound is precipitated:

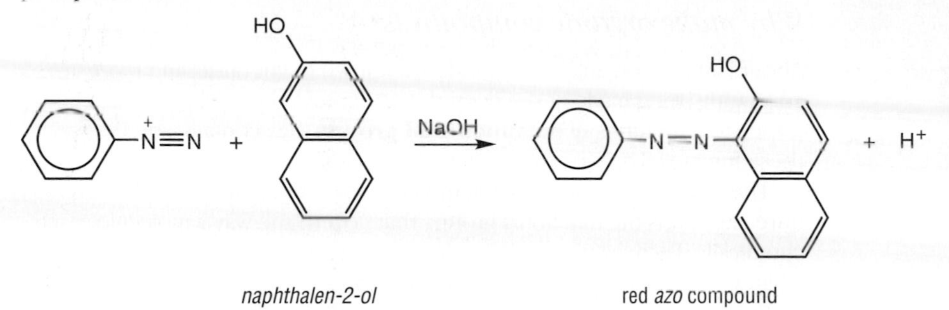

Coupling with amines

Diazonium salts also couple with arylamines such as phenylamine.

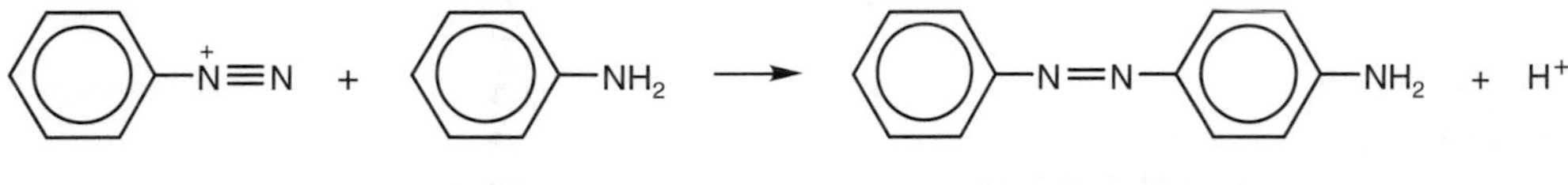

Many different azo compounds can be formed by coupling different diazonium salts with a whole range of coupling agents. Unlike diazonium salts, the azo compounds are stable so their colours do not fade.

Problems for 13.10

1 Write the structures of the products you would expect to be formed at each stage when:
  a phenylamine is dissolved in sulfuric acid
  b sodium nitrite solution is added to the cooled solution in part **a**
  c the product from part **b** is added to an alkaline solution of phenol.

2 a Draw the structure of the diazonium salt that would form when 4-aminophenol is treated with sodium nitrite in the presence of ice-cold hydrochloric acid.
  b Write equations for the coupling reaction of this diazonium salt with
    i phenol
    ii naphthalen-2-ol
    iii phenylamine.

3 Give the structures of the diazonium compounds and coupling agents that you would need to make each of the azo compounds shown here. (The coupling agent usually contains a phenol group or an amine group attached to an arene ring system.)

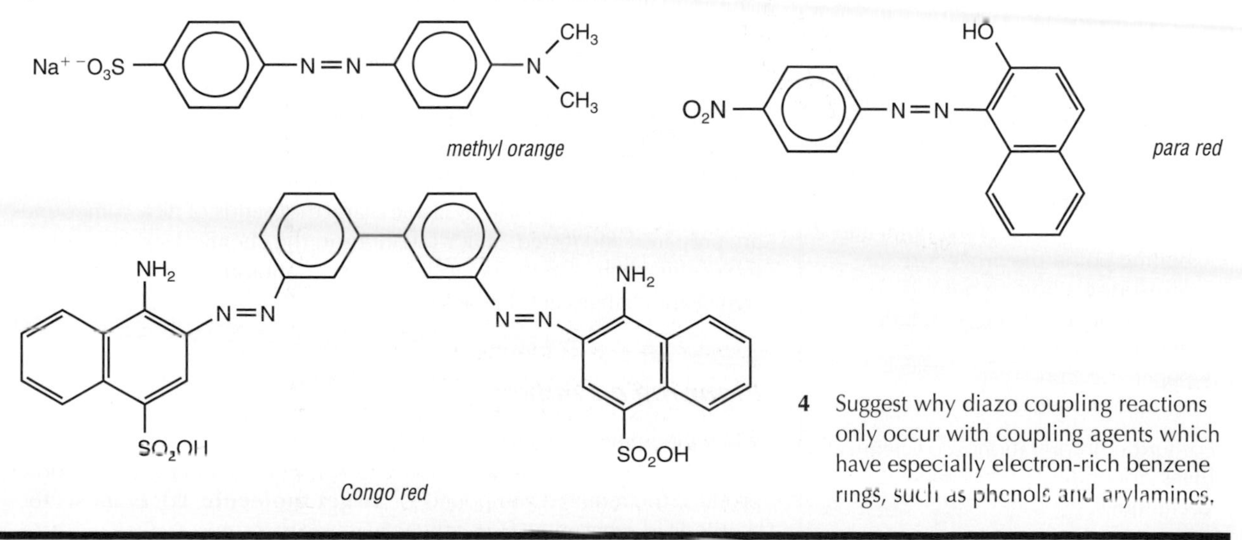

4 Suggest why diazo coupling reactions only occur with coupling agents which have especially electron-rich benzene rings, such as phenols and arylamines.

14

ORGANIC SYNTHESIS

14.1 *Planning a synthesis*

Why make organic compounds?

About 7 million organic compounds are known to chemists, and more are constantly being made or discovered. Fortunately, their behaviour can be understood in terms of the **functional groups** they contain, and the way these functional groups are arranged in space.

The carbon framework gives a molecule its shape, but is usually fairly unreactive. It is the functional groups that govern the way a molecule reacts chemically. Each group usually undergoes the same type of reactions, whatever carbon framework it is attached to.

For example, the hydroxyl group, –OH, and the alkene group, C=C, react in characteristic ways whether they are found in simple molecules, such as ethanol and ethene, or in the more complex steroid, *cholesterol* (Figure 1).

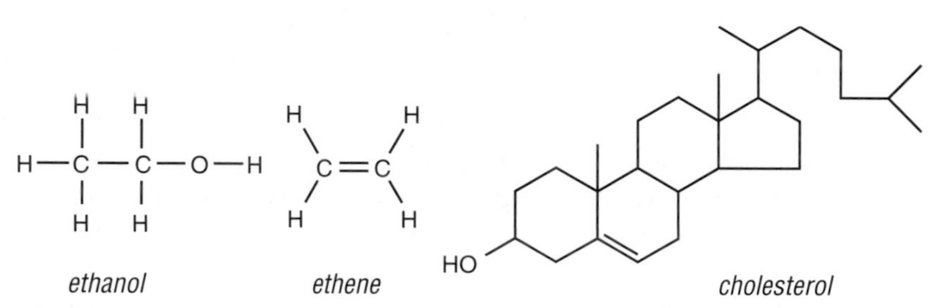

▲ **Figure 1** Ethanol, ethene and cholesterol. Cholesterol is a natural product, but can be synthesised in the laboratory. It was first synthesised by the American Nobel Prizewinner R. B. Woodward in 1952 – the synthetic route involved more than 40 steps!

New compounds are made by a process of **synthesis**, in which more complex structures are built up from simpler starting materials. The synthetic chemist is a sort of molecular architect, who plans and carries out strategies for making new and useful substances. There may be many steps in the synthesis, each involving the preparation of a new compound from the previous compound. However, the basic principles are the same in each preparation (see the margin box).

Chemists make new compounds for many different reasons. At one time, it was the only sure way to confirm the structure of a compound. Many complex syntheses of natural products were undertaken for this reason. Once they thought they knew the structure of a natural compound, chemists would synthesise it, and then check that the synthetic compound had the same properties as the natural one.

Nowadays, most new substances are made in the hope that they will be useful in everyday life. In the pharmaceutical industry, for every medicine that becomes commercially available many thousands of new compounds are prepared and tested. Minor variations in the chemical structure of molecules, such as in the penicillin range of antibiotics, can result in significant changes in their biological activity.

Planning a synthesis

When planning an organic synthesis, several steps may well be needed and a number of **intermediates** may have to be prepared and purified in order to produce the required compound, or **target molecule**. For example, the synthesis of ethylamine from ethene involves two steps:

Carrying out an organic preparation

There are normally three stages involved in preparing an organic compound.

Organic reactions rarely give one pure product – there are usually two or more products from competing side-reactions, and unreacted starting material. The reaction mixture must be 'worked up' to extract the desired product in impure form – this is then purified.

The three stages are:

1 Reaction
2 Extraction of product
3 Purification of product.

After purification, the product is usually tested to confirm its purity.

The preparation of a halogenoalkane includes all these stages.

The **experimental techniques** that are commonly used in organic preparations – which you should know about – are:

- heating under reflux
- distillation
- vacuum filtration
- recrystallisation
- measurement of melting point and boiling point (the procedure for measuring a boiling point is not on the specification but is included here for completion)
- paper and thin-layer chromatography.

For further information, see **Chemical Ideas Appendix 1: Experimental Techniques**.

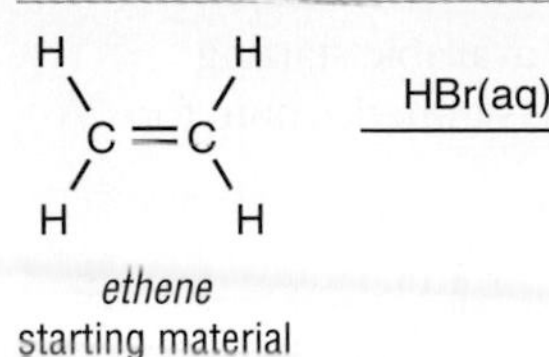

HBr(aq) → CH_3-CH_2-Br → c. NH_3(aq) → $CH_3-CH_2-NH_2$

ethene starting material — *bromoethane* intermediate — *ethylamine* target molecule

Acceptable starting materials should ideally be cheap and readily available. Most hydrocarbons containing six or fewer carbon atoms are obtained from petroleum refining. These, and simple compounds made from them, are good starting points for synthesis. Sometimes a readily available natural product from a plant or animal source may be used.

The starting point in planning a synthesis is to examine the required compound itself – this is called the target molecule. By looking at the functional groups it contains, chemists can work backwards, through a logical sequence of reactions, until suitable **starting materials** can be found. This process is known as **retrosynthesis.**

Consider a very simple example – the preparation of methylphenyl ketone:

methyl phenyl ketone

Our first step would be to look for an obvious disconnection that would give rise to two fragments (synthons) which could feasibly be used to produce the target molecule. The obvious disconnection in this molecule would be between the benzene ring and the ketone group:

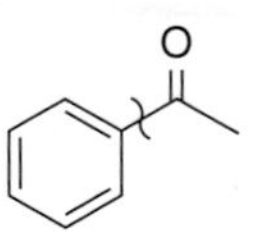

disconnection between benzene ring and ketone group

This would give rise to the following two synthons:

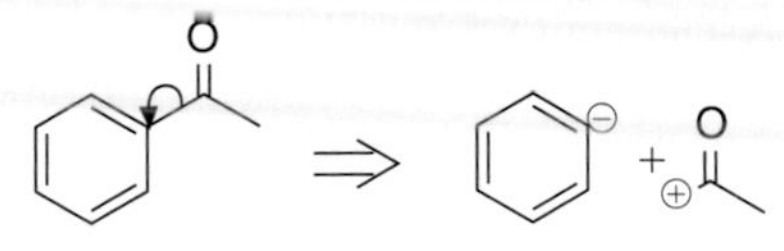

target molecule — synthons

Although neither synthon exists (you couldn't go and get a jar of either from a chemical store), they correspond to two actual compounds that could be used produce the target molecule. If they didn't, you would have to look for a different disconnection.

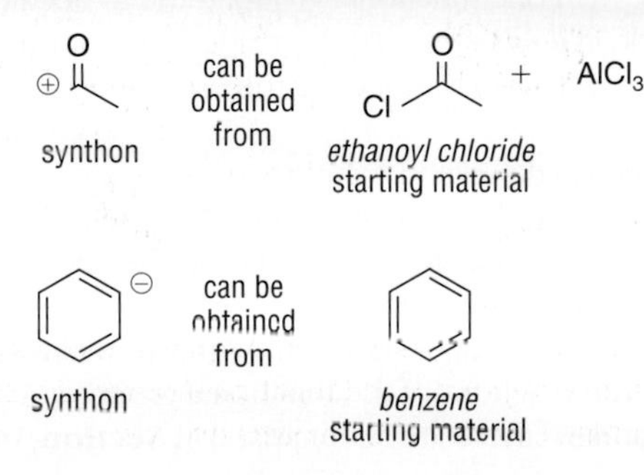

The language of retrosynthesis

- **Target molecule** – the desired compound.
- **Retrosynthetic analysis** – the process of *working backwards* from the target molecule in order to devise a suitable synthetic route (or routes) on paper.
- **Readily available starting materials** – cheap, commercially available compounds that can be modified to produce the target molecule.
- **Disconnection** – a paper operation involving an imagined breaking of a bond to suggest two fragments (or **synthons**) that could be reacted together to make the target molecule.
  Note: the arrow symbol ⇒ is used to indicate a disconnection. When planning any synthesis, ensure that a good chemical reaction corresponds to the reverse of the disconnections.
- **Synthon** – an idealised fragment, usually a cation or anion, resulting from a disconnection. Often synthons don't exist as such, but help in the correct choice of reagent.
- **Synthetic equivalent** – the actual compounds that are used to function as synthons.

Both benzene and ethanoyl chloride are readily available starting materials, so we can reverse the situation to see the synthetic route for methyl phenyl ketone.

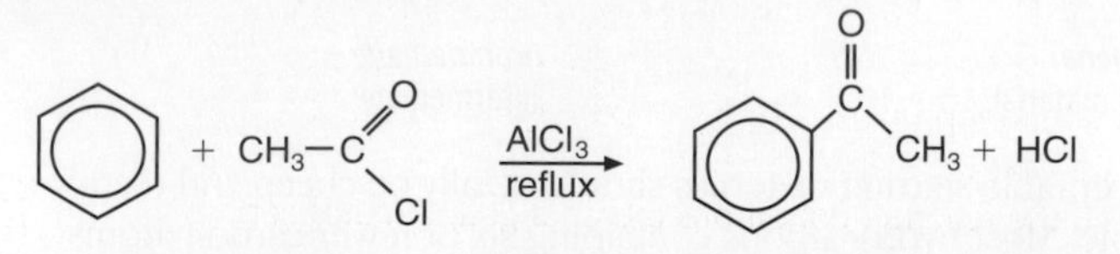

In a more complex synthesis involving several steps, there will often be more than one possible route to the target molecule. Choices will have to be made then. The preferred route is *usually* the one with the fewest steps – but this may not always be the case. Sometimes other factors are more important, such as the cost of the starting materials and the reagents, the time involved, disposal of waste materials and possible safety and health hazards.

Other important factors to consider are the **overall yield** and **atom economy** of the synthesis. (See **Section 15.7** for more information on these factors.)

Getting the right isomer

The yield can be reduced when there are several by-products. Sometimes isomers of a product molecule are formed, and only one is wanted.

This is always a problem in substitution reactions involving benzene rings in which there is already a group present. In the nitration of methylbenzene, for example, three isomeric products are formed:

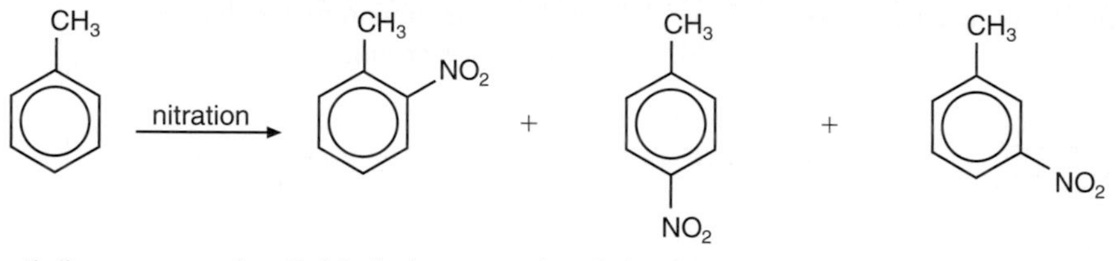

In this case, the products are all solids and the required isomer can be separated by fractional crystallisation (separation of a number of compounds using crystallisation) or by chromatography.

Separating isomers can sometimes be difficult and time consuming. This is particularly true in the case of **optical isomers** (see **Section 3.5**), so a pure D- or L-form of a molecule is usually more difficult to prepare.

Choosing the reagents

An organic chemist chooses the reactions needed for a synthesis from a vast 'tool-kit' of reactions of functional groups.

If there is more than one functional group in the molecule, it is important to check, in *each* step, that the reagents do not react with the other groups present. This may influence the order in which the steps are carried out.

There is often no single correct solution to a synthetic problem, and several alternatives may well be equally viable. The preferred route will take into account all the relevant factors – and may not be one which seemed most likely at first.

Before you try your hand at designing a synthesis, make sure that you are familiar with the main reactions of the functional groups you have met throughout the course – these are summarised in **Section 14.2**.

Problems for 14.1

1 a Ethanal, CH_3CHO, can be prepared by the mild oxidation of ethanol. In an experiment, 15.0 g of ethanol produced 10.6 g of pure ethanal. What is the percentage yield of ethanal?

b If the reaction conditions are changed, ethanol can be further oxidised to ethanoic acid. Starting from the same mass of ethanol, 14.3 g of pure ethanoic acid were obtained. What was the percentage yield of ethanoic acid?

2 Compound Z can be prepared from the starting material, compound W, by two alternative routes. Route I (W → X → Y → Z) involves three steps. Route II (W → V → Z) involves two steps. The yields for each step are shown below. Which route has the highest overall yield?

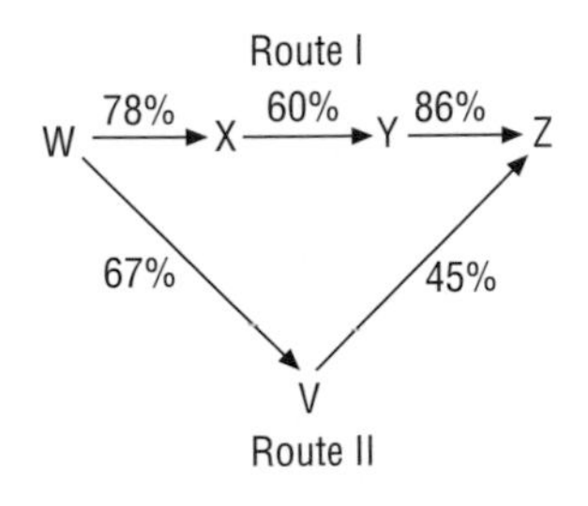

3 Phenylamine can be made in two steps from benzene:

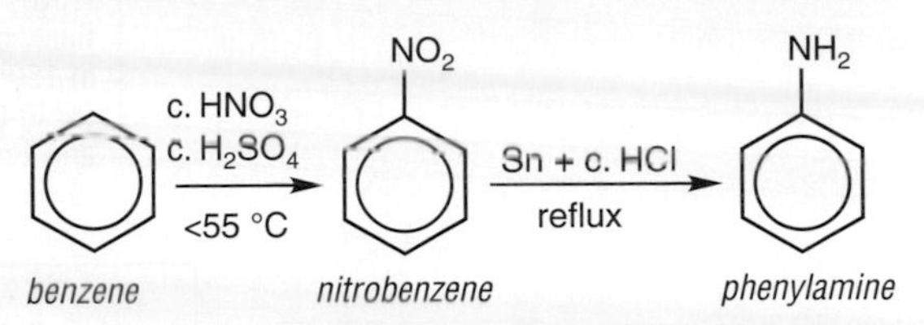

Nitration of benzene gives an 85% yield of nitrobenzene. The second step gives an 80% yield of phenylamine. If you carry out the synthesis, starting with 20.0 g benzene, what mass of phenylamine would you obtain?

4 a List the three stages of any organic preparation – for example, preparation of ethanal in problem **1a.**

b Even when there are no side-reactions in an organic reaction, a 100% yield of product is rarely obtained. Suggest reasons why this might be.

14.2 *A summary of organic reactions*

The following sections summarise the important reactions of the various functional groups you have met throughout the course – *these are the reactions you should remember*. You should be able to use them correctly to convert one functional group into another, or to design the synthesis of an organic molecule.

Alkenes

Simple alkenes, such as ethene and propene, are obtained by cracking petroleum fractions. They are much more reactive than the corresponding alkanes and make good starting materials for a synthesis. You can see the main reactions of alkenes in Figure 1 (page 336).

All the reactions of alkenes in Figure 1 are **addition** reactions. The reaction may proceed by an *ionic mechanism* or it may involve *radicals* (**Section 12.2**).

For example, addition polymerisation reactions involve radicals, whereas the addition of bromine is an **electrophilic addition** reaction in which the alkene reacts with a polarised bromine molecule:

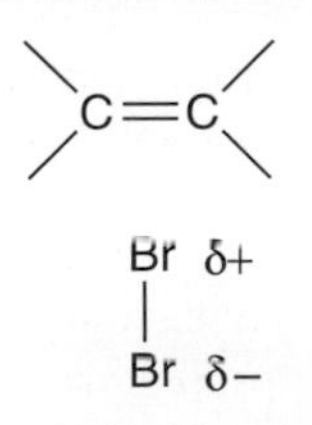

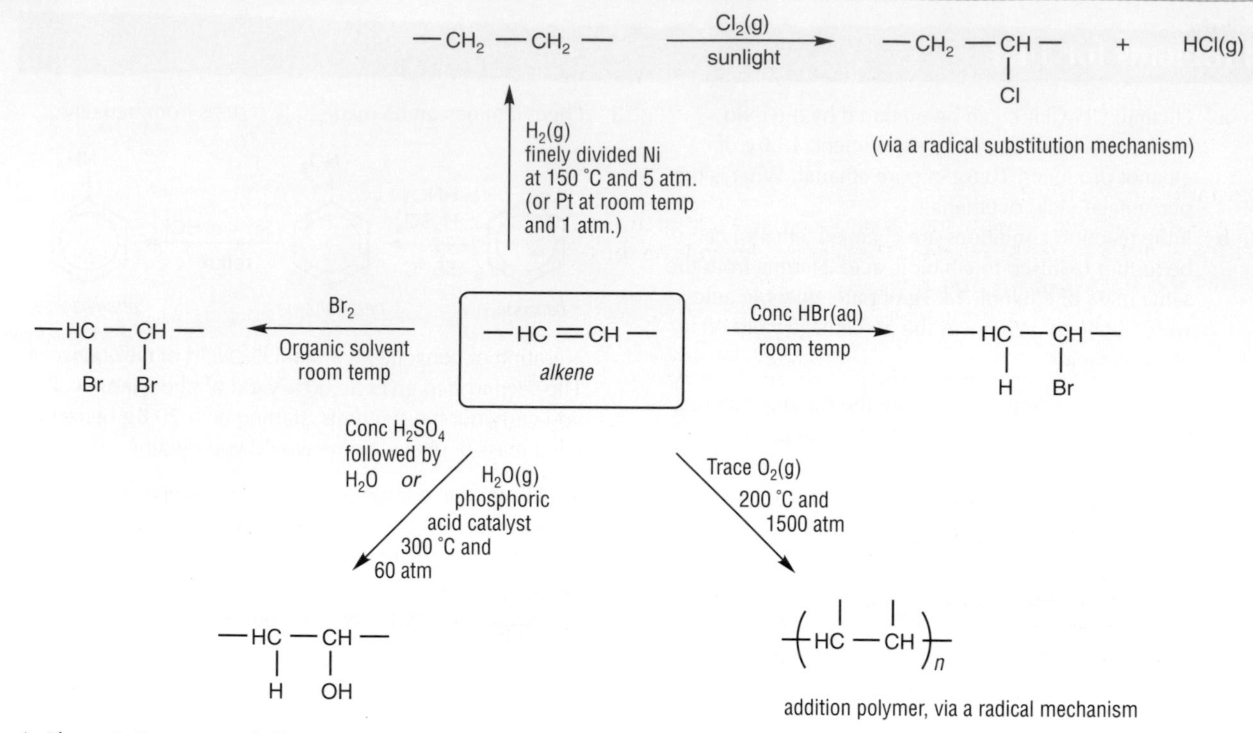

▲ **Figure 1** Reactions of alkenes.

Addition of an unsymmetrical molecule, such as HBr, to an alkene may lead to two possible isomeric products. For example, when propene reacts with HBr in a polar solvent, two addition products are formed:

$$CH_3{-}CH{=}CH_2 \xrightarrow{HBr} CH_3{-}CH_2{-}CH_2{-}Br + CH_3{-}CH(Br){-}CH_3$$

propene main product

The isomer with a central Br atom is the main product, but it must be separated from small amounts of the other isomer.

Halogenoalkanes

Halogenoalkanes are useful synthetic intermediates because they readily undergo a large variety of **nucleophilic substitution** reactions (**Section 13.1**). Three important examples are shown in Figure 2.

These reactions are useful because they allow one functional group to be converted into another.

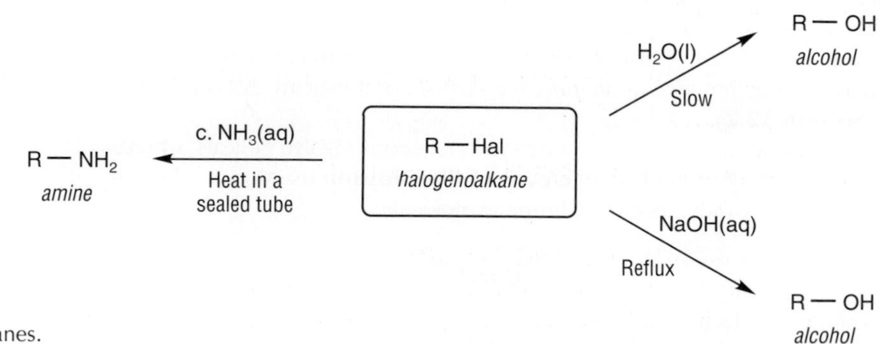

► **Figure 2** Reactions of halogenoalkanes.

Alcohols

Alcohols undergo a variety of reactions that can be useful in synthesis (see **Section 13.2**). The important reactions of alcohols are summarised below.

Oxidation of alcohols

Alcohols are oxidised by warming with an oxidising agent such as acidified potassium dichromate(VI) solution. The product depends on the type of alcohol and on the way the oxidation is carried out.

Primary alcohols give first aldehydes, and then carboxylic acids:

$$\underset{\textit{primary alcohol}}{RCH_2OH} \xrightarrow[\text{distil}]{Cr_2O_7^{2-}/H^+(aq)} \underset{\textit{aldehyde}}{RCHO} \xrightarrow[\text{reflux}]{Cr_2O_7^{2-}/H^+(aq)} \underset{\textit{carboxylic acid}}{RCOOH}$$

To make the aldehyde, the oxidising agent is dripped slowly into the hot alcohol and the aldehyde is distilled off as soon as it is formed, before it has time to be oxidised further to the acid.

To make the carboxylic acid the alcohol is heated under reflux with an excess of the oxidising agent to make sure the reaction goes to completion.

With *secondary alcohols*, the oxidation stops at the ketone. The ketone has no hydrogen atoms on the carbon atom of the ketone group, so it cannot easily undergo further oxidation:

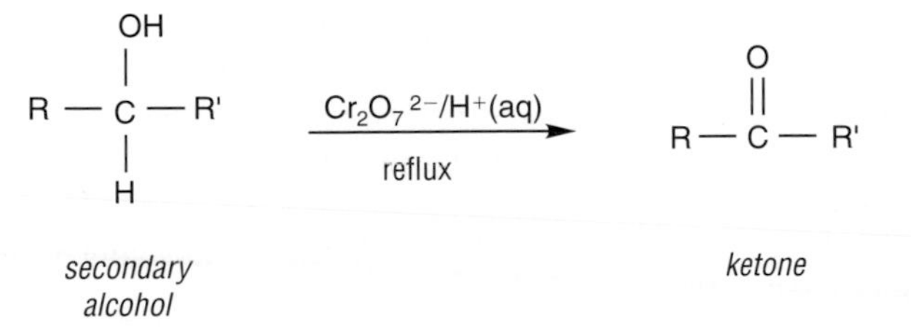

Tertiary alcohols have no hydrogen atoms on the carbon atom attached to the –OH group and are much more difficult to oxidise.

Under more vigorous oxidising conditions, both ketones and tertiary alcohols will react but the oxidation then involves the breaking of carbon–carbon bonds and usually leads to a mixture of products.

Dehydration of alcohols

Many alcohols readily eliminate a molecule of water to give an alkene. This is called a **dehydration** reaction and is an example of **elimination**. You can think of an elimination reaction as being the reverse of an addition reaction.

For example, ethene is formed when ethanol vapour is passed over alumina (Al_2O_3) at 300 °C:

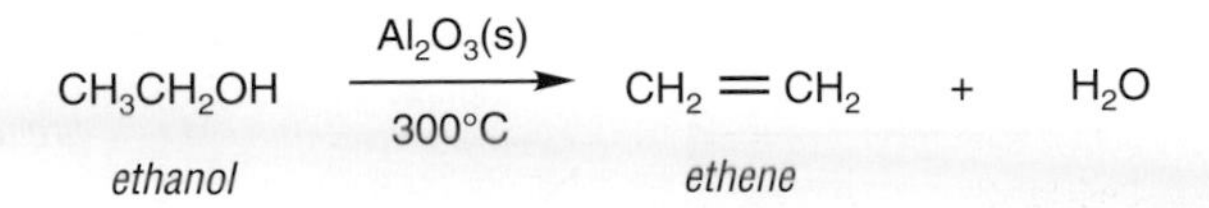

Alcohols can also be dehydrated by heating with concentrated sulfuric acid.

Reaction of alcohols with HCl and HBr

Alcohols can be converted to chloroalkanes or bromoalkanes by reaction with hydrochloric acid, HCl(aq), or with hydrobromic acid, HBr(aq), respectively. The hydrobromic acid used to make bromoalkanes is prepared in situ (in the reaction mixture) by the reaction of NaBr with concentrated sulfuric acid.

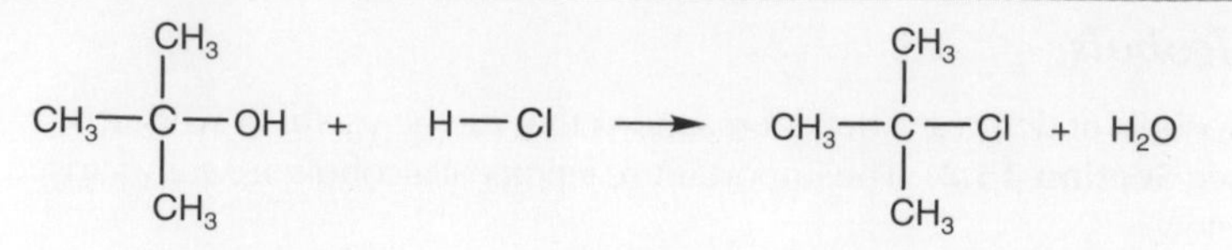

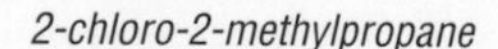

This is a **nucleophilic substitution** reaction. The alcohol group must first be protonated in the strongly acid solution before it can be displaced by the Cl^- nucleophile:

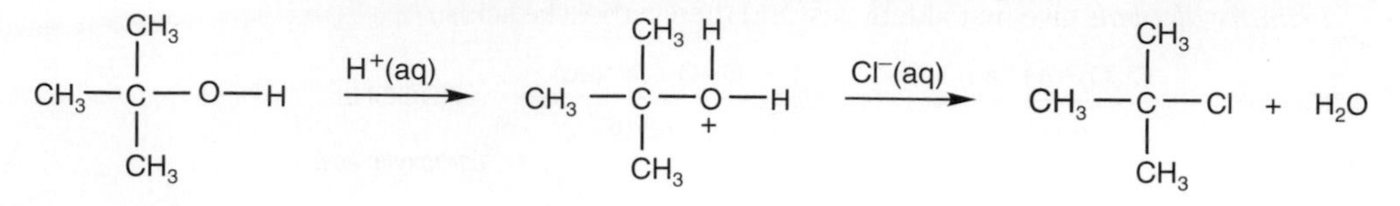

Esterification

An alcohol reacts with a carboxylic acid in a **condensation** reaction to give an ester. The alcohol and the carboxylic acid are heated under reflux in the presence of a few drops of concentrated sulfuric acid, acting as a catalyst. To get the best yield of ester, the water is distilled off as it forms, driving the equilibrium to the right:

$$R{-}OH + R'{-}COOH \underset{}{\overset{c.\ H_2SO_4\ catalyst}{\rightleftharpoons}} R{-}O{-}C({=}O){-}R' + H_2O$$

Esters can also be made by the reaction of alcohols with acylating agents such as acyl chlorides R′–COCl and acid anhydrides $(R'CO)_2O$ (see **Section 13.5**).

The main reactions of primary alcohols are shown in Figure 3 – you could draw similar diagrams for secondary and tertiary alcohols.

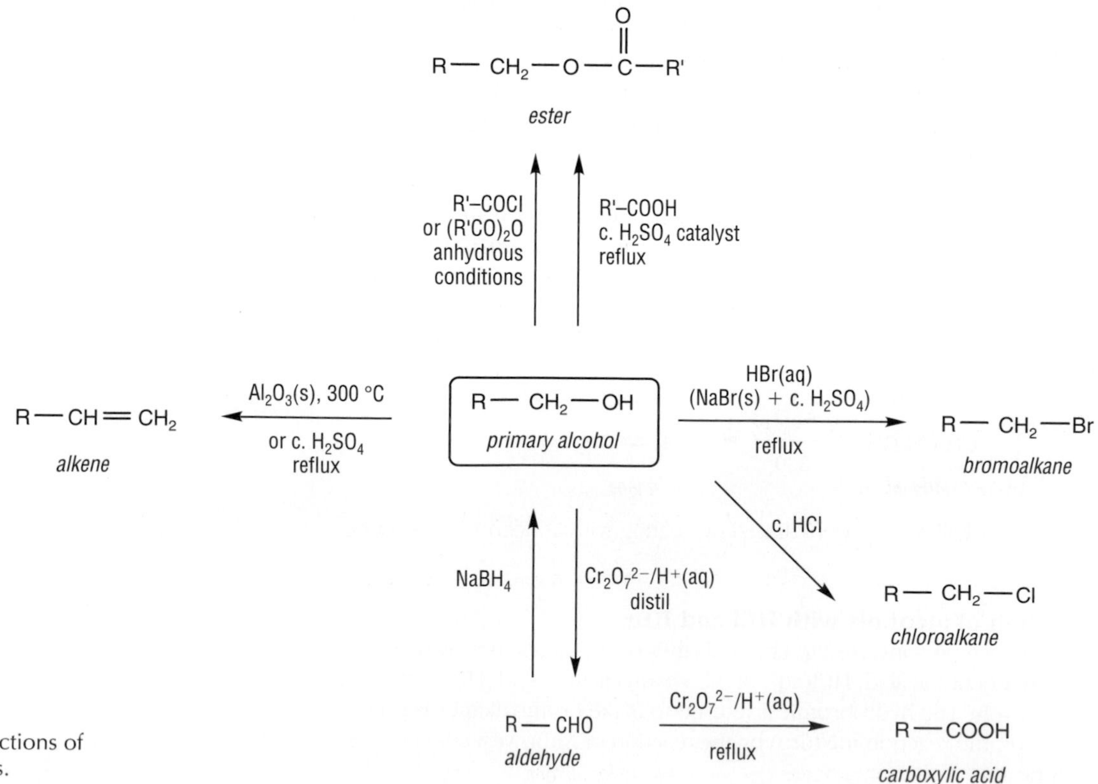

► **Figure 3** Reactions of primary alcohols.

Aldehydes and ketones

Aldehydes and ketones are useful intermediates in organic synthesis (see **Section 13.7**). Their reactions are summarised in Figures 4 and 5.

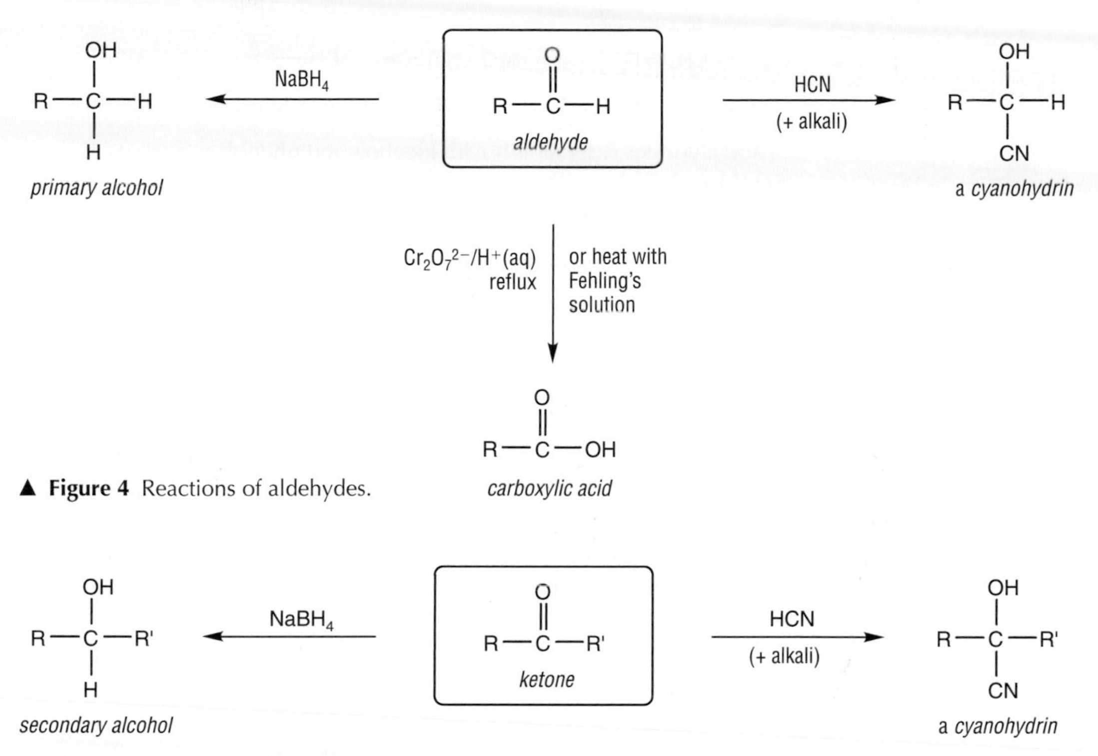

▲ **Figure 4** Reactions of aldehydes.

▲ **Figure 5** Reactions of ketones.

Ketones can be distinguished from aldehydes because they are not easily oxidised. They do not give a red precipitate of copper(I) oxide when heated with Fehling's solution.

The addition of H–CN to the C=O bond in aldehydes and ketones is a **nucleophilic addition** reaction.

Carboxylic acids and related compounds

Carboxylic acids themselves are not particularly reactive, but they can be converted into the more reactive *acyl chlorides* and *acid anhydrides*. These compounds are very useful as synthetic intermediates or as reagents in synthesis.

You can find more information about acids and their derivatives in **Sections 13.3**, **13.4**, **13.5** and **13.8**.

Acyl (acid) chlorides

Acyl chlorides, such as ethanoyl chloride, are readily prepared from carboxylic acids. They are more reactive than chloroalkanes, such as chloroethane – the chlorine is much more readily displaced as chloride ion by nucleophiles.

$CH_3-C(=O)Cl$ — *ethanoyl chloride*

CH_3CH_2-Cl — *chloroethane*

Acyl chlorides undergo a variety of **nucleophilic substitution** reactions. Most are readily hydrolysed by cold water. Ethanoyl chloride, for example, fumes in moist air as droplets of ethanoic acid and hydrochloric acid are formed:

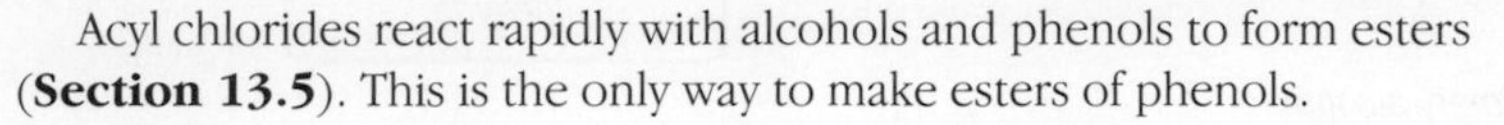

Acyl chlorides react rapidly with alcohols and phenols to form esters (**Section 13.5**). This is the only way to make esters of phenols.

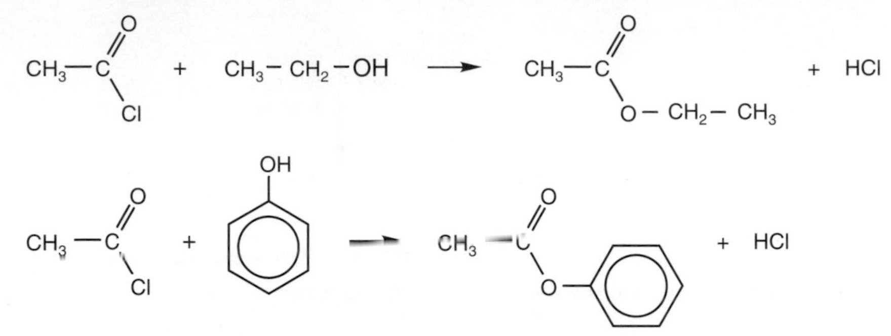

Acyl chlorides also react with ammonia and amines to form amides (**Section 13.8**). These reactions take place readily at room temperature.

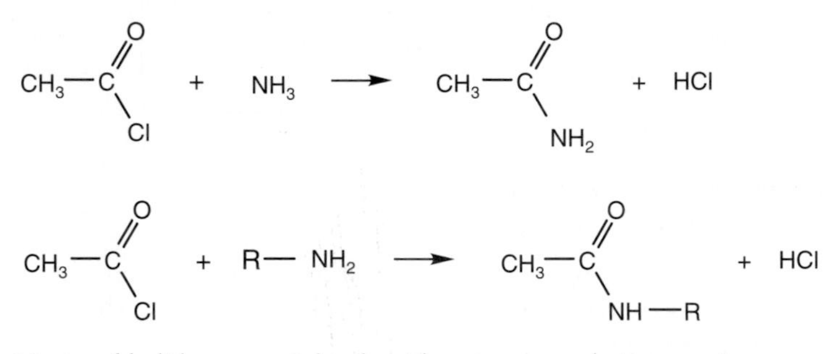

Most acyl halides react violently with water, so acylation reactions are usually carried out under strictly **anhydrous conditions**. Benzoyl chloride is an exception – it is much less reactive than ethanoyl chloride and can be used in aqueous solutions.

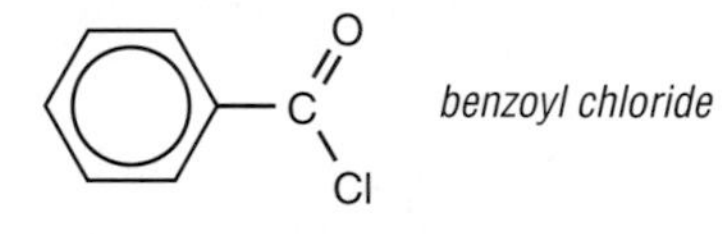

Acid anhydrides

Acid anhydrides are also acylating agents and react in the same way as acyl halides with water and alcohols, and with amines to form amides, although not quite so vigorously. Like acyl chlorides, they must be used under **anhydrous conditions**. Acylation reactions using acid anhydrides often require heating under reflux.

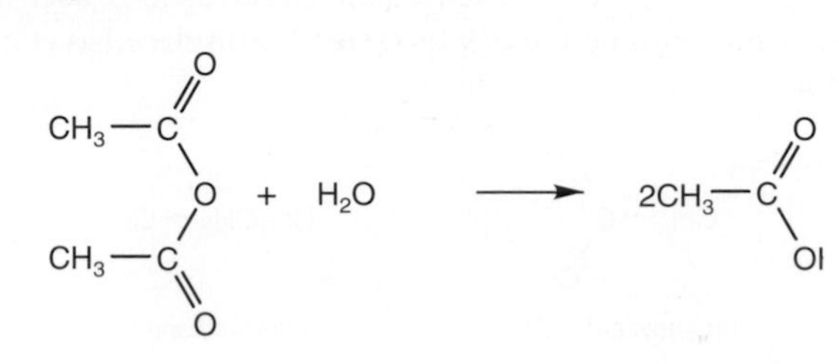

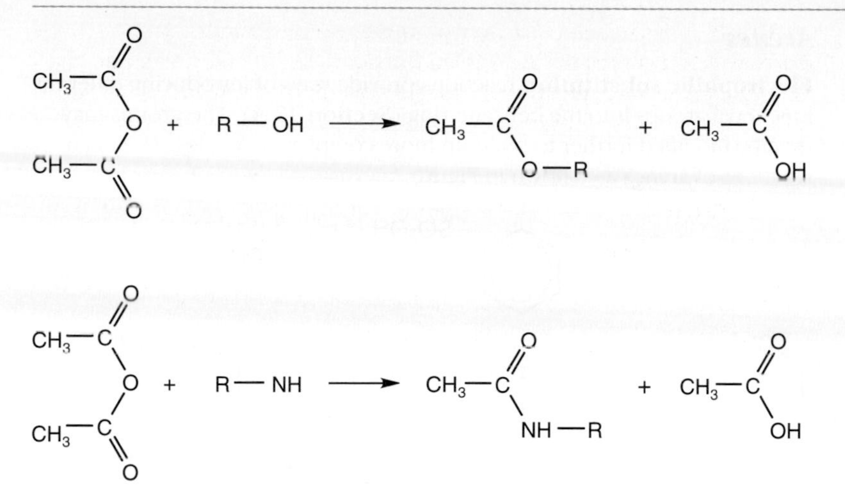

You can see the main reactions of carboxylic acids and their derivatives in Figure 6.

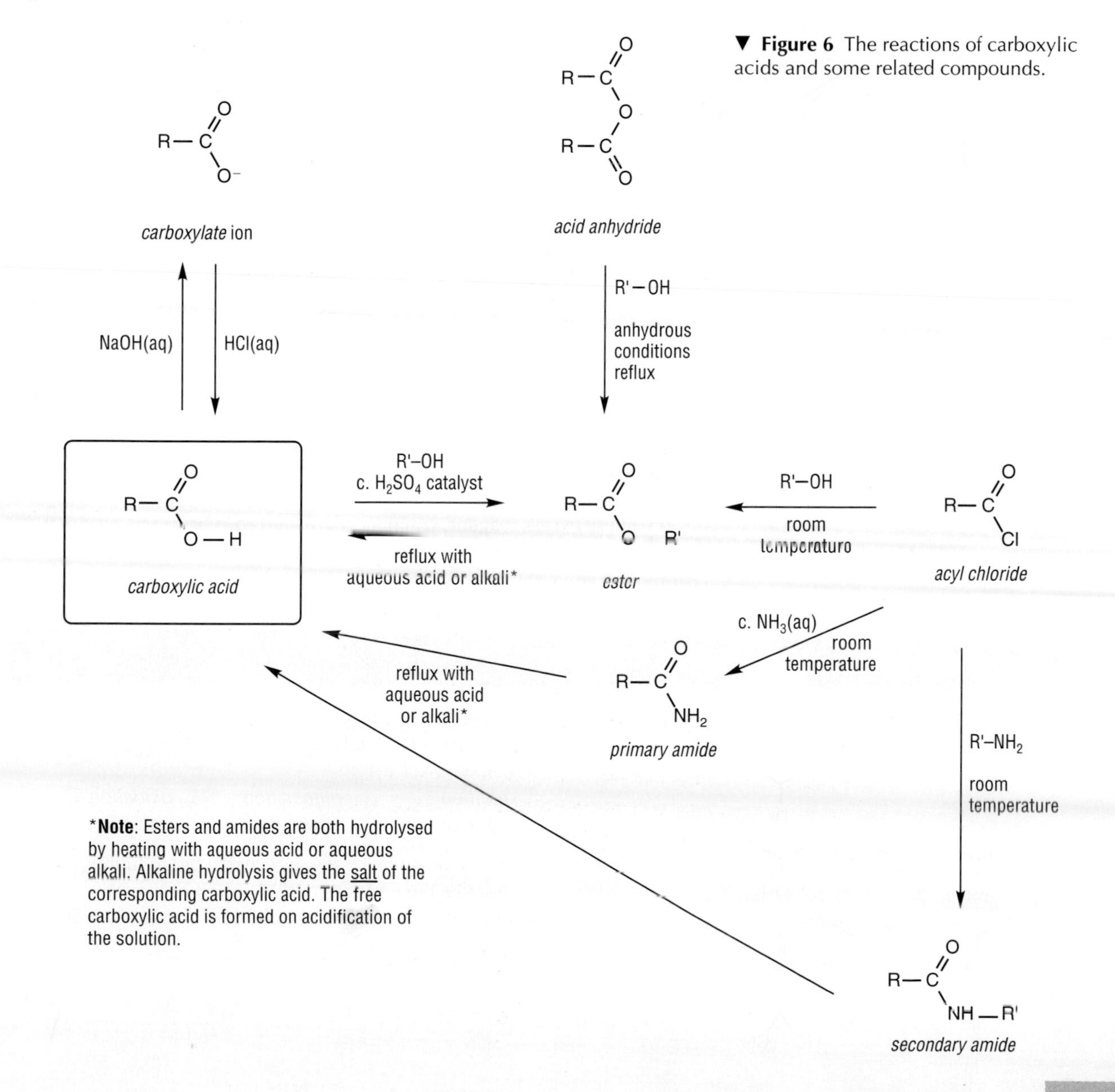

▼ **Figure 6** The reactions of carboxylic acids and some related compounds.

Arenes

Electrophilic substitution reactions provide ways of introducing different functional groups into the benzene ring (**Section 12.4**). The groups may then be modified further to build up more complex molecules. The main reactions of arenes are shown in Figure 7.

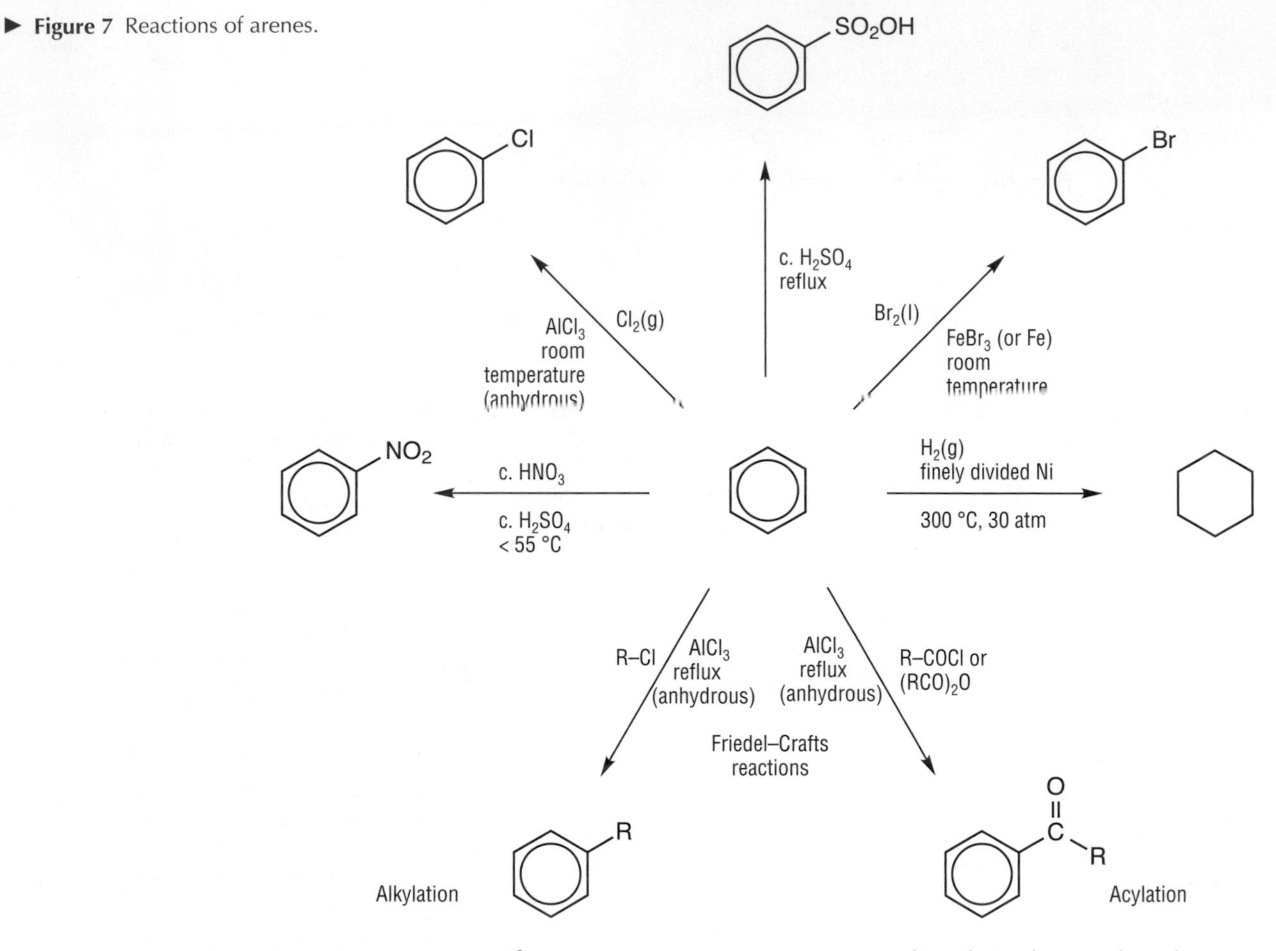

▶ **Figure 7** Reactions of arenes.

If you use excess reagent, you may get di- and tri-substituted products. Often a mixture of products is formed and you will have to separate the compound you want.

Problems for 14.2

1 Name the reagent and give the conditions you would use to bring about the following conversions:

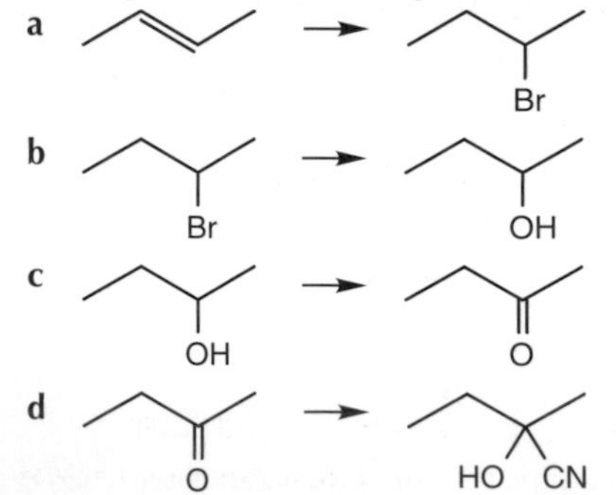

2 For each of the reactions in problem 1, state the type of reaction occurring by choosing a word from the list below:

addition *elimination* *oxidation*
substitution *condensation* *reduction*

Where possible, choose a second word to further describe the reaction mechanism. Choose from:

radical *nucleophilic* *electrophilic*

3 Treatment of methylbenzene with chlorine gas *in the presence of sunlight* at room temperature gives chloromethylbenzene.

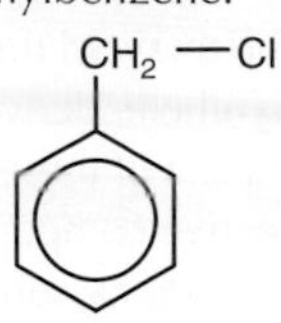

a What *type* of reaction mechanism do you think is involved?
b Suggest a possible mechanism for the reaction.
c Describe how you would convert chloromethylbenzene into

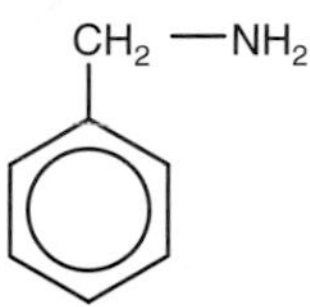

4 a What products would be formed by treating methylbenzene with chlorine in the presence of anhydrous aluminium chloride?
b What type of reaction mechanism is involved?
c Explain the mechanism you have suggested in part **b**.

5 a Complete the following reaction schemes by inserting the structures of the missing reactants or products, or by writing the reaction conditions on the arrow.

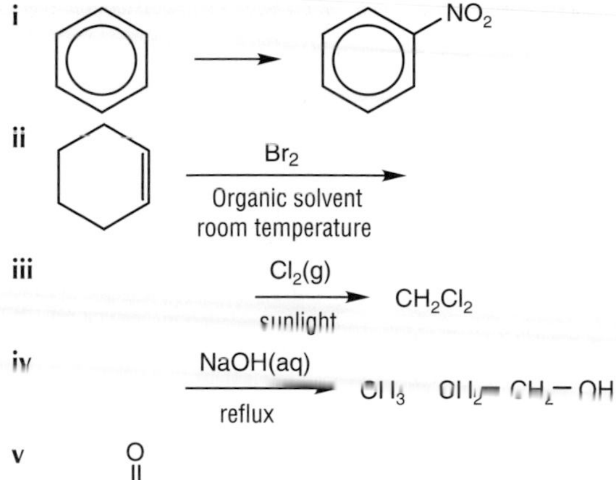

v

$$\begin{array}{l} CH_2-C(=O)-Cl \\ | \\ CH_2-C(=O)-Cl \end{array} \xrightarrow[\text{room temperature}]{H_2O(l)}$$

vi

$$CH_3-CH_2-CH(CH_3)-C(=O)Cl \xrightarrow[\text{AlCl}_3,\ \text{reflux}]{C_6H_6}$$

b Choose *two* words, one from list **A** and one from list **B**, to describe the mechanism of each of the reactions **i–vi** in part **a**.

A	**B**
electrophilic	*substitution*
nucleophilic	*addition*
radical	*elimination*

6 a Complete the following equations by filling in the missing products or reactants:

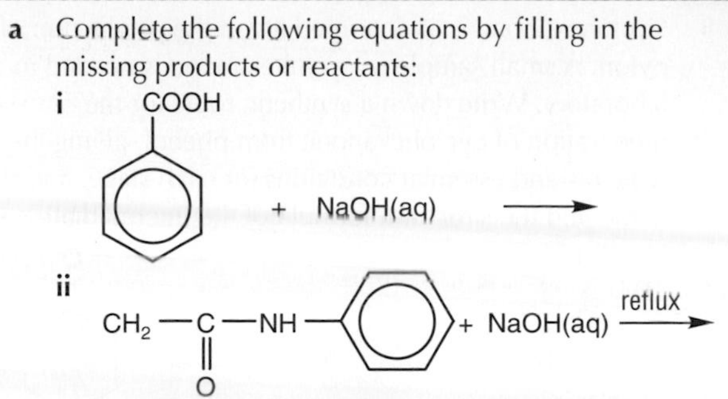

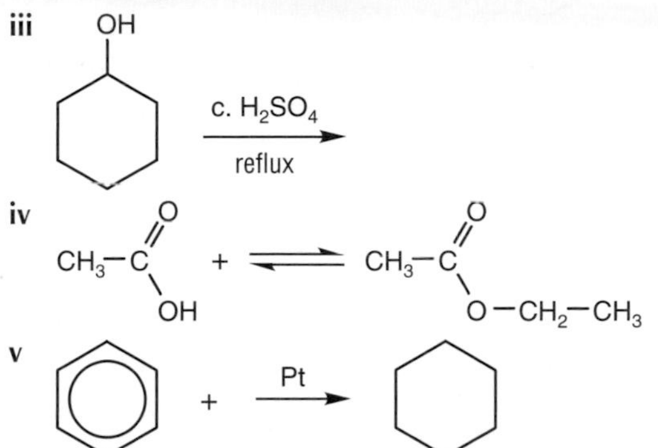

b For each of the reactions **i–v** in part **a**, choose a word to describe the *type* of reaction from:

reduction	*hydrolysis*	*oxidation*
dehydration	*acid-base*	*esterification*

c Give a second example, in the form of a balanced equation, for each type of reaction you select.

7 As a pharmaceutical chemist, you are interested in making minor modifications to the structure of penicillin F, in the hope of extending its range of effectiveness. Penicillins can be made in the laboratory by reacting 6-aminopenicillanic acid (6-APA) with a suitable acyl chloride. For penicillin F, the acyl chloride is $CH_3–CH_2–CH=CH–CH_2–COCl$, which can be made from the corresponding carboxylic acid.
Your aim is to synthesise a range of acyl chlorides to react with 6-APA to make some modified penicillins. You decide to work with the acid, $CH_3–CH_2–CH=CH–CH_2–COOH$, rather than the more reactive acyl chloride.
Use your knowledge of organic reactions to show how you would convert the acid into the following compounds. In each case, give the essential conditions and write a balanced equation for the reaction.

a The saturated acid
$CH_3-CH_2-CH_2-CH_2-CH_2-COOH$

b The halogenoacid
$CH_3-CH_2-CH(Br)-CH_2-CH_2-COOH$

c The hydroxyacid
$CH_3-CH_2-CH(OH)-CH_2-CH_2-COOH$

d The unsaturated acid
$CH_3-CH=CH-CH_2-CH_2-COOH$
(This may not be the only product of your reaction).

8 Cyclohexanone is an intermediate in the production of nylon. A small sample of cyclohexanone is needed in the laboratory. Write down a synthetic route for the preparation of cyclohexanone from phenol, giving the reagents and essential conditions for each stage, and the name and the structural formula of the intermediate.

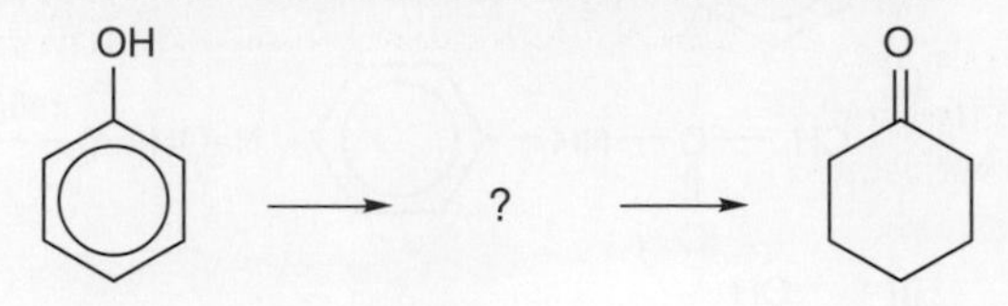

9 Cocaine was the first effective local anaesthetic to be used in minor surgery. Unfortunately, it can over-stimulate a patient and it is dangerously addictive. Modern local anaesthetics have to avoid these problems. Some of them are synthetic derivatives of 4-aminobenzoic acid. The simplest of these is benzocaine. This is often the active component of ointments to relieve the pain caused by sunburn.

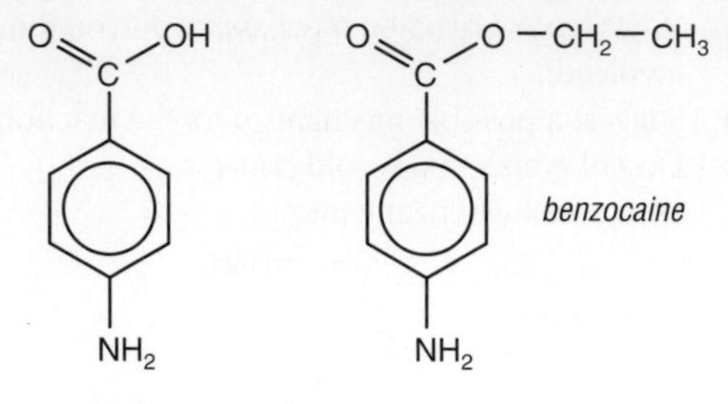

Design a synthesis of benzocaine, starting from 4-nitrobenzoic acid.

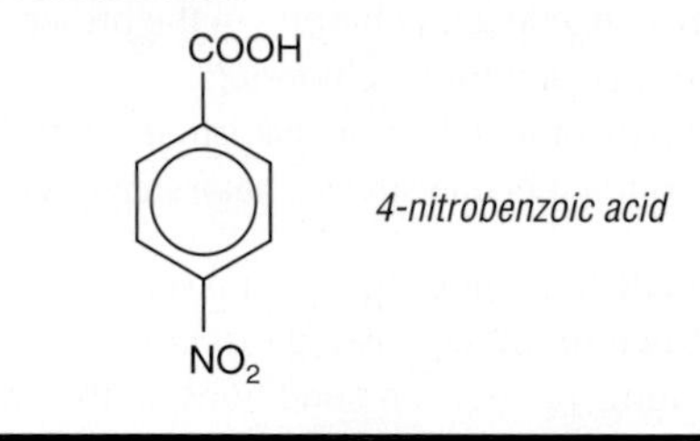

15 GREENER INDUSTRY

15.1 *The operation of a chemical manufacturing process*

Why study green chemistry?

The chemical industry is a major contributor to both the quality of our lives and to our national economy. In EU countries alone, it is estimated that the chemical industry provides employment for over 4 million people, with total sales valued at 613 billion euros.

The chemical industry earns its money by carrying out chemical conversions and selling products either for further chemical reactions or for formulation into final products. These final products are varied and include cleaners, paints, inks, medicines, pesticides, polymers, fertilisers, fuels, fuel additives, dyes and soaps.

All the principles of chemistry you learn in your studies are applied in one way or another in the chemical industry. Studying the industry enables you to recognise the practical importance of chemistry.

The first few sections of this chapter look at aspects of the operation of a chemical manufacturing process, while the latter half of the chapter focuses on ways in which the chemical industry has begun to consider and improve its 'green' credentials.

◀ **Table 1** EU chemical sales in 2005.

Country	Percentage share of EU market
Denmark	24.9%
France	15.6%
Italy	12.5%
UK	9.4%
Spain	7.1%
Netherlands	6.5%
Belgium	6.4%
Ireland	5.6%
other	12%

Source: Facts and Figures, CEFIC, 2006

Batch or continuous?

The chemical manufacturing process takes place within a chemical plant. It is the process by which a new product is made. Chemical manufacture includes specific instructions relating to the use of site equipment (called the plant), quantities and qualities of raw materials, mixing sequences, reaction conditions (including temperatures and pressures) and purification processes. Most chemical processes involve a sequence of events (called *unit operations*) which are represented in Figure 1. Input or removal of energy may be required at any stage in a process.

▼ **Figure 1** Sequence of unit operations in a chemical plant.

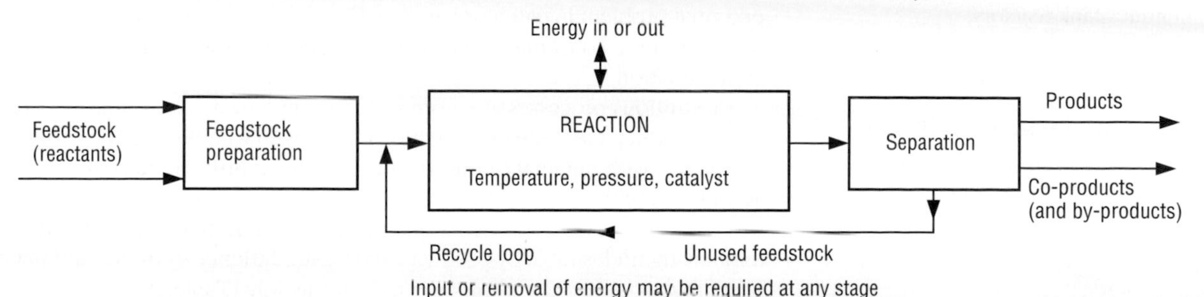

For a particular process, this sequence may be organised in one of two ways – **batch** or **continuous.** Sometimes a mixture of batch and continuous unit operations may be applied to achieve a given product. The organic preparations you have performed in the laboratory are batch processes.

In a *batch process*, the starting materials are put into a vessel and allowed to react together. The reaction is monitored and, when complete, the reaction phase is terminated. The product is then separated from the reaction mixture. This process is repeated in an identical manner, batch by batch, until the required amount of product is manufactured (Figure 2a).

In a *continuous process*, the starting materials are fed in at one end of the plant and product is withdrawn at the other end of the plant in a continuous flow (Figure 2b). The process may proceed through various unit operations in specifically designed parts, each dedicated to a particular step in the sequence.

Plants capable of operating batch processes are generally more versatile than continuous plants. It is usually possible to carry out a variety of chemical reactions in batch operations using the same vessel with no, or only minor, modifications. Batch processes are most cost effective when relatively small amounts of products are required. Batch processes are also more applicable to slow reactions, where a long residence time in the reaction vessel makes adaptation to continuous operation difficult.

There are drawbacks to batch processes relating to safety and contamination. In a batch reaction, relatively large reacting masses might not be easy to cool if a reaction is highly exothermic. Continuous processes tend to operate with relatively low volumes of reactants together at any time, allowing faster removal of thermal energy and better control.

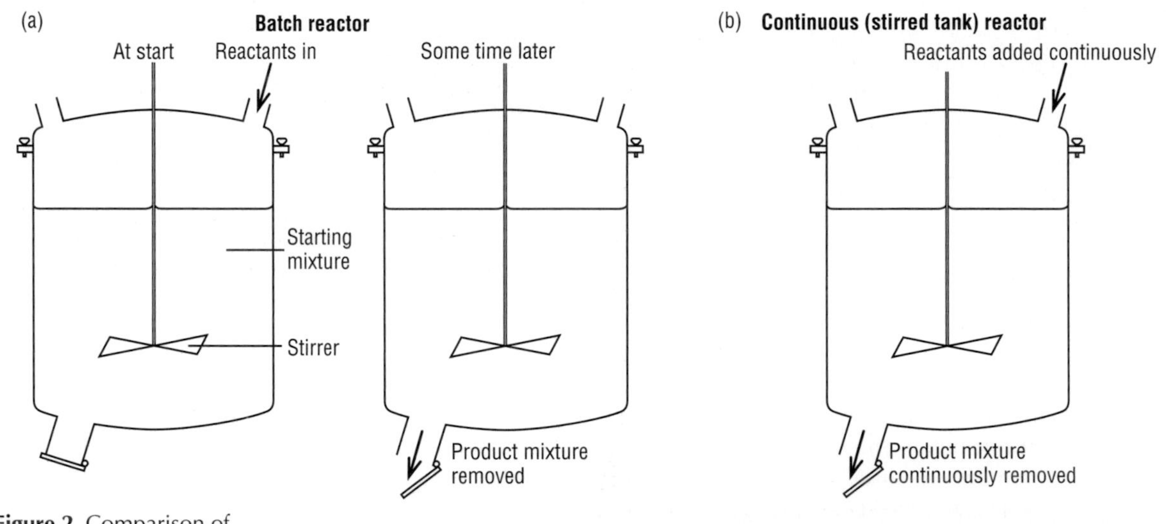

▲ **Figure 2** Comparison of **a** batch and **b** continuous tank reactors.

Contamination from batch to batch and product to product is also a potential problem in batch processing. Detailed clean-out procedures, which are time consuming and costly, must be followed to minimise cross-contamination.

Continuous processes are more suited to high-tonnage production of a single product. Once steady conditions have been established, the process can be run with minimal labour, relying heavily on instrumental and automatic control. To operate a continuous process, the plant is designed to produce a product of consistent quality under optimum conditions, thereby minimising undesirable by-products and waste. Ethene, ammonia and many other high-tonnage chemicals are made in this way (Table 2).

Batch	Continuous
copper refining	ethene, propene
organic bromine compounds	poly(ethene)
steel by BOS process	ammonia
pharmaceuticals, e.g. aspirin	bromine
dyes	sulfuric acid
pesticides	

◀ **Table 2** Examples of chemicals manufactured by batch or continuous processes.

However, there are often drawbacks to continuous processes. In particular, they require tailor-made plant which is less flexible. Nevertheless, there are new ingenious ways in which some flexibility can be incorporated by changing the feedstock or the conditions of the process. Table 3 summarises some of the advantages and disadvantages of the two types of process.

Advantages	Disadvantages
Batch	
• cost effective for small quantities; capital cost of plant much lower – possible to buy plant 'off the shelf' • slow reactions can be catered for • range of products can be made in the same vessel • greater percentage conversion is achieved compared with the same-sized continuous reactor for the same time	• charging and emptying the reactor is time consuming; no more product is being formed in this **shut-down** time • larger workforce needed • contamination more likely • exothermic reactions can be more difficult to control
Continuous	
• suited to high-tonnage production • greater throughput; shut-down may not occur for months or even years • contamination risk is low when used for one product • controlled because fine adjustments are possible • consistent quality ensured • requires minimal labour • more easily automated	• very much higher capital cost before any production can occur • not cost effective when run below full capacity • contamination risk is higher when used for two or more products

◀ **Table 3** Advantages and disadvantages of batch and continuous processes.

Plant construction materials

It is essential to choose construction materials which do not react with the feedstocks, catalysts, solvents and products involved in the process. The wrong choice can lead to lower efficiency, hazardous reactions and product contamination – not to mention holes in the reaction vessels or pipes. Glass-lined vessels, alloys or glass-reinforced plastics are widely used in place of steel components when there is a corrosion risk.

Problems for 15.1

1 Consider a typical school day. Make a list of 10 manufactured items that you usually come into contact with during the day between waking up and going to sleep.
Next to each item write down what contribution you think the chemical industry has made to its production. You may prefer to do this with a group of two or three other students.
Now sort your list of items into different categories of chemical industry – for example the polymer industry. How many items do you have left which have not been dependent on a contribution from the chemical industry?

2 Classify the following processes as batch or continuous:
- **a** the removal of carbon monoxide, oxides of nitrogen and unburnt hydrocarbons in a catalytic converter on a petrol-driven motor vehicle
- **b** cake making in the home
- **c** the distillation of crude oil in industry to provide a range of fractions for different purposes
- **d** the conversion of glucose into a mixture of glucose and fructose by passing the glucose through a bed of the immobilised enzyme, glucose isomerase. (The product is called 'high fructose syrup'. This syrup is sweeter than either fructose or glucose. The food and soft drinks industry use millions of tonnes of high fructose syrup every year.)

15.2 *Raw materials*

Raw materials are the starting point for any industrial chemical process. They are the materials from which **feedstocks** are made – the reactants that go into a chemical process. The raw materials usually have to be 'prepared' or treated to ensure that they are sufficiently pure and present in the correct proportions to use as feedstock. For example, by far the largest part of an ammonia plant is concerned with feedstock preparation, making the nitrogen and hydrogen mixture for direct conversion to ammonia. Methane is reacted with steam to produce hydrogen, and nitrogen is obtained by fractional distillation of liquefied air. The raw materials for the manufacture of ammonia are therefore air, water and methane.

An important part of feedstock preparation is getting the feedstock into a form in which it is easy to handle. Transferring gases and liquids is relatively easy because they can be transported by pipes within the chemical plant, or even across the country. Even so, the cost of pumping may be high and so every effort is made to keep the number of pumps and length of piping down to a minimum.

Solids are expensive to handle. Sometimes they are melted and maintained as hot liquids to reduce the transportation costs. For example, the sulfur used in the manufacture of sulfuric acid is often delivered by ship to the plant as a molten liquid, and is then also used in the plant as a liquid. Another way of handling solids is to mix them with a liquid to form a **slurry** – they can then be pumped for many kilometres along pipelines.

Often there is little choice of feedstock and raw materials for a chemical process. You can see some examples in Table 1.

A vast proportion of the organic chemicals produced today are derived from oil and natural gas. Natural gas is mainly methane but ethane, propane and butane are often present as well. These are steam cracked to produce ethene and propene (Figure 1).

Distillation of crude oil produces a variety of fractions, in particular LPG (liquefied petroleum gas), naphtha and gas (or diesel) oil. These fractions are converted into a variety of building blocks from which many chemicals are made. These building blocks are often alkenes. However, branched-chain alkanes, cycloalkanes and aromatic hydrocarbons are also produced for use in unleaded petrol (Figure 1).

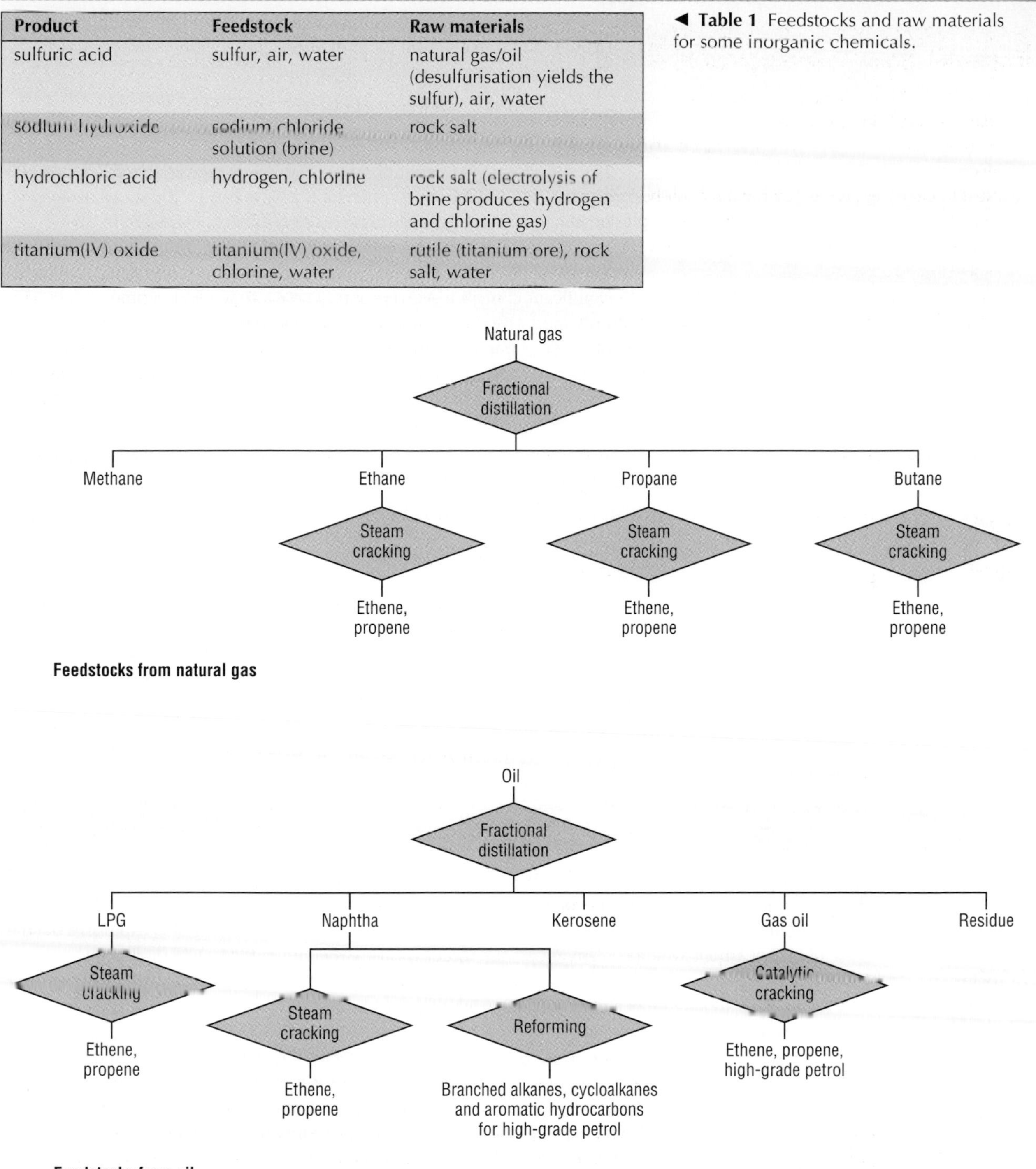

Product	Feedstock	Raw materials
sulfuric acid	sulfur, air, water	natural gas/oil (desulfurisation yields the sulfur), air, water
sodium hydroxide	sodium chloride solution (brine)	rock salt
hydrochloric acid	hydrogen, chlorine	rock salt (electrolysis of brine produces hydrogen and chlorine gas)
titanium(IV) oxide	titanium(IV) oxide, chlorine, water	rutile (titanium ore), rock salt, water

◀ **Table 1** Feedstocks and raw materials for some inorganic chemicals.

▲ **Figure 1** Feedstocks from oil and natural gas.

Co-products and by-products

When feedstock is passed through a reactor, a number of things might happen to it.

1 The reaction may form only one product. For example:

$$N_2(g) + 3H_2(g) \rightleftharpoons 2NH_3(g)$$

2 *The reaction may form two products.* For example, phenol is manufactured from (1-methylethyl)benzene – often called cumene. The overall reaction can be represented as

$$\underset{cumene}{C_6H_5CH(CH_3)_2} + O_2 \rightarrow \underset{phenol}{C_6H_5OH} + \underset{propanone}{(CH_3)_2CO}$$

The propanone formed in this reaction is an example of a **co-product**. The ratio of product to co-product is always fixed – the more desired product produced, the more co-product is also produced. In this example 6 tonnes of propanone are produced at the same time as 10 tonnes of phenol. Proceeds from the sale of propanone make a significant contribution to profits. The route would become uncompetitive if demand for propanone was to fall.
Another example is in the electrolysis of brine when chlorine and sodium hydroxide are co-products in the production of hydrogen.

3 *A reaction other than the one that was intended may occur.* For example epoxyethane, $(CH_2)_2O$, can be made in a one-step process in which ethene is mixed with oxygen and passed over a silver catalyst at 300 °C and a pressure of about 3 atmospheres.

$$2C_2H_4 + O_2 \rightarrow 2(CH_2)_2O \qquad \text{desired reaction}$$

Under these conditions, there is also the possibility of the ethene being completely oxidised:

$$C_2H_4 + 3O_2 \rightarrow 2CO_2 + 2H_2O \quad \text{unwanted side reaction}$$

The carbon dioxide and water formed in this unwanted reaction are called **by-products**. Chemists and chemical engineers work to try to reduce the amount of by-product by reducing the amount of side reaction occurring.

4 Some feedstock may remain unreacted.

Problem for 15.2

1 Until recently in the UK, naphtha was used as the feedstock for making ethanoic acid. Naphtha is one of the fractions obtained in refining petroleum. It is principally a mixture of hydrocarbons which contain between 6 and 10 carbon atoms per molecule.
Naphtha is oxidised by air at a temperature of about 200 °C and a pressure of about 50 atmospheres. No catalyst is needed and the reaction takes place in one step. The product contains about 50% ethanoic acid together with a range of compounds including propanone, methanoic acid, propanoic acid and butane-1,4-dioic acid. Some of these co-products are useful; others are simply burned at the chemical plant to provide energy.
Naphtha oxidation has been replaced by a process in which methanol reacts with carbon monoxide in the presence of a rhodium/iodine catalyst:

$$CH_3OH(l) + CO(g) \rightarrow CH_3COOH(l) \qquad \Delta H = -135\,kJ\,mol^{-1}$$

This process takes place at a similar temperature to that used in naphtha oxidation but the pressure is lower (about 30 atmospheres). The catalyst is expensive but the process is very specific. A 99% yield of ethanoic acid is obtained. Methanol is cheap and readily available from a number of sources. It can be made from crude oil or natural gas.
One way of investigating the success and efficiency of a chemical process is to look at the advantages and disadvantages associated with factors such as:

- the conditions employed
- the feedstock
- the product
- the co-products.

a Taking the four factors in turn, suggest one advantage in each case which the methanol/carbon monoxide process has over the naphtha oxidation.

b Suggest one disadvantage which the methanol/carbon monoxide process has compared to naphtha oxidation.

AS

15.3 *Costs and efficiency*

Fixed and variable costs

Many factors contribute to the cost of a chemical process. Major costs – such as research and development, plant design and construction, and initial production – are incurred before any product is sold into the market. Sales of the product have to generate enough return to offset these initial costs and generate a profit for the company (see Figure 1).

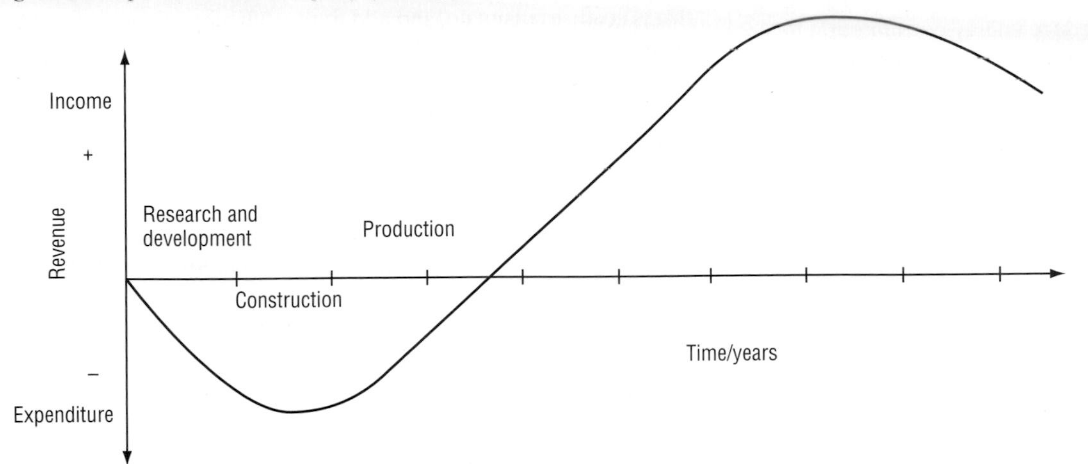

▲ **Figure 1** In the early years there is considerable expense with no income. Later in the life of the product, profits may reduce because demand for the product is reduced or because maintenance costs increase as the plant gets older.

The profit generated is the difference between the selling price of the product and the costs of production. Production costs are made up of two elements – **fixed costs** and **variable costs**.

Fixed (or indirect) costs are those incurred by the company whether they produce one tonne or tens of thousands of tonnes of product. For example, as soon as it has been built the production plant starts to lose value, or depreciate, regardless of how much product is made. Other fixed costs include labour costs, land purchase or rental, sales expenses, telephone bills, etc. Hence, the fixed cost element in the production cost is calculated by spreading the total annual charges over the number of units produced per year. If only 1 tonne of product is produced per year then the fixed cost element of production cost will be significantly higher than if 100 tonnes are produced with the same fixed costs. As fixed costs are allocated against a product, it is best to divide the cost against the maximum output of product possible. This will only occur when the plant is operating at maximum capacity.

A significant element of fixed costs are **capital costs**. They relate to establishing chemical plant, buildings and infrastructures around which manufacture is based. These capital costs are recovered as part of the fixed cost element (depreciation). In general, for normal accountancy purposes, the life of a plant is assumed to be 10 years, after which it may be said to be 'written off'. In practice, many plants have a perfectly satisfactory life beyond 10 years. In estimating the profitability of a plant, it will be necessary to recoup the investment in capital costs (as well as covering the fixed and variable costs) before the plant is written off.

Variable (or direct) costs relate specifically to the unit of production. Raw materials are the most obvious variable cost, along with costs of effluent treatment and disposal, and the cost of distributing the product. If no production occurs then these variable costs will not be incurred, whereas fixed costs will still have to be paid.

Efficiency

The efficiency of a chemical process will also affect total costs. The efficiency of a chemical process depends on various physical factors such as temperature, pressure and rate of mixing. When choosing the temperature and pressure for a particular reaction step, it might seem appropriate to go for the highest possible to maximise the rate at which the product is formed. However, it is not always as simple as this and it is necessary to find the conditions that give the most economical conversion. For example, very high temperatures and pressures require very specialised, expensive plant that is costly to maintain, and add to the difficulty of controlling the chemical reactions.

Green chemistry aims to reduce the use of feedstocks to a minimum, recycle unused reactants and solvents, and reduce energy consumption to a minimum. This usually results in minimum waste. Cost control often utilises the ideas of green chemistry.

Even more importantly, choice of reaction conditions can have a significant impact on *yield*.

Industry aims to maximise the output of product in a given time in order to maximise profits. Achieving this involves a balancing act between rate and position of equilibrium.

The rate of a reaction will be influenced by a number of factors including temperature. You can see more detail of this in **Section 10.2**. Generally, the higher the temperature then the faster the rate of reaction. However, the amount of product obtained is affected by the position of equilibrium. The further the position of equilibrium lies to the right in the desired reaction (i.e. favours product formation) then the greater the percentage conversion to products. You can read more about this in **Sections 7.1** and **7.2**. Both of these factors need to be taken into account when selecting reaction conditions.

The costs of a process can be reduced considerably if an effective catalyst can be found. Huge savings can also be made by recycling unreacted feedstock, and in saving energy that would otherwise be lost to the environment. Sometimes the ability to sell on a co-product can be the difference between making a profit or not.

Recycling

The reclamation and recycling of unreacted materials is an important aspect of process development for both economic and environmental reasons.

Recycling means separating unreacted feedstock from the reaction mixture, and this is not always easy. It is important that impurities do not get recycled along with any unreacted feedstock. If an impurity is recycled but does not react, its concentration will build up in the reaction step and it will interfere with the reaction. For example, argon from the air is recycled along with unreacted nitrogen and hydrogen in the manufacture of ammonia. Argon is unreactive and it is difficult to remove, so it is not possible to recycle continuously. The argon-enriched nitrogen and hydrogen mixture is bled off at intervals.

Saving energy

Efficient use of energy is a significant aim in most chemical processes. During the past few years, often in response to high energy costs and environmental considerations, chemical companies have markedly reduced their energy consumption.

Many chemical reactions involve the release of thermal energy. This thermal energy can be conserved by lagging pipes and by using heat exchangers (Figure 2). Energy from exothermic parts of the process can be used to supply energy to endothermic parts. The energy released in an exothermic reaction can be used to raise the temperature of the reactants.

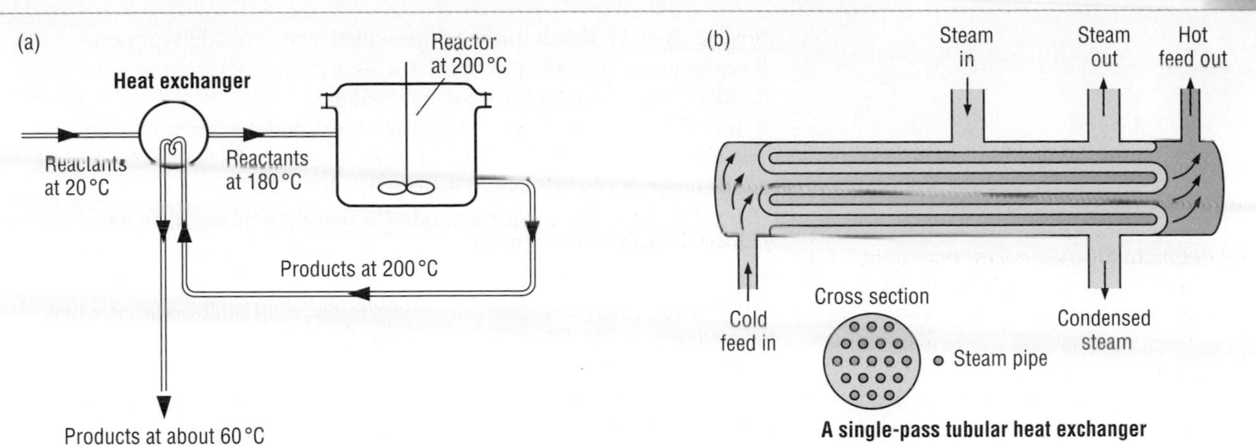

▲ **Figure 2 a** An example of how energy can be used in a chemical process. **b** Diagram of a single-pass tubular heat exchanger.

Water or steam is often used to transfer energy within a chemical plant. Steam, which is generated in a separate plant away from the process, is a much safer alternative to oil and gas, or to using electricity where flammable substances are involved.

In an integrated plant, energy (transferred in steam, hot liquid or gas streams) from one process can be used in a completely different part of the plant. This reduces the quantity of oil, gas and electricity that has to be purchased. By contrast, a small site would have to raise steam in a special boiler using purchased fuel oil, gas or electricity. The price of energy in the small plant would therefore be considerably greater in relative terms than that in an integrated plant.

Problems for 15.3

1 Explain why:

a the capital cost of a chemical plant designed to produce 200 tonnes of product a day is less than twice that of a plant designed to produce 100 tonnes a day.

b if you double the output of a chemical plant, you may double the variable costs but you don't double the fixed costs.

2 Ethanol is made by the addition of steam to ethene. The equation for this exothermic reaction is

$$CH_2{=}CH_2(g) + H_2O(g) \rightleftharpoons C_2H_5OH(g) \qquad \Delta H = -87.5\,kJ\,mol^{-1}$$

In one plant the conditions for this process are 300°C and 70 atmospheres pressure. The catalyst is a solid impregnated with phosphoric acid. When the ratio of reactants is 1 mole of ethene to 2 moles of steam, the proportion of ethene converted at equilibrium is 40%. However, the process is operated so that a 10% conversion of ethene is achieved following one pass of the reactants through the catalyst.

a With the aid of a diagram, suggest how some of the thermal energy released in the reaction can be put to good use in the process.

b Explain why it is more desirable to achieve a 10% conversion of ethene than a figure closer to the equilibrium value at this temperature and pressure.

15.4 *Plant location*

The chemical industry in the UK grew up around the sources of its major raw materials, near the customer industry that it served or in a location with good transport links.

With increasing specialisation in the UK and a much improved communications network, these factors are no longer the primary considerations in locating manufacturing sites.

New chemical plants are often built near to existing works. This may be because a district specialising in the manufacture of a particular product is more likely to be able to provide the skilled labour needed for new developments. For example, specialist steel producers were sited in Sheffield, a traditional centre for steelmaking.

It may be that the feedstock for the new process is already being produced at the existing site. There may be less opposition to building a new plant at an existing works than to the development of a new site.

In addition to these considerations, it may save money if facilities such as a canteen, medical centre and administration block are shared between existing and new plants.

However, many well-established locations are in built-up areas where the increasing concerns of the local population give rise to problems in extending plant or building new facilities.

Increasingly, chemical companies are seeking more remote locations for their operations. But then another problem arises because they are accused of spoiling the countryside.

The actual site chosen should be served by a good communications network – building this from scratch would add significantly to the cost of opening the plant. Road, rail and water are all important means of transport, and deep sea access is useful for the import and export of bulk materials such as oil. Ideally the site would be level, free from danger of subsidence and have scope for further expansion. Waste treatment and disposal are always concerns to be borne in mind. Often the reasons for the location of a particular industry are complex, and each case needs to be considered on its own merits.

Problem for 15.4

1 Explain the reasons for the location of the following chemical industries:
   a the Fawley oil refinery located on the south coast near Southampton
   b sulfuric acid is often manufactured in vicinity of an oil refinery
   c Speciality Minerals' high grade calcium carbonate production plant is situated on the canal system in Birmingham
   d the UK's chlor–alkali industry is situated in and around Cheshire.

AS

15.5 *Health and safety*

On 11th December 2005, the Buncefield fuel depot was the site of the UK's biggest peacetime blaze. The depot, situated outside the town of Hemel Hempstead 40 km (25 miles) north west of London, is just off the M1 motorway. Unleaded motor fuel was being pumped into a storage tank. Safeguards on the tank failed to stop more fuel being pumped when the full level was reached, resulting in the tank overflowing. The overflow from the tank led to the rapid formation of a rich fuel and air vapour, which caused an initial explosion. Further explosions followed and a large fire took hold, eventually engulfing 20 large storage tanks. Emergency services declared a major emergency and at the peak of the incident 25 fire engines, 20 support vehicles and 180 firefighters were on site. The blaze was finally extinguished on 15th December. Some 2000 homes had to be evacuated.

Fortunately events like this are very rare. The safety record of the chemical industry in the United Kingdom is very good, and the chances of a major accident occurring are very low. Unfortunately, if an accident does happen the consequences can be severe, as at Buncefield. We need to balance the risk, and the likelihood of that risk occurring, against the enormous benefits the chemical industry can bring to our lives.

Safety legislation

Safety is a major consideration in all operations in the chemical industry. All aspects of safety are affected by national and European Union legislation. Legislation and planning are essential, but the key factor in ensuring safety is for everyone within the chemical plant to recognise that it is in their interests to work safely. Some of the main pieces of legislation affecting the chemical industry are:

1 *The Health and Safety at Work Act, 1974*
UK legislation places responsibility for health and safety with the employer. Personal safety is rated very highly and on a typical visit to a production site you may see eye-baths, showers, toxic gas refuges, breathing apparatus, emergency control rooms and, on larger sites, the company's fire brigade, ambulance service and a well-equipped medical centre with its own qualified doctors and nurses.

2 *The Control of Substances Hazardous to Health (COSHH) Regulations, 2002*
These regulations control the amount of exposure that employees have to hazardous chemicals. There are various means of minimising exposure, depending on the chemical process concerned. The production of these hazardous materials will already have been minimised by altering the reaction conditions – for example, altering the temperature might reduce the amount of a toxic by-product being produced. Steps taken might also include the use of extractors on site and the implementation of safe handling and storage of hazardous chemicals. These could be raw materials, reactants, products or by-products. Table 1 shows some simplified examples of the risks caused by certain materials that are commonly found on chemical sites and the methods that can be taken to reduce these risks occurring. 'Hazard' is the potential that a hazardous substance has to cause harm. 'Risk' is the likelihood the hazardous substance will cause harm under the conditions of its use.

Hazard	Risk	Methods taken to reduce the risk occurring
flammable gases	explosion/ fire	stored in flameproof, pressurised cylinders; incinerate under controlled conditions; extractor fans
acidic gases	burns	neutralise by passing through a scrubber containing alkaline material; regular checking of plant for leaks
toxic emissions	various, depending on the material	check protocol for process to minimise emissions; monitor levels of emissions; ensure that all personnel are familiar with evacuation procedure

◀ **Table 1** Examples of how the occurrence of risks can be minimised.

3 *The Control of Major Accident Hazards (COMAH) Regulations, 1999*
Some reagents used by chemical companies could become a hazard to people living near the plant in the event of an accident. This applies particularly to poisonous gases or volatile liquids. If a plant requires the use of a reagent such as chlorine then the company works with the local authority and emergency services. Emergency procedures are carefully planned and rehearsed at regular intervals. These plans are outlined in leaflets, and a video may be produced to brief people on what to do if the plant siren sounds continuously. The siren is tested at regular intervals and this test ensures that everyone recognises

the warning. The most common emergency procedure requires everyone to go indoors and close all windows and doors until the all-clear is given.

Green chemistry

The principles of green chemistry aim to minimise the impact of chemical production on both living organisms and the environment. The main aims are to:

- minimise waste (for example, through recycling or by finding ways to make waste products useful)
- reduce energy consumption (for example, through use of enzymes to reduce high temperatures needed for reactions, increasing atom economy – see **Section 15.7**)
- reduce feedstock consumption (for example, by minimising solvent use or by reducing the number of steps needed in a process).

In many cases, the steps taken to achieve these aims also improve the profit margin of a process. You will have seen in the preceding sections of this chapter that many of the efforts made by the chemical industry to maximise profit also involve acting in a 'greener' way. The two ideas are not mutually exclusive – in fact, they often complement one another. Green chemistry is a sustainable way forward, meeting the needs of everyone.

AS

15.6 Waste disposal

The environment is an important consideration for chemical companies. It is not acceptable to allow harmful substances to escape into the environment. In fact, the pollution resulting from the accidental escape of materials from a plant, or the disposal of untreated waste products, not only damages the environment but also jeopardises the future of the company itself.

In the past, waste has been dumped in the nearest convenient place – directly into the atmosphere, into a disused quarry or into rivers, lakes and the sea. Alternatively, it was contained in purpose-built ponds or tips which have caused problems with toxic materials leaching out into nearby streams and waterways. These methods are no longer acceptable and are now illegal unless special permission is given. Waste must now be treated and it can only be disposed of when in a state that is not harmful to the environment. Liquid waste from chemical works has to meet legal requirements, against such criteria as pH level and metal ion content, before being released into natural waterways or sewage systems. It must therefore be treated appropriately – for example, neutralising any acidic waste streams. Water containing organic waste cannot be discharged into rivers or canals if it would significantly reduce the oxygen content of the water, as this would cause fish to die from lack of oxygen.

Gases which contain contaminants are purified by bubbling them through neutralising solutions to remove soluble contaminants. Particles of dust can be removed by filtration or other methods.

Sulfur dioxide is a good example of a potential pollutant that can be successfully removed from waste gases, enhancing atmospheric environment protection. The ores used in the manufacture of many metals (copper, lead and zinc for example) are sulfides. These are roasted in air to form an oxide as one of the first stages in the manufacture of a metal. For example:

$$2PbS(s) + 3O_2(g) \rightarrow 2PbO(s) + 2SO_2(g)$$

The sulfur dioxide produced is not allowed to escape but is led away from the furnace to a nearby plant, which converts it into sulfuric acid using the contact process:

$$2SO_2(g) + O_2(g) \rightarrow 2SO_3(g)$$
$$SO_3(g) + H_2O(l) \rightarrow H_2SO_4(aq)$$

The sulfuric acid can then be sold. An ingenious way of converting an unwanted co-product into profit!

Problem for 15.6

1 Another gas which causes great problems in the atmosphere is nitrogen(II) oxide (NO), which is discharged by cars, lorries, factories and power stations – in fact, from anywhere where air is heated to high temperatures and nitrogen and oxygen react. The NO discharged is oxidised to form nitrogen(IV) oxide, NO_2, in the cool atmosphere.

a Write an equation for the formation of nitrogen(II) oxide, NO, from oxygen and nitrogen.

b Write an equation for the formation of nitrogen(IV) oxide, NO_2, from nitrogen(II) oxide and oxygen.

These oxides are removed catalytically in car exhausts. However, this process is too expensive for power stations because very large quantities of gases must be treated. An ingenious process, known as gas reburn, is being tried at Longannet power station in Scotland. Natural gas (CH_4) is fed into the flue gases (containing nitrogen(II) oxide) above the flames of the burning coal. The methane is oxidised by the nitrogen(II) oxide to form nitrogen, carbon dioxide and water.

c Write an equation for this process.

d What do you think happens to any excess methane?

15.7 *Percentage yield and atom economy*

AS

It would be unusual these days if you lived in a household that recycled no waste at all. Increased environmental awareness and government targets for recycling ensure that nearly all of us sort out at least some of our rubbish before it is collected. In addition, many people are attempting to live a much more sustainable lifestyle, actively working at consuming less of the Earth's resources and consequently producing far less waste for disposal.

Unsurprisingly, this movement towards waste reduction is mirrored in the chemical industry, and is widely referred to as 'green chemistry'. The Green Chemistry Program was established in the United States in 1991 with the aim of encouraging the design of chemical products and processes that reduce or eliminate the use or generation of hazardous substances in the design, manufacture and use of chemical products e.g. polymer production. It is probably the fastest growing area in chemistry today, and it is forcing chemical manufacturers to ensure that their products and processes are sustainable.

In this section you will look at some of the principles of green chemistry and consider how they are changing the way that the chemical industry works today.

Percentage yield

For many years, chemists have considered **percentage yield** to be the deciding factor when determining whether or not a particular reaction is economically viable. In other words, can enough be produced to make enough money to make the whole process worthwhile? To calculate the percentage yield of a particular reaction you need to use the percentage yield calculation:

$$\%\text{ yield} = \frac{\text{actual mass of product}}{\text{theoretical maximum mass of product}} \times 100$$

It is possible to work out the percentage yield for any reaction. Consider the conversion of a ketone to an alkene – a reaction commonly used in the manufacture of many vitamins and pharmaceuticals. This type of reaction is known as a Wittig reaction.

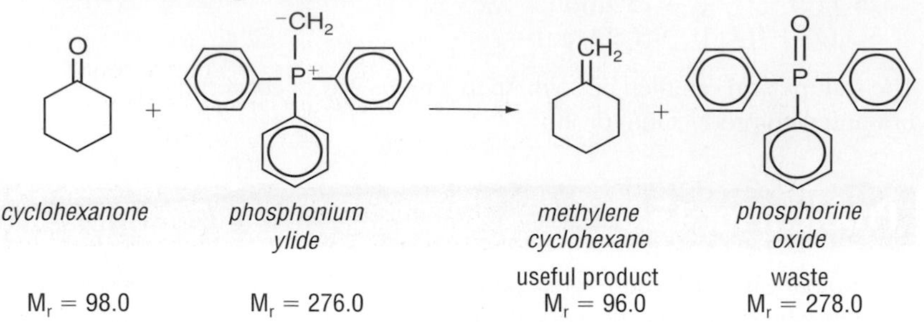

When this reaction is performed on an industrial scale, 1000 tonnes of ketone produce 842 tonnes of methylenecyclohexane. What is the percentage yield of this reaction?

Calculating percentage yield

To calculate the percentage yield you need to use the equation and the relative formula masses for the reactant and product given above.

Step 1 Calculate the maximum possible yield given the amount of starting material. From the equation, 98 tonnes of cyclohexanone could produce 96 tonnes of methylenecyclohexane. 1000 tonnes of cyclohexanone would therefore produce

$$\frac{96}{98} \times 1000 = 979.6 \text{ tonnes}$$

Step 2 Calculate the percentage yield given that 842 tonnes of methylenecyclohexane are produced.

$$\% \text{ yield} = \frac{\text{actual mass of product}}{\text{theoretical maximum mass of product}} \times 100$$

$$= \frac{842}{979.6} \times 100$$

$$= 86\%$$

You may have already made a chloroalkane from an alcohol in your practical work and calculated the percentage yield of your reaction. If you have, you will realise that a yield of 86% makes this reaction very efficient indeed for an organic reaction, which is one of the reasons why it is so widely used in industry.

But is it as efficient as it first appears? What the yield calculation completely overlooks is the quantity of phosphine oxide waste product also generated. In fact, when you look at the relative formula masses of the useful product (methylenecyclohexane = 96.0) and the waste product (phosphine oxide = 278.0) you can see that for every 96 tonnes of useful product, 278 tonnes of waste are generated – nearly three times the amount by mass. Now that doesn't look so efficient after all!

Atom economy

We evidently need to consider how efficiently all the reactant atoms are used in a chemical reaction, and take account of the amount of reactant that ends up in the desired product and the amount lost as waste.

For this reason, industrial chemists have started to consider the **atom economy** of a process, rather than just the percentage yield:

$$\% \text{ atom economy} = \frac{\text{relative formula mass of useful product}}{\text{relative formula mass of the reactants used}} \times 100$$

Let us look again at the production of methylenecyclohexane:

M_r of all reactants = 98.0 + 276.0 = 374.0
M_r of useful product = methylenecyclohexane = 96.0

$$\% \text{ atom economy} = \frac{96.0}{374.0} \times 100 = 26\%$$

Not so efficient after all! In fact, when atom economy is taken into account, many of the reactions traditionally used to manufacture chemicals are not very efficient. Some of these reactions are covered in the next section.

Problems for 15.7

1 Ethyl ethanoate is produced by the esterification of ethanol with ethanoic acid in the presence of concentrated sulfuric acid:

$$CH_3COOH(l) + C_2H_5OH(l) \longrightarrow CH_3COOC_2H_5(l) + H_2O(l)$$

ethanoic acid *ethanol* *ethyl ethanoate* *water*

a Calculate the relative formula masses of ethanol, ethanoic acid and ethyl ethanoate.
b If the starting mass of ethanol is 23 g, calculate the expected mass of ethyl ethanoate (assume that excess ethanoic acid is used).
c Calculate the actual percentage yield if 36 g of ethyl ethanoate are produced.

2 Ethanol can also be oxidised to produce ethanoic acid by heating it with an oxidising agent such as acidified potassium dichromate(VI) solution:

$$CH_3CH_2OH \xrightarrow[\text{reflux}]{Cr_2O_7^{2-}/H^+(aq)} CH_3COOH$$

ethanol *ethanoic acid*

Calculate the percentage yield of this reaction if the starting amount of ethanol was 23.0 g and 20.0 g of ethanoic acid were produced.

3 Iron is produced from iron ore in the blast furnace by reducing iron ore with carbon monoxide.

$$Fe_2O_3 + 3CO \longrightarrow 2Fe + 3CO_2$$

iron ore *carbon monoxide* *iron* *carbon dioxide*

Calculate the percentage yield of this process if 10 tonnes of the ore yield 0.67 tonnes of iron.

4 Alcohols can be dehydrated to produce alkenes by heating the alcohol with aluminium oxide. For example, propanol can be dehydrated to produce propene and water:

$$CH_3CH_2CH_2OH \xrightarrow[300°C]{Al_2O_3} CH_3CH{=}CH_2 + H_2O$$

propanol *propene* *water*

a Calculate the relative formula mass of the starting material, propanol.
b Calculate the relative formula mass of the useful product, propene.
c Calculate the atom economy of this reaction.

5 Organic chemists sometimes want to increase the chain length of a hydrocarbon during an organic reaction. The easiest way to do this is to react an aldehyde, such as ethanal, with hydrogen cyanide:

$$CH_3{-}\overset{\overset{O}{\|}}{C}{-}H + H{-}CN \longrightarrow CH_3{-}\overset{\overset{OH}{|}}{CH}{-}CN$$

ethanal *hydrogen cyanide* *2-hydroxypropanenitrile*

a Calculate the atom economy of this reaction.
b Explain why this reaction has such good atom economy.

6 Alcohols can be prepared by replacing the halogen of a halogenoalkane with an –OH group, provided by a water molecule. For example bromobutane will react (rather slowly) with water to produce butanol:

$$CH_3CH_2CH_2CH_2Br + H_2O \longrightarrow CH_3CH_2CH_2CH_2OH + HBr$$

bromobutane *water* *butanol* *hydrogen bromide*

a Calculate the atom economy of this reaction.
This reaction can be speeded up by using sodium hydroxide, NaOH, instead of water to provide the hydroxide ions. In this case, the waste product of the reaction is not hydrogen bromide, HBr, but sodium bromide, NaBr.
b Write an equation for this reaction.
c What effect would changing the reactant in this way have on the atom economy?

7 Synthetic chemists often use the following reaction to make a chemical called 2-*t*-butyl-*p*-cresol:

p-cresol (OH and CH_3 on benzene ring) + MTBE ($CH_3{-}C(CH_3)(OCH_3){-}CH_3$) $\xrightarrow{\text{cat}}$ 2-t-butyl-p-cresol (ring with OH, $C(CH_3)_3$ and CH_3) + CH_3OH

p-cresol *MTBE* *2-t-butyl-p-cresol* *methanol*

a Calculate the atom economy for this reaction.
b Calculate the percentage yield if 19.61 g of *p*-cresol was used and 13.0 g of 2-*t*-butyl-*p*-cresol was produced.
c Use the percentage yield and atom economy of this reaction to calculate how much of the starting material is actually converted into useful product.

15.8 *Which reactions have the highest atom economy?*

Most of the reactions used in chemical synthesis can be classified as one of the following types:

- rearrangement
- addition
- substitution
- elimination.

This section will cover each of these reaction types to see how they differ in terms of atom economy.

Rearrangement reactions

The isomerisation reaction of hydrocarbons, to improve the efficiency of fuels, is an example of a rearrangement reaction – you studied these in the **Developing Fuels** module. The following equation shows the isomerisation of hexane to 3-methylpentane. Note that the relative formula mass of the useful product is the same as the reactant.

$$CH_3CH_2CH_2CH_2CH_2CH_3 \longrightarrow CH_3CH_2CH(CH_3)CH_2CH_3$$

hexane $M_r = 86.0$ *3-methylpentane* $M_r = 86.0$

Because rearrangement reactions effectively reorganise the constituent atoms within a particular molecule, no atoms are lost as waste and so rearrangements are, by definition, 100% atom economical.

Addition reactions

The bromination of cyclohexene is an example of an addition reaction – you may have used this reaction to test for an alkene in the **Polymer Revolution** module.

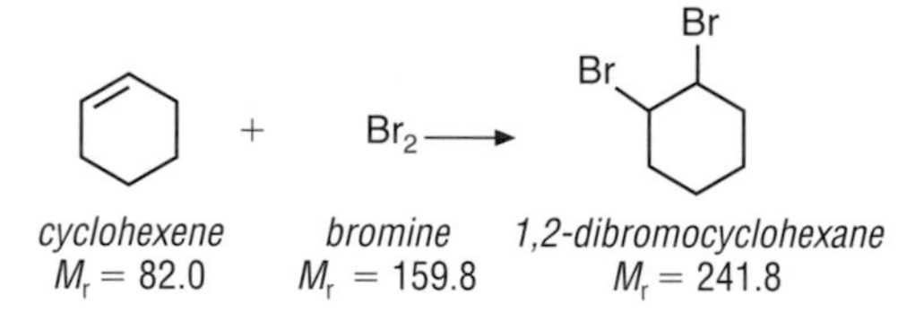

cyclohexene $M_r = 82.0$ *bromine* $M_r = 159.8$ *1,2-dibromocyclohexane* $M_r = 241.8$

Total M_r of reactants = 241.8
M_r of useful product = 241.8

$$\%\ \text{atom economy} = \frac{M_r \text{ of useful product}}{M_r \text{ of the reactants used}} \times 100$$

$$= \frac{241.8}{241.8} \times 100 = 100\%$$

Again, you can see from this example that addition reactions are always 100% atom economical as the reactants are added together and there are no leaving atoms or molecules.

Substitution reactions

In the **Elements from the Sea** module you used a substitution reaction to prepare a halogenoalkane from an alcohol:

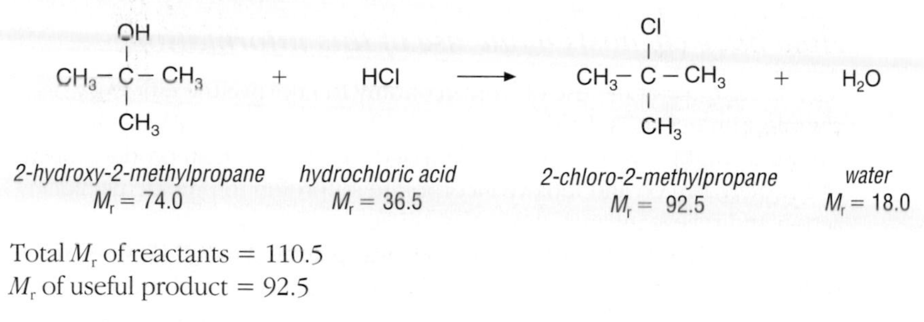

Total M_r of reactants = 110.5
M_r of useful product = 92.5

$$\% \text{ atom economy} = \frac{92.5}{110.5} \times 100 = 84\%$$

Substitution reactions involve substituting one group for another group. The leaving group is not incorporated in the final product, and so the atom economy of these reactions will always be less than 100%. The exact value will depend on the formula masses of the substituted groups. In fact, with an atom economy of 84%, this substitution reaction is relatively efficient. However, the Wittig reaction that you came across in **Section 15.7** had an atom economy of only 26% because of the large mass of the leaving group.

Elimination reactions

In the **Polymer Revolution** module you came across elimination reactions where alcohols are converted to alkenes, for example

$$CH_3CH_2CH_2OH \longrightarrow CH_3CH{=}CH_2 + H_2O$$

propan-1-ol $M_r = 60.0$ *propene* $M_r = 42.0$ *water* $M_r = 18.0$

Total M_r of reactants = 60.0
M_r of useful product = 42.0

$$\% \text{ atom economy} = \frac{42.0}{60.0} \times 100 = 70\%$$

Elimination reactions involve removing a group from a molecule, so there will always be another product as well as the useful product. In addition, any reagent used will not be incorporated in the final product and so will also be a waste product. For this reason elimination reactions are always less than 100% efficient.

Which type of reaction is the most atom efficient and environmentally friendly?

The different types of reaction used in chemical synthesis have very different atom economies. What the examples above show is that rearrangement and addition reactions are 100% atom economic. Substitution reactions are far less atom economic – the actual degree of atom economy depending on the group substituted. Elimination reactions are the least economic of all. Table 1 summarises this information.

▼ **Table 1** Summary of atom economy for different types of reaction.

Reaction types in order of decreasing atom economy
rearrangement, addition
substitution
elimination ↓

Some reactions, such as condensation reactions, can appear to be more difficult to categorise. However, on closer inspection, a condensation reaction is actually an addition reaction followed by an elimination reaction.

How have chemists made use of this information?

A good example of the use of atom economy to improve the efficiency of chemical processes is the development of ibruprofen, an analgesic or painkiller within a very important class of drugs. Well-known products such as Nurofen, Brufen and Ibuleve incorporate ibuprofen to provide painkilling action and relieve inflammation or swelling.

The traditional synthesis of ibuprofen was patented in the 1960s by the Boots Company plc, and until 1990 industrial production of this drug was almost exclusively by one method. The problem was that this method consists of six steps, many of which were substitution or elimination reactions with low atom economies. The overall atom economy of the Boots process was 32%, indicating that more than two-thirds (by mass) of the materials used were not ending up in the target molecule (the desired product).

In the mid 1980s, Boots' manufacturing patent expired, and other companies were free to produce ibuprofen. New, more atom-efficient methods of production had to be found. Boots teamed up with a company called Hoescht Celanese (together they were called BHC) and initiated research into new methodology. In 1992 they launched full scale production of ibuprofen using novel methodology at a new plant in Texas. The new method featured only three steps. Moreover, only one of these was an inefficient substitution reaction – the other two were atom-efficient addition reactions. Atom economy was increased to 77%, profits were maintained in spite of greater market competition and BHC earned themselves a Presidential Green Chemistry Award.

Problems for 15.8

1 Put the following sets of reactions in order of increasing atom economy (least economic first):

Set 1

a $(CH_3)_2C(CH_3)\text{–}OH + HCl \longrightarrow (CH_3)_2C(CH_3)\text{–}Cl + H_2O$

b $CH_3CH_2OH \longrightarrow CH_2=CH_2 + H_2O$

c $CH_3CH_2CH_2CH_2CH_2CH_3 \longrightarrow CH_3CH_2CH(CH_3)\text{–}CH_2CH_3$

d

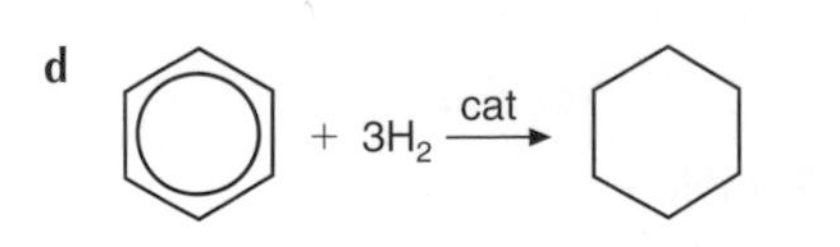

Set 2

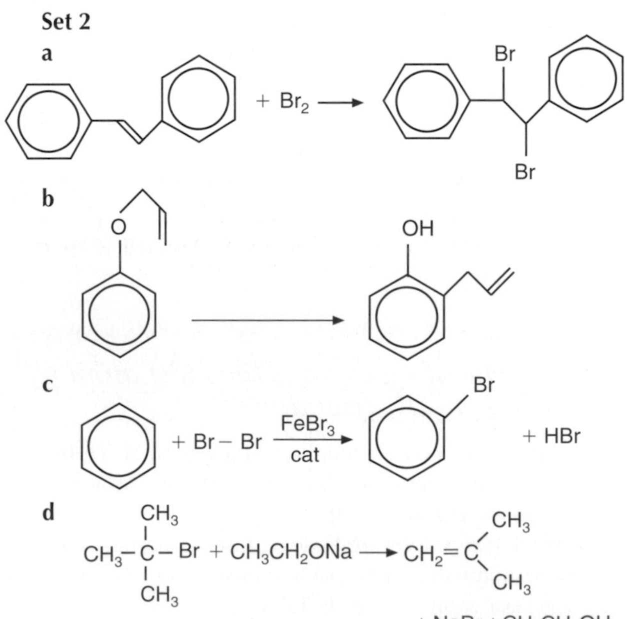

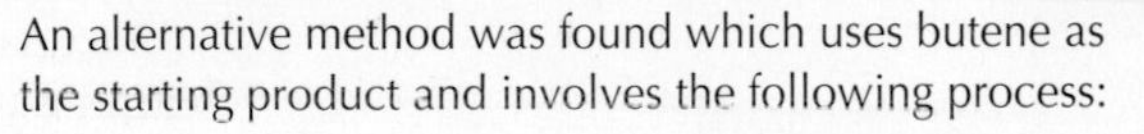

2 In the 1960s, the usual method for producing a chemical called malic anhydride was through the oxidation of benzene:

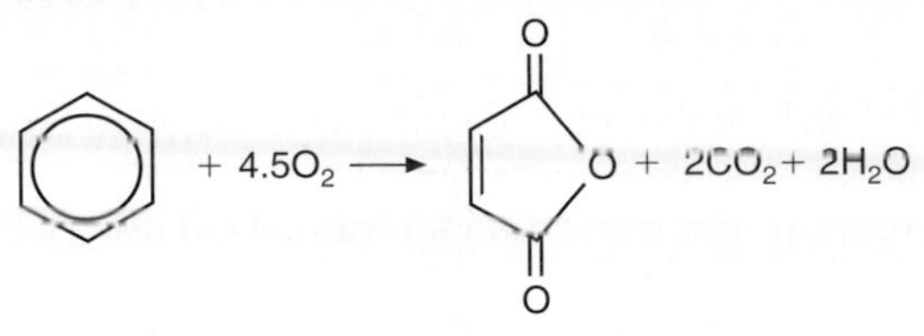

The oil crisis of the 1970s caused the price of benzene to rise to such an extent that this method was no longer competitive.

An alternative method was found which uses butene as the starting product and involves the following process:

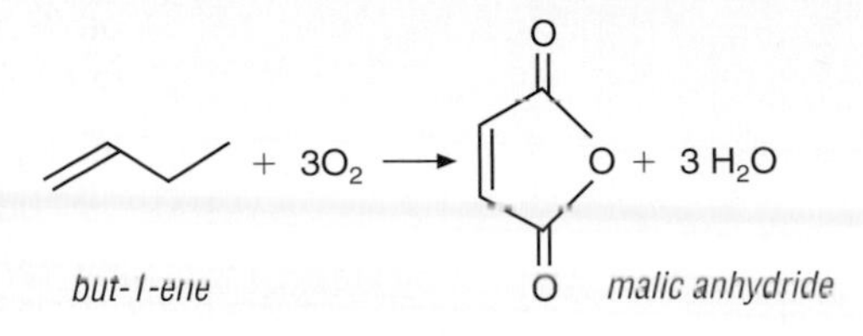

a Calculate the atom economy of both processes.

b The change from using the benzene process to the butene process was actually driven by the relative costs of the two starting products, but what was the effect on the atom economy?

15.9 *Atom economy and enzymes*

As explored in the previous section, synthetic chemists know that they need to improve the atom economy of their processes in order to produce less waste material and to increase their profitability. This usually means reducing the number of steps in a synthesis and shifting from atom-uneconomic reaction types, such as elimination and substitution, to atom-economic reaction types, such as rearrangement and addition.

The solution to the challenge of improving atom economy often lies in the use of catalysts – particularly biological catalysts, or enzymes. For many years, biochemists have been aware that whenever a living organism needs to convert one substance to another, enzymes are used to speed up the process. A reaction requiring many steps in the laboratory (plus extensive heating and use of reagents) might be accomplished by an enzyme at room temperature, in water, very quickly. This increases atom economy by reducing the number of steps – and also serves to reduce energy requirements. In addition, unlike many traditional reagents, enzymes can be reused and their large size makes it easy to separate them from the product of the reaction.

One area where enzymes have been used very successfully is in paper recycling. For recycling, the paper material (consisting of newspapers, cardboard and magazines) is mixed with water to form a slurry. Unfortunately, recycled paper does not just contain paper, but all sorts of materials such as adhesives, plastics, inks and other additives (known in the trade as 'stickies'). These tend to build up on the processing equipment, forcing production to stop periodically for cleaning, which is traditionally carried out using organic solvents. Halting production costs the company money – as does the purchase and disposal of high volumes of solvent.

A solution to this problem was found by using an enzyme to break down the 'stickies'. The enzyme is capable of hydrolysing the insoluble organic sticky molecules and converting them to water-soluble materials which are easily removed with the waste water from the process. Usage couldn't be easier – the enzyme is simply added at several different points during the process, negating the requirement for plant down-time and organic solvent use. The estimated savings are significant – the use of such enzymes at a paper mill producing 1000 tonnes of paper per day reduces solvent use by 275 000 kg per year and increases production by 6%, leading to greatly enhanced profits.

Success stories like this are driving research into the enzyme catalysis of reactions throughout the chemical industry, and there are already many other processes in which enzymes are used to increase efficiency. Some examples are:

- Use of enzymes to break down corn and other cereal starch to produce corn syrup or fructose, a material used to sweeten literally thousands of food products. Previously this conversion was carried out using an acid process at high temperature and pressure. Replacement with an enzyme process has reduced energy costs and cut the amount of by-products by 50%.
- Use of enzymes to increase nutrient uptake and decrease phosphate by-products in the livestock industry.
- Use of enzymes in the energy industry to manufacture cleaner biofuels from agricultural waste materials.
- Increasing use of enzymes in waste treatment – for example, to destroy cyanide ions left over from gold extraction and in the production of some polymers.

Problems for 15.9

1. Describe how enzymes catalyse reactions.
2. List the ways in which enzymes increase the efficiency of an industrial process.

15.10 *Green chemistry and recycling*

Polymer production requires significant quantities of resources, both as a raw material and to deliver energy for the manufacturing process. It is estimated that 4% of the oil extracted worldwide each year is used as a feedstock for the production of plastics, and an equal amount during manufacture. In addition, plastics manufacture produces waste and emissions, although their environmental impact varies according to the type of plastic being produced and the production method employed.

The disposal of plastic products also contributes significantly to their environmental impact. Because most plastics are non-biodegradable they take a long time to break down, possibly hundreds of years – although no one knows for certain because plastics haven't existed for long enough. With more and more plastics products (particularly plastic packaging) being disposed of soon after purchase, the landfill space required for plastics waste is a growing concern.

▲ **Figure 1** Plastics can take hundreds of years to degrade.

Recycling

In addition to reducing the amount of plastic waste requiring disposal, recycling of plastics can also serve to:

- conserve fossil fuel feedstocks
- reduce consumption of energy
- reduce the amount of solid waste going to landfill
- reduce carbon dioxide (CO_2), nitrogen(II) oxide (NO) and sulfur dioxide (SO_2) emissions.

How are polymers recycled?

Plastic process scrap recycling

Currently most plastic recycling in the UK is of 'process scrap' from industry – i.e. polymers left over from the production of plastics. These are relatively simple and economical to recycle and the material is relatively uncontaminated. Process scrap represents some 250 000 tonnes of the plastic waste produced in the UK and approximately 95% of this is recycled. This is usually described as reprocessing rather than recycling.

Post-use plastic recycling

Post-use plastic can be described as plastic material recycled from products that have undergone a full service life prior to being recovered. Most of this is household waste, and one of the major problems here lies with collection. In 2003 an estimated 24 000 tonnes of plastic bottles were collected – only 5.5% of all plastic bottles sold. The sorting of plastics prior to mechanical recycling is another problem. Mechanical recycling of plastics refers to processes which involve the melting, shredding or granulation of waste plastics. At the moment, most sorting for mechanical recycling in the UK is undertaken by trained staff who manually sort the plastics into polymer type and/or colour. This reliance on manual labour makes post-use plastic recycling an expensive process – technology is being introduced to sort plastics automatically using various techniques such as X-ray fluorescence, infrared spectroscopy, electrostatics and flotation. Following sorting, the plastic is either melted down directly and moulded into a new shape, or melted down after being shredded into flakes and then processed into granules called re-granulate.

Chemical or feedstock recycling

Feedstock recycling describes a range of recovery techniques used in making plastics. These techniques break down polymers into their constituent monomers, which in turn can be used again in refineries or in petrochemical and chemical production.

Use of bioplastics

Besides recycling, another way in which polymer manufacturers can adhere to the principles of green chemistry is by examining their feedstocks and moving towards use of renewable raw materials. Significantly, a number of manufacturers have been exploring alternatives to plastics made from non-renewable fossil fuels. Such alternative 'bioplastics' include polymers made from plant sugars and plastics grown inside genetically modified plants or microorganisms.

Recycling other materials

Recycling of other materials besides plastics can also benefit the environment in terms of reducing the amount of waste to be disposed of, and reducing the energy consumed in their production. For example, recycling steel saves up to 74% of the energy needed to make steel from new raw materials. Recycling 1 tonne of steel not only reduces water pollution in steel manufacture by 76%, but also saves 1.5 tonnes iron ore, 0.5 tonnes of coke and 1.28 tonnes of solid waste.

Thanks to its magnetic properties, steel is one of the easiest packaging materials to recover from the waste stream. Steel packaging can be automatically extracted from non-sorted refuse, or separated from other recyclable materials using efficient, low-cost magnets. All steel packaging can be recycled, with the exception of aerosols. They are sorted using magnets and cleaned by incineration.

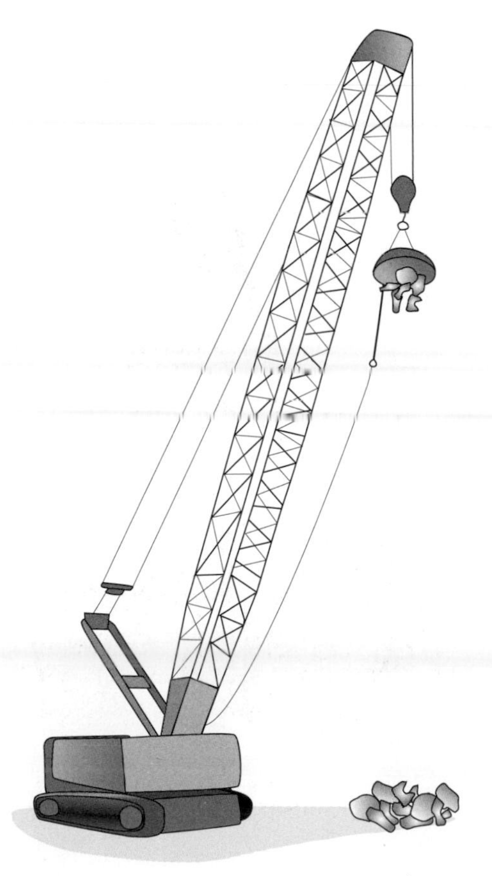

▲ **Figure 2** Steel is much easier to sort than plastic.

Reducing carbon dioxide emissions through recycling

Reduced use of feedstocks and reduced production of waste material are not the only advantages of recycling. Because the energy consumption per tonne of recycled material is so much lower than the energy needed to produce a tonne of the same material from new raw materials, less fossil fuel is burned during production. This offers another environmental advantage in the form of reduced carbon emissions. Table 1 summarises the savings in carbon dioxide emissions to the environment per tonne of material recycled.

▶ **Table 1** Carbon dioxide emissions saved through recycling.

Recycled material	CO_2 emissions saved per tonne of material recycled/kg
aluminium	15 420
steel	2095
plastics	1661
rubber	651
glass	330

A 'cradle to grave' approach to looking at carbon emissions resulting from the complete life cycle of a polymer should lead to a greener way of manufacturing, using recycling, and eventually disposing of polymers. This, in turn, should lead to significant reductions in carbon emissions.

APPENDIX 1 EXPERIMENTAL TECHNIQUES

Being a good chemist isn't just about knowing and understanding all the chemical knowledge in this book. It's also about understanding and being familiar with experimental techniques. The main way that experimental techniques are examined in this course is through AS and A2 coursework. Written examinations might also ask for details of experimental techniques. This appendix contains an outline of the techniques you need to know about for coursework and examinations.

A *Synthesis*

1 Heating under reflux

Use of technique

This technique is used for reactions involving volatile liquids. It ensures that reactants and/or products do not escape while the reaction is in progress. This is a particular problem since many organic liquids are flammable.

a Put the reactants into a pear-shaped or round-bottomed flask and add a few anti-bumping granules to prevent the reaction mixture from boiling and 'bumping' out of the flask.
Do not stopper the flask – doing this would cause pressure to build up and the glassware could crack or the stopper could fly out. In either case, a serious accident could result.

b Attach a condenser vertically to the flask to convert escaping vapours back to liquids. This will ensure that any volatile liquids return to the reaction flask and that no products will be lost. Connect to a water supply as shown in Figure 1 – this maximises the efficiency of the condenser.

c Heat so that the reaction mixture boils gently, using a Bunsen flame or heating mantle. When refluxing correctly, the condensate should reach no more than half way up the condenser, and the condensed vapour should drip back into the reaction flask steadily.

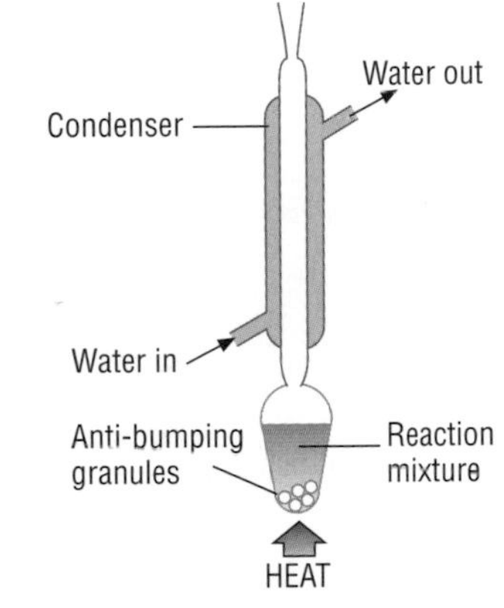

▲ **Figure 1** Reflux apparatus.

2 Purifying an organic liquid product

Use of technique

This technique is used to purify an organic liquid product after it has been synthesised.

a The first step is to separate the impure organic product from the reaction mixture. If the main product has been obtained with another immiscible liquid (often an aqueous liquid – for example, in the preparation of a halogenoalkane) the two layers must be separated. This can be done using a separating funnel. The layers will separate, with the denser liquid forming the lower layer. Allow the layers to settle and then run off and dispose of the aqueous layer – take care to keep the correct layer. Run the organic (product) layer into a clean conical flask.

b This next stage is needed only if there are acidic or alkaline impurities present. Add sodium hydrogencarbonate solution and shake well to remove acidic impurities. If the crude product is alkaline and needs neutralising then add a dilute acid until the mixture is neutral.

c Finally, the crude product has to be dried. Add anhydrous sodium sulfate and swirl the mixture. It is possible to use other anhydrous salts, such as calcium chloride, to dry organic compounds.

The pure product can then be separated by distillation (see **Technique 3**).

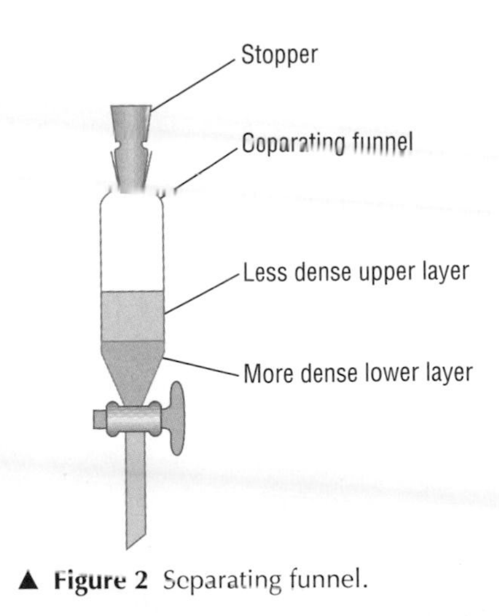

▲ **Figure 2** Separating funnel.

B Purification

3 Simple distillation

Use of technique

This technique relies on the fact that in a mixture of miscible liquids, each component has its own unique boiling point. By heating the mixture, each pure component can be vaporised, recondensed and collected. The components will evaporate in the order of their boiling points – the one with the lowest boiling point will evaporate first.

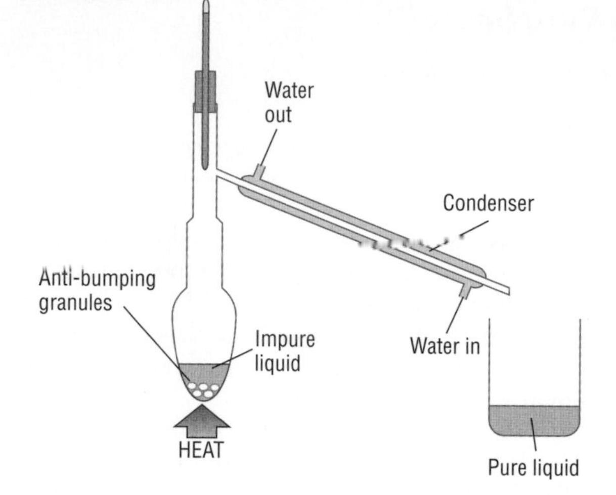

► **Figure 3** Simple distillation apparatus.

a Put the mixture to be separated into a pear-shaped flask and add a few anti-bumping granules. These granules burst the bubbles in the boiling mixture and reduce the chance of boiling over. Set up the distillation apparatus as shown in Figure 3 – remember the following points:
- the bulb of the thermometer needs to be positioned exactly as indicated in the diagram – this ensures an accurate reading of the vapour temperature as the distillation progresses;
- ensure that the water supply is connected exactly as indicated in the diagram – this makes sure that the condenser is always full of cold water.

b Heat the mixture until it boils gently, using a Bunsen flame or heating mantle. Heating mantles do not have a naked flame and so reduce the risk of fires with flammable liquids. If a heating mantle is used it is advisable to support it on a laboratory jack. This allows the heat source to be removed when needed.

c Check the thermometer reading. When the vapour temperature is approximately two degrees below the boiling point of the liquid you are about to collect, put the collecting beaker in place. Continue to heat the mixture, collecting the distilled condensate until the temperature of the vapour rises above the boiling point of the liquid you are collecting. Stop heating.

d If you require another compound of a higher boiling point from the mixture then repeat step **c** using a clean collecting beaker.

4 *Thin-layer chromatography*

Use of technique

This technique is used to separate small quantities of organic compounds, and also to purify organic substances. It can also be used to follow the progress of a reaction over time. A suitable solvent must be chosen. The method relies on the fact that different organic compounds have different affinities for a particular solvent, and so will be carried through the chromatography medium (plate) at different rates. When chromatography is carried out using a silica plate, it is known as thin-layer chromatography. A similar technique can be carried out using paper rather than a silica plate. This is known as paper chromatography.

a Spot the test mixture and reference sample(s) on a pencil line 1 cm from the base of the thin-layer chromatography plate. Pencil is used because it will not run into the solvent.

b Suspend the plate in a beaker containing the solvent, as shown in Figure 4, and cover the beaker with a watch glass to prevent the solvent from evaporating.

c Remove the plate when the solvent front is near the top. Mark how far the solvent has reached. Allow the plate to dry.

d Locate any spots with iodine, ninhydrin or under an ultraviolet lamp.

e Match the heights reached, or R_f values, with those of known compounds.

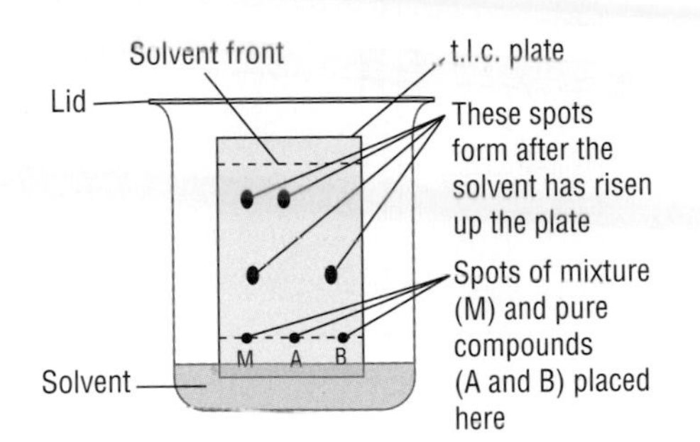

▲ **Figure 4** Thin-layer chromatography.

5 *Recrystallisation*

Use of technique

This technique is used to purify solid crude organic products. The mixture should contain mainly one product, with small amounts of impurities. It works on the principle that only the desired compound will dissolve to an appreciable extent in a suitable hot solvent. When cooled, the pure organic compound will drop out of solution (recrystallise) – any other soluble impurities stay in solution.

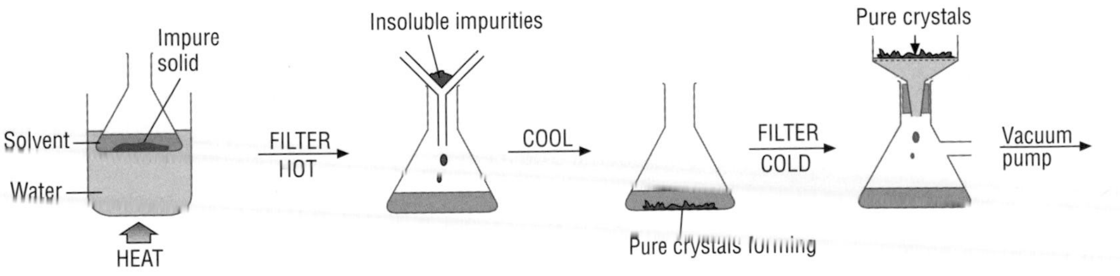

▲ **Figure 5** The process for recrystallising an impure solid.

a Select a solvent in which the desired substance is very soluble at higher temperatures, and insoluble, or nearly so, at lower temperatures. This ensures that the maximum amount of pure solid is obtained when the mixture is cooled at the end of the process.

b Dissolve the mixture to be treated in the minimum quantity of hot solvent – the smaller the amount of solvent used the better the yield of the desired substance.

c Filter to remove any insoluble impurities and retain the filtrate. It is best to preheat the filter funnel and conical flask to prevent any solid crystallising out at this stage.

d Leave the filtrate to cool until crystals form.

e Collect the crystals by vacuum filtration (see **Technique 6**).

f Dry the crystals by leaving them in the open, covered with an inverted filter funnel to prevent contamination by impurities from the atmosphere. Alternatively, the crystals can be dried in an oven set at an appropriate temperature.

6 Vacuum filtration

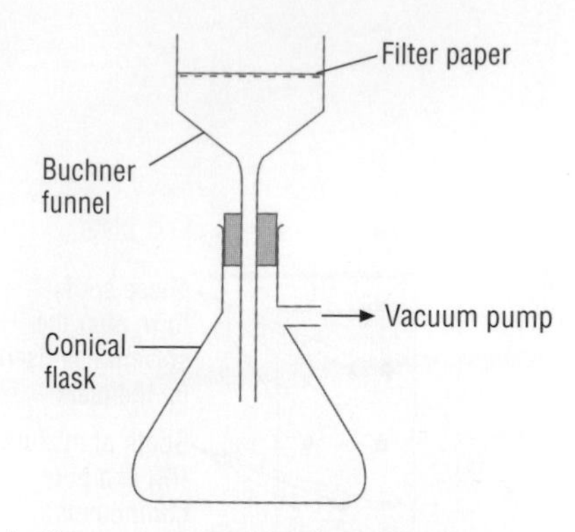

▲ **Figure 6** Apparatus for vacuum filtration.

Use of technique

This is a method for separating a solid from a filtrate rapidly.

a Connect a conical flask to a vacuum pump via the side arm, as shown in Figure 6. The pump will create a partial vacuum so that the filtrate gets 'pulled through' quickly. Do not switch the pump on yet.

b Dampen a piece of filter paper and place it flat in the vacuum funnel. Switch the vacuum pump on and then carefully pour in the mixture to be filtered.

c Disconnect the flask from the vacuum pump before turning the pump off – this avoids 'suck back'. Solids may be dried as described in **Technique 5**, step **f**.

C Analysis

7 Determining melting points

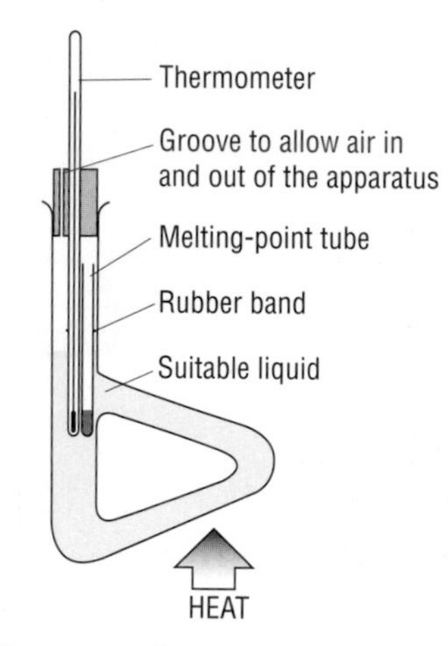

▲ **Figure 7** Melting point apparatus.

Use of technique

This technique is used to determine the melting point of, usually, organic solids. The melting point can then be used as evidence of a product's identity and purity.

a Seal the end of a glass melting point tube by heating it to melting in a Bunsen flame.

b Introduce a small amount of the dry, crushed solid into the melting point tube – tap the tube so that the solid falls to the bottom of the sealed end.

c Fix the tube in the melting point apparatus and heat the surrounding liquid gently, stirring to ensure even heating throughout. The temperature should rise very slowly.

d Note the temperature at which the solid starts and finishes melting. The difference between the highest and lowest temperatures recorded is known as the melting range.

This value can then be compared to the published value for the melting point of the solid to see how pure the substance is – the wider the melting range, the more impure the substance. A pure compound will melt within 0.5 °C of the true melting point.

8 Making a standard solution

Use of technique

A standard solution is made up fresh whenever a concentration has to be accurate. Here it is assumed that 250 cm^3 of solution is being made up.

a Calculate the mass of solute required. In a weighing bottle, weigh out approximately this amount accurately, to the nearest 0.01 g. Make a note of this mass.

b Pour a suitable volume, say 100 cm^3, of deionised water into a 250 cm^3 beaker. Carefully transfer the weighed solute into the water from the weighing bottle.

c Reweigh the weighing bottle. The difference between the mass of the weighing bottle and solute and the weighing bottle once emptied is the mass of solute transferred. This step takes account of any residue left in the weighing bottle, which will not be dissolved in the standard solution.

d Stir the mixture in the beaker to dissolve the solute. This should be done thoroughly to ensure complete dissolving.

e Transfer the solution to a clean, rinsed 250 cm³ volumetric flask, using a filter funnel. Rinse the beaker and stirring rod well with deionised water, making sure that all the washings go into the volumetric flask.

f Add deionised water to the solution, swirling at intervals to mix the contents, until the level is within about 1 cm of the mark on the neck of the volumetric flask.

g Using a dropping pipette, add deionised water so that the bottom of the meniscus is level with the mark on the neck of the flask – when you are looking at it at eye level.

h Insert the stopper in the flask and invert it, shaking thoroughly to ensure complete mixing.

9 Acid–base titration

Use of technique

This technique is used to determine the concentration of an acid or an alkali very accurately.

The method described below assumes that you have made up a standard solution of an alkali, such as sodium hydroxide, and that you are titrating it against an acid, such as hydrochloric acid, in order to calculate the concentration of the acid.

a Using a funnel, rinse a burette with some of the hydrochloric acid solution that is to be used. Then fill it with the hydrochloric acid. Run a little of the acid through the burette into a waste beaker to fill the tip. Take the initial burette reading to the nearest 0.05 cm³ and make a note of the measurement (see Figure 8).

b Rinse a clean beaker with the standard sodium hydroxide solution to be used, and then half fill it. Use a pipette filler to rinse a clean 25.0 cm³ pipette with some of the standard sodium hydroxide solution. Then fill the pipette to the mark – take care to make sure that the bottom of the meniscus rests exactly on the mark on the pipette stem. Wipe the outside of the pipette with a paper towel.

c Carefully run the alkaline solution into a clean 250 cm³ conical flask. Once the pipette is empty, touch its tip against the inside of the conical flask – the pipette has then delivered exactly the volume it is designed to.

d Add 2 or 3 drops of a suitable indicator and swirl to mix – if you add too much it will reduce the accuracy of the results since the indicator may react with acids and alkalis.

e Run hydrochloric acid from the burette into the flask. Swirl the flask continually and watch for the first hint of the solution changing colour. This first titration should be used as a trial run – you may well overshoot the end point by a little, but it will give a rough indication of the amount of acid required. Record the final burette reading – the volume of acid used is called the 'titre'.

f If necessary, refill the burette and record the initial burette reading.

g Using the pipette filler, transfer 25.0 cm³ of the sodium hydroxide solution to a clean conical flask – if you reuse the previous conical flask then it must be rinsed thoroughly with deionised water. Add 2 to 3 drops of the indicator and swirl to mix.

h Run in hydrochloric acid solution to 1 cm³ below the rough titre. Then add the acid drop by drop, swirling after each, until the colour of the indicator changes.

i Repeat steps **f**, **g** and **h** until there are three concordant results – that is, three results within 0.10 cm³ of each other.

The volume and concentration of the standard sodium hydroxide solution required for neutralisation can be used to calculate the concentration of the hydrochloric acid – details of these calculations are given in **Section 1.5**.

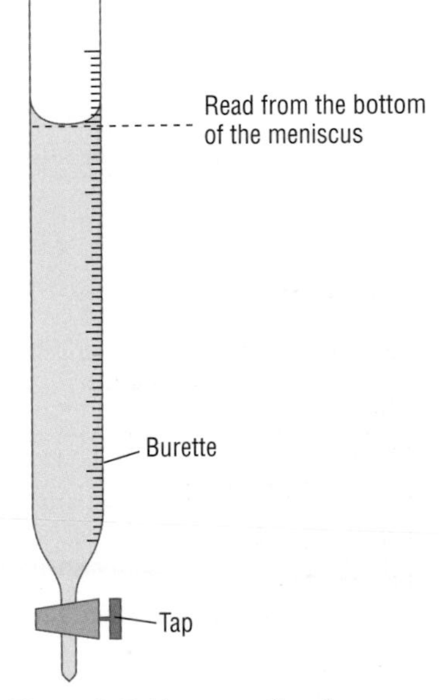

▲ **Figure 8** Taking a reading from a burette.

10 Redox titration

Use of technique

The procedure for a redox titration is similar to that used for acid–base titrations. The difference is in the type of reaction occurring, which is a redox reaction – electrons are transferred from one species to another. Often, there is no need for an indicator since one of the reactants or products is coloured – for example, when manganate(VII) ions are in the conical flask and are being reduced, the reaction is over when the last of the purple colour disappears.

Alternatively, if the manganate(VII) ions are being added from a burette to a colourless solution, the end point is when the first pale pink colour appears.

11 Measuring a cell e.m.f.

Use of technique

This technique is used to determine the potential of an electrochemical cell. Standard electrode potentials can be measured by connecting any half-cell to a standard hydrogen half-cell, or a calibrated reference half-cell.

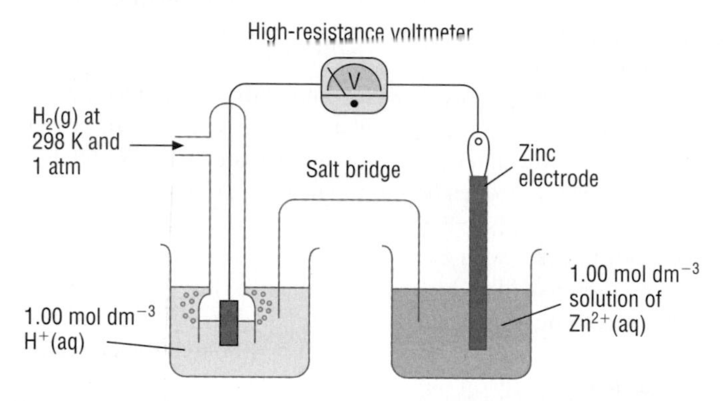

► **Figure 9** An example of a standard electrochemical cell.

a Construct the half-cell whose electrode potential is to be measured. Ensure that all solutions have a concentration of 1.0 mol dm^{-3} and are at 25°C.

b Connect the half-cell to a standard hydrogen half-cell – or other reference cell (often Cu^{2+}/Cu) – using a high-resistance voltmeter and salt bridge as shown in Figure 9.

c Check that the reading on the voltmeter is positive – if it is then the half-cell connected to the positive terminal of the voltmeter is the positive electrode. If the reading is negative, change the connections round on the voltmeter to give a positive reading.

d Record the voltmeter reading – this is the required cell e.m.f. ($E^{\ominus}_{cell}$).

12 Using a colorimeter

Use of technique

This technique is used to determine the concentration of a coloured solution. It works on the principle that coloured solutions absorb certain wavelengths of light (e.g. a purple solution will absorb orange light strongly).

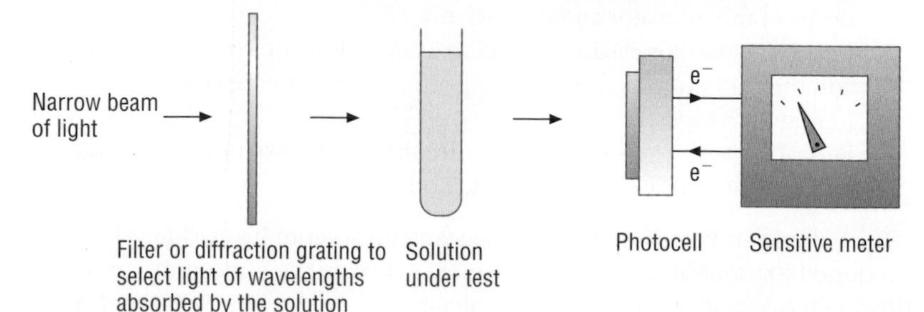

► **Figure 10** A simplified diagram of a colorimeter.

The amount of this light that is either absorbed or transmitted by a solution can be measured. This is proportional to the concentration of the solution.

a Select a filter with the complementary colour to the solution being tested – for example, choose an orange filter for a purple solution. This allows only those wavelengths absorbed most strongly by the solution to pass through to the sample.

b Make up a range of standard solutions of the test solution – there should be solutions both above and below the concentration of the unknown solution.

c Zero the colorimeter using a tube/cuvette of pure solvent – this will be water in most cases.

d Measure the absorbance of each of the standard solutions, and plot a calibration curve of concentration against absorbance.

e Measure the absorbance of the unknown sample and use the calibration curve to determine the concentration of the unknown solution.

13 Measuring the enthalpy change of combustion of a fuel

Use of technique

This technique is used to determine the enthalpy change of combustion when a fuel is burned. The values obtained experimentally can then be compared with theoretical values.

a Using a measuring cylinder, pour a known volume of water into a copper calorimeter, record its temperature and make a note of it.

b Support the calorimeter over a spirit burner containing the fuel to be tested. Surround it with a draught excluder to help to reduce energy losses, as shown in Figure 11.

c Weigh the spirit burner – keep the cap on the burner to reduce loss of the fuel by evaporation.

d Put the burner under the calorimeter, remove the cap and light the wick.

e Use the thermometer to stir the water all the time it is being heated – carry on heating until the temperature has risen by 15 to 20 °C.

f Extinguish the burner and put the cap back in place. Keep stirring the water and make a note of the highest temperature reached.

g Weigh the burner again.

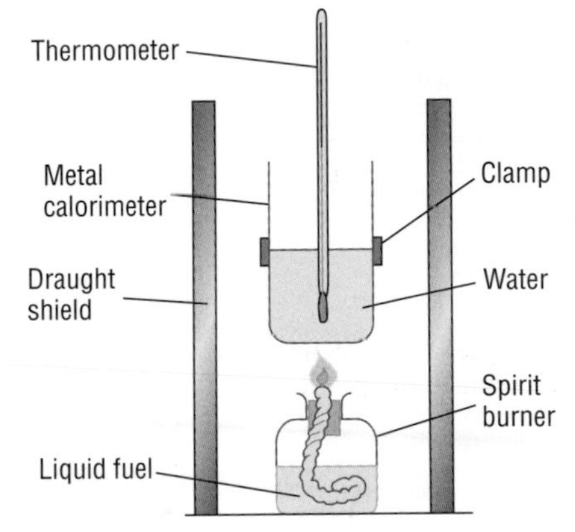

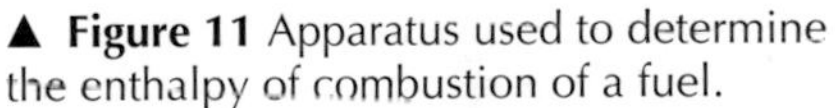
▲ **Figure 11** Apparatus used to determine the enthalpy of combustion of a fuel.

The results can be used to calculate the enthalpy of combustion for the fuel under test. You can see how to do this in **Section 4.1**.

INDEX